MW00883741

Brigade Combat Team

January 2021

United States Government
US Army

Contents

DISTRIBUTION RESTRICTION: Approved for public release; distribution is unlimited.

*This publication supersedes FM 3-96, dated 8 October 2015.

Figures

Tables

This page intentionally left blank.

Preface

Army FM 3-96 provides doctrine for the brigade combat team (BCT). This manual describes how the BCT, as part of a joint team, shapes operational environments, prevents conflict, conducts large-scale ground combat, and consolidates gains against a peer threat. FM 3-96 describes relationships, organizational roles and functions, capabilities and limitations, and responsibilities within the BCT. *Tactics*, the employment, ordered arrangement, and directed actions of forces in relation to each other (ADP 3-90), are discussed in this manual and are intended to be used as a guide. They are not prescriptive. FM 3-96 applies to the three maneuver BCT types: Infantry, Stryker, and Armored. This manual supersedes FM 3-96, dated 8 October 2015.

The principal audience for FM 3-96 is the commanders, staffs, officers, and noncommissioned officers of the brigade, battalions, and squadron within the BCT. The audience also includes the United States Army Training and Doctrine Command institutions and components, and the United States Army Special Operations Command. This manual serves as an authoritative reference for personnel developing doctrine, materiel and force structure, institutional and unit training, and standard operating procedures for the BCT. For lower maneuver echelon specific discussions, see the appropriate Army techniques publication for that organization.

To comprehend the doctrine contained in this manual, readers must first understand the characteristics of the Army Profession (trust, honorable service, military expertise, stewardship, and esprit de corps) as described in ADP 1 and the principles of Army leadership as described in ADP 6-22 and FM 6-22. Readers also must understand the principles and tenets of unified land operations as well as decisive action, and the links between the operational and tactical levels of war described in JP 3-0; ADP 3-0, and FM 3-0; FM 3-94; ATP 3-91, and ATP 3-92. In addition, readers should understand the fundamentals of the operations process found in ADP 5-0, associated with offensive and defensive operations contained in ADP 3-90, FM 3-90-1, and reconnaissance, security and tactical enabling tasks contained in FM 3-90-2. The reader must comprehend how stability operations tasks described in ADP 3-07 and FM 3-07, carry over and affect offensive and defensive operations and vice versa. Readers must understand how the operation process fundamentally relates to the Army's design methodology, military decision-making process, and troop leading procedures and the exercise of command and control as described in ADP 6-0, FM 6-0, and ATP 6-0.5.

Commanders, staffs, and subordinates ensure their decisions and actions comply with applicable U.S., international, and in some cases, host-nation laws and regulations. Commanders at all levels ensure that their Soldiers operate in accordance with the law of war and the rules of engagement. (See FM 6-27.)

FM 3-96 uses joint terms where applicable. Selected joint and Army terms and definitions appear in both the glossary and the text. Terms for which FM 3-96 is the proponent publication (the authority) are marked with an asterisk (*) in the glossary. Definitions for which FM 3-96 is the proponent publication are boldfaced in the text and the term is italicized. For other definitions shown in the text, the term is italicized, and the number of the proponent publication follows the definition.

FM 3-96 applies to the Active Army, the Army National Guard/the Army National Guard of the United States, and the United States Army Reserve unless otherwise stated.

The proponent for FM 3-96 is the United States Army Maneuver Center of Excellence. The preparing agency is the Maneuver Center of Excellence, Directorate of Training and Doctrine, Doctrine and Collective Training Division. Send comments and recommendations on a DA Form 2028, (*Recommended Changes to Publications and Blank Forms*) to: Commanding General, Maneuver Center of Excellence, Directorate of Training and Doctrine, ATTN: ATZK-TDD, 1 Karker Street, Fort Benning, GA 31905-5410; by email to usarmy.benning.mcoe.mbx.doctrine@mail.mil; or submit an electronic DA Form 2028.

This page intentionally left blank.

Introduction

The Army provides readily available, trained and equipped, and globally responsive forces to shape the operational environment, prevent conflict, and prevail in large-scale ground combat while consolidating gains as part of unified action. Army forces, which consist of trusted Army professionals of character, competence, and commitment, maintain proficiency in the fundamentals of unified land operations, the Army's operational concept, and possess capabilities to meet specific geographic combatant command requests. Army forces provide combatant commanders with BCTs—a combined arms, close combat force that can operate as part of a division or a joint task force. BCTs, with unified action partners, conduct land operations to shape security environments, prevent conflict, prevail in ground combat, and consolidate gains. BCTs provide the Army with multiple options for responding to and resolving crises. The BCT, within the division or corps scheme of maneuver, defeats enemy forces, controls terrain, secures populations, and preserves joint force freedom of action.

FM 3-96 defines the employment and ordered arrangement of forces within the BCT during the conduct of decisive action across the range of military operations. The tactics addressed in this manual include the ordered arrangement and *maneuver*—movement in conjunction with fires (ADP 3-0)—of units in relation to each other, the terrain, and the enemy. Tactics vary with terrain and other circumstances; they change frequently as the enemy reacts and friendly forces explore new approaches. Applying tactics usually entails acting under time constraints with incomplete information. Tactics always require judgment in application; they are always descriptive, not prescriptive. FM 3-96 addresses the tactical application of tasks associated with the offense, the defense, and operations focused on stability. FM 3-96 does not discuss defense support of civil authorities.

Employing tactics addressed in FM 3-96 may require using and integrating techniques and procedures. Echelon-specific Army techniques publications address *techniques*, non-prescriptive ways or methods used to perform missions, functions, or tasks (CJCSM 5120.01B) and *procedures*, standard, detailed steps that prescribe how to perform specific tasks (CJCSM 5120.01B).

This manual incorporates the significant changes in Army doctrinal terminology, concepts, constructs, and proven tactics developed during recent operations. It also incorporates doctrinal changes and terms based on ADP 3-0, FM 3-0, and ADP 3-90.

Note. This manual is written based on the current structure of the BCT and its subordinate units. Future changes to the organizational structures of the BCT will be published as change documents to the manual.

The following is a brief introduction and summary of changes by chapter.

Chapter 1 – Organization

Chapter 1 provides the doctrinal foundation for the three types of BCTs: the Infantry brigade combat team, the Stryker brigade combat team, and the Armored brigade combat team. The chapter addresses the mission, capabilities, limitations, and internal organization of each type of BCT.

Chapter 2 – The Brigade Combat Team and the Operational Environment

Chapter 2 discusses the BCT's role in military operations and its interactions with operational environments. The chapter addresses key doctrinal concepts on how the Army fights regardless of which element of decisive action (offense, defense, or stability) currently dominates the BCT's area of operations.

Chapter 3 – Threat

Chapter 3 addresses threats as a fundamental part of an overall operational environment. The chapter identifies a threat as any combination of actors, entities, or forces that have the capability and intent to harm U.S. forces, U.S. national interests, or the homeland. This includes individuals, groups of individuals (organized or not organized), paramilitary or military forces, nation states, or national alliances. The chapter provides the understanding for today's forces to deal with symmetrical threats as seen in Operation Desert Storm, as well as asymmetrical threats seen during Operation Iraqi Freedom and Operation Enduring Freedom. In addition, chapter 3 discusses—

- Threat characteristics and organization.
- Threat countermeasures.
- Countering adaptations and retaining the initiative.

Chapter 4 – Mission Command

Chapter 4 addresses the fundamentals of mission command as the Army's approach to command and control. The approach requires the commander, as a trusted Army professional, to make decisions and to take actions consistent with the Army Ethic and Army Values. It requires the commander to lead from a position that allows timely decisions based on an assessment of the operational environment and application of judgment. In addition, chapter 4—

- Addresses the command and control warfighting function as it assists the commander with combining the art and science of command and control.
- Emphasizes the human aspects of mission command.
- Discusses BCT command and staff operations.
- Describes how the commander cross-functionally organizes the staff into cells and working groups.
- Describes the establishment of centers and meetings to assist with coordinating operations.
- Describes the types and composition of command posts at brigade echelon.
- Addresses air-ground operations and intelligence support team considerations.
- Discusses cyberspace electromagnetic activities, with major emphasis directed towards electromagnetic warfare operations (electromagnetic warfare replaces the term electronic warfare).

Chapter 5 – Reconnaissance and Security

Chapter 5 discusses reconnaissance and security as continuous and essential to support the conduct of offense, defense, and stability. Chapter 5 provides—

- The doctrinal basis for reconnaissance and security forces.
- A discussion of information collection.
- An overview of reconnaissance fundamentals and reconnaissance operations.
- An overview of security fundamentals and security operations.
- An overview of surveillance and intelligence operations.

Chapter 6 – Offense

Chapter 6 discusses offensive actions to defeat, destroy, or neutralize the enemy. The chapter addresses the characteristics of a BCT offense and describes the four offensive operations: movement to contact, attack, exploitation, and pursuit. Chapter 6 also discusses—

- Common offensive planning considerations and offensive control measures.
- Forms of maneuver (flank attack removed as a form; frontal attack now frontal assault).
- Transitions to other tactical operations.

Chapter 7 – Defense

Chapter 7 discusses defensive actions to defeat enemy attacks, gain time, control key terrain, protect critical infrastructure, secure the population, and economize forces. The chapter addresses BCT defense characteristics and describes the three defensive operations: area defense, mobile defense, and retrograde. Chapter 7 also discusses—

- Common defensive planning considerations and defensive control measures.
- Forms of the defense.
- Forms of defensive maneuver.
- Transitions to other tactical operations.

Chapter 8 – Stability

Chapter 8 addresses BCT support to operations focused on stability operations tasks. This chapter encompasses various military missions, tasks, and activities conducted outside the United States in coordination with other instruments of national power. In addition, chapter 8—

- Addresses the foundation (principles and framework), and environment during stabilization.
- Discusses the BCT's responsibilities and roles when supporting stability operations tasks (added sixth stability operations task–conduct security cooperation).
- Addresses area security operations and security force assistance missions.
- Discusses the transition from stability to other tactical operations.

Chapter 9 – Sustainment

Chapter 9 discusses the process that sustainment planners and operators use to anticipate the needs of the maneuver units. Chapter 9 also discusses—

- Fundamentals of sustainment.
- Sustaining the BCT.
- Staff and unit responsibilities and relationships.
- Echelon support.
- Echelons above brigade sustainment (specifically, corps and division).
- Brigade support area.

This page intentionally left blank.

Chapter 1

Organization

Brigade combat teams (BCTs) organize to conduct *decisive action*—the continuous, simultaneous execution of offensive, defensive, and stability operations or defense support of civil authorities tasks (ADP 3-0). BCTs are the Army's primary combined arms, close combat force. BCTs often operate as part of a division or joint task force. The division or joint task force acts as a tactical headquarters that typically directs the operations of between two to five BCTs across the range of military operations. The tactical headquarters assigns the BCT its mission, area of operations, and supporting elements. The headquarters coordinates the BCT's actions with other BCTs in the formation. The BCT might be required to detach subordinate elements to other brigades attached or assigned to the division or task force. Usually, this tactical headquarters assigns augmentation elements to the BCT. Field artillery, maneuver enhancement, sustainment, and combat aviation brigades can all support BCT operations. (See ATP 3-91 for additional information on division operations.)

Note. This FM does not address defense support of civil authorities. (See ADP 3-28 and ATP 3-28.1 for information.)

BCTs include capabilities across the command and control, movement and maneuver, intelligence, fires, sustainment, and protection warfighting functions. These capabilities are scalable to meet mission requirements. All BCTs include maneuver; field artillery; intelligence; signal; engineer; chemical, biological, radiological, and nuclear (CBRN); and sustainment capabilities. Higher commanders augment BCTs with additional combat power for specific missions. Augmentation might include aviation, Armor, Infantry, field artillery, air defense, military police, civil affairs, a tactical psychological operations (PSYOP) company, engineers, additional CBRN capabilities, cyberspace, and information systems. Organizational flexibility enables the BCT to accomplish missions across the range of military operations.

Chapter 1 provides the doctrinal foundation for the three types of BCTs: the Infantry brigade combat team (IBCT), the Stryker brigade combat team (SBCT), and the Armored brigade combat team (ABCT). The chapter addresses the mission, capabilities, limitations, and internal organization of each BCT.

SECTION I – INFANTRY BRIGADE COMBAT TEAM

1-1.   The IBCT is an expeditionary, combined arms formation optimized for dismounted operations in *complex terrain*—a geographical area consisting of an urban center larger than a village and/or of two or more types of restrictive terrain or environmental conditions occupying the same space (ATP 3-34.80). The IBCT can conduct entry operations by ground, airland, air assault, or amphibious assault into austere areas of operations with little or no advanced notice. Airborne IBCTs can conduct vertical envelopment by parachute assault. The IBCT's dismounted capability in complex terrain separates it from other functional brigades and maneuver BCTs.

1-2.   Mission variables, categories of specific information needed to conduct operations, help to determine the task organization and required augmentation for the IBCT. For example, if additional *tactical mobility*—the ability of friendly forces to move and maneuver freely on the battlefield relative to the enemy (ADP 3-90)—is required, the higher tactical headquarters can temporarily augment the IBCT with aviation assets to conduct air movements or air assault operations (see FM 3-99). Augmentation can include wheeled assets such as the mine-resistant ambush protected family of vehicles (see ATP 3-21.10).

1-3.   The role of the IBCT is to close with the enemy by means of fire and movement to destroy or capture enemy forces, or to repel enemy attacks by fire, close combat, and counterattack to control land areas, including populations and resources. ***Fire and movement* is the concept of applying fires from all sources to suppress, neutralize, or destroy the enemy, and the tactical movement of combat forces in relation to the enemy (as components of maneuver applicable at all echelons). At the squad level, fire and movement entails a team placing suppressive fire on the enemy as another team moves against or around the enemy.**

1-4.   The IBCT performs complementary missions to SBCTs and ABCTs. The IBCT optimizes for the offense against conventional, hybrid, and irregular threats in severely restrictive terrain. The IBCT performs missions such as reducing fortified areas, infiltrating and seizing objectives in the enemy's rear, eliminating enemy force remnants in restricted terrain, and securing key facilities and activities. The IBCT conducts stability operations tasks in the wake of maneuvering forces.

1-5.   IBCTs configure for area defense and as the fixing force component of a mobile defense. The IBCT's lack of heavy combat vehicles reduces its logistic requirements. Not having heavy combat vehicles gives higher commanders greater flexibility when adapting various transportation modes to move or maneuver the IBCT. Airborne IBCTs conduct airborne assault-specific missions. All IBCTs can conduct air assault operations. (See FM 3-99 for information on airborne and air assault operations.)

1-6.   The IBCT is a combined arms force organized around dismounted Infantry. Cavalry, field artillery, engineer, intelligence, signal, sustainment, and CBRN reconnaissance units are organic to the IBCT (see figure 1-1). Unique to the IBCT is the weapons company in each Infantry battalion, composed of four mounted assault platoons and provides those battalions with the capability to defeat light enemy armor threats with organic mounted tube launched, optically tracked, wire guided/wireless guided Improved Target Acquisition System, M2 series heavy machine gun, and MK-19 40-millimeter (mm) grenade machine gun weapon systems (see paragraph 1-9). Higher commanders augment the IBCT for a specific mission with additional capabilities. Augmentation can include aviation, Armor, field artillery, air defense, military police, civil affairs, a tactical PSYOP element, engineers, CBRN, and additional information systems assets. Three Infantry battalions and the Cavalry squadron serve as the IBCT's primary maneuver forces.

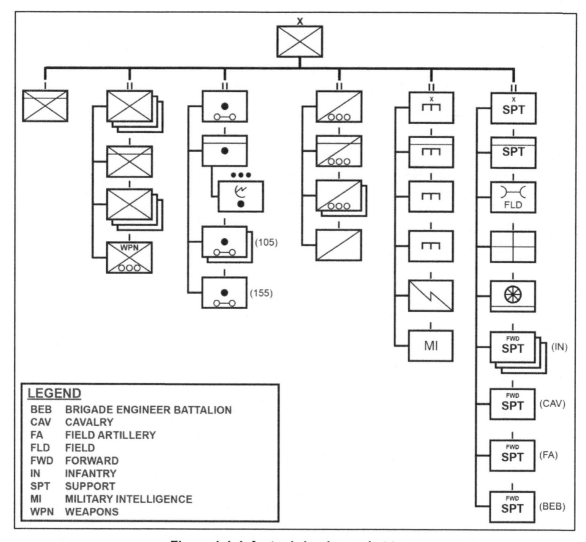

Figure 1-1. Infantry brigade combat team

1-7.   The Infantry battalions organize with a headquarters and headquarters company, three Infantry rifle companies, and a weapons company (see figure 1-2 on page 1-4). The headquarters and headquarters company provides planning and intelligence, signal, and fire support to the battalion. The headquarters company has a battalion command section, a battalion staff section, a company headquarters, battalion medical, scout, and mortar platoons, a signal section, and a sniper squad. The headquarters company mortar platoon is equipped with 120-mm mortars (trailer towed) and 81-mm mortars (ground mounted). The battalion receives a forward support company (FSC) for sustainment purposes (see chapter 9), normally in a direct support relationship. (See ATP 3-21.20 for additional information.)

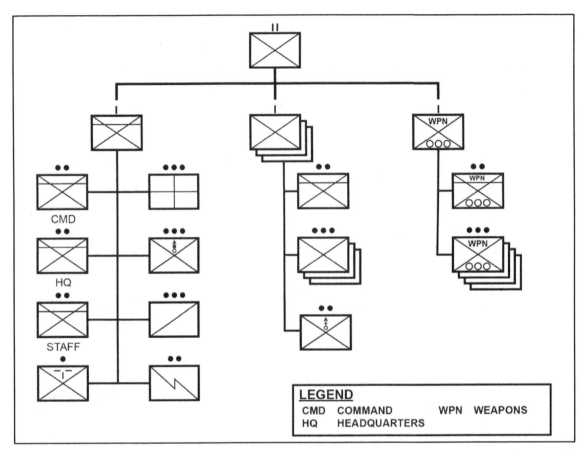

Figure 1-2. Infantry battalion

1-8.   Infantry rifle companies have three Infantry rifle platoons, a mortar section, a Raven unmanned aircraft system (UAS) team, and a headquarters section. Each rifle platoon has three Infantry rifle squads and a weapons squad. The mortar section has two squads, each with a 60-mm mortar. Habitual attachments to the Infantry rifle company include a fire support team at the company level and forward observer teams at the platoon level, medics assigned to the rifle platoons, and a senior medic at the company level. (See ATP 3-21.10 and ATP 3-21.8 for additional information.)

1-9.   The Infantry weapons company has a company headquarters and four assault platoons. Each assault platoon has two sections of two squads and a leader's vehicle. Each squad contains four Soldiers and a vehicle mounting the heavy weapons. The heavy weapons can be tailored to a mission based on the commander's mission analysis. Infantry weapons companies are equipped with the following weapons: the tube launched, optically tracked, wire guided/wireless guided Improved Target Acquisition System, the MK19, the M2, and the M240 series machine gun. While all of the weapons vehicles can mount the MK19 and the M2, only two vehicles per platoon are equipped to mount the Improved Target Acquisition System. Habitual attachments for the weapons company include a fire support team at the company level and medics. (See ATP 3-21.20, appendix D for additional information.)

Note. The Infantry battalion scout platoon and IBCT Cavalry squadron organize, train, and equip to conduct reconnaissance, security operations, and surveillance. However, reconnaissance, security operations, and surveillance remain a core competency of the Infantry rifle company, platoon, and squad.

1-10.   The IBCT Cavalry squadron's mission focuses on *information requirements*—in intelligence usage, those items of information regarding the adversary and other relevant aspects of the operational environment that need to be collected and processed in order to meet the intelligence requirements of a commander (JP 2-0)—tied to the execution of tactical missions (normally reconnaissance, security operations, and surveillance). The squadron's information collection effort answers the commander's priority intelligence requirements. Information acquired during collection activities about the threat and the area of interest allows the IBCT commander to focus combat power, execute current operations, and prepare for future operations simultaneously.

1-11. The Cavalry squadron (see figure 1-3) has four troops: a headquarters and headquarters troop, two mounted Cavalry troops, and one dismounted Cavalry troop. (See ATP 3-20.96.) The headquarters troop organization includes a command section, the troop headquarters section, the squadron primary staff, a medical section, a sniper section, a retransmission (known as RETRANS) section, an attached fire support cell, and a tactical air control party (TACP). The two mounted Cavalry troops (three scout platoons each) are equipped with wheeled vehicles (each with a crew and scout team for dismounted operations), tube launched, optically tracked, wire guided/wireless guided Improved Target Acquisition Systems, the Long-Range Advance Scout Surveillance Systems, a mortar section (120-mm trailer towed), and a Raven UAS team. The dismounted Cavalry troop (two dismounted scout platoons each) enables dismounted infiltration and rotary-wing aircraft insertion and has a mortar section (60-mm ground mounted), a Raven UAS team, and a sniper squad. Habitual attachments to the Cavalry troop include a fire support team at the troop level and forward observer teams at the platoon level, medics assigned to each platoon, and a senior medic at the troop level. (See ATP 3-20.97 and ATP 3-20.98.) The squadron receives an FSC for sustainment purposes (see chapter 9), normally in a direct support relationship.

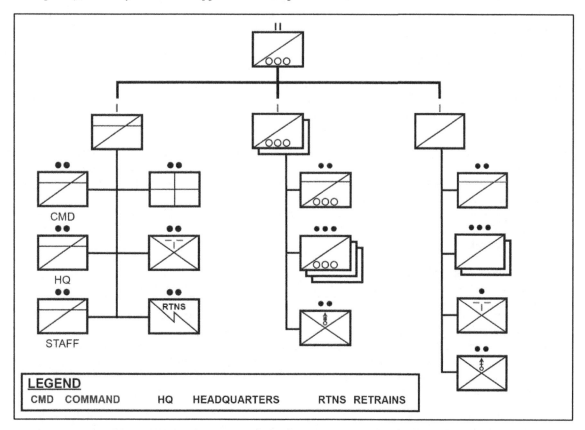

Figure 1-3. Infantry brigade combat team Cavalry squadron

1-12. The IBCT field artillery battalion has four batteries: a headquarters and headquarters battery, two 105-mm firing batteries (six-gun M119 series towed howitzer battery), and one 155-mm firing battery (six-gun M777 series towed howitzer battery). The firing batteries in a battalion have two 3-gun firing platoons. The field artillery battalion provides massing fires in space and time on single or multiple targets with precision, near precision, and area fires to support IBCT operations. The IBCT field artillery battalion has a target acquisition platoon (counterbattery and countermortar radars) organized and equipped to quickly detect, and accurately locate, classify, and report indirect fire from enemy mortars, artillery, and rockets to permit their immediate engagement with counterfire. The information provided includes the point of origin, predicted point of impact, radar cross section, and velocity. The battalion receives an FSC for sustainment purposes, normally in a direct support relationship. The battalion receives an FSC for sustainment purposes (see chapter 9), normally in a direct support relationship. (See ATP 3-11.23 for additional information.)

> *Note.* Paragraphs 1-13 through 1-27 in this section discuss the brigade engineer battalion (BEB) and the brigade support battalion (BSB) in the BCT. This discussion includes the differences in these battalion formations for each type of BCT (IBCT, SBCT, and ABCT).

1-13. The BEB provides a baseline of combat capabilities to the BCT that can be augmented with specialized units from echelons above brigade (see FM 3-34). The BEB has a headquarters and headquarters company, two engineer companies, a signal company, military intelligence company, a tactical unmanned aircraft system (known as TUAS) platoon (located in the military intelligence company), and a CBRN reconnaissance platoon (located in the headquarters and headquarters company). The command and support relationship between these units dictate whether the BEB logistically supports or coordinates support with the BCT, the BSB, or other unit higher headquarters. Unless the BCT directs otherwise, the BEB retains command and support relationships with organic and attached units, regardless of location on the battlefield. Companies may be task organized further to maneuver task forces or a subordinate company or troop. The battalion receives an FSC for sustainment purposes (see chapter 9), normally in a direct support relationship.(See ATP 3-34.22 for additional information.)

1-14. The headquarters and headquarters company of the BEB consists of a battalion headquarters, company headquarters, CBRN reconnaissance platoon, medical section, and unit ministry team. The headquarters and headquarters company commander assists the engineer battalion commander in designating the location of the headquarters and headquarters company operations center. The battalion commander directs the location of the company. The company units (not including detachments) receive their missions from the battalion commander.

1-15. The engineer battalion headquarters consists of a command section and staff sections. Staff sections consist of the personnel staff officer (S-1), intelligence staff officer (S-2), operations staff officer (S-3), logistics staff officer (S-4), and signal staff officer (S-6). The battalion's operations section is responsible for training, operations, and plans for the battalion. The operations section includes combat (see ATP 3-34.22), general (see ATP 3-34.40), and construction surveyor (see ATP 3-34.40) engineers. A CBRN platoon, under the headquarters and headquarters company, is responsible for providing CBRN technical advice to the BEB.

> Note. Geospatial engineers are assigned to the BCT headquarters and headquarters company, overseen by the assistant brigade engineer (known as ABE) and the BCT S-2.

1-16. In some instances, the BCT commander may direct the BEB to secure one or both of the BCT's command posts (CPs), assign the BEB to their own area of operations, or give the BEB responsibility for a base perimeter or area defense. A significant change to the engineer battalion mission may affect its ability to provide engineer support to the BCT. The BCT staff weights the level of risks associated with these missions and may recommend additional engineer augmentation from echelons above brigade units to mitigate potential negative effects.

1-17. The two engineer companies of the BEB provide the BCT with the minimum capability to perform mobility, countermobility, and survivability tasks during the conduct of decisive action. These tasks include bypassing, marking, and breaching obstacles, assisting in the assault of fortified positions, emplacing obstacles to shape terrain, constructing or enhancing survivability positions, conducting route reconnaissance and information collection, and identifying and clearing explosive hazards. Supporting these tasks maintains the BCT's freedom of maneuver and inhibits the enemy's ability to mass and maneuver. Each company is slightly different, but the company's primary focus is to support the combat engineering discipline with breaching, gap crossing, survivability assets, and route clearance capabilities.

1-18. Engineer company A's (see figure 1-4) organizational structure is identical in the Infantry and Armored BCTs. Company A, in the IBCT and ABCT, provides combat engineer support and consists of a company headquarters and headquarters platoon, two combat engineer platoons, and one engineer support platoon. In a Stryker BCT, company A provides combat engineer support and has a company headquarters and headquarters platoon and two combat engineer platoons. Instead of an engineer support platoon, the SBCT has a bridge section and a horizontal squad. Company A provides mobility, countermobility, and survivability; and limited construction support to the three type of BCTs. Combat engineer platoons provide the three types of BCTs with capabilities for breaching and obstacle emplacement. Horizontal squads within the three types of BCTs provide specialized engineer equipment to support limited general engineering tasks assigned to the company. Rapidly emplaced bridge system capabilities reside within company A, Stryker and Armored types of BCTs and airborne IBCTs. Infantry and Stryker BCTs have mine clearing line charges. Stryker and Armored BCTs have Volcano Mine Systems able to construct scatterable antitank mine systems with self-destruct capabilities.

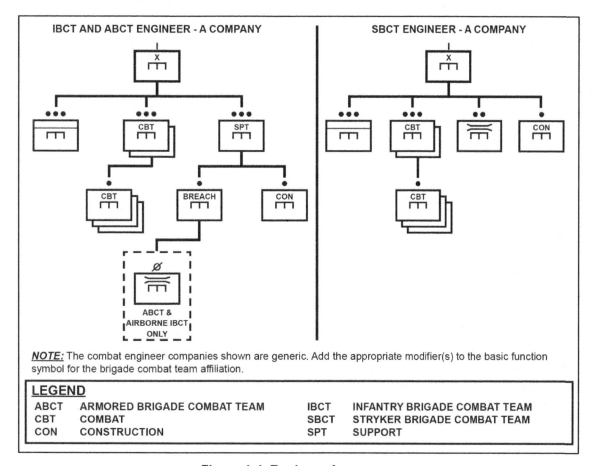

Figure 1-4. Engineer A company

1-19. Engineer company B is slightly different in the Infantry, Armored, and Stryker BCTs. Company B is generally of the same composition as engineer company A, but it has an additional route clearance platoon (see figure 1-5 on page 1-8). This platoon provides the detection and neutralization of explosive hazards and reduces obstacles along routes that enable force projection and logistics. This route clearance platoon can sustain lines of communications as members of the combined arms team or independently in a permissive environment. The Infantry and Armored BCT organizations for this company are organized the same; however, the breach section contains different equipment and capabilities. The Stryker and Armored breach section consist of bridging, whereas the IBCT breach section consists of mine clearing line charges. The IBCT currently does not have a bridging capability and requires augmentation from echelons above brigade engineers if bridging capability is required. The airborne IBCT has a rapidly emplaced bridge system. (See ATP 3-34.22 for additional information on the engineer companies within the BEB.)

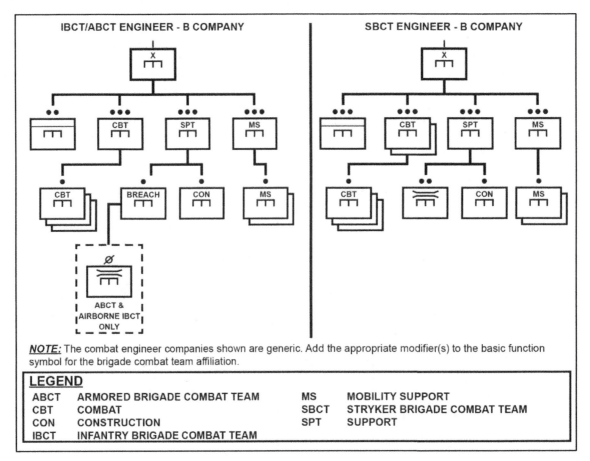

Figure 1-5. Engineer B company

1-20. The brigade signal company provides communication and information support for the BCT. It connects the BCT to the Department of Defense information network-Army (DODIN-A). The company has a headquarters and network support platoon, and two network extension platoons.

1-21. The headquarters and network support platoon consist of the company headquarters section, a small CP support team, and a RETRANS team (see figure 1-6). The company headquarters section provides command and control, logistics, and administrative support for the unit.

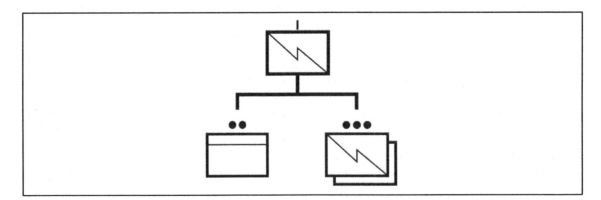

Figure 1-6. Brigade signal company

1-22. Network extension platoons' resources to provide connectivity to their assigned CPs and consist of a Joint Network Node/Secure Mobile Anti-Jam Reliable Tactical Terminal team, a high capacity line of sight section, and a RETRANS team. The Joint Network Node/Secure Mobile Anti-Jam Reliable Tactical Terminal team provides network equipment that enables CPs to use line of sight or beyond-line of sight systems. Joint network node equipment provides the connectivity to the DODIN-A satellite and terrestrial network transport systems. The Joint Network Node system connects BCT CPs, brigade support areas, higher headquarters, Army forces, and joint task forces. Each system maintains the interface capability to terminate network circuits, provide data and battlefield video teleconference services, and interface with special circuits (such as the Defense Switched Network). The Joint Network Node system provides network planning and monitoring for the BCT wide-area network. The Secure Mobile Anti-Jam Reliable Tactical Terminal provides an alternate protected satellite communications connection, the DODIN-A. The network extension section has traditional RETRANS teams and gateway systems for enhanced position location and reporting system units. RETRANS teams are managed by the BCT S-3, with S-6 coordination, to ensure specific BCT and subordinate frequencies have coverage and are adjusted as operations changes with expanding or collapsing terrain. Alternate RETRANS options include three specific frequencies uploaded into the TUAS located within the military intelligence company of the BEB.

> *Note.* The Department of Defense information network (DODIN) operations section is located within the BCT headquarters and headquarters company signal platoon, under the control of the BCT S-6. The DODIN operations section and the signal company of the BEB are the critical command and control systems during distributed operations within the BCT.

1-23. Usually, one network-extension support platoon locates at the BCT main CP. The other platoon locates in an area where best aligned to support the BCT tactical CP (when established) or at the BEB or BSB main CP. The users supported by the BCT signal company use Army command and control software and hardware capabilities to collaborate, decide, and lead the BCT's operations. (See FM 6-02 for additional information.)

1-24. The military intelligence company mission is to conduct analysis, full motion video, signals intelligence, geospatial intelligence, and human intelligence (HUMINT) activities. The military intelligence company must task organize with the BCT intelligence cell to form the brigade intelligence support element. The military intelligence company must frequently task organize its collection platoons based on the mission. Personnel from the military intelligence company maintain the threat portion of the common operational picture (COP [see paragraph 2-27]); integrate intelligence operations as part of the information collection effort (see chapter 5); and execute signals intelligence, HUMINT, and geospatial intelligence. The military intelligence company has a headquarters section, an information collection platoon, an intelligence and electromagnetic warfare (EW) systems integration (maintenance section), a TUAS platoon, multifunctional platoon (flexible design-signals intelligence, HUMINT, and site exploitation task capable), and an United States Air Force weather team (see figure 1-7 on page 1-10). Intelligence operations, conducted by the military intelligence company, collect information about the intent, activities, and capabilities of threats and relevant aspects of the operational environment to support the BCT commanders' decision-making across the range of military operations. The military intelligence company provides analysis and intelligence production support to the BCT S-2 and supports the BCT and its subordinate commands through collection, analysis, and dissemination of information and intelligence. (See ATP 2-19.4 for additional information.)

> *Note.* The military intelligence company (K series) within the BCT distributes intelligence support teams regardless of which element of decisive action (offense, defense, or stability) currently dominates. The BCT can employ anywhere from two intelligence analysts, for example to a maneuver company, or a large team of intelligence analysts as an intelligence support team to support, based on the situation, an Infantry battalion, BEB, field artillery battalion, Cavalry squadron, BSB, or to further augment the BCT intelligence cell or brigade intelligence support element (see chapter 4).

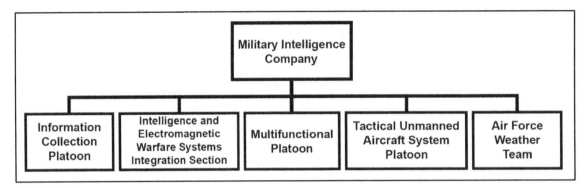

Figure 1-7. Military intelligence company

1-25. The TUAS platoon, within the military intelligence company, provides the BCT commander with an organic aerial reconnaissance, surveillance, security, and communications relay capability. (See ATP 3-04.64.) The TUAS platoon comprises of a mission planning and control section and a launch and recovery section and is equipped with four RQ-7 Shadow aircraft. TUAS platform and sensor payloads include electro-optical and infrared cameras. Onboard Global Positioning System (GPS) instruments provide navigational information. Sensor capabilities are based on a variety of factors, including altitude, field of view, depression angle, sensor payload, and standoff range. The air vehicle provides coverage for up to 7 hours at 50 kilometers (31 miles) from the launch and recovery site. The maximum range, which is limited by the data link capability, is 125 kilometers (78 miles). Imagery collection from the UAS platoon assists commanders and planners primarily by—

- Providing situational awareness of the terrain, both natural and manmade, to support the creation of products by the geospatial intelligence cell to support the staff's conduct of intelligence preparation of the battlefield (IPB) via—
  - Various baseline geospatial intelligence-based studies, such as helicopter landing zones.
  - Port and airfield studies.
  - Gridded reference graphics.
- Using imagery as a confirming source of intelligence for another intelligence discipline, such as signals intelligence or HUMINT.
- Supporting the targeting effort, including information for combat assessment through the detection and tracking of targets before and after an attack.

1-26. The CBRN reconnaissance and surveillance platoons that support the BCT are organized and equipped in two capability variations, a light platoon and a heavy platoon. The light platoon primarily conducts dismounted reconnaissance and surveillance using dismounted sets, kits, and outfits and up-armored vehicles. The platoon provides subject matter expertise on CBRN environments and is a force multiplier for survivability in dense urban terrain or subterranean environments. The heavy platoon conducts mounted reconnaissance using a CBRN reconnaissance vehicle. The platoon possesses the speed and protection to rapidly identify and mark routes and large areas while maintaining tempo with armored and mechanized forces. These two platoon variations primarily perform two information collection tasks (reconnaissance and surveillance) directed at CBRN targets. (See ATP 3-11.37 for additional information.) The CBRN reconnaissance platoons have the following capabilities:

- Detect and provide field confirmatory identification of CBRN hazards.
- Locate, identify, mark, and report contaminated areas.
- Collect CBRN samples as required in the overall sample management plan and coordinate for sample evacuation.
- Assess and characterize hazards to confirm or deny the presence of CBRN material and in support of site exploitation.

1-27. The BSB is the organic sustainment unit of the BCT. The BSB plans, prepares, executes, and assesses replenishment operations to support brigade operations. The BSB ensures the BCT can conduct self-sustained operations. The BSB's command group consists of the commander, command sergeant major, executive officer (XO), and unit ministry team. The BSB staff includes the S-1, S-2, S-3, S-4, support operations, Sustainment Automation Support Management Office, and S-6 sections. The six forward support companies provide each battalion and squadron commander within the BCT with dedicated logistic assets, less class VIII (medical supplies), that meet the supported unit's requirements. The BSB also has an assigned distribution company, a field maintenance company, and a Role 2 medical company. The Role 2 medical company provides Army Health System (AHS) (health service support and force health protection) and class VIII support. The BSB within the SBCT and the ABCT provides the same function and has the same general configuration as the BSB within the IBCT, with the most significant differences in the maintenance capabilities. (See chapter 9 and ATP 4-90 for additional information.)

SECTION II – STRYKER BRIGADE COMBAT TEAM

1-28. The SBCT is an expeditionary combined arms force organized around mounted Infantry. SBCT units operate effectively in most terrain and weather conditions. The role of the SBCT is to close with the enemy by means of fire and movement to destroy or capture enemy forces, or to repel enemy attacks by fire, close combat, and counterattack to control land areas, including populations and resources. The SBCT can gain the initiative early, seize and retain *key terrain*—an identifiable characteristic whose seizure or retention affords a marked advantage to either combatant (ADP 3-90), and conduct *massed fire*—fire from a number of weapons directed at a single point or small area (JP 3-02), to stop the enemy.

1-29. The SBCT is task organized to meet specific mission requirements. All SBCTs include maneuver, field artillery, intelligence, signal, engineer, CBRN, and sustainment capabilities (see figure 1-8 on page 1-12). This organizational flexibility enables SBCTs to function across the range of military operations. Unique to the SBCT is the weapons troop (with three antitank guided missile (ATGM) platoons and three mobile gun system (known as MGS) platoons) that provides the SBCT the ability to defeat light-skinned enemy armor or task organize those assets to maneuver battalions based on mission requirements (see paragraph 1-33). Higher commanders augment the SBCT for a specific mission with additional capabilities such as aviation, Armor, field artillery, air defense, military police, civil affairs, a tactical PSYOP element, engineers, CBRN, and information systems assets.

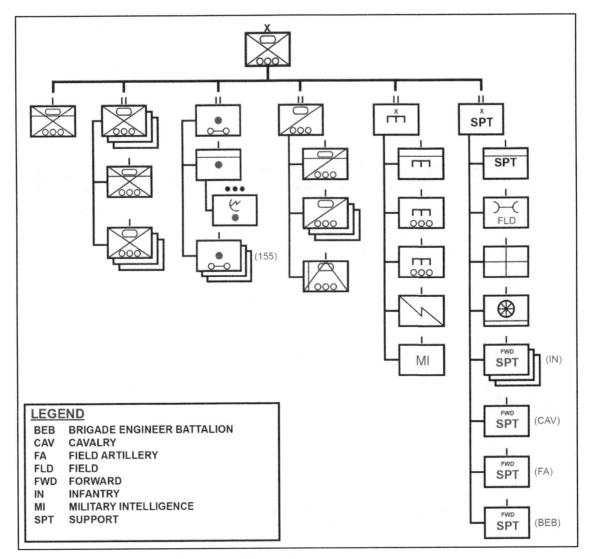

Figure 1-8. Stryker brigade combat team

1-30. SBCTs balance combined arms capabilities with significant mobility. The SBCT primarily fights as a dismounted Infantry formation that includes three SBCT Infantry battalions. The SBCT Infantry battalion has a headquarters and headquarters company, and three SBCT Infantry rifle companies each with three SBCT Infantry rifle platoons (see figure 1-9). The headquarters and headquarters company provides planning and intelligence, signal, and fire support to the battalion. The headquarters company has a battalion command section, a battalion staff section, a company headquarters, battalion medical, scout, and mortar platoons, a signal section, and a sniper squad. The headquarters company mortar platoon is equipped with 120-mm Stryker mortar carrier vehicles that have an 81-mm mortar dismounted capability. Each SBCT Infantry rifle company has a section of organic 120-mm Stryker mortar carrier vehicles that have a 60-mm mortar dismounted capability and a Raven UAS team. Habitual attachments to the SBCT Infantry rifle company include a fire support team at the company level and forward observer teams at the platoon level, medics assigned to the rifle platoons, and a senior medic at the company level. The battalion receives an FSC for sustainment purposes (see chapter 9), normally in a direct support relationship. (See ATP 3-21.21 and ATP 3-21.11 for additional information.)

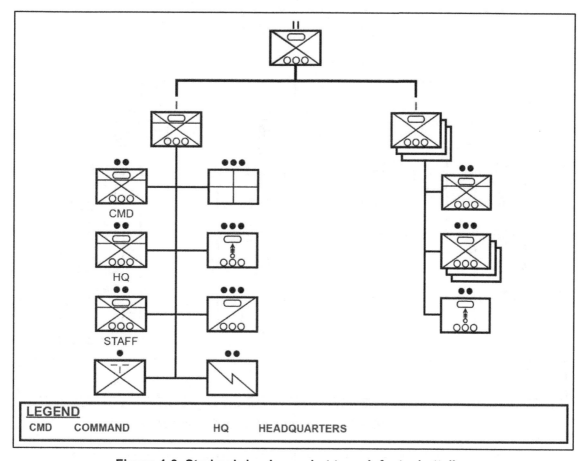

Figure 1-9. Stryker brigade combat team Infantry battalion

1-31. The Cavalry squadron of the SBCT is extremely mobile. The Cavalry squadron is composed of five troops, one headquarters and headquarters troop, three Cavalry troops equipped with Stryker reconnaissance vehicles, and weapons troop equipped with Stryker ATGM vehicles and Stryker MGS vehicles (see figure 1-10 on page 1-14). The headquarters troop organization includes a command section, the troop headquarters section, the squadron primary staff, a medical section, a sniper section, a RETRANS section, an attached fire support cell, and a TACP. The squadron receives an FSC for sustainment purposes (see chapter 9), normally in a direct support relationship. (See ATP 3-20.96.)

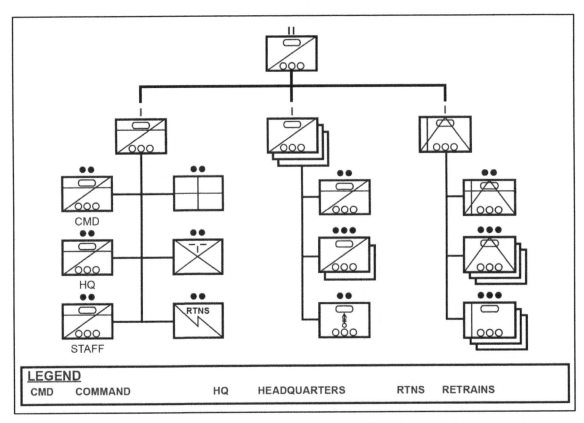

Figure 1-10. Stryker brigade combat team Cavalry squadron

1-32. Each Cavalry troop includes headquarters section, two scout platoons, a Raven UAS team, and a mortar section. The two scout platoons contain four reconnaissance vehicles, each with a crew and scout team for dismounted operations. The mortar section consists of two 120-mm mounted mortar carrier vehicles led by a sergeant first class. Habitual attachments to the Cavalry troop include a fire support team at the troop level and forward observer teams at the platoon level, medics assigned to each platoon, and a senior medic at the troop level. (See ATP 3-20.97 and ATP 3-20.98.)

1-33. The weapons troop combat power resides within its three ATGM platoons and three MGS platoons. It has a headquarters section with an assigned Infantry carrier vehicle. The ATGM platoon engages the enemy by means of long-range antiarmor fires and maneuvers to destroy or to repel the enemy's assaults by fire, and counterattack. The platoon consists of three ATGM vehicles. The MGS platoon provides precise long-range direct fire to destroy or suppress hardened enemy bunkers, machine gun positions, sniper positions, and long-range threats. It also creates Infantry breach points in urban, restricted, and open rolling terrain. The MGS 105-mm main gun provides the platoon with limited antiarmor, self-defense capabilities. The platoon consists of four MGS vehicles. Attachments include a fires support team with a fire support vehicle from the field artillery battalion to support with fires and medics with a medical support vehicle from the medical platoon of the headquarters and headquarters troop of the Cavalry squadron. (See ATP 3-21.91 for additional information.)

1-34. The SBCT field artillery battalion has four batteries: a headquarters and headquarters battery and three six-gun lightweight M777-series 155-mm towed howitzer batteries. The SBCT field artillery battalion organizes each howitzer battery with two firing platoons of three guns each. The battalion supports SBCT operations with precision, near precision, and area fires. The field artillery battalion has two AN/TPQ-53 counterfire radars and four AN/TPQ-50 lightweight countermortar radars for target acquisition. (See ATP 3-09.42 for additional information.)

1-35. Section I of this chapter discusses the BEB and the BSB within the BCT. This discussion includes the differences in these battalion formations for each type of BCT (IBCT, SBCT, and ABCT).

SECTION III – ARMORED BRIGADE COMBAT TEAM

1-36. The ABCT's role is to close with the enemy by means of fire and movement to destroy or capture enemy forces, or to repel enemy attacks by fire, close combat, and counterattack to control land areas, including populations and resources. The ABCT organizes to concentrate overwhelming combat power. Mobility, protection, and firepower enable the ABCT to conduct offensive operations with great precision and speed. The ABCT performs complementary missions to the IBCT and SBCT.

1-37. The ABCT conducts offensive operations to defeat, destroy, or neutralize the enemy. The ABCT conducts defensive operations to defeat an enemy attack, buy time, economize forces, and develop favorable conditions for offensive actions. During stability, the ABCT's commitment of time, resources, and forces establish and reinforce diplomatic and military resolve to achieve a safe, secure environment and a sustainable peace.

1-38. The ABCT conducts sustained and large-scale combat operations within the foundations of unified land operations through decisive action. The ABCT seizes, retains, and exploits the initiative while synchronizing its actions to achieve the best effects possible. During combat operations, the ABCT can fight without additional combat power but can be task organized to meet the precise needs of its missions. The ABCT conducts expeditionary deployments and integrates its efforts with unified action partners.

1-39. The ABCT (figure 1-11 on page 1-16) is a combined arms organization consisting of three combined arms battalions of Armor and mechanized Infantry companies. Cavalry, field artillery, engineer, intelligence, signal, sustainment, TUAS, and CBRN reconnaissance units are organic to the ABCT. Higher commanders augment the ABCT for a specific mission with additional capabilities. Augmentation can include aviation, Infantry, field artillery, air defense, military police, civil affairs, tactical PSYOP element, engineers, CBRN, and additional information systems assets.

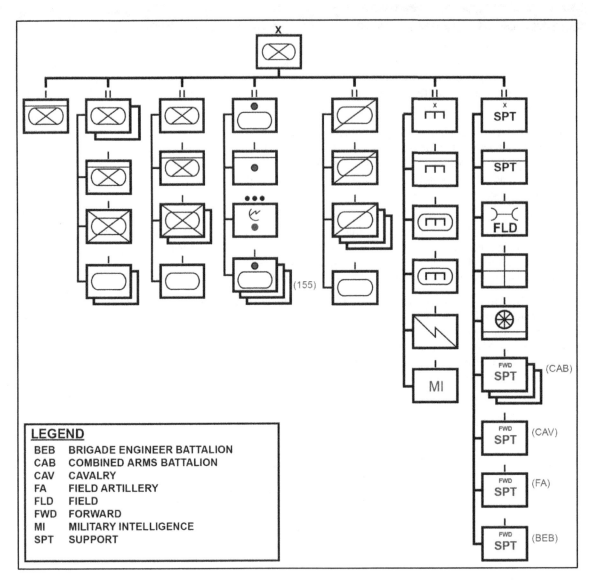

Figure 1-11. Armored brigade combat team

1-40. Three combined arms battalions are the ABCT's primary maneuver force. Each combined arms battalion conducts sustained combined arms and close combat operations as an essential part of the ABCT formation. The combined arms battalions of the ABCT serve as a deterrent to armed conflict; they can deploy worldwide in the conduct of decisive action. Combined arms battalions execute operations within their assigned areas of operations in support of the commander's scheme of maneuver. The combined arms battalion receives an FSC for sustainment purposes (see chapter 9), normally in a direct support relationship.

1-41. Combined arms battalions combine the efforts of their Armor companies and mechanized Infantry companies along with their headquarters and headquarters company to execute tactical missions as part of a combined arms operation. Within the ABCT, two combined arms battalions (see figure 1-12) have two Armor companies (each with three tank platoons and a headquarters section) and one mechanized Infantry company (with three mechanized Infantry platoons, a headquarters section, and a Raven UAS team); and one combined arms battalion has two mechanized Infantry companies (each with three mechanized Infantry platoons, a headquarters section, and a Raven UAS team) and one Armor company (with three tank platoons and a headquarters section). The headquarters and headquarters company of each combined arms battalion provides planning and intelligence, signal, and fire support to the battalion. Each headquarters company has a battalion command section, a battalion staff section, a company headquarters, battalion medical, scout, and mortar platoons, a signal section, and a sniper squad. The headquarters company mortar platoon is equipped with 120-mm mortar carrier vehicles that have a 120-mm mortar dismounted capability. Habitual attachments to the maneuver companies include a fire support team at the company level and forward observer teams at the platoon level, medics assigned to the rifle platoons, and a senior medic at the company level. (See ATP 3-90.5 and ATP 3-90.1 for additional information.)

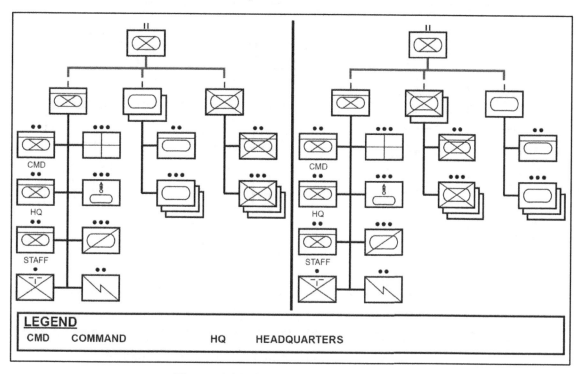

Figure 1-12. Combined arms battalion

1-42. The fundamental purpose of the Cavalry squadron is to perform reconnaissance and security operations in close contact with the enemy and civilian populations, often in conjunction with fighting for information to support the ABCT commander. The conduct of security operations by the squadron provides an economy of force while allowing the ABCT commander the flexibility to conserve combat power for engagements where better desired.

1-43. The Cavalry squadron has a headquarters and headquarters troop, three ground Cavalry troops, and an Armor company (see figure 1-13 on page 1-18). The headquarters troop organization includes a command group, the troop headquarters section, the squadron primary staff that is; personnel, intelligence, operations, logistics, signal, the medical platoon, an attached fire support cell, and a TACP. The squadron has 120-mm self-propelled mortars (see ATP 3-20.96). The squadron receives an FSC for sustainment purposes (see chapter 4), normally in a direct support relationship. The ground Cavalry troops have two platoons with six Bradley fighting vehicles and a Raven UAS team. The Armor company has three platoons with four M1 Abrams main battle tanks. Habitual attachments to the Cavalry troop tank and company include a fire support team at the troop/company level and forward observer teams at the platoon level, medics assigned to each platoon, and a senior medic at the troop/company level. (See ATP 3-20.15, ATP 3-20.97, and ATP 3-20.98.)

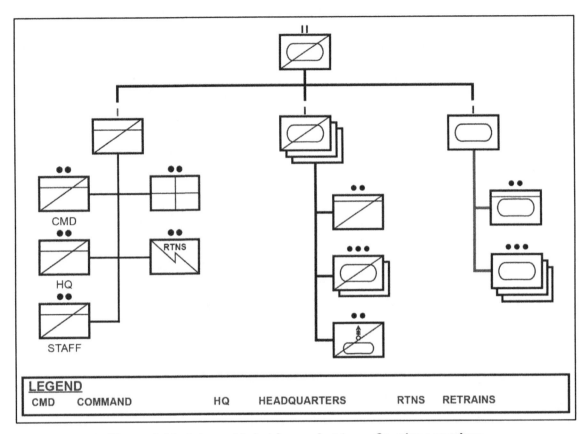

Figure 1-13. Armored brigade combat team Cavalry squadron

1-44. The ABCT field artillery battalion has four batteries, a headquarters and headquarters battery and three batteries of six M109 (family of vehicles) Paladin self-propelled 155-mm howitzers. The batteries are manned and equipped to operate as two separate firing platoons of three guns. The field artillery battalion provides massing fires in space and time on single or multiple targets with precision, near precision, and area fires to support ABCT operations. The field artillery battalion has two AN/TPQ-53 counterfire radars and four AN/TPQ-50 lightweight countermortar radars for target acquisition. (See ATP 3-09.42 for additional information.)

1-45. Section I of this chapter discusses the BEB and the BSB within the BCT. This discussion includes the differences in these battalion formations for each type of BCT (IBCT, SBCT, and ABCT).

Chapter 2

The Brigade Combat Team and the Operational Environment

This chapter discusses the brigade combat team's (BCT) role in military operations and its interactions with operational environments. The chapter addresses key doctrinal concepts on how the Army fights regardless of which element of decisive action (offense, defense, or stability) currently dominates the BCT's area of operations.

SECTION I – OPERATIONAL OVERVIEW

2-1. Threats to United States interests throughout the world are countered by the ability of U.S. forces to respond to a wide variety of challenges along a competition continuum that spans cooperation to war. Threats seek to mass effects from multiple domains quickly enough to impede joint operations. Threats attempt to impede joint force freedom of action across the air, land, maritime, space, and cyberspace domains. Understanding how threats present multiple dilemmas to Army forces from these domains help Army commanders identify (or create), seize, and exploit opportunities during operations to achieve a position of advantage relative to the enemy. This section briefly covers key doctrinal concepts on how the Army fights and how operational environments shape the nature and affect the outcome of military operations. (See ADP 3-0 and FM 3-0 for a complete discussion of these doctrinal concepts.)

ARMY STRATEGIC ROLES

2-2. The Army's primary mission is to organize, train, and equip its forces to conduct prompt and sustained land combat to defeat enemy ground forces and seize, occupy, and defend land areas. The Army accomplishes its mission by supporting the joint force and unified action partners in four strategic roles: shape operational environments, prevent conflict, prevail in large-scale ground combat, and consolidate gains. The strategic roles clarify the enduring reasons for which the Army is organized, trained, equipped, and employed. Strategic roles are not tasks assigned to subordinate units. (See ADP 3-0 for additional information.)

MILITARY OPERATIONS

2-3. The range of military operations is a fundamental construct that relates military activities and operations in scope and purpose against the backdrop of the competition continuum (see figure 2-1 on page 2-2). Rather than a world either at peace or at war, the competition continuum describes a world of enduring competition conducted through a mixture of cooperation, competition below armed conflict, and armed conflict and war. The potential range of military activities and operations extends from military engagement and security cooperation, up through large-scale combat operations in war. Whether countering terrorism as part of a crisis response or limited contingency operation, deterring an adversary or enemy from taking undesirable actions, or defeating a peer threat in large-scale ground combat, the nature of conflict is constant.

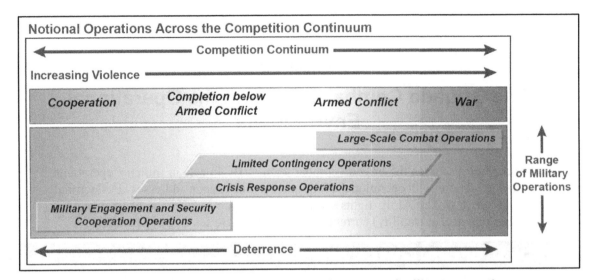

Figure 2-1. Competition continuum and the range of military operations

2-4.   Given the complex and dynamic nature of an operational environment, the competition continuum of force does not proceed smoothly from cooperation, through competition below armed conflict, to armed conflict and war and back. Within the backdrop of the competition continuum, unstable peace may erupt into an insurgency that quickly sparks additional violence throughout a region, leading to a general state of war.

2-5.   Military engagement and security cooperation activities build networks and relationships with partners, shape regions, keep day-to-day tensions between nations or groups below the threshold of armed conflict, and maintain U.S. global influence. Crisis response and limited contingency operations typically focus in scope and scale to achieve specific strategic or operational-level objectives in an operational area. Large-scale combat operations occur as major operations and campaigns aimed at defeating an enemy's armed forces and military capabilities in support of national objectives.

2-6.   Deterrence applies across the competition continuum. The purpose of deterrence is to dissuade an adversary from taking undesirable actions because of friendly capabilities and the will to use them. Deterrence takes different forms according to the situation and where Army forces are on the competition continuum. Many of the operations listed in the range of military operations may serve as deterrents in certain situations to deter further unwanted actions. (See ADP 3-0 for additional information.)

OPERATIONAL ENVIRONMENT

2-7.   An *operational environment* is a composite of the conditions, circumstances, and influences that affect the employment of capabilities and bear on the decisions of the commander (JP 3-0). Operational environments shape the nature and affect the outcome of military operations. The complex and dynamic nature of an operational environment and the threats that exist within an operational environment make determining the relationship between cause and effect difficult and contributes to the uncertainty of the military operation. BCTs and subordinate units must make every effort to understand the specific operational environment in which they operate in.

ANTICIPATED OPERATIONAL ENVIRONMENTS

2-8.   The operational environment encompasses physical areas of the air, land, maritime, space, and cyberspace domains (see ADP 3-0). Included as well in the operational environment is the *information environment*—the aggregate of individuals, organizations, and systems that collect, process, disseminate, or act on information (JP 3-13) and the electromagnetic spectrum (EMS)—the range of frequencies of electromagnetic radiation from zero to infinity (see ATP 6-02.70). Included within these are adversary, enemy, friendly, and neutral actors that are relevant to a specific operation (see chapter 3). Although an operational environment and information environment are defined separately, they are interdependent and integral to each other. In fact, any activity that occurs in the information environment simultaneously occurs in and affects one or more of the operational environment domains.

2-9. The information environment is comprised of three dimensions: physical, informational, and cognitive. Within the physical dimension of the information environment is the connective infrastructure that supports the transmission, reception, and storage of information. Also, within this dimension are tangible actions or events that transmit a message in and of themselves, such as reconnaissance patrols, ground and aerial surveillance, civil affairs projects, and intelligence operations efforts. Within the informational dimension is the content or data itself. The informational dimension refers to content and flow of information, such as text or images, or data that staffs can collect, process, store, disseminate, and display. The informational dimension provides the necessary link between the physical and cognitive dimensions. Within the cognitive dimension are the minds of those who are affected by and act upon information. These minds range from friendly commanders and leaders, to foreign audiences affecting or being affected by operations, to enemy, threat or adversarial decision makers. (See FM 3-13, chapter 1 for additional information about each dimension.)

2-10. The information element of combat power (see paragraph 2-79) is integral to optimizing combat power, particularly given the increasing relevance of operations in and through the information environment to achieve decisive outcomes. Information operations and the information element of combat power are related but not the same. Information is a resource. As a resource, it must be obtained, developed, refined, distributed, and protected. Information operations, along with knowledge management and information management, are ways the BCT can harness this resource and ensure its availability, as well as operationalize and optimize it. (See paragraphs 4-88 to 4-96 for a discussion of knowledge and information management).

2-11. The operational environment of the BCT includes all enemy, adversarial, friendly, and neutral systems across the range of military operations and is part of the higher commander's operational environment. The BCT's operational environment includes the physical environment, the state of governance, technology, and local resources, and the culture of the local populace. As the operational environment for each operation is different, it also evolves as the operation progresses. Commander and staff continually assess and reassess the operational environment as they seek to understand how changes in the nature of threats and other variables affect not only their force but other actors as well. The commander and staff use the Army design methodology (see paragraph 4-47), operational variables, and mission variables to analyze an operational environment in support of the operations process.

Operational and Mission Variables

2-12. When alerted for deployment, redeployment within a theater of operations, or assigned a mission, the BCT's assigned higher headquarters (generally the division) provides an analysis of the operational environment. That analysis includes *operational variables*, a comprehensive set of information categories used to describe an operational environment (ADP 1-01). The categories are political, military, economic, social, information, infrastructure, physical environment, and time (PMESII-PT). The purpose of operational variables is to provide a broad, general set of information categories that assist commanders and staffs in analyzing and developing a comprehensive understanding of an operational environment.

2-13. Upon receipt of a mission, commander and staff filter information categorized by operational variables into *relevant information*—all information of importance to the commander and staff in the exercise of command and control (ADP 6-0)—with respect to the assigned mission. The commander uses *mission variables*—the categories of specific information needed to conduct operations (ADP 1-01), to focus on specific elements of an operational environment during mission analysis. This analysis enables the commander and staff to combine operational variables and tactical-level information with knowledge about local conditions relevant to their mission. Mission variables are mission, enemy, terrain and weather, troops and support available, time available, civil considerations (METT-TC). (See FM 6-0 for additional information.)

Threat and Hazards

2-14. For every operation, threats and hazards are a fundamental part of the operational environment. Commanders at all levels must understand threats, criminal networks, enemies, and adversaries, to include both state and nonstate actors, in the context of their operational environment. When the commander understands the threat, the commander can visualize, describe, direct, lead, and assess operations to seize, exploit, and retain the initiative. Threats include any combination of actors, entities, or forces that have the capability and intent to harm U.S. forces, U.S. national interests, or the homeland. Threats may include individuals, groups of individuals (organized or not organized), paramilitary or military forces, nation-states, or national alliances. When threats execute their capability to do harm to the United States, they become categorized as an enemy. (See chapter 3 for a detailed discussion of the possible threats within the BCT's operational environment.)

2-15. A *hazard* is a condition with the potential to cause injury, illness, or death of personnel; damage to or loss of equipment or property; or mission degradation (JP 3-33). Hazards include disease, extreme weather phenomena, solar flares, and areas contaminated by toxic materials. Hazards can damage or destroy life, vital resources, and institutions, or prevent mission accomplishment. Understanding hazards and their effects on operations allows the commander to understand better the terrain, weather, and various other factors that best support the mission. Understanding hazards also helps the commander visualize potential impacts on operations. Successful interpretation of the environment aids in correctly opposing threat courses of action (COAs) within a given geographical region. Possible hazards within the BCT's operational environment are addressed throughout this manual.

SPECIFIC OPERATIONAL ENVIRONMENTS

2-16. Specific operational environments include urban, mountain (includes cold weather regions), desert, and jungle. Subsurface areas are conditions found in all four operational environments. Offensive, defensive, and stability operations tasks in these environments follow the same planning process as operations in any other environment, but they do impose specific techniques and methods for success. The uniqueness of each environment may affect more than their physical aspects but also their informational systems, flow of information, and decision-making. As such, mission analysis must account for the information environment and cyberspace within each specific operational environment. Each specific operational environment has a specific manual because of their individual characteristics.

Urban Terrain

2-17. Operations in urban terrain are Infantry-centric combined arms operations that capitalize on the adaptive and innovative leaders at the squad, platoon, and company level. Plans must be flexible to promote disciplined initiative by subordinate leaders. Flexible plans are characterized by a simple scheme of maneuver and detailed control measures that enable combined arms operations. When assigning areas of operation to subordinate units, commanders must consider the size and density of the civilian population or noncombatants within urban terrain to prevent units from culminating early due to large numbers of civilians and noncombatants in their area of operations. In the offense, task organizing battalion and company-combined arms team at the right place and time is key to achieving the desired effects. In the defense, the combined arms team turns the environment's characteristics to its advantage. Urban areas are ideal for the defense because they enhance the combat power of defending units. (See ATTP 3-06.11 and ATP 3-06 for additional information.)

Mountainous Terrain and Cold Weather Environments

2-18. Operations in mountainous terrain are conducted for three primary purposes: to deny an enemy a base of operations; to isolate and defeat enemy; and to secure lines of communication. Enemy tactics commonly involve short violent engagements followed by a hasty withdrawal through preplanned routes. The enemy often strikes quickly and fights only as long as the advantage of the initial surprise is in their favor. Attacks may include direct fires, indirect fires, or improvised explosive devices (IEDs) and may be against stationary or moving forces. The design of the landscape, coupled with climatic conditions, creates a unique set of mountain operations characteristics that are characterized by close fights with dismounted Infantry, decentralized small-unit operations, degraded mobility, increased movement times, restricted lines of communications, and operations in thinly populated areas. (See ATP 3-90.97, ATP 3-21.50, and ATP 3-21.18 for additional information.)

Desert Terrain

2-19. Operations in desert terrain require adaptation to the terrain and climate. Equipment must be adapted to a dusty and rugged landscape with extremes in temperature and changes in visibility. The BCT orients on primary enemy approaches but prepares for an attack from any direction. Considerations for operations in desert terrain include lack of concealment and the criticality of mobility; use of obstacles to site a defense, which are limited; strong points to defend choke points and other key terrain; and mobility and sustainment. (See FM 90-3 for additional information.)

Jungle Terrain

2-20. Operations in jungle terrain combine dispersion and concentration. For example, a force may move out in a dispersed formation to find the enemy. Once the force makes contact, its subordinate forces close in on the enemy from all directions. Operations are enemy-oriented, not terrain-oriented. Forces should destroy the enemy wherever found. If the force allows the enemy to escape, the force will have to find the enemy again, with all the risks involved. Jungle operations use the same defensive fundamentals as other defensive operations. Considerations for offensive and defensive operations in a jungle environment include limited visibility and fields of fire, ability to control units, and limited and restricted maneuver. (See ATP 3-90.98 for additional information.)

Subsurface Areas

2-21. A subsurface area is a condition found in all four operational environments described in paragraphs 2-16 through 2-20. Subsurface areas are below the surface level (ground level) that may consist of underground facilities, passages, subway lines, utility corridors or tunnels, sewers and storm drains, caves, or other subterranean spaces. This dimension includes areas both below the ground and below water. Additional subterranean areas include drainage systems, cellars, civil defense shelters, mines, military facilities, and other various underground utility systems. In older cities, subsurface areas can include pre-industrial age hand-dug tunnels and catacombs. (See ATP 3-21.51 for information on threat and hazardous subterranean structures existing or operating in concealment or hidden or when utilized in secret by an enemy or adversary.)

2-22. Subsurface areas may serve as secondary and, in fewer instances, primary avenues of approach at lower tactical levels. Subsurface areas are used for cover and concealment, troop movement, command functions, and engagements, but their use requires intimate knowledge of the area. When thoroughly reconnoitered and controlled, subsurface areas offer excellent covered and concealed lines of communications for moving supplies and evacuating casualties. Attackers and defenders can use subsurface areas to gain surprise and maneuver against the rear and flanks of an enemy and to conduct ambushes. However, these areas are often the most restrictive and easiest to defend or block. The commander may need to consider potential avenues of approach afforded by the subsurface areas of rivers and major bodies of water that border urban areas.

2-23. Knowledge of the nature and location of subsurface areas is of great value to both friendly and enemy forces. The effectiveness of subsurface areas depends on superior knowledge of their existence and overall design. A thorough understanding of the environment is required to exploit the advantages of subsurface areas. Maximizing the use of these areas could prove to be a decisive factor while conducting offensive and defensive operations, and stability operations tasks. (See ATTP 3-06.11, TC 2-91.4, ATP 3-06, and ATP 3-34.81 for additional information on subsurface areas.)

SECTION II – UNDERSTAND, SHAPE, AND INFLUENCE

2-24. The BCT commander must understand the operational objectives and tactical situation, cultural conditions, and ethical challenges of the operational environment, in order to shape the operational environment through action, influence the population and its leaders, and consolidate gains (see section III) to seize, retain, and exploit disciplined initiative throughout the range of military operations (see paragraph 2-3). Regardless of which element of decisive action (offense, defense, or stability) currently dominates (see paragraph 2-73), the commander conducts multiple missions assessing and balancing the need for decisive action, judicious restraint, security, and protection to shape the operational environment and seeks to achieve a common goal and end state that nests with higher command objectives. The BCT commander must understand competing interests within the operational and information environment to determine what is of value to competitive parties (to include identified adversaries or enemies) and entities within the BCT's area of operations. Understanding competing interests helps the commander develop ethical, effective, and efficient COAs that influence the populace and political structure, enhance the security situation, and lead to mission success. Additionally, the commander anticipates ethical risks associated with the BCT's operational environment and understands how adherence to the Army Ethic provides moral basis for decisive action and how it becomes a force multiplier in all operations (see ADP 1).

It Is All About the Information

1 In both Joint and Army doctrine, information operations is defined as "the integrated employment, during military operations, of information-related capabilities in concert with other lines of operation to influence, disrupt, corrupt, or usurp the decision-making of adversaries and potential adversaries while protecting our own." 2 In more general terms, information operations support the commander's ability to achieve a position of relative advantage through activities in the information environment (the physical, informational, and cognitive dimensions) to influence the adversary's will to fight; disrupt, corrupt, or usurp its capabilities to collect, process, and disseminate information; and ultimately manipulate (deceive) or disrupt an adversary decision-maker's understanding of the operational environment.

Commanders visualize and understand the operational environment through information. As an element of combat power, information enables decision-making, and its transmission aids decisive operations. Today, modern technology has significantly increased the speed, volume, and access to information. Concurrently, technology has enabled significant means to disrupt, manipulate, distort, and deny information—technology that adversaries have already demonstrated a willingness to use with great effect.

"It is all about the information;" that whoever controlled the information could dominate competition and conflict. 3 In large-scale combat operations, this remains as true as ever. Commanders direct resources toward intelligence collection in order to develop the situation and gain the sufficient information required to make a timely and informed decision. Just as importantly, measures must be put into place to protect friendly information while simultaneously developing and executing means in all domains to attack the adversary's ability to access, process and disseminate information. In this way information operations enable an accurate understanding of the operational environment while disrupting or manipulating that of the adversary. Through information operations, the adversary/enemy decision-maker's reality should be that which best supports achieving a position of relative advantage. That said, more needs to be done to fully garner the true potential of information as an element of combat power in a large-scale combat operations context. Common sense dictates that information absent accompanying action does not resonate cognitively in the same way when both are present and complementary. The duality of the relationship between action and information must become a constant theme of operations in the Information Age" of the 21st Century.

In Iraq and Afghanistan, the Taliban and al Qaeda staged countless engagements against United States and its partners, less for the physical effects in the immediate operational environment, but rather to gain an informational advantage around the world. Videotaped improvised explosive device attacks, while devastating, worked well to promote an image of organizational credibility, bolster adherents' will to fight, radicalize vulnerable populations, and increase financial support. More importantly with respect to large-scale combat operations, Russian information confrontation activity preceding, during, and following its illegal annexation of Crimea and invasion of eastern Ukraine demonstrates the power of integrated operations in the information environment, in this case more appropriately termed information warfare. Russia successfully sowed disinformation causing the international community to distrust the information it was receiving while also crippling the Ukrainian response through cyberspace operations, electronic warfare, and psychological operations. The confusion and misdirection caused by Russian information warfare had a paralytic effect on Western decision-makers—so much so that Russia was able to achieve its strategic and political objectives before Western leaders could mount a credible response.

Various

UNDERSTANDING THE OPERATIONAL ENVIRONMENT

2-25. Interests are motivations that provide insight to perceived rights, influences, responsibilities, and power. Interests influence how populations perceive complexity, physical security, political systems, economic influence, tribal and religious identity, self-serving, or a combination of two or more. The BCT commander and staff develop an understanding of operational variables—PMESII-PT and mission variables—METT-TC through information collection to enhance situational awareness and understanding of competing interests. At the tactical level, intelligence operations, reconnaissance, security operations, and surveillance are the four-primary means of information collection. The commander and staff can frame a problem if they understand competing interests within the area of operations. They seek to understand the motivations and recognize that each interest has multiple perspectives. The commander and staff consider political interests from multiple perspectives to operate effectively under conditions of complexity and in close contact with enemies and populations. Understanding interests assist the commander and staff to synchronize information-related capabilities that shape the information environment and to modify behaviors to further sustainable objectives.

2-26. Understanding interests requires analysis of operational variables and mission variables within a particular region. Understanding requires an appreciation of the operational environment's complex, humanistic, and political environs within the context of war as a contest of wills. The BCT commander and staff must develop an understanding of the local audience's cultural communication techniques to communicate with them effectively. They also must understand that the most important aspect of cultural communication is how the population receives the information rather than how the unit transmits the information. Determination of valued interests within an area of operations provides options for the BCT to establish programs that incentivize cooperation leading to mission accomplishment. Comprehension of interests and anticipation of ethical risks allows understanding to implement disincentives that seek to coerce and persuade adversaries, enemies, and neutral parties with interest's counter to the objectives the BCT and higher have established. The understanding and acknowledgement of interests help to frame information-related capabilities in future operations.

2-27. Efforts to understand interests begin before deployment (see paragraph 2-141). Country studies, analysis of the social demographics, constructs of local, sub-national and national governance, and understanding of key personalities and organizations within the BCT's future area of operations provide baseline knowledge to increase situational awareness and identify potential areas of friction before deployment. The BCT commander and staff consider operational variables and mission variables within their area of operations to gain an understanding of the interests and motivations particular to different groups and individuals to enhance situational awareness. Unified action partners and Army special operations forces are key resources that the BCT uses to develop situational understanding to shape efforts that lead to a sustainable, secure environment. Analysis of these resources allows informed leaders to identify information gaps and develop COAs that increase their situational understanding within their area of operations. The *common operational picture*—a display of relevant information within a commander's area of interest tailored to the user's requirements and based on common data and information shared by more than one command (ADP 6-0)—is key to achieving and maintaining situational understanding. Although the common operational picture (COP) is ideally a single display, it may include more than one display and information in other forms, such as graphical representations or written reports. BCTs primarily leverage the Command Post of the Future (known as CPOF), or its replacement the Command Post Computing Environment (known as CPCE) to establish a digital COP.

Notes. The COP is holistic, with layers that include for example, friendly forces, enemy forces, medical, sustainment, fire support, and engineer assets (to include obstacles) positioning. Over digital systems, the individual users depending on what information is required at any given moment can turn information systems on or off. This ensures the user has accurate information when required, for example, critical medical or sustainment nodes, and clean/dirty route locations.

CPCE, replacing CPOF is the primary computing environment under the common operating environment. CPCE is the central computing environment developed to support command posts (CPs) and combat operations and will be interoperable with mounted and mobile/handheld systems. CPCE provides an integrated, interoperable, and cyber-secure computing infrastructure framework for multiple warfighting functions. CPCE provides Army units with a core infrastructure, including a COP tool, common data strategy, common applications such as mapping and chat, common hardware configurations and common look and feel (user interface). CPCE provides an integrated command and control capability across CP and platforms, through all echelons, and provides simplicity, intuitiveness, core services and applications, and warfighter functionality in the areas of fires, logistics, intelligence, airspace management, and maneuver.

2-28. The BCT conducts information collection through intelligence operations, reconnaissance, security operations, and surveillance that focus on intelligence requirements to bridge information gaps. Gaps identified during intelligence preparation of the battlefield (IPB) develop into information requirements through aggressive and continuous operations to acquire information. The BCT staff considers operational variables and mission variables, with emphasis on civil considerations, to understand the interests within the area of operations. *Civil considerations* are the influence of manmade infrastructure, civilian institutions, and attitudes and activities of the civilian leaders, populations, and organizations within an area of operations on the conduct of military operations (ADP 6-0). Information requirements that develop situational understanding of the interests within an area of operations are defined and collected by focusing civil considerations within the characteristics of areas, structures, capabilities, organizations, people, and events, (ASCOPE). The commander uses various capabilities (civil affairs, psychological operations [PSYOP] forces, intelligence professionals, academic studies, and U.S. Embassy staff) to understand the nuances and particulars of organizations and people within the area of operations.

2-29. The commander and staff consider culture and pillar organizations that influence the operational environment's civil considerations. Culture is the shared beliefs, values, customs, behaviors, and artifacts members of a society use to cope with the world and each other. Pillar organizations are organizations or systems on which the populace depends for support, security, strength, and direction. Examination of a culture gives insight to the motivations and interests of people and organizations. Consideration of a culture is imperative to successful shaping operations that set conditions for future successes. Thorough understanding of the interests of groups and individuals allow for informed and viable COAs that seek to favorably shape the environment and contribute to positive outcomes and objectives within the BCT's area of operations.

2-30. Host-nation security organizations and political partners provide invaluable insight into values, beliefs, and interests. As organizations are comprised of the people, they secure and govern, their native fluency in the customs, courtesies, cultures, beliefs, interests, and ideals provide the partnering BCT cultural perspective and intelligence that develop understanding of the operational environment. Close positive relationships with host-nation partners breed trust, which leads to an enhanced understanding of the operational environment.

SHAPE THE ENVIRONMENT

2-31. Army operations to shape bring together all the activities intended to promote regional stability and to set conditions for a favorable outcome in the event of a military confrontation. Army operations to shape help dissuade adversary activities designed to achieve regional goals short of military conflict. Shaping activities, although not all inclusive, include security cooperation and forward presence to promote U.S. interests and develop allied and friendly military capabilities for self-defense and multinational operations. Regionally aligned and engaged Army forces are essential to accomplishing objectives to strengthen the global network of multinational partners and preventing conflict.

Regionally Aligned and Engaged Army Forces

1 As part of a response to Russian aggression in Ukraine, the deployment of the 3rd Armored brigade combat team (ABCT), 4th Infantry Division (ID) in January 2017, marked the start of back-to-back, nine-month rotations of U.S. troops and equipment to Poland. The 3rd ABCT consisted of approximately 3,500 Soldiers, 80 plus M1 Abrams tanks, 140 plus M2 Bradley fighting vehicles, 15 plus M109A6 Paladins, and 400 plus high mobility multipurpose wheeled vehicles. The deployment of both personnel and equipment from a continental United States location forced theater level sustainment and transportation functions to exercise their systems for the first time and increased the overall ability of the United States to project combat power into Eastern Europe. Following the ABCT's consolidation in Poland near the Drawsko, Pomorskie, and Zagan training areas, the ABCT dispersed across seven locations in Eastern Europe for training and exercises with European allies.

Along with 3rd ABCT, the 10th Combat Aviation Brigade from Fort Drum, New York, deployed to Europe in February 2017 and headquartered in Illesheim, Germany, with forward-positioned aircraft in task forces in Latvia, Romania, and Poland. The 10th Combat Aviation Brigade consisted of approximately 2,200 Soldiers (including 400 Soldiers from an additional attached aviation battalion), 24 AH-64 Apaches, 50 UH-60 Blackhawks, and 10 CH-47 Chinooks. 2

Once 3rd ABCT completed reception, staging, onward movement, and integration, the BCT conducted various training, security cooperation, and security assistance missions across seven counties from Estonia to Bulgaria. These rotations enhanced deterrence capabilities, increased the ability to respond to potential crises, and defended North Atlantic Treaty Organization and its allies. The ABCT rotations remained under U.S. command and focused on strengthening capabilities, interoperability with coalition partners, and sustaining readiness through bilateral and multinational training and exercises. 3 Ultimately the actions taken by 3rd ABCT, 4th ID, created greater understanding of the unique conditions within the European operating environment and helped shape the operational environment for future operations.

Operation Atlantic Resolve 17 shaped the European operational environment in several discreet ways. It established routine deployment of forces that expanded to nearly 6,000 personnel, 2,500 wheeled vehicles and pieces of equipment, 385 tracked vehicles, and almost 100 rotary-wing aircraft into the European theater. It also increased cooperation and interoperability between the United States and more than a dozen counties to include Belgium, Bulgaria, Czech Republic, Denmark, Estonia, Georgia, Germany, Greece, Hungary, Italy, Lithuania, Netherlands, Norway, Poland, Romania, Slovakia, and the United Kingdom. It established key logistical nodes and infrastructure capable of supporting rapid deployment of combat credible forces. Finally, it deterred further Russian aggression into Eastern Europe and provided a credible threat of force in case of future aggression. 4

Various

2-32. Shaping the operational environment requires a deep understanding of competing dynamics across all domains within the BCT's area of operations. The commander builds mutual trust within the BCT and externally with both coalition partners and indigenous population, to support shaping the operational environment. Mutual trust developed through commander and staff actions, shape and set conditions that facilitate actions consistent with the understanding of competing interest and the moral principles of the Army Ethic.

2-33. The BCT commander and staff consider the competitive environment of the area of operations and shape the operational environment to set conditions to seize, retain, and exploit the initiative. Different political entities, different personalities, tribal dynamics, religious interests, economic motivations, sources of security, and potential havens of refuge for enemies all contribute to the competitive nature of the operational environment. Not all of these interests are parallel and mutually supportive of the objectives and end state for a particular region. The commander and staff develop situational understanding and influences personalities and organizations to achieve objectives to shape the environment. Shaping the environment includes persuading and empowering other personalities and organizations to modify behaviors and actions consistent with the BCT commander's intent and objectives. Setting conditions is an enduring process throughout all *phases*—a planning and execution tool used to divide an operation in duration or activity (ADP 3-0)—of an operation.

2-34. Shaping an area of operations requires integration of a broad variety of missions across multiple domains that can range from raids, civil reconnaissance, tactical PSYOP, and engaging with local leaders. In order to plan and execute these missions the BCT as well as its subordinate units must understand how the civil considerations affect the other mission variables. In particular, commanders may need to emphasize understanding organizations and people to include the relationships between them. When conditions are not favorable for future operations or are not aligned to the desired end state, the BCT conducts missions to shape the area of operation, changing those conditions, and setting conditions for future operations. The BCT establishes countermeasures to counter cross-domain threats and influencers attempting to shape the battlefield themselves.

2-35. The BCT commander and staff understand through analysis of operational and mission variables, enhanced and developed thorough information collection to understand the competing dynamics within the area of operations. The commander and staff seek to understand the populations' interests and motivations and to identify pillar organizations that provide guidance, inspiration, and strength to the population. The BCT must understand who is influential in the area of operations to engage leaders, influence behaviors, and persuade neutral and fringe groups to synthesize with BCT objectives, and to plan and execute limited offensive operations that set conditions for future successes. Ultimately, greater understanding of operational and mission variables is essential to the development, planning, and execution of information-related capabilities that shape the operational environment.

2-36. The BCT commander seeks to understand the competitive interests within an area of operations and how these interests influence desired outcomes and objectives. Some interests and motivations are supportive of the BCT's objectives and others conflict, counter, and disrupt efforts of the desired end state. The BCT's ability to shape and set conditions for favorable outcomes relates to the BCT's ability to understand the influence of different competitive interests. The commander and staff seek answers to information gaps through the development of intelligence requirements that are satisfied through active information operations (see ATP 3-13.1) within a given area. Staffs develop options for the commander through information collection and analysis so the commander can inform the populace and influence various actors to shape the environments.

2-37. Analysis of the motivations and interests of personalities and organizations provide insight to future information operations seeking to modify behaviors counter to friendly force objectives. Subordinate commanders and staffs develop plans and operations that support the BCT commander's intent and desired end state. Supporting efforts empower key influencers and organizations, and persuade neutral audiences, to bolster legitimacy and secure vital interests and objectives. Coercive efforts attack to neutralize the enemy's narrative. The BCT uses coercive efforts to counter enemy propaganda and isolates adversaries or enemies from their support base to begin the psychological breakdown of adversary or enemy organizations. The BCT shapes conditions for favorable objectives in line with the interests of the host nation contributing to the enemy's defeat through use of military deception, engagements, and communication mediums.

2-38. Activities that shape the operational environment derive success in how effectively they persuade the populace and empower the host-nation government. All efforts focus on bolstering the legitimacy of the rule of law and the host nation's ability to provide for effective governance. Persuasion and empowerment demand the BCT use engagement strategies to make connections and form relationships with pillar organizations and individuals who control and influence the local community. Engagements secure common and clearly defined goals and ideals that provide a common reference point for future engagements and activities. Engagements seek to reinforce the authority of legitimate leaders and pillars and to restore or solidify confidence in host-nation security forces, governance, and rule of law. Persuasive efforts utilize a compelling narrative that justifies and explains friendly actions while delegitimizing motivations and behaviors of those who are counter to positive gains within the area of operations. Additionally, persuasive efforts specifically target neutral or fringe entities with the goal of tipping neutrality to a favorable alliance.

2-39. Defensive operations will likely be constant during BCT operations, to include the protection of information, friendly forces, or civil populations. Offensive operations of limited scope, duration, and objectives targeting enemy, capabilities, groups, or individuals seize initiative and opportunities contributing to enduring success. BCTs identify opportunities to seize, retain, and exploit the initiative to destroy, dislocate, disintegrate, or isolate enemy organizations and discredit enemy actions as trust builds and information is collected. Offensive operations shape the operational environment within operational frameworks and establish conditions for future operations. Effective offensive operations retain initiative through actions and coherent and compelling themes and messages to inform and influence audiences.

INFLUENCE AUDIENCES

2-40. The BCT commander ensures actions, themes, and messages complement and reinforce each other to accomplish objectives. An information theme is a unifying or dominant idea or image that expresses the purpose for an action. A message is a verbal, written, or electronic communication that supports an information theme focused on an audience. A message supports a specific action or objective. Actions, themes, and messages are inextricably linked. The commander ensures actions, themes, and messages complement and reinforce each other and support operational objectives. The commander keeps in mind that every action implies a message, and avoids contradictory actions, themes, or messages.

2-41. Throughout operations, the commander informs and influences audiences inside and outside of the BCT. The commander informs and influences by conducting Soldier and leader engagements, radio programs, command information programs, operations briefs, and unit website posts or social media. Soldiers and leaders build trust and legitimacy by demonstrating their adherence to the moral principles of the Army Ethic. That trust is earned by the actions of every Soldier and leader assigned or attached to the BCT. Commanders and subordinate leaders convey the importance of that message to their subordinates at every opportunity. The BCT staff assists the commander to create shared understanding and purpose inside and outside of the BCT, and among all affected audiences. Shared understanding synchronizes the words and actions of Soldiers and leader to achieved mutual trust that informs and influences audiences inside and outside the BCT (see ADP 1).

2-42. Influence is central to shaping the operational environment. All activities conducted by the BCT directly or indirectly contribute to or detract from the BCT's ability to influence the populace and environment. Information related tools, techniques, and activities are the integration of designated information-related capabilities to synchronize themes, messages, and actions with operations to inform U.S. and global audiences, influence foreign audiences, and affect adversary and enemy decision-making. Information related capabilities clarify intentions through common narratives, counter enemy propaganda, expose corruption within competing groups or entities, and bolster the legitimacy of host nation power and governance. Information related capabilities modify behaviors and efforts through persuasion, cooperation, or coercion that leads to successful operations that secure the populace and provide order to the social structure.

2-43. Narratives provide a communication mechanism and are the unifying structures between action and communication with the populace. Simple narratives tie together the actions of the BCT with unit objectives. Simple narratives provide a basis for informing and influencing leaders and pillars as to the purpose behind actions and activities conducted by host nation forces and the BCT. Compelling narratives seek to address concerns and interests of the populace while explaining the methodologies endeavored by the host nation government and security forces in partnership with the BCT. All BCT leaders must understand the narrative as they play a central role in key leader engagements and all information-related capabilities. Narratives explain and justify friendly actions while delegitimizing enemy and adversary actions. Narratives simultaneously serve as both communication mechanisms and counterpropaganda instruments that gain the populaces' favor. Narratives seek to neutralize or disable the support structures provided to adversary or enemy groups and factions. BCTs must know the multiple narratives within a given information environment. The BCT gains valuable insights from competing narratives to determine the multiple and disparate interests and motivations of the population and its subsets. BCT leaders rely on attached PSYOP and other key personnel to identify the actors and analyze the narratives to determine competing narratives; staffs then articulate these competing narratives to inform the commander's decisions. Key staff members informing the commander's narrative guidance and actions include the information operations officer, public affairs officer, brigade civil affairs operations staff officer (S-9), PSYOP staff planner, and brigade judge advocate.

2-44. The BCT addresses enemy propaganda efforts by preempting and countering enemy propaganda to neutralize their effects on friendly actions and objectives. The BCT maintains credibility with the host-nation populace and counters enemy propaganda that seeks to delegitimize host-nation government and friendly forces actions to maintain the initiative. Use of mainstream media, social media, community meetings, key leader engagements, and other messaging mechanisms provide multiple means to counter enemy propaganda and address accusations and misinformation before the local, regional, national, and global audience perceives deceit and lies as truth and fact. The BCT must actively collect information and intelligence that allows unhindered observation of enemy messaging and propaganda platforms to identify enemy information campaigns that seek to degrade the effectiveness of friendly actions and activities. The BCT employs specifically trained capabilities to construct narratives, identify enemy messaging and propaganda efforts through information collection, and aggressively deliver countermessages that discredit enemy propaganda. They also assess the impact of friendly and enemy influences upon the populace at the local, regional, national, and international level to mitigate the effects of enemy propaganda.

2-45. Networks of systematic crime and corruption that undermine progress for their own political or economic gains require transparency, accountability, and combined oversight with host-nation partners. Political environments and security organizations allow opportunists to infiltrate legitimate systems and pursue agendas outside the interests, aims, and objectives that support sustainable and favorable outcomes. Political subversion undermines legitimacy and gives enemies and adversaries insider information about friendly motivations and operations.

2-46. The BCT commander and staff must understand the external and internal influences of corruption within host nation political, economic, and security systems. The BCT, in partnership with the host nation, must identify corrupt officials, discredit enemy influence in legitimate systems, and eliminate subversive elements that promote negative influences on legitimate governmental processes or other pillar organizations. When the host nation denies enemy organizations sanctuary in pillar organizations, they are forced to seek support elsewhere or retire from a given area, thereby making themselves vulnerable to friendly forces that can identify transitions, seize initiative, exploit weakness, and neutralize or destroy enemy forces.

2-47. Above all, the BCT supports efforts designed to bolster host-nation partner legitimacy among the populace and global audience. Legitimacy takes on varying forms depending upon the social, cultural, and political systems of a particular society. Rule of law is fundamental to legitimate governance. Partnered security operations between the BCT and host-nation security forces are essential to gaining and maintaining the rule of law and a sustainable security environment. The populace decides whether the governance mechanisms within their society are legitimate, because local and cultural norms define legitimacy and acceptance by the people. Measurable and noticeable progress, however slight, enhances legitimacy that improves the security, law and order, economic situation, and social structure over time.

2-48. The BCT commander and staff that exhibit an understanding of the information environment are prepared to synchronize information-related capabilities to enhance the effectiveness of operations. Information operations communicate action and intent to the populace, encourage cooperation through persuasion and relationships, effectively counter enemy propaganda, expose and defeat corruption, and bolster the legitimacy of host nation partners. Effective information operations create effects in the commander's portion of the information environment and enable sustainable outcomes that lead to rule of law, effective governance, address the needs of the people, and enhance mission accomplishment.

INFLUENCE OUTCOMES

2-49. The BCT commander and staff employ information-related capabilities within the BCT's area of operations to empower the successful accomplishment of objectives. Influence alters public opinion garnering support for military and diplomatic operations. Well planned and executed, information operations lead to diplomatic and political conclusions that can minimize or eliminate the need for military operations. All assets and capabilities at a commander's disposal have the capacity to achieve objectives and inform and influence to varying degrees. Some examples of resources the commander may use include combat camera, human intelligence (HUMINT), maneuver, and network operations. Objectives encapsulate the results of activities and the expected or desired conclusion of missions and tasks. Use of information-related capabilities nested within tactical, operational, and strategic objectives reinforce narratives that inform and promote influence.

2-50. Culture, history, religion, politics, tradition, and needs hierarchies contribute to interpretation and acceptance of the narratives presented to adversaries, host-nation forces, and indigenous populations. Competing narratives clash within the operational environment concurrently with lethal, nonlethal, and ancillary capabilities within the operational environment. The commander and staff work with information-related capabilities such as civil affairs, public affairs, military intelligence, and other capabilities. Capabilities such as PSYOP forces, draft, implement, distribute, and monitor the effectiveness of narratives. Unintended or unconsidered consequences impacts from activities and actions of entities outside of the commander's sphere of control, and adversary or enemy competing narratives struggle for acceptance or rejection of the narrative within the operational environment. Using environmental metrics, civil considerations, intelligence, monitoring of media (external and social), and constant attention to all competing narratives increases the commander's development of influence within an area of operations.

2-51. Influence and outcomes are inextricably linked so commanders can consolidate the elements of combat power resulting in mission success and end state accomplishment. Subsets within influence and its attainment are concepts and actions such as conflict resolution, negotiation, accommodation, reconciliation, compromise, and release of authority and responsibility to host-nation military and political forces and entities. Continuous information collection and intelligence analysis of the area of operations are essential to gain and implement influence. The BCT commander relies on organic, specifically trained staff and attached capabilities to analyze the information environment and recommend adjustments to physical actions and the commander's narrative to maintain narrative dominance. The commander and staff achieve desired effects by synchronizing operations and the narrative.

2-52. Building trust and legitimacy are key information influence objectives. As trusted Army professionals, the BCT commander and staff conduct risk assessment and take into account ethical considerations in planning and execution. (See ADP 1.) In population-centric operational environments, earning and maintaining the trust of either the host nation or indigenous population are essential to mission accomplishment.

SECTION III – CONSOLIDATION OF GAINS

2-53. The BCT consolidates gains and favorable milestones to seize and exploit weaknesses, capitalize on opportunities, and further the allies' interests to secure stable political settlements and objectives complimentary to desired outcomes. *Consolidate gains* are activities to make enduring any temporary operational success and to set the conditions for a sustainable security environment allowing for a transition of control to other legitimate authorities (ADP 3-0). Consolidate gains is an integral part of winning and achieving success across the competition continuum and the range of military operations. It is the follow-through to achieve the commander's intent, and essential to retaining the initiative over determined enemies and adversaries.

2-54. Consolidate gains is not a mission. It is an Army strategic role defined by the purpose of the tasks necessary to achieve enduring political outcomes to military operations and, as such, represents a capability that Army forces provide to the joint force commander. Consolidating gains enables a transition from the occupation of a territory and control of populations by Army forces—that occurred because of military operations—to the transfer of control to legitimate authorities. Activities to consolidate gains occur across the range of military operations and often continue through all phases of a specific operation.

2-55. To consolidate gains, the BCT commander reinforces and integrates the efforts of all unified action partners within the area of operations. The BCT staff deliberately plans and prepares, in coordination with higher headquarters, for consolidating gains to capitalize on successes before an operation begins. Planning should address changes to task organization and the additional assets required in a specific situation. Additional engineer, military police, civil affairs, and medical capabilities typically support the security and stability of large areas. In some instances, the BCT is in charge of integrating and synchronizing their activities, in others the BCT is in a supporting role. The provision of minimum-essential stability operations tasks (see paragraph 8-48) within the BCT's area of operations can include providing security, food, water, and medical treatment.

2-56. The BCT consolidates gains within its capability by seizing, retaining, and exploiting initiative and opportunities resulting from information collection, interaction with people and organizations, offensive and defensive operations, information-related capabilities, civil-military operations, and cyberspace electromagnetic activities (CEMA). The BCT consolidates gains through decisive action, executing offense, defense, and stability to defeat the enemy or adversary in detail and begin to set security conditions that support the desired end state (see ADP 3-0). The BCT develops and reassesses the situation, perception, and opportunities through continuous information collection to maintain positive momentum and tactical gains.

2-57. The BCT commander and staff influence host partners and populace through compelling narratives that explain actions, discredit enemy propaganda, and highlight common goals, themes, and messages. The BCT develops information and intelligence to understand, shape, and influence the operational environment and consolidate positive gains leading towards desired objectives. The commander and staff analyze operational and mission variables to provide understanding of the operational environment and to influence the people and organizations within the BCT's area of operations. The BCT influences, persuades, and empowers people and organizations to shape the environment and support sustainable objectives. Setting conditions to shape transcends phases and is continuous throughout all operations.

2-58. The BCT commander seeks opportunities to maintain pressure on enemy forces, highlight and promote positive contributions in rule of law and governance, and exploit weaknesses in enemy narratives to consolidate successes. Executing tasks to accomplish objectives that are consistent with the higher commander's intent achieves consolidation. Consolidation of gains capitalizes on the positive actions and objectives through information collection, offensive and defensive operations, information operations, narratives, themes, messages, and host-nation partnerships to bridge tactical success with operational and strategic objectives. In essence, the consolidation of gains (demonstrated in the discussion below) links positive, contributing tactical actions with operational and strategic objectives.

Necessity to Consolidate Gains and Establish Area Security

1 FM 3-0 states that consolidation of gains occurs in portions of an area of operations, where large-scale combat operations are no longer occurring. Nevertheless, enemy forces will likely continue to fight and exploit any kind of friendly weaknesses across all domains. Thus, the constant awareness of the necessity to consolidate gains, to plan accordingly, and to allocate sufficient resources must become a constant staff consideration.

Before units can appropriately consolidate gains, they must successfully establish security. The actions of the 82d Airborne Division on 3 and 4 February 1945 provide a clear example of how units should first establish security in an area before beginning deliberate consolidation of gains.

[3 February 1945] The division strengthened and consolidated defensive positions; eliminated scattered groups of enemy remaining in rear areas; repulsed strong counterattacks and inflicted heavy casualties on the enemy. [4 February 1945] The division maintained defensive positions and patrolled aggressively to the East. The 99th ID commenced relief of front-line units of the division.

2 Only one day after the 82d Airborne Division successfully breached the German Siegfried Line against "insane opposition," the division not only consolidated the newly captured position as part of the actions on the objective, but continued clearing bypassed enemy remnant elements by patrolling into enemy territory. This created the tactical conditions for relief-in-place by follow-on units and further offensive operations. Such successful examples create vivid and informative lessons which sharpen and define the Army's understanding of how to consolidate gains on the future battlefield.

Various

2-59. Ultimately, the host nation must have the capability to ensure a safe and secure environment and must likewise develop the capacity to maintain acceptable conditions related to good governance, the rule of law, social well-being, and economic development. The BCT commander builds partner capacity through collaboration and empowerment that enhances the legitimacy of host-nation forces and government (see chapter 8, section I). Partner capacity must be sustainable and eventually independent of the BCT's influence to maintain legitimate authority and perception of the rule of law and governance. (See ADP 3-0 for further discussion.)

Note. Consolidating gains is not the same as unit consolidation. BCT subordinate units routinely conduct consolidation upon occupying a new position on the battlefield or achieving mission success. (See paragraph 6-30 for information on unit consolidation, to include reorganization.)

SECTION IV – LARGE-SCALE COMBAT OPERATIONS

2-60. *Large-scale combat operations* are extensive joint combat operations in terms of scope and size of forces committed, conducted as a campaign aimed at achieving operational and strategic objectives (ADP 3-0). They are at the far right of the competition continuum and associated with war. Large-scale combat operations are intense, lethal, and brutal. Conditions include complexity, chaos, fear, violence, fatigue, and uncertainty. Battlefields will include operations across the entire expanse of the land, air, maritime, space, and cyberspace domains and noncombatants crowded in and around large cities. Enemies will employ conventional military tactics, terror, criminal activity, and information warfare to complicate operations further. Enemy activities in the information environment will be inseparable from ground operations. As in the past, large-scale combat operations present the greatest challenge for Army forces across the range of military operations, and as expressed by U.S. Army LTG (retired) Michael D. Lundy:

Large-Scale Combat Operations

Since the Soviet Union's fall in 1989, the specter of large-scale ground combat against a peer adversary was remote. During the years following, the U.S. Army found itself increasingly called upon to lead multinational operations in the lower to middle tiers of the range of military operations and competition continuum. The events of 11 September 2001 led to more than 15 years of intense focus on counterterrorism, counterinsurgency, and stability operations in Iraq and Afghanistan. An entire generation of Army leaders and Soldiers were culturally imprinted by this experience. We emerged as an Army more capable in limited contingency operations than at any time in our nation's history, but the geopolitical landscape continues to shift, and the risk of great power conflict is no longer a remote possibility.

While our Army focused on limited contingency operations in the Middle East and Southwest Asia, other regional and peer adversaries scrutinized U.S. military processes and methods and adapted their own accordingly. As technology has proliferated and become accessible in even the most remote corners of the world, the US military's competitive advantage is being challenged across all the warfighting domains. In the last decade, we have witnessed an emergent China, a revanchist and aggressive Russia, a menacing North Korea, and a cavalier Iranian regime. Each of these adversaries seeks to change the world order in their favor and contest U.S. strategic interests abroad. The chance for war against a peer or regional near-peer adversary has increased exponentially, and we must rapidly shift our focus to successfully compete in all domains and across the full range of military operations.

Over the last several years, the U.S. Army has rapidly shifted the focus of its doctrine, training, education, and leader development to increase readiness and capabilities to prevail in large-scale ground combat operations against peer and near-peer threats. Our new doctrine dictates that the Army provide the joint force four unique strategic roles: shaping the security environment, preventing conflict, prevailing in large-scale combat operations, and consolidating gains to make temporary success permanent. To enable this shift of focus, the Army is now attempting to change its culture shaped by over 15 years of persistent limited-contingency operations. Leaders must recognize that the hard-won wisdom of the Iraq and Afghanistan wars is important to retain but does not fully square with the exponential lethality, hyperactive chaos, and accelerated tempo of operations across multiple domains when facing a peer or near-peer adversary.

U.S. Army LTG (retired) Michael D. Lundy

ARMY FORCES IN LARGE-SCALE GROUND COMBAT OPERATIONS

2-61. *Large-scale ground combat operations* are sustained combat operations involving multiple corps and divisions (ADP 3-0). Army forces constitute the preponderance of land combat forces, organized into corps and divisions, during large-scale combat operations. Army forces seize the initiative, gain and exploit positions of relative advantage in multiple domains to dominate an enemy force, and consolidate gains. Corps and divisions execute decisive action (offense, defense, and stability), where offensive and defensive operations make up the preponderance of activities conducted during combat operations. Corps and division commanders must explicitly understand the lethality of large-scale combat operations to preserve combat power and manage risk. Commanders use ground maneuver and other land-based capabilities to enable maneuver in the air, land, cyberspace, and maritime domains. Commanders leverage operations in the cyberspace and space domains, and the information environment to support ground maneuver.

2-62. Corps and division commanders use fires to create effects in support of Army and joint operations. *Cross-domain fires*—fires executed in one domain to create effects in a different domain (ADP 3-19)—provide commanders with the flexibility to find the best system to create the required effect and to build redundancy into their plan. Cross-domain fires also present a more complex problem to the adversary or enemy than fires within a single domain. *Multi-domain fires*—fires that converge effects from two or more domains against a target (ADP 3-19)—converge surface-based fires with other effects across domains to create multiple dilemmas, taxing the enemy's ability to effectively respond. A commander may employ offensive cyberspace operations to attack an enemy air defense network while surface-to-surface fires destroy enemy air defense radars and air-to-surface fires destroy the air defense command and control nodes.

2-63. During large-scale combat operations, the performance of offensive operations is traditionally associated with a favorable combat-power ratio. Combat multipliers often provide positions of relative advantage, even when Army forces are outnumbered. A numerically superior force is not a precondition for performing offensive operations. Rather, a commander must continuously seek every opportunity to seize the initiative through offensive operations, even when the force as a whole is on the defense. This requires the commander to perform economy of force measures to adequately resource the force's main effort. The offensive plan allows the corps or division to shift and synchronize combat power where necessary to reinforce the main effort. Mobility, surprise, and aggressive execution are the most effective means for achieving tactical success when performing both offensive and defensive operations. Bold, aggressive tactics may involve significant risk; however, greater gains normally require greater risks. A numerically inferior force capable of bold and aggressive action can create opportunities to seize and exploit the initiative. (See chapter 6 for additional information on the offense.)

2-64. Defending corps and division commanders seek to push the enemy off-balance in multiple domains and the information environment when that enemy initially has the initiative. The key to a successful corps or division defense is the orchestration and synchronization of combat power across all available domains and the information environment to converge effects. Commanders decide where to concentrate combat power and where to accept risk as they establish engagement areas (EAs). Corps and division defensive planning normally calls for the decisive operation to culminate in the main battle area (MBA) with the attacking enemy's defeat. The plan allows the corps or division to shift and synchronize combat power where necessary to reinforce MBA units. Spoiling attacks and counterattacks designed to disrupt the enemy and to prevent the enemy from massing or exploiting success are part of MBA operations. The headquarters future operations and plans integrating cells conduct contingency planning to counter potential enemy penetrations of forward defenses within the MBA. The key consideration before diverting any assets from the corps or division's decisive operation is if the threat to its sustainment capabilities jeopardizes mission accomplishment. Although a corps or division may sustain the temporary loss of sustainment from its support area, it loses the battle if defeated in the MBA. (See chapter 7 for additional information on the defense.)

2-65. Corps and division offensive and defensive plans must address how preparations for, and the conduct of, operations impact the civilian population of the area of operations. This includes the conduct of noncombatant evacuation operations for U.S. civilians and other authorized groups. The BCT commander's legal and moral obligations to that civilian population must be met as long as meeting those obligations does not deprive the defense of necessary resources. The commander is legally and morally responsible for the decisions made by, and actions of the BCT and must plan and have the foresight to mitigate and reduce the risk of unintended effects such as excessive collateral damage and negative psychological impacts on noncombatant populace. Those effects may be positive or negative—which create or reinforce instability in the area of operations. Improper planning could lead to severe consequences that adversely affect efforts to gain or maintain legitimacy and impede the attainment of both short term and long-term goals for the U.S. forces commander. (See ADP 1 for additional information.)

2-66. Ideally, the host-nation government will have the capability to provide area security for its population and conduct the six stability operations tasks. To the extent that a host-nation government is unable to conduct the immediately necessary minimum-essential stability operations tasks, the defending unit will perform stability operations tasks within its capability and request further support. Corps and division commanders analyze the situations they face to determine the minimum essential stability operations tasks and the priority associated with each task. This analysis includes a plan to consolidate gains in operational areas once large-scale combat operations culminate. (See chapter 8 for additional information on stability.)

2-67. As the consolidation of gains is an integral part of all operations, corps and division commanders assign purposefully task organized forces consolidation areas (see figure 2-2 on page 2-18) within their assigned area of operations to begin consolidate gains activities concurrent with large-scale combat operations. Consolidate gains activities provide freedom of action and higher tempo for those forces committed to the close and deep support areas. Forces begin consolidate gains activities after achieving a minimum level of control and when there are no on-going large-scale combat operations in a specific portion of their area of operations. Designating a maneuver force responsible for consolidation areas enables freedom of action for units in the other corps and division areas by allowing them to focus on their assigned tasks and expediting the achievement of the overall purpose of the operation.

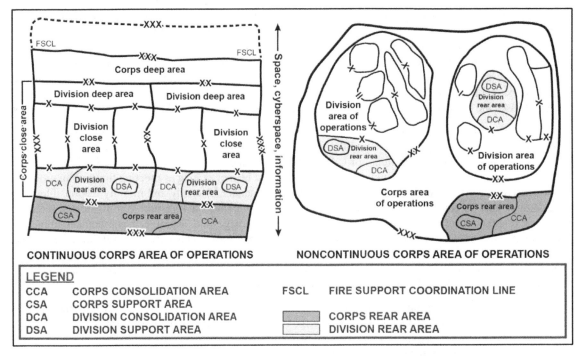

Figure 2-2. Notional corps and division areas of operations

2-68. Forces assigned the mission of consolidating gains execute area security and stability operations tasks. Initially the focus is on combined arms operations against bypassed enemy forces, defeated remnants, and irregular forces to defeat threats against friendly forces in the consolidation area(s), as well as those short of the rear boundaries of the BCTs in the close area. Friendly forces may eventually create or reconstitute an indigenous security force through security cooperation activities (see chapter 8) as the overall focus of operations shifts from large-scale combat operations to consolidating gains. Optimally, a division commander would assign a BCT to secure a consolidation area. A division is the preferred echelon for this mission in a corps area of operations. The requirement for additional forces to consolidate gains as early as possible should be accounted for early during planning with appropriate force tailoring by the theater army. (See ADP 3-0, ATP 3-92, and ATP 3-91 for additional information.)

THE BRIGADE COMBAT TEAM IN LARGE-SCALE COMBAT OPERATIONS

2-69. During large-scale combat operations, maneuver BCTs conduct offensive and defensive operations, and stability operations tasks. The BCT and its subordinate echelons concentrate on performing offensive and defensive operations and necessary tactical enabling tasks. The BCT performs only those minimal-essential stability operations tasks—providing civil security, food, water, shelter, and emergency medical treatment—to civilians located within their area of operations in accordance with the laws of war and international standards. Commanders balance the provision of those minimum-essential stability tasks with their capability to conduct the offense or defense. The BCT does not conduct operationally significant consolidate gains activities unless tasked to do so, usually within a division consolidation area. The BCT commander orchestrates rapid offensive maneuver to provide the commander with an opportunity to compel, persuade, or deter an enemy decision or action. Defending headquarters develop plans to find, fix, and destroy enemy forces conducting distributed enemy operations in, and major penetrations into support and consolidation areas.

2-70. During the execution of close operations, the BCT commander employs the appropriate offensive form of maneuver to close with an enemy to mitigate any disadvantage in capabilities. This typically requires rapid movement through close or complex terrain during periods of limited visibility. Subordinate unit movement formations move in as dispersed a manner as possible while retaining the capability to mass effects against enemy forces at opportune times and places. Joint enablers become more effective when an enemy has no time to focus on singular friendly capabilities in the five domains. Units perform attacks that penetrate enemy defenses or attack them frontally or from a flank. Depending on the situation, they also infiltrate enemy positions, envelop them, or turn enemy forces out of their current positions. Those units then exploit success to render enemy forces incapable of further resistance. (See chapter 6 for additional information on the offense.)

2-71. In the defense, the division and BCT commanders' intent and concept of operations may be to use available fires to defeat, deter, or delay an enemy before major enemy forces come into direct fire range within the MBA. Corps and division commanders' concept may be for field artillery cannon and rockets, and offensive cyberspace operations to conduct suppression of enemy air defenses to enable combat aviation brigade attack reconnaissance assets and joint fires to delay or disrupt the approach of enemy second echelon or reserve forces. Thus, allowing those MBA BCTs to complete their defeat of the enemy's initial attack with their organic assets before enemy second echelon or reserve forces join close combat operations. Defending commanders direct the delivery of effects in multiple domains to establish positions of relative advantage necessary for a successful counterattack. Defense plans at each echelon retain a reserve regardless of the defensive operation assigned. The reserve must be an uncommitted force available for commitment at the decisive moment during the operation. The division or BCT tasked to provide the defensive forward security force (see chapter 5) might conduct either a cover or guard mission, or other offensive operation to set conditions to regain the initiative and transition to the offense. (See chapter 7 for additional information on the defense.)

Note. Controlling commander coordinates for the fire support coordination line to be closer to the forward edge of the battle area to better facilitate the employment of joint fires. BCT commanders establish coordinated fire lines to facilitate the employment of surface-to-surface fires.

2-72. When the stability element of decisive action currently dominates actions within a BCT's area of operations, the responsibility for providing for the needs of the civilian population generally rests with the host-nation government or designated civil authorities, agencies, and organizations. The BCT performs minimal-essential stability operations to provide security, food, water, shelter, and medical treatment when there is no civil authority present. Under these conditions, the BCT commander assesses available resources against the mission to determine how best to conduct these minimum-essential stability operations tasks and what risks must be accepted. The priorities and effort given to stability operations tasks will vary within each subordinate unit's area of operations. Within this stability environment, area security may be the predominant method of protecting the civilian population and support and consolidation areas that are necessary to facilitate the positioning, employment, and protection of resources required to sustain, enable, and control forces and the civilian population. (See chapter 8 for additional information on stability.)

DECISIVE ACTION

2-73. Operations conducted outside the United States and its territories simultaneously combine three elements of decisive action—offense, defense, and stability. Through decisive action, commanders seize, retain, and exploit the initiative while synchronizing their actions to achieve the best effects possible. As a single, unifying idea, decisive action provides direction for the entire operation. Decisive action begins with the commander's intent and concept of operations. The commander's intent includes the operation's purpose, key tasks, and the conditions that define the end state (see chapter 4). Commanders and staffs refine the concept of operations during planning and preparation and determine the proper allocation of resources and tasks. Throughout an operation, they may adjust the allocation of resources and tasks as conditions change.

2-74. The *simultaneity*—the execution of related and mutually supporting tasks at the same time across multiple locations and domains (ADP 3-0)—of the decisive action is not absolute. The simultaneity of decisive action varies by echelon and span of control. The higher the echelon, the greater the possibility of simultaneous offensive, defensive, and stability operations. At lower echelons, an assigned task may require all the echelons' combat power (see paragraph 2-77) to execute a specific task. For example, a division always performs offensive, defensive, and stability operations simultaneously. Subordinate BCTs perform some combination of offensive, defensive, and stability operations, but they may not perform all three simultaneously.

2-75. While BCTs perform some combination of offensive, defensive, and stability operations, they generally are more focused by their immediate priorities on a specific element, particularly during large-scale ground combat operations. While an operation's primary element is offense, defense, or stability, different subordinate units involved in that operation may be conducting different types and subordinate variations of operations. During decisive action, commanders rapidly shift emphasis from one element to another to maintain tempo and keep enemy forces off balance. Maintaining tempo and flexibility through transitions contributes to successful operations. Commanders perform tactical enabling operations to help in the planning, preparation, and execution of any of the elements of decisive action. Tactical enabling operations are never decisive operations. Commanders use tactical enabling operations to complement current operations or to transition between phases or element of decisive action.

2-76. BCTs employ mutually supporting lethal and nonlethal capabilities in multiple domains to generate overmatch, present multiple dilemmas to the enemy, and enable freedom of movement and action. BCT commanders and staffs use their situational understanding to choose the right combinations of combined arms (see paragraph 2-88) to place the enemy at the maximum disadvantage. As trusted Army professionals, commanders and staffs are expected to make decisions (for example, ethical, effective, and efficient) and take actions consistent with the moral principles of the Army Ethic. Decisive action requires that they implement judicious use of lethal and nonlethal force balanced with restraint, tempered by professional judgment. (See ADP 3-0 and ADP 1 for additional information.)

COMBAT POWER

2-77. Commanders conceptualize capabilities in terms of combat power. *Combat power* is the total means of destructive, constructive, and information capabilities that a military unit or formation can apply at a given time (ADP 3-0). Combat power includes all capabilities provided by unified action partners that are integrated, synchronized, and converged with the commander's objectives to achieve unity of effort in sustained operations. The eight elements of combat power are leadership, information, command and control, movement and maneuver, intelligence, fires, sustainment, and protection. Commanders apply leadership and information throughout to multiply the effects of the other six elements of combat power. The other six elements—command and control, movement and maneuver, intelligence, fires, sustainment, and protection—are collectively known as warfighting functions. A *warfighting function* is a group of tasks and systems united by a common purpose that commanders use to accomplish missions and training objectives (ADP 3-0).

2-78. *Leadership* is the activity of influencing people by providing purpose, direction, and motivation to accomplish the mission and improve the organization (ADP 6-22). Leaders adhere to and uphold a shared identity as trusted Army professionals by fulfilling their sworn oaths to support and defend the US Constitution. Trusted Army professionals, demonstrate character, competence, and commitment, by making decisions and taking actions in adherence to the moral principles of the Army Ethic, including Army Values, which reflect American values and the expectations of the American people. (See ADP 6-22 and ADP 1 for more information.)

2-79. Information alone rarely provides an adequate basis for deciding and acting. Effective command and control require further developing information into knowledge so commanders can achieve understanding. Information operations is the commander's primary means to optimize the information element of combat power and supports and enhances all other elements to gain an operational advantage over an enemy or adversary (see ATP 3-13.1). *Information operations* is the integrated employment, during military operations, of information-related capabilities in concert with other lines of operation to influence, disrupt, corrupt, or usurp the decision making of adversaries and potential adversaries while protecting our own (JP 3-13). (See FM 6-0 and FM 3-13 for more information.)

2-80. The *command and control warfighting function* is the related tasks and a system that enable commanders to synchronize and converge all elements of power (ADP 3-0). (See ADP 6-0 for more information.)

2-81. The *movement and maneuver warfighting function* is the related tasks and systems that move and employ forces to achieve a position of relative advantage over the enemy and other threats (ADP 3-0). (See ADP 3-90 for more information.)

2-82. The *intelligence warfighting function* is the related tasks and systems that facilitate understanding the enemy, terrain, weather, civil considerations, and other significant aspects of the operational environment (ADP 3-0). Specifically, other significant aspects of the operational environment include threats, adversaries, the operational variables, and can include other aspects depending on the nature of operations. (See ADP 2-0 and FM 2-0 for more information.)

2-83. The *fires warfighting function* is the related tasks and systems that create and converge effects in all domains against the adversary or enemy to enable operations across the range of military operations (ADP 3-0). (See ADP 3-19 and FM 3-09 for more information.)

2-84. The *sustainment warfighting function* is the related tasks and systems that provide support and services to ensure freedom of action, extend operational reach, and prolong endurance (ADP 3-0). (See ADP 4-0 for more information.)

2-85. The *protection warfighting function* is the related tasks and systems that preserve the force so the commander can apply maximum combat power to accomplish the mission (ADP 3-0). (See ADP 3-37 for more information.)

2-86. Commanders employ three means to organize combat power: force tailoring, task organizing, and mutual support, which are defined below:

- *Force tailoring* is the process of determining the right mix of forces and the sequence of their deployment in support of a joint force commander (ADP 3-0).
- *Task-organizing* is the act of designing a force, support staff, or sustainment package of specific size and composition to meet a unique task or mission (ADP 3-0).
- *Mutual support* is that support which units render each other against an enemy, because of their assigned tasks, their position relative to each other and to the enemy, and their inherent capabilities (JP 3-31).

Note. Task organization is a temporary grouping of forces designed to accomplish a particular mission (ADP 5-0).

2-87. Commanders consider mutual support when task organizing forces, assigning areas of operations, and positioning units. The two aspects of mutual support are supporting range and supporting distance. *Supporting range* is the distance one unit may be geographically separated from a second unit yet remain within the maximum range of the second unit's weapons systems (ADP 3-0). *Supporting distance* is the distance between two units that can be traveled in time for one to come to the aid of the other and prevent its defeat by an enemy or ensure it regains control of a civil situation (ADP 3-0). (See ADP 3-0 and ADP 3-90 for additional information.)

COMBINED ARMS

2-88. Applying combat power depends on combined arms to achieve its full destructive, disruptive, informational, and constructive potential. *Combined arms* is the synchronized and simultaneous application of arms to achieve an effect greater than if each element was used separately or sequentially (ADP 3-0). Through combined arms, the BCT commander integrates leadership, information, and each of the warfighting functions and their supporting systems. Used destructively, combined arms integrate different capabilities so that counteracting one makes the enemy vulnerable to another. Used constructively, combined arms multiply the effectiveness and efficiency of Army capabilities used in stability operations tasks.

2-89. Combined arms use all Army, joint, and multinational capabilities (when available)—in the air, land, maritime, space, and cyberspace domains—in complementary and reinforcing ways. Complementary capabilities protect the weaknesses of one system or organization with the capabilities of a different warfighting function. For example, commanders use artillery (fires) to suppress an enemy bunker complex pinning down an Infantry unit during tactical movement (movement). The infantry unit then closes with (maneuver) and destroys the enemy. In this example, the fires warfighting function complements the movement and maneuver warfighting function.

Notes. In the context of Army tactics, *movement* is the positioning of combat power to establish the conditions for maneuver (ADP 3-90). To direct movement, BCT forces use movement techniques, use movement formations, and conduct battle drills to mitigate the risk of making contact with the enemy before maneuvering. Commanders and subordinate leaders must avoid confusing tactical movement with maneuver. Tactical movement is movement in preparation for contact; maneuver is movement while in contact. Actions on contact are the process by which a unit transitions from tactical movement to maneuver. (See chapter 6.)

In both the offense and defense, contact occurs when a unit encounters any situation that requires an active or passive response to a threat or potential threat. The eight forms of contact are visual; direct; indirect; nonhostile; obstacles; aircraft; chemical, biological, radiological, and nuclear (CBRN), and electromagnetic warfare (EW). The conduct of tactical offensive and defensive operations most often involves conduct using the visual, direct, and indirect forms.

2-90. Reinforcing capabilities combine similar systems or capabilities within the same warfighting function to increase the function's overall capabilities. In urban operations, for example, Infantry, aviation, Armor, and special forces (movement and maneuver) often operate close to each other. This combination reinforces the protection, maneuver, and direct fire capabilities of each. The Infantry protects tanks from enemy Infantry and antitank systems; tanks provide protection and firepower for the Infantry. Army aviation attack and reconnaissance units maneuver above buildings to observe and fire from positions of advantage, while other aircraft may help sustain the ground elements. Special forces units enable indigenous forces, which can provide combat information, intelligence, target locations, and provide a layer to the COP of the BCT that cannot be realized without an indigenous view. Army space-enabled capabilities and services such as communications and global positioning satellites enable communications, navigation, situational awareness, protection, and sustainment of land forces.

2-91. Other capabilities—such as close air support (see ATP 3-09.32) and Army special operations forces (see FM 3-05)—can complement or reinforce the BCT's capabilities. For example, close air support planning and execution in support of the BCT is tightly integrated, and focused on providing timely and accurate fires in close proximity to the enemy. Army special operations forces such as PSYOP units are attached to the BCT and provide unique capabilities to influence populations, facilitate graduated response, and protect noncombatants among other force multiplying effects.

2-92. Combined arms multiply Army forces' effectiveness in all operations. Units operating without support of other capabilities generate less combat power and may not accomplish their mission. Employing combined arms requires highly trained Soldiers, skilled leadership, effective staff work, and integrated information systems. As stewards of the Army Profession, the BCT commander and staff fully leverage their military expertise and work effectively, efficiently, and ethically to optimize the use of available resources. Commanders synchronize combined arms operations utilizing command and control systems to apply the effects of combat power to the best advantage. They conduct simultaneous combinations of offensive, defensive, and stability operations to defeat an opponent on land and establish conditions that achieve the commander's end state.

HASTY VERSUS DELIBERATE OPERATIONS

2-93. Army forces are task organized specifically for an operation to provide a fully synchronized combined arms team. That combined arms team conducts extensive rehearsals while also conducting shaping operations to set the conditions for the conduct of the force's decisive operation. Most operations lie somewhere along a continuum between two extremes—hasty operations and deliberate operations. A *hasty operation* is an operation in which a commander directs immediately available forces, using fragmentary orders, to perform tasks with minimal preparation, trading planning and preparation time for speed of execution (ADP 3-90). A *deliberate operation* is an operation in which the tactical situation allows the development and coordination of detailed plans, including multiple branches and sequels (ADP 3-90). Determining the right choice involves balancing several competing factors.

2-94. The decision to conduct a hasty or deliberate operation is based on the commander's current knowledge of the enemy situation and assessment of whether the assets available (including time) and the means to coordinate and synchronize those assets are adequate to accomplish the mission. If they are not, the commander takes additional time to plan and prepare for the operation or bring additional forces to bear on the problem. The commander makes that choice in an environment of uncertainty, which always entails some risk. Ongoing improvements in command and control systems continue to assist in the development of a COP of friendly and enemy forces while facilitating decision-making and communicating decisions to friendly forces. These improvements can help diminish the distinction between hasty and deliberate operations; they cannot make that distinction irrelevant.

2-95. The commander may have to act based only on available *combat information*—unevaluated data, gathered by or provided directly to the tactical commander which, due to its highly perishable nature or the criticality of the situation, cannot be processed into tactical intelligence in time to satisfy the user's tactical intelligence requirements (JP 2-01)—in a time-constrained environment. The commander must understand the inherent risk of acting only on combat information, since it is vulnerable to enemy deception operations and can be misinterpreted. The commander's intelligence staff helps assign a level of confidence to combat information used in decision-making.

2-96. A commander cannot be successful without the capability of acting under conditions of uncertainty while balancing various risks and taking advantage of opportunities. Although a commander strives to maximize knowledge of available forces, the terrain and weather, civil considerations, and the enemy, a lack of information cannot paralyze the decision-making process. A commander who chooses to conduct hasty operations must mentally synchronize the employment of available forces before issuing fragmentary orders. This includes using tangible and intangible factors, such as subordinate training levels and experience, a commander's own experience, perception of how the enemy will react, understanding of time-distance factors, and knowledge of the strengths of each subordinate and supporting unit to achieve the required degree of synchronization. (See ADP 3-90 for additional information.)

CLOSE COMBAT

2-97. Only on land do combatants routinely and in large numbers come face-to-face with one another. *Close combat* is that part of warfare carried out on land in a direct fire fight, supported by direct and indirect fires and other assets (ADP 3-0). Close combat destroys or defeats enemy forces. It encompasses all actions that place friendly forces in immediate contact with the enemy where the commander uses fire and movement in combination. Our forces or the enemy can initiate close combat.

2-98. The primary mission of subordinate elements of the BCT is to close with the enemy by means of fire and movement to destroy, defeat, or capture the enemy, to repel the enemy assault by fire, close combat, and counterattack, or all of these. Units involved in close combat may—

- Employ direct and indirect fires.
- Execute combined arms maneuver to obtain positions of relative advantage.
- Receive effective enemy direct and indirect fires.
- Have no or only a limited ability to maneuver.
- Have a battalion/squadron or one or more of its companies/troops decisively engaged.

2-99. Close combat places a premium on leadership, positive command and control, and clear and concise orders. During close combat, leaders have to think clearly, give concise orders, and lead under great stress. Key terms used within this section and throughout this publication include the following:

- *Defeat*—to render a force incapable of achieving its objectives (ADP 3-0).
- *Destroy*—a tactical mission task that physically renders an enemy force combat-ineffective until it is reconstituted. Alternatively, to destroy a combat system is to damage it so badly that it cannot perform any function or be restored to a usable condition without being entirely rebuilt (FM 3-90-1).

- *Direct fire*—fire delivered on a target using the target itself as a point of aim for either the weapon or the director (JP 3-09.3).
- *Fires*—the use of weapon systems or other actions to create specific lethal or nonlethal effects on a target (JP 3-09).
- Indirect fire—the fire delivered at a target not visible to the firing unit; the fire delivered to a target that is not itself used as a point of aim for the weapons or the director (see TC 3-09.81).
- *Neutralize*—a tactical mission task that results in rendering enemy personnel or materiel incapable of interfering with a particular operation (FM 3-90-1).
- *Suppress*—a tactical mission task that results in the temporary degradation of the performance of a force or weapon system below the level needed to accomplish its mission (FM 3-90-1).
- *Suppression*—the temporary or transient degradation by an opposing force of the performance of a weapons system below the level needed to fulfill its mission objectives (JP 3-01).

OPERATIONS STRUCTURE

2-100. The operations structure—the operations process, warfighting functions, and operational framework—is the Army's common construct for *unified land operations*—simultaneous execution of offense, defense, stability, and defense support of civil authorities across multiple domains to shape operational environments, prevent conflict, prevail in large-scale ground combat, and consolidate gains as part of unified action (ADP 3-0). The operations structure allows Army leaders to organize the effort rapidly and effectively and, in a manner, commonly understood across the Army. The operations process provides a broadly defined approach to developing and executing operations. The warfighting functions provide an intellectual organization for common critical functions (see paragraph 2-77). The operational framework provides Army leaders with basic conceptual options for visualizing and describing operations. (See ADP 3-0 for additional information.)

OPERATIONS PROCESS

2-101. The Army's framework for exercising command and control is the *operations process*—the major command and control activities performed during operations: planning, preparing, executing, and continuously assessing the operation (ADP 5-0). The operations process is a commander-led activity to plan, prepare, execute, and assess military operations (see chapter 4 for details). These activities may be sequential or simultaneous. In fact, they are rarely discrete and often involve a great deal of overlap. The commander, assisted by the staff, uses the operations process to drive the conceptual and detailed planning necessary to understand their operational environment; visualize and describe the operation's end state and operational approach; make and articulate decisions; and direct, lead, and assess military operations. (See ADP 5-0 for additional information.)

OPERATIONAL FRAMEWORK

2-102. The commander and staff use an operational framework, and associated vocabulary, to help conceptualize and describe the concept of operations in time, space, purpose, and resources. An *operational framework* is a cognitive tool used to assist commanders and staffs in clearly visualizing and describing the application of combat power in time, space, purpose, and resources in the concept of operations (ADP 1-01). An operational framework establishes an area of geographic and operational responsibility for the commander and provides a way to visualize how the commander will employ forces against the enemy. To understand this framework is to understand the relationship between the area of operations and operations in *depth*—the extension of operations in time, space, or purpose to achieve definitive results (ADP 3-0). Proper relationships allow for simultaneous operations and massing of effects against the enemy.

2-103. The operational framework has four components. First, the commander is assigned an area of operations for the conduct of operations. Second, the commander can designate deep, close, rear, support, and consolidation areas to describe the physical arrangement of forces in time and space. Third, within this area, the commander conducts decisive, shaping, and sustaining operations to articulate the operation in terms of purpose. Finally, the commander designates the main and supporting efforts to designate the shifting prioritization of resources.

Note. The BCT does not conduct operationally significant consolidate gains activities unless tasked to do so, usually within a division consolidation area.

Area of Operations

2-104.   An *area of operations* is an operational area defined by a commander for land and maritime forces that should be large enough to accomplish their missions and protect their forces (JP 3-0). In operations, the commander uses *control measures*—a means of regulating forces or warfighting functions (ADP 6-0)—to assign responsibilities, coordinate maneuver, and control combat operations. Within the area of operations, the commander integrates and synchronizes combat power across multiple domains. To facilitate this integration and synchronization, the commander designates targeting priorities, effects, and timing within the assigned area of operations. The loss or severe degradation of combat power within the BCT's area of operations by enemy attacks in any domain (see chapter 4) can prevent the successful execution of missions. Responsibilities within an assigned area of operations include—

- Terrain management.
- Information collection, integration, and synchronization.
- Civil-military operations.
- Movement control.
- Clearance of fires.
- Security.
- Personnel recovery.
- Airspace management of assigned airspace users.
- Minimum-essential stability operations tasks.

2-105.   The commander considers the BCT's area of influence when assigning an area of operations to subordinate commanders. An *area of influence* is a geographical area wherein a commander is directly capable of influencing operations by maneuver or fire support systems normally under the commander's command or control (JP 3-0). Understanding the area of influence helps the commander and staff plan branches to the current operation in which the force uses capabilities outside the area of operations. An area of operations should not be substantially larger than the unit's area of influence. Ideally, the area of influence would encompass the entire area of operations. An area of operations that is too large for a unit to control can allow sanctuaries for enemy forces and may limit joint flexibility.

2-106.   An *area of interest* is that area of concern to the commander, including the area of influence, areas adjacent thereto, and extending into enemy territory (JP 3-0). This area also includes areas occupied by enemy forces who could jeopardize the accomplishment of the mission. An area of interest for stability operations tasks (see chapter 8) may be much larger than that area associated with the offense and defense (see chapters 6 and 7, respectively). The area of interest always encompasses aspects of the air, cyberspace, and space domains since capabilities residing in all three enable and affects operations on land.

2-107.   Areas of operations may be contiguous or noncontiguous. When they are contiguous, a boundary separates them. When areas of operations are noncontiguous, subordinate commands do not share a boundary. The higher headquarters retains responsibility for the area not assigned to subordinate units. (See ADP 3-0 for additional information.)

2-108.   Commanders and staff are responsible to coordinate and integrate the actions of Army airspace users over an area of operations regardless of whether they have been assigned airspace control responsibility for a volume of airspace (see FM 3-52 and ATP 3-52.2). Commanders exercise airspace management through control of airspace users, which is inherent in mission command to control assigned or supporting forces in all domains.

Deep, Close, Rear, Support, and Consolidation Areas

2-109. Commanders can designate deep, close, rear, support, and consolidation areas to describe the physical arrangement of forces in time, space, and focus. (See figure 2-3.) The BCT may have established areas or may be operating in a higher headquarters designated area. A description of each area follows—

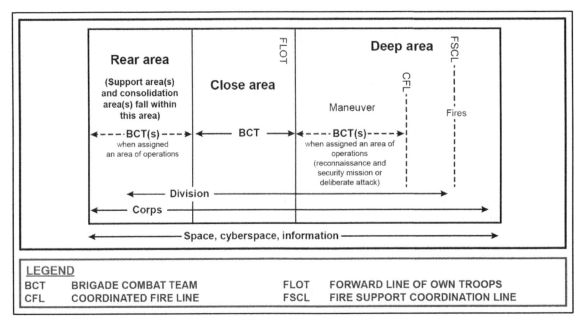

Figure 2-3. Deep, close, rear, support, and consolidation areas

Deep Area

2-110. A *deep area* is where the commander sets conditions for future success in close combat (ADP 3-0). Operations in the deep area involve efforts to prevent uncommitted enemy forces from being committed in a coherent manner. The commander's deep area generally extends beyond subordinate unit boundaries out to the limits of the commander's designated area of operations. The purpose of operations in the deep area is frequently tied to other events distant in time, space, or both time and space. BCT operations in the deep area might disrupt command and control systems, sustainment, and follow-on forces. While division and corps capabilities allow for operations in the deep area to disrupt the operational movement of reserves; cannon, rocket, or missile; and follow-on forces. In an operational environment where the enemy recruits insurgents from a population, deep operations might focus on interfering with the recruiting process, disrupting the training of recruits, or eliminating the underlying factors that enable the enemy to recruit.

Close Area

2-111. The *close area* is the portion of a commander's area of operations where the majority of subordinate maneuver forces conduct close combat (ADP 3-0). Operations in the close area are operations within a subordinate commander's area of operations. The BCT commander plans to conduct decisive operations using maneuver in the close area, and positions most of the maneuver force within it. Within the close area, one unit may conduct the decisive operation while others conduct shaping operations. A close operation requires speed and mobility to rapidly concentrate overwhelming combat power at the critical time and place and to exploit success.

Rear Area

2-112. The rear area is that area within a unit's area of operations extending forward from its rear boundary to the rear boundary of the area assigned to the next lower level of command. It is an area where most forces and assets locate that support and sustain forces in the close area. Rear operations include—

- Security.
- Sustainment.
- Terrain management.
- Movement control.
- Protection.
- Infrastructure development.

Support Area

2-113. In operations, the BCT commander may refer to a support area. The *support area* is the portion of the commander's area of operations that is designated to facilitate the positioning, employment, and protection of base sustainment assets required to sustain, enable, and control operations (ADP 3-0). The commander assigns a support area as a subordinate area of operations to support functions. It is where most sustaining operations occur. Within a division or corps support area, a designated BCT or maneuver enhancement brigade provides area security, terrain management, movement control, mobility support, clearance of fires, and tactical combat forces for security. Corps and divisions may have one or multiple support areas, located as required to best support the force. These areas may be noncontiguous to the other areas, in the close area, or in the rear area. (See chapter 4 for additional information.)

Consolidation Area

2-114. The *consolidation area* is the portion of the land commander's area of operations that may be designated to facilitate freedom of action, consolidate gains through decisive action, and set the conditions to transition the area of operations to follow on forces or other legitimate authorities (ADP 3-0). Corps and division commanders may establish a consolidation area, particularly in the offense as the friendly force gains territory, to exploit tactical success while enabling freedom of action for forces operating in the other areas. A consolidation area has all the characteristics of a close area, with the purpose to consolidate gains through decisive action once large-scale ground combat has largely ended in that particular area of operations.

2-115. The consolidation area requires a purposefully task organized, combined arms unit to conduct area security and stability operations tasks (see chapter 8) and employ and clear fires. (See ATP 3-21.20, chapter 4 for illustrations of area security and stability operations tasks performed by an Infantry brigade combat team [IBCT].) For a division, the BCT assigned responsibility for the consolidation area will initially focus primarily on security operations tasks that help maintain the tempo of operations in other areas, and it is likely to conduct offensive operations to defeat or destroy enemy bypassed units in order to protect friendly forces positioned in or moving through the area. The division consolidation area grows as the BCTs in close operations advance. When division boundaries shift, as is likely during the offense, the corps/division consolidation area will grow, and the balance of security and stability operations tasks may shift towards more of a stability focus, as conditions allow. The division responsible for the corps consolidation area conducts tasks designed to set conditions for the handover of terrain to host-nation forces or legitimate civilian authorities. (See ADP 3-0 for additional information.)

Decisive, Shaping, and Sustaining Operations

2-116. Decisive, shaping, and sustaining operations lend themselves to a broad conceptual orientation. The *decisive operation* is the operation that directly accomplishes the mission (ADP 3-0). The decisive operation determines the outcome of a major operation, battle, or engagement. The decisive operation is the focal point around which the commander designs an entire operation. Multiple subordinate units may be engaged in the same decisive operation across multiple domains. Decisive operations lead directly to the accomplishment of the commander's intent. The commander typically identifies a single decisive operation (see figure 2-4 on page 2-28).

2-117. A *shaping operation* is an operation at any echelon that creates and preserves conditions for success of the decisive operation through effects on the enemy, other actors, and the terrain (ADP 3-0). In combat, synchronizing the effects of aircraft, artillery fires, and obscurants to delay or disrupt repositioning forces illustrates shaping operations. Information operations, for example, may integrate Soldier and leader engagement tasks into the operation to reduce tensions between subordinate units within the BCT and different ethnic groups through direct contact between subordinate leaders and local leaders. Shaping operations may occur throughout the area of operations and involve any combination of forces and capabilities. Shaping operations set conditions for the success of the decisive operation and may be conducted prior to or simultaneously with the decisive operation. The commander may designate more than one shaping operation.

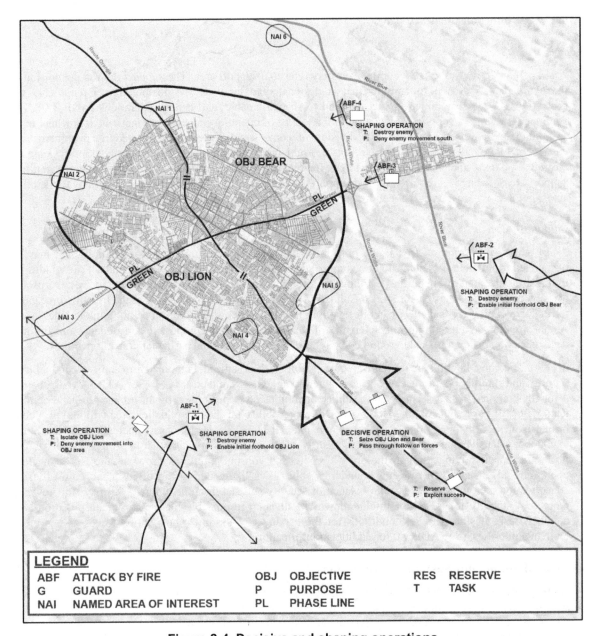

Figure 2-4. Decisive and shaping operations

2-118. A *sustaining operation* is an operation at any echelon that enables the decisive operations or shaping operations by generating and maintaining combat power (ADP 3-0). Sustaining operations differ from decisive and shaping operations in that they focus internally (on friendly forces) rather than externally (on the enemy or environment). Sustaining operations include personnel and logistics support, support area security, movement control, terrain management, and infrastructure development. Sustaining operations occur throughout the area of operations, not just within a support area. Failure to sustain may result in mission failure. Sustaining operations determine how quickly the force can reconstitute and how far the force can exploit success.

2-119. Throughout decisive, shaping, and sustaining operations, the commander and staff ensure that—

- Forces maintain positions of relative advantage.
- Operations are integrated with unified action partners.
- Continuity is maintained throughout operations.

Position of Relative Advantage

2-120. A *position of relative advantage* is a location or the establishment of a favorable condition within the area of operations that provides the commander with temporary freedom of action to enhance combat power over an enemy or influence the enemy to accept risk and move to a position of disadvantage (ADP 3-0). Positions of relative advantage provide the commander with an opportunity to compel, persuade, or deter an enemy decision or action. The commander maintains the momentum through exploitation of opportunities to consolidate gains and continually assess and reassess friendly and enemy effects for further and future opportunities.

2-121. The commander understands that positions of advantage are temporary, and may be planned for or spontaneous. The commander seeks positions of relative advantage before combat begins, and exploits success throughout operations. As the commander recognizes and gains positions of relative advantage, enemy forces will attempt to regain a position of advantage. As such, subordinate units of the BCT leverage terrain to their advantage and pit their strength against a critical enemy weakness. Subordinate units maneuver to a position that provides either positional advantage over the enemy for surveillance and targeting, or a position from which to deliver fires in support of continued movement towards an advantageous position; or to break contact.

Integration in Operations

2-122. The commander integrates BCT operations within the larger effort. The commander, assisted by the staff, integrates numerous processes and activities (see chapter 4) within the headquarters and across the force. Integration involves efforts to operate with unified action partners and efforts to conform BCT capabilities and plans to the larger concept. The commander extends the depth of operations through joint integration across multiple domains, including air, land, maritime, space, and cyberspace.

> *Note.* Army forces conduct operations across multiple domains, as part of a joint force, to seize, retain, and exploit control over enemy forces. For example, Army forces use aviation and unmanned aircraft systems (UASs) in the air domain, and protect vital communications networks in cyberspace, while retaining dominance in the land domain. (See ADP 3-0 and ADP 3-19 for additional information.)

2-123. When determining an operation's depth, the commander considers the BCT's own capabilities as well as available joint capabilities and limitations. The commander sequences and synchronizes operations in time and space to achieve simultaneous effects throughout an area of operations. The commander seeks to use capabilities within the BCT that complement those of unified action partners. Effective integration requires the staff to plan for creating shared understanding and purpose through collaboration with unified action partners.

Maintaining Continuity in Operations

2-124. Decision-making during operations is continuous; it is not a discrete event. The commander balances priorities carefully between current and future operations. The commander seeks to accomplish the mission efficiently while conserving as many resources as possible for future operations. To maintain continuity of operations, the commander and staff ensure they—

- Make the fewest changes possible.
- Facilitate future operations.

2-125. The commander makes only those changes to the plan needed to correct variances. The commander keeps as much of the current plan the same as possible. This presents subordinates with the fewest possible changes. The fewer the changes, the less resynchronization needed, and the greater the chance changes will be executed successfully.

2-126. When possible, the commander and staff ensure changes do not preclude options for future operations. The staff develops options during planning, or the commander infers them based on the staff assessment of the current situation. Developing or inferring options depends on validating earlier assumptions and updating planning factors and staff estimates. The concept of future operations may be war-gamed using updated planning factors, estimates, and assumptions. (See chapter 4.) The commander projects the situation in time, visualizes the flow of battle, and projects the outcomes of future operations and consolidating gains.

Main and Supporting Efforts

2-127. The commander designates main and supporting efforts to establish clear priorities of support and resources among subordinate units. The *main effort* is a designated subordinate unit whose mission at a given point in time is most critical to overall mission success (ADP 3-0). The main effort is usually weighted with the preponderance of combat power or the operation is designed where the effort is singularly focused. Task centric execution may be by event or phase. Typically, the commander shifts the main effort one or more times during execution. Designating a main effort temporarily prioritizes resource allocation. When the commander designates a unit as the main effort, it receives priority of support and resources in order to maximize combat power. The commander establishes clear priorities of support, and shifts resources and priorities to the main effort as circumstances and the commander's intent require. The commander may designate a unit conducting a shaping operation as the main effort until the decisive operation commences. However, the unit with primary responsibility for the decisive operation then becomes the main effort upon the execution of the decisive operation.

2-128. A *supporting effort* is a designated subordinate unit with a mission that supports the success of the main effort (ADP 3-0). The commander resources supporting efforts with the minimum assets necessary to accomplish the mission. The force often realizes success of the main effort through the success of the supporting effort(s). (See ADP 3-0 for additional information.)

FORCE PROJECTION

2-129. *Force projection* is the ability to project the military instrument of national power from the United States or another theater, in response to requirements for military operations (JP 3-0). Future conflicts will place a premium on promptly deploying landpower and constantly adapting to each campaign's unique circumstances as they occur and change. Army forces combine expeditionary capability and campaign quality to contribute crucial, sustained landpower to unified action.

2-130. Expeditionary capability is the ability to promptly deploy combined arms forces worldwide into any area of operations and conduct operations upon arrival. The BCT's ability to alert, mobilize, rapidly deploy with little notice, and operate immediately on arrival enables it to shape conditions early within the operational area and exploit successes while consolidating gains at the tactical level.

2-131. Campaign quality is the ability to sustain operations as long as necessary and to conclude operations successfully. The Army's campaign quality extends its expeditionary capability well beyond deploying combined arms forces that are effective upon arrival. It is an ability to conduct sustained operations for as long as necessary, adapting to unpredictable and often-profound changes in an operational environment as the campaign unfolds. The BCT is organized, trained, and equipped for endurance with an appropriate mix of combat forces together with maneuver support and sustainment units. (See ADP 3-0 and FM 3-0 for more information.)

PREPARATIONS

2-132. The success of the BCT in combat operations begins with its preparations for combat. These preparations include a mission-oriented training program, pre-mobilization and predeployment plans that support the BCT's specific regional contingencies. Given these contingencies, the commander and the staff derive the critical tasks and missions the BCT will most likely be called upon to execute. They express these tasks and missions as the BCT's mission-essential tasks. (See FM 7-0.) The commander develops and executes a mission-oriented training program that incorporates the entire mission-essential task. This training must be mentally challenging, physically demanding, and as realistic as common sense, safety, and resources permit. Such training allows units to deploy rapidly in accordance with an N-hour sequence or deploy in accordance with an X-hour sequence when not involved in a rapid, short notice deployment.

2-133. The X-hour/N-hour sequences for deployment are developed and followed to ensure all reports, actions, and outload processes are accomplished at the proper time during marshalling. They aid in developing air and deployment schedules and are flexible to allow for modifications based on the mission and the unit commander's concept of the operation.

2-134. X-hour is the unspecified time that commences unit notification for planning and deployment preparation in support of potential contingency operations that do not involve rapid, short notice deployment. X-hour sequence is an extended sequence of events initiated by X-hour that allow a unit to focus on planning for a potential contingency operation, to include preparation for deployment.

2-135. N-hour is the time a unit is notified to assemble its personnel and begin the deployment sequence. The N-hour sequence starts the reverse planning necessary after notification to have the first assault aircraft en route to the objective area for commencement of the parachute assault or to begin movement to a port of embarkation (POE) in accordance with the order for execution. An N-hour sequence may be employed by any unit when it is notified for a contingency deployment and assigned a date/time for first element departures either by wheels up by strategic airlift or a designated start time for convoy or line-haul to a POE.

2-136. In anticipation of an order for execution the BCT staff and its key leaders begin preparing or updating an operation order or plan. The length of X-hour planning varies based on the contingency planning or crisis action planning situation and the specific operation order or plan. It normally ceases with either the designation of N-hour, or if political or military events warrant, no further action. Deployment planning sequences fall into one of three scenarios:

- Unconstrained X-hour sequence. Used primarily for deliberate planning or crisis-action planning that is not under a time constraint.
- Constrained X-hour sequence. Used for crisis action planning.
- N-hour sequence. May be proceeded by an X-hour sequence.

2-137. The X-hour and N-hour sequence and deployment procedures are covered in the unit's tactical standard operating procedures (SOPs). Both sequences are just part of the force projection picture. (See ATP 3-35 for additional information.)

Note. ATP 3-35.1 provides the framework for commanders and their staff at all levels and deploying units on the employment of Army pre-positioned stocks to support force projection and the combatant commanders. It describes the missions, duties, and responsibilities of all organizations involved in moving Army pre-positioned stocks to an operational area and handing it off to designated Army units. It also describes planning and executing pre-positioned operations as well as supporting the combatant commander in a theater.

PROCESS

2-138. Force projection encompasses a range of processes including mobilization, deployment, employment, sustainment, and redeployment, and is inherently joint requiring detailed planning and synchronization. Each force projection activity influences the other having overlapping timelines repeated continuously throughout an operation. Deployment, employment, and sustainment are inextricably linked so one cannot be planned successfully without the others.

2-139. The operational speed and tempo reflect the ability of the deployment pipeline to deliver combat power where and when the joint force commander requires it. A disruption in the deployment will inevitably affect employment. Decisions made early in the process directly affect the success of the operation. (See ATP 3-35 for additional information.)

Mobilization

2-140. *Mobilization* is the process by which Armed Forces of the United States or part of them are brought to a state of readiness for war or other national emergency (JP 4-05). Whether deploying as part of a division or in an independent operation, the BCT will generally conduct the following sequence of activities for mobilization: planning, alert, home station, mobilization station, and POE:

- Planning. The BCT assists the division by maintaining and improving its combat readiness, preparing mobilization plans and files as directed by higher headquarters (including logistics and family support plans), providing required data to various mobilization stations (through division headquarters) as appropriate, ensuring unit movement data accuracy, conducting required mobilization and deployment training.
- Alert. This phase begins when the unit receives notice of a pending order. Commanders complete administrative and personnel processing actions begun during the planning phase. The alert phase concludes with preparation for deployment. When directed, BCT commanders may exchange liaison teams with the gaining command.
- Home station. Home station activities bring the reserve units onto active status; augmentation forces are identified and positioned. During this phase, the BCT takes the necessary steps to clear installation accounts and hand receipts, if required. The BCT also dispatches an advance party (and under certain conditions an early-entry CP, see chapter 4) to the mobilization station.
- Mobilization station. In this phase, the BCT plans for and provides the specific support called for in the applicable mobilization plan or as tasked by its parent division. Throughout this phase, the unit continues to train to mission-essential task in preparation for deployment.
- POE. BCT actions at the air or the sea POE include preparing and loading equipment and manifesting and loading personnel. This phase ends when the BCT departs from the POE.

Deployment

2-141. *Deployment* is the movement of forces into and out of an operational area (JP 3-35). It is composed of activities required to prepare and move a force as it task organizes, tailors itself for movement based on the mission, concept of operations, available lift, and other resources. The employment concept is the starting point for deployment planning. Proper planning establishes what, where, and when forces are needed and sets the stage for a successful deployment. Consequently, how the commander intends to employ forces is the basis for orchestrating the deployment structure. All deployment possibilities must be examined as they dramatically influence employment planning. Deployment directly affects the timing and amount of combat power that can be delivered to achieve the desired effects.

2-142. The joint deployment process is divided into four phases—deployment planning; predeployment activities; movement; and joint reception, staging, onward movement, and integration. The terminology used to describe the Army deployment phases is in synch with the joint process. The joint process includes a planning phase at the outset whereas the Army considers planning to be woven through all the phases. The Army deployment process consists of four distinct but interrelated phases that are addressed in the following paragraphs. A successful deployment requires implementation of each process with seamless transitions and interactions among all of them. The phases are not always sequential and could overlap or occur simultaneously. (See ATP 3-35 for additional information.)

Deployment Planning

2-143. Deployment planning is a logical process that focuses on Soldiers, supplies, and equipment, ways to deploy them, and the required information to track them. In particular, deployment plans require detailed information. Knowing the right details will help to guide the unit through an effective deployment. The heart of deployment planning is an accurate list of Soldiers and equipment that will deploy—the unit deployment lists.

2-144. Deployment and employment planning decisions are based on the anticipated operational environment to be encountered in the operational area. Understanding the operational environment helps the commander anticipate the results of various friendly, adversary, and neutral actions and how they affect operational depth and reach of force employment, as well as mission accomplishment. The operational environment is generally described by three conditions: permissive, uncertain, or hostile. (See JP 3-35 for more information.)

2-145. Force employment plans and schedules drive deployment planning, and execution preparation requirements and movement timelines. Deployment operations provide forces ready to execute the supported commander's orders. Deployment plans and schedules align with the operation plans/orders and support the associated force movement requirements, allowing for predeployment preparations.

2-146. During planning, four principles apply to the broad range of activities encompassing deployment. They are—precision, synchronization, knowledge, and speed:

- Precision applies to every activity and piece of data.
- Synchronization of those activities required to close the force successfully.
- Knowledge upon which decisions are made.
- Speed, the proper focus is on the velocity of the entire force projection process, from planning to force closure.

2-147. The end state of the deployment plan is to synchronize deployment activities to facilitate employment execution. The steps used in planning and preparation include analyze the mission, structure forces, refine deployment data, prepare the force, and schedule movement. Successful deployment planning will require knowledge of the unit's deployment responsibilities, an understanding of the total deployment process, and an intellectual appreciation of the link between deployment and employment.

Predeployment Activities

2-148. When ordered to deploy, the BCT task organizes, echelons, and tailors its units based on the assigned mission and available lift and other resources. Higher echelon plans determine the command, communications, intelligence, and logistics relationships. Some modifications to existing operations plans will normally be necessary. These plans should also specify any joint and combined operations relationships, if known. Within the division plan, the BCT commander prioritizes lift requirements consistent with the mission variables of METT-TC and establishes the sequence in which the BCT's units deploy relative to the movement of other forces and other Services. Maximum use of in-theater intelligence sources is essential. Sources include special operations forces area assessments, the country team, and higher headquarters.

2-149. Echeloning is organizing and prioritizing units for movement. Echelons are often divided into elements such as advance parties, initial combat forces, follow-on forces, and closure forces. Each echelon has a designated echelon commander. Tailoring is the adding to or subtracting from planned task organizations based upon a mission analysis, available transportation, pre-positioned assets, and host-nation support. Task organizing is the temporary grouping of forces to accomplish a certain mission. Task organizing and echeloning occur during initial planning.

2-150. Force tailoring is the process of determining the right mix of forces and the sequence of their deployment in support of a joint force commander. Force tailoring is situational dependent and occurs after mission analysis by the higher headquarters commander and staff. The BCT commander tailors the force after mission analysis, identifying initial strategic lift, pre-positioned assets, and host nation or contract services or assets.

2-151. Following the receipt of a mission, the BCT prepares its personnel and equipment for deployment through preparation for overseas movement activities. These activities ensure deploying units meet all requirements associated with deployment into another theater of operation as directed by Army regulations and local authorities.

Movement

2-152. Deployment includes preparing or moving the BCT, its equipment, and supplies to the area of operations in response to a crisis or natural disaster. The movement phase in the Army process is discussed in two segments—fort-to-port and port-to-port. The Army relies on United States Transportation Command to provide the strategic lift or surface movement to and from the POE:

- Fort-to-port. Units deploying with the BCT complete their preparation for overseas movement based on the mobilization plan and the joint task force's time phased deployment list. Units update their automated unit equipment lists to deployment equipment lists and submit them to the installation transportation office for transmission to United States Transportation Command. Based on information given to the joint operations planning and execution system, United States Transportation Command provides movement guidance for the BCT's movement to the POE through its component command.

- Port-to-port. This phase begins with the departure of BCT elements on strategic lift or surface movement from the POE. It ends with the BCT's closure in theater. The commander and staff must be prepared to update intelligence and, as necessary, modify plans while in transit.

Reception, Staging, Onward Movement, and Integration

2-153. This process applies only to unopposed entry operations. It begins with the arrival of BCT units (may include an early-entry CP, see chapter 3, at the port of debarkation [POD]) in the theater, and ends when the BCT departs the POD. Except in opposed entry operations, the BCT can expect reception, staging, onward movement, and integration support, whether provided by theater support contracts, external support contracts (primarily the Army logistics civil augmentation program [LOGCAP], see ATP 4-10.1), or regionally available commercial host-nation support, or military assets, must be sufficient to immediately support the arrival of deploying units. The primary requirement is coordinating the BCT's onward movement to its first destination.

2-154. Effective reception, staging, onward movement, and integration matches personnel with their equipment, minimizes staging and sustainment requirements while transiting the PODs, and begins onward movement as quickly as possible. Onward movement begins with personnel and equipment linkup, sustainment, the receipt of pre-positioned systems or logistics stocks at designated marshaling areas, and the reconfiguration of BCT forces. A plan to accomplish integration and maintain combat readiness must be understood, trained, and ready to implement upon arrival. Onward movement ends when the BCT arrives at the gaining command's staging areas where preparations for military operations occur.

Employment

2-155. *Employment* is the strategic, operational, or tactical use of forces (JP 5-0). Entry requirements following deployment vary. The BCT's entry into an area of operations can be either opposed or unopposed. In both cases, the BCT may use an intermediate staging base (see FM 3-99) to complete preparations and shorten lines of communications.

Unopposed Entry

2-156. Unopposed entry operations generally support host nation or forward presence forces. Hostilities may be underway or imminent, but the POD is secure and under friendly control. The commander sequences combat forces and supporting structure into the contingency area to gain and sustain the initiative and protect the force. Actions include the following:

- Link up with in-theater forces.
- Prepare to assist the host nation or forward presence forces.
- Protect the BCT and other collocated units, if required.
- Build up the combat abilities through training, familiarization, and acclimatization of the troops to the operational environment.
- Support to humanitarian and disaster relief.
- Facilitate the arrival of follow-on forces.

Opposed Entry

2-157. Opposed entry requires the integration and synchronization of multiservice capabilities in a concerted military effort against a hostile force. It is an extremely complex and hazardous operation that risks the assault force's defeat. Natural forces such as unfavorable weather and sea states represent hazards that are not normally such dominant factors. The assault force's key advantages lie in its mobility, flexibility, ability to concentrate balanced forces, and the ability to strike with great power at a selected point in the hostile defense system. Opposed entry operations exploit the element of surprise. They also capitalize on enemy weaknesses by applying the required type and degree of force at the most advantageous times and places.

2-158. The existence of a forcible entry capability induces the enemy to disperse forces, and in turn may result in making wasteful efforts to defend everything. The typical sequence in this type operation is to gain, secure, and expand a lodgment as part of a larger force before continuing operations. As an assault force, the BCT deploys by various means (parachute assault [airborne Infantry BCT only], air landing force, helicopter-borne air assault, and amphibious assault) into the objective area to seize initial assault objectives, neutralize enemy units, prepare obstacles, and secure additional landing zones. The intent is to introduce additional forces as quickly as possible to secure the initial lodgment area. (See FM 3-99 and JP 3-18 for additional information.)

Sustainment

2-159. Sustainment is the provision of logistics, financial management, personnel services, and health services support necessary to maintain operations until successful mission accomplishment. Sustainment of force projection operations is a complex process involving the geographic combatant commander, strategic and joint partners such as U.S. Transportation Command, and transportation component commands like Air Mobility Command, Military Sealift Command, Surface Deployment and Distribution Command, United States Army Material Command, Defense Logistics Agency, Service Component Commands, and Army generating forces. The Joint Network Node system connects the BCT to support area high headquarters, other sustainment Army forces and joint task forces during the force projection process. Sustainment of force gives Army forces its operational reach, freedom of action and endurance. (See ADP 4-0 for additional information.)

Redeployment

2-160. *Redeployment* is the transfer of forces and/or materiel to home or demobilization stations for reintegration and/or out-processing (ATP 3-35). This process for the BCT includes two major functions: deployment back to home station or to another theater and consolidation and reorganization as part of higher echelon reconstitution.

2-161. Redeployment is the preparation for and movement of the BCT from a theater to its designated follow-on continental United States or outside the continental United States base or to any other location. Commanders must contend with the same challenges as in deployment. Protection remains critical. Redeployment activities must be planned and executed to optimize the readiness of redeploying forces and material to meet new contingencies or crises. Redeployment phases include reconstitution for strategic movement, movement to the redeployment assembly areas, movement to the POE, strategic lift, reception at the POD, and onward movement.

2-162. *Reconstitution* is actions that commanders plan and implement to restore units to a desired level of combat effectiveness commensurate with mission requirements and available resources (ATP 3-21.20). Reconstitution activities include rebuilding unit integrity and accounting for Soldiers and equipment. These activities continue after the force arrives in the continental United States or in the home theater. The focus is on reconstituting units and their assigned equipment to premobilization levels of readiness, regenerating logistic stockpiles, and accounting for mobilized equipment and supplies.

Note. In large-scale ground combat operations, reconstitution will be a task conducted in theater by units in order to maintain lethality, freedom of action, operational reach and prolong endurance.

This page intentionally left blank.

Chapter 3

Threat

Threats are a fundamental part of an overall operational environment. A *threat* is any combination of actors, entities, or forces that have the capability and intent to harm United States forces, United States national interests, or the homeland (ADP 3-0). Threats may include individuals, groups of individuals (organized or not organized), paramilitary or military forces, nation states, or national alliances. When threats execute their capability to do harm to the United States, they become enemies.

SECTION I – UNDERSTANDING THE THREAT

3-1.   In general, the various actors in any area of operations can qualify as a threat, an enemy, an adversary, a neutral actor, or a friend. An *enemy* is a party identified as hostile against which the use of force is authorized (ADP 3-0). An enemy is also called a combatant and is treated as such under the law of war. An *adversary* is a party acknowledged as potentially hostile to a friendly party and against which the use of force may be envisaged (JP 3-0). A neutral is a party identified as neither supporting nor opposing friendly or enemy forces. Land operations often prove complex because a threat, an enemy, an adversary, a neutral, or a friend intermix, often with no easy means to distinguish one from another.

Today's operational environment presents threats to the Army and joint force that are significantly more dangerous in terms of capability and magnitude than those faced in recent operations in Iraq and Afghanistan. FM 3-0 presents a two-fold major paradigm shift: The shift from the industrial to the information age and from non-state to peer and near peer threats leading to transitions, such as from counterinsurgency to large-scale combat operations. For the first time in U.S. Army doctrinal history, FM 3-0 names the enemies and adversaries, which possess the capabilities to contest and degrade the battlefield across all domains as the "4+1:" Russia, China, North Korea, and Iran as well as radical ideologues and transnational criminal organizations, such as Islamic State of Iraq and the Levant or al-Qa'ida. These threats continuously challenge the US in multiple domains and in most cases purposefully below the threshold of open conflict.

THREAT COMPOSITION, DISPOSITION, AND INTENTION

3-2.   Leaders must understand that not all potential state adversaries seek to avoid U.S. forces or strengths, particularly those state adversaries with overwhelming numbers combined with favorable ground, and those with chemical, biological, radiological, and nuclear (CBRN) weapons of mass destruction (WMD) capability—North Korea and Iran are examples. Today's forces must prepare to deal with symmetrical threats as seen in Operation Desert Storm, as well as asymmetrical threats seen during Operation Iraqi Freedom and Operation Enduring Freedom.

UNDERSTANDING WITHIN THE CONTEXT OF THE OPERATIONAL ENVIRONMENT

3-3.   The BCT commander must understand threats, criminal networks, enemies, and adversaries, to include both state and nonstate actors, in the context of the operational environment. When the BCT commander understands the threat, the commander can visualize, describe, direct, and assess operations to seize, exploit, and retain the initiative and consolidate tactical gains. The commander and staff must develop and maintain running estimates (see chapter 4) of the situation. To develop and maintain running estimates of the situation as the basis for continuous adaptation, the commander and staff must consider their own forces within the realm of emerging threats as well as the mission, terrain, friendly forces, and civilian populations.

3-4. The BCT engages in close combat while operating in complex terrain in close proximity to civilian populations. Current and future battlefields require the BCT to fight and win in mountainous, urban, jungle, cold weather, and desert environments and subsurface areas. The physical challenges presented by complex terrain, and the continuous interactions of numerous actors, each with their own agendas, objectives, interests, and allegiances, influence the operational environment and mission accomplishment.

3-5. The impact of operational and mission variables on the operational environment produces additional layers of complexity to BCT operations. As a result, the BCT commander and staff must understand the complicated relationships and the complex interactions between the various actors that produce tactical challenges and opportunities. In the context of close combat, the BCT focuses on the assigned area of operations and use the mission variables to conduct analyses in order to gain the required understanding. Understanding is critical to seizing, retaining, and exploiting the initiative over enemies and adversaries. Understanding is equally critical to the consolidation of tactical gains to achieve sustainable political outcomes consistent with the mission.

STATE AND NONSTATE ACTORS

3-6. The BCT must be prepare to defeat determined state and nonstate actors that combine conventional and unconventional tactics to avoid our strengths (such as mobility, long-range surveillance, and precision fires capabilities) while attacking our perceived vulnerabilities (such as our difficulty identifying the enemy among civilian populations). Current and future threats use a variety of means, including conventional combined arms operations, terrorism, insurgency, political subversion, and information operations to evade our forces and disrupt tactical and combined arms capabilities. (See FM 3-24.2.) Enemies and adversaries will attempt to seize the initiative and dictate the terms and tempo of operations in their favor while relying on their established sources of strength. These sources of strength include networks that facilitate the undetected movement of logistics, finances, people, and weapons areas within complex terrain to exploit U.S. and unified action partner military, political, social, economic, and information vulnerabilities.

3-7. The enemy employs tactical countermeasures to limit the BCT's ability to develop the situation, to avoid decisive engagements, and to initiate contact under advantageous conditions. The enemy also employs technological countermeasures to reduce their signature on the battlefield and degrade our force's ability to detect, engage, and destroy them. Many hostile nation states continue to procure conventional capabilities such as tanks, antitank guided missiles (ATGM), manned aircraft, and air defense systems. These conventional weapons systems are increasingly available to nonstate enemy organizations. Enemy forces also integrate emerging technology such as robotics, unmanned aircraft systems (UASs), cyber, and nanotechnologies. Enemies and adversaries combine conventional and unconventional tactics to counter, evade, or disrupt the BCT's efforts across the range of military operations.

3-8. *Weapons of mass destruction* are chemical, biological, radiological, or nuclear weapons capable of a high order of destruction or causing mass casualties, excluding the means of transporting or propelling the weapon where such means is a separable and divisible part from the weapon (JP 3-40). The use of WMD in future conflict is inevitable. Many threat organizations already possess WMD and their delivery systems (for example, rockets and artillery). Enemies employ these WMD to obtain a relative advantage over U.S. forces to achieve their objectives. Threat organizations that do not currently possess WMD consistently seek opportunities to acquire them. The potential catastrophic effects associated with the threat or use of WMD adds greater uncertainty to an already complex environment. The BCT commander must anticipate and plan for the conduct of countering WMD (see paragraphs 60-80 and 8-92).

3-9. In current and future conflicts, the BCT commander and staff must rapidly develop a detailed and adaptable understanding of the threat, as it exists within the context of local conditions. Such a contextualized understanding allows the commander and staff to determine the nature of the conflict and to gain visibility of the enemy's structure and methods of operation. This determination allows commanders to identify emerging opportunities to seize, retain and exploit the initiative, exert influence over local actors, and consolidate tactical gains into operational and strategic successes. By understanding the internal workings of current and future enemies and adversaries, the BCT commander can exploit possibilities to disrupt the enemy and then rapidly dislocate, isolate, disintegrate, and destroy enemy forces.

NETWORKS

3-10. The BCT commander and staff must determine an enemy's strategies, objectives, and the multiple dimensions, (physical, psychological, informational, and political) in which the enemy operates to defeat the enemy. The BCT identifies and depicts networks (such as criminal, financial, terrorist, security forces) as friendly, enemy, or neutral based on how they affect the mission. The BCT supports friendly networks, influences neutral networks, and disrupts, neutralizes, or defeats enemy networks. Network assessment is continuous and collaborative, integrating unified action partners whenever possible. Unified action partners supply much of the information needed for an accurate assessment. At the tactical level, units develop an understanding of various networks through reconnaissance, intelligence operations, and surveillance (see chapter 5) in close contact with the enemy and civilian populations. Network assessment considerations include—

- Objectives and strategy.
- Key individuals, groups, nodes, and their roles within a network.
- Relationships between key individuals and networks.
- Means and methods of communicating.
- Resources that flow across, into, and out of networks (such as people, money, weapons, and narcotics).
- Network intersections where illicit networks connect to legitimate institutions and leaders.
- Network strengths and vulnerabilities.

POTENTIAL THREAT GROUPS

3-11. Complex operational environments may include joint transregional, all-domain, and multifunctional threats and conflicts, calling for Army operations across multiple domains, including air, land, maritime, space, and cyberspace. Threats include nation state militaries, insurgent organizations, transnational criminal organizations, and terrorist groups. These threat groups may align or partner with each other based on mutual goals, self-interest, convenience, required capabilities, and common interests. As a result, the BCT commander must prepare to defeat a complicated and often shifting array of enemies and threats. Understanding threat and enemy capabilities, as well as their political, economic, or ideological aims, is an essential element of seizing, retaining, and exploiting the initiative.

STATES

3-12. States are sovereign governments that control a defined geographic area. Although social movements and global real time communications reduce the relative power of some states, the state remains the entity that generates, sustains, and employs combat power. States have a number of advantages over organizations. These advantages include the recognition and support of other states, the authority to create laws governing the population and the authority to enforce laws through the control of institutions such as their standing armies and internal security forces, and the ability to raise money through taxation. Using their military forces, states have access to the institutions required to generate doctrinal, organizational, training, and materiel components of combined arms teams and their associated combat power. As a result, the BCT commander must understand a sovereign government's combat capabilities to work with or fight against that sovereign government.

NONSTATE ORGANIZATIONS

3-13. Nonstate organizations are groups that operate within states, but who act outside of the system to support or achieve their own political goals. Such groups may vary in size and organizational structure, changing over time and environments. Frequently, organizations consist of a predominant tribal, ethnic, national, or religious group, but there are corporate, criminal, and transnational organizations as well. Threat organizations may vary in capabilities and in the goals, they pursue. Often enemies and adversaries seek alliances of convenience by combining criminal networks, terrorists, state and nonstate actors, insurgents, transnational groups, proxies, and paramilitaries to attain short- or near-term objectives. For example, during the Iraq war, a variety of organizations operated in the country, some of whom posed threats to the U.S. mission. Nonstate organizations included Al Qaida, the Islamic State of Iraq, Jaysh al-Mahdi, Asaab al Haq, Khattaib Hizballah, the Sons of Iraq, and a variety of Kurdish militia groups. Energy Logistics Iraq Ltd and other corporations with their private security forces operated inside the country, also. At times, the United Nations and other transnational organizations or nongovernment organizations operated within the country. Each of the organizations that operated in Iraq had different, frequently opposing goals. Many of the organizations were directly opposed to U.S. forces, but even the organizations that were not overt enemies had separate goals that did not align with U.S. interests. The BCT commander, therefore, must understand and prepare to work with and fight against a wide variety of organizations, many of which may be tied directly to sovereign states.

CRIMINAL NETWORKS AND OPPORTUNISTS

3-14. Criminal networks are often stakeholders in state weaknesses. The government institutions' weaknesses allow criminal networks to have freedom of movement and to divert state resources without repercussions from law enforcement and rule of law. Criminal networks often ally other state and nonstate organizations to engage in and facilitate a range of illicit activities (intimidation and coercion) to capture and subvert critical state functions and institutions. These networks often align regionally and ethnically. The networks build alliances with political leaders, financial institutions, law enforcement, foreign intelligence, and security agencies to pursue political and criminal agendas. Many networks operate with impunity, consistently avoiding meaningful investigations and prosecution, by exerting influence within law enforcement, investigative and judicial institutions within a nation state government.

> *Note.* The *rule of law* is a principle under which all persons, institutions, and entities, public and private, including the state itself, are accountable to laws that are publicly promulgated, equally enforced, and independently adjudicated, and that are consistent with international human rights principles (ADP 3-07).

3-15. Opportunists often take advantage of unstable conditions to pursue their personal goals and agendas. Opportunists can work with, for, or against an insurgency. Their interests determine their actions, operations, and conduct. An opportunist can work both sides to gain a positional advantage, to maximize influence, to maximize profits, or to avoid retribution. Opportunists can facilitate movement of insurgents while providing intelligence to counterinsurgents. Counterinsurgent or insurgent objectives do not restrict or govern opportunists.

3-16. The BCT commander and staff must identify the presence of criminal networks and opportunists. The commander and staff assess criminal networks and opportunists' impact on the mission and protection while planning and executing actions to mitigate those negative impacts. The BCT works with local, federal, U.S. Army, unified action partners, and law enforcement personnel to mitigate the threat of these groups and individuals. The BCT integrates law enforcement personnel into their operations and synchronizes their operations to facilitate the reduction and elimination of criminal networks and the threat posed by opportunists, ultimately creating an environment where local law enforcement agencies can assume responsibility in this effort.

> *Note.* Military police possess the capabilities to identify, deter, mitigate, and defeat criminal actors and networks, crime-conducive conditions, and other factors from within the criminal environment that can destabilize an area or threaten short- and long-term operational success (see FM 3-39). The BCT provost marshal is responsible for planning, coordinating, requesting, and employing military police assets.

INDIVIDUALS

3-17. Identifying the threat posed by states and organizations is relatively easy when compared to the challenge of identifying the threat posed by a single individual. Although U.S. forces have not historically focused on neutralizing the threat of a single person, the growing interconnectivity of states, organizations, and individuals increases the ability of an individual with sufficient computer technical skills to attack U.S. interests and forces using an army of computers. The BCT must be prepared to defend its command and control system against cyber-attacks, whether initiated by a state, organization, or individual. In addition to fighting and defeating states and organizations, the BCT commander and staff must retain the ability to identify, disrupt, and isolate individuals within the political, social, and tactical context of the operational environment.

SECTION II – THREAT CHARACTERISTICS AND ORGANIZATION

3-18. The BCT possesses the capability to fight and win against regular and irregular forces (may involve nation-states using proxy forces or nonstate actors such as criminal and terrorist organizations). The term hybrid threat captures the complexity of operational environments, the multiplicity of actors involved, and the blurring of traditionally regulated elements of conflict.

3-19. Regular forces are part of nation states that employ military capabilities and forces in military competition and conflict. Normally, regular forces conduct operations to accomplish the following objectives, defeat an enemy's armed forces, destroy an enemy's war making capacity, and seize or retain territory.

3-20. Regular forces often possess technologically advanced weapon systems integrated into mechanized and motorized combined arms formations and light Infantry forces. Military equipment that the BCT may encounter in combat include armored fighting vehicles, antiarmor systems, air defense systems, ballistic missiles, manned and unmanned aircraft, indirect-fire systems, mines, EW, and digital communications systems. Regular force organizations are hierarchical (companies, battalions, brigades, and so forth) with a centralized command and control structure. Regular forces can conduct long-term conventional and unconventional operations. Examples of regular forces include—

- Islamic Republic of Iran Army.
- Peoples Liberation Army of China.
- Russian Army.
- North Korean People's Army.

3-21. Irregular forces may be armed individuals or groups who are not members of the regular armed forces, police, or other internal security forces. Irregular forces employ unconventional, asymmetric methods to counter U.S. advantages. Unconventional methods may include terrorism, insurgency, and guerrilla warfare. A weaker threat or enemy often uses unconventional methods to exhaust the U.S. collective will through protracted conflict. Economic, political, informational, and cultural initiatives usually accompany and may even be the chief means of an attack on U.S. influences. Irregular forces or complex threats include paramilitaries, terrorists, guerillas, and criminal organizations and networks.

3-22. Irregular forces or complex threats have political, ideological, or grievance related objectives tied to their motivation. These grievances may be real or perceived. Identifying these insurgent objectives and motivations can be difficult for a number of reasons, such as (these same issues occur in other irregular threats such as paramilitary, guerrillas, and criminal organizations)—

- Multiple insurgent groups with differing goals and motivations may be present.
- Insurgent leaders may change, and the movement's goals change with them.
- Organizations may fracture into two or more new entities with different or opposing goals.
- Movement leaders may have different motivations from their followers.
- Insurgents may hide their true motivations and make false claims.
- Goals of the insurgency may change due to operational environment changes.

3-23. Irregular forces customarily operate in small, dispersed, decentralized formations or cells (team and squad size) within a decentralized command and control structure. Irregular forces are often highly motivated with established local, regional and worldwide support networks. Irregular forces threat capability is limited to small arms weapons, antitank weapons, man portable air defense missiles, mortars, short-range rockets, homemade radio frequency weapons, rudimentary robotics, counter UASs, and improvised explosive devices (IEDs). However, some irregular threats possess the financial means to acquire advanced weapon systems and technologies. Examples of irregular forces in armed conflicts include—

- Revolutionary Army Forces of Columbia People's Army (1964).
- Mujahidin in Afghanistan (1979).
- Palestine Liberation Organization in the West Bank (2001).
- Al Qaeda in Iraq (2007).
- Taliban in Afghanistan (2009).
- Islamic State in Iraq and the Levant (2013).
- Iran's Quds Force support to nonstate actors in foreign countries.

3-24. The term hybrid threat evolved to capture the seemingly increased complexity of operations, the multiplicity of actors involved, and the blurring among traditional elements of conflict. A *hybrid threat* is the diverse and dynamic combination of regular forces, irregular forces, terrorist, or criminal elements acting in concert to achieve mutually benefitting effects (ADP 3-0). Hybrid threats combine traditional forces governed by law, military tradition, and custom with unregulated forces that act without constraints on the use of violence. These may involve nation-states using proxy forces or nonstate actors such as criminal and terrorist organizations that employ sophisticated capabilities traditionally associated with states. Hybrid threats may include nation state actors that employ protracted forms of warfare, possibly using proxy forces to coerce and intimidate, or nonstate actors employing capabilities traditionally associated with states. Hybrid threats can operate under a centralized or decentralized command and control structure. Hybrid threats are most effective when they exploit friendly constraints, capability gaps, and lack of situational awareness.

3-25. Combat experiences in Afghanistan, Iraq, and other recent conflicts in Lebanon, Mali, Syria, Gaza, Northern Nigeria, and Southern Thailand demonstrate a migration of capabilities, tactics, and techniques previously only associated with military forces of nation states to state sponsored and nonstate entities. This migration of capabilities presents friendly maneuver forces with a challenge that extends beyond defeating an enemy's regular force. Current and future threats do and can combine and transition between regular and irregular forces adopting strategies, tactics, and techniques to evade and disrupt U.S. advantages and gain tactical advantages within the physical, psychological, informational, and political dimensions of armed conflict. As a result, the BCT must prepare to counter lethal evasion and disruption capabilities from a variety of forces (regular, irregular, and hybrid) in current and future operational areas.

SECTION III – THREAT COUNTERMEASURES

3-26. To predict threat countermeasures in time and space, the BCT commander and staff (as conducted during the intelligence preparation of the battlefield [IPB]) must first understand threat capabilities, tactics, and techniques. Current and future enemies employ a series of integrated tactical and technical countermeasures to counter friendly operational and tactical advantages. Countermeasures are deception operations, dispersion, concealment, and the intermingling with civilians in urban terrain. The enemy also employs technological countermeasures, such as cyber-attacks and Global Positioning System (GPS) jamming, to evade and disrupt the friendly force's ability to develop the situation, seize the initiative, and consolidate tactical gains into favorable outcomes.

EVASION

3-27. Enemies operate within complex terrain to evade friendly weapon systems, advanced combined arms and air-ground capabilities. They operate in and among the population to evade detection, preserve their combat power, and retain their freedom of movement. The enemy often establishes relationships with local, regional, and transnational criminal organizations, and violent extremist organizations to finance their operations and gain access to illicit trafficking networks to move illegal weapons, munitions, WMD, people, narcotics, or money.

3-28. Enemy forces use deception, cover and concealment, smoke, or other obscurant when conducting operations. They move in small, dispersed units, formations, groups, or cells to avoid detection. They conduct short engagements with three to ten person elements that break contact before friendly forces can bring indirect fire or airborne strike platforms to bear. The enemy creates false battlefield presentations and reduces signatures through deliberate and expedient means of deception to frustrate friendly information collection efforts. The enemy uses hardened and buried facilities and multispectral decoys to mask the signatures of high-value systems (such as short-range ballistic missiles and surface to air missiles). The enemy also exploits safe havens within hostile states or in ungoverned areas and takes advantage of subsurface means to avoid detection (for example, tunnels, underground facilities, sewers, drainage systems, and other subterranean spaces). As enemies evade U.S. and coalition forces, they seek to expand their freedom of movement through intimidation and coercion. The enemy exploits civilian populations and cultural sites to hide key weapon systems.

DISRUPTION

3-29. Enemy forces employ combinations of lethal and other actions to disrupt BCT efforts to shape the operational environment. Lethal actions can be offensive and defensive in nature through decisive force. Other actions can be agitation, propaganda, and exploitation of the local population. Enemies employ integrated and networked combined arms teams to offset friendly capabilities. They employ small, dispersed, squad-sized teams armed with technologically advanced lethal weapons. Lethal weapons include rocket-propelled grenades, ATGMs, and man-portable air defense systems to conduct short engagements, and to defend against friendly armored, and counter-manned and UASs capabilities. Enemies seek opportunities to mass fires and forces against vulnerable targets, such as small combat outposts, dismounted patrols, and logistic convoys where they believe they can achieve quick victories with little risk of decisive engagements. When available, enemy forces employ armored or technical vehicles to increase their tactical mobility, protection, and firepower. Enemies integrate indirect fires such as rockets, mortars, and artillery into their operations.

3-30. Enemies augment their small combined arms teams' tactical capabilities by employing inexpensive countermeasures such as IED, Molotov cocktails, suicide bombers, civilians as obstacles (demonstrators and crowds to incite riots), CBRN weapons and materials, and fire and smoke as weapon systems. Enemies use these countermeasures to impede friendly forces' ability to move and maneuver, or to prevent and delay friendly forces from conducting operations. At the same time, enemies seek to acquire technologies such as UASs (that may be weaponized for precision strike capability), satellite imagery, forward looking infrared, and EW systems or platforms.

3-31. The presence of civilians in the area of operations can interfere with military operations. Capabilities such as, engineers civil affairs, military police, and psychological operations (PSYOP) forces can plan and perform populace control measures that ensure freedom of maneuver by mitigating civil interference, which enables commanders to maintain tempo and preserve combat power.

3-32. The enemy is proficient at establishing and maintaining communications and at disrupting BCT command and control systems. The enemy disrupts combined arms capabilities through combinations of jamming electromagnetic frequencies, cyber-attacks, data pirating, and satellite neutralization. Developing and maintaining these capabilities requires extensive recruitment, training, and outsourcing for personnel with the required skill set to conduct such attacks.

3-33. The loss of space-based communications due to enemy activity remains a major concern for friendly forces conducting operations. Whether the enemy action against satellites or with intermittent jamming and spoofing causes the communication interruptions, the resulting black out requires friendly forces to adapt and adjust until the restoration of communications. Short-term losses or disruptions of satellite communications will be mitigated through alternative communications methods, courier networks, and complete understanding and execution of the commander's intent and concept of operations.

3-34. Regular, irregular, and hybrid forces present formidable tactical challenges to the BCT when combined with area denial weapons. Area denial weapons included area denial systems, artillery munitions, land mines, and WMD. Enemy operations emphasize deception, cover, mobility, and most importantly, depth in the defense. In the offense, enemy operations emphasize deception, cover, mobility, and most importantly, infiltration techniques. Taken together, regular, irregular, and hybrid forces on the current and future battlefield employ significant combined arms capabilities that seek to disrupt BCT operations and dislocate BCT combined arms capabilities.

PUBLIC PERCEPTION

3-35. Enemies recognize the importance of public perception and its impact on the conduct of operations. The enemy attempts to influence the will of the American people, key allies, and the populations among whom there are conflicts, through propaganda, disinformation, and attacks on U.S. and allies' assets at home and abroad. The enemy conducts propaganda and disinformation operations to shape local and international public opinion and perception against the U.S., host nation, or coalition forces. The enemy undermines ongoing stabilization efforts, marginalizing successes, exploiting instances of friendly force missteps, and fabricating or exaggerating friendly force cultural shortcomings. Enemy organizations attempt to manipulate local, regional, and worldwide news and social media outlets to achieve their ends and solicit new recruits to their cause. For example, mobile phones can activate IEDs with the results captured on digital cameras, transmitted via satellite phones, and posted on internet chat rooms for a worldwide audience. Additionally, the enemy operating within urban terrain uses tactics that increase the potential for civilian casualties and collateral damage to undermine the resolve of both the United States and the local populace.

POLITICAL CONSIDERATIONS

3-36. Politics considerations, and in particular, competition for power, resources, and survival drive conflicts and are key to their resolution. Understanding the political dynamics at the local level allows the BCT commander and staff to identify the enemy's strategy, capabilities, and potential weaknesses within the political environment. This understanding aids in identifying targets that undermine or counter U.S. and coalition efforts that consolidate gains and achieve a sustainable political outcome consistent with U.S. vital interests.

3-37. The enemy exploits societal divisions along political, economic, ethnic, tribal, and religious lines. The enemy offers benefits to favored groups and disenfranchises opposing groups within the population to exploit societal divisions. These activities protect their sources of strength, consolidate their power, and assist in establishing political legitimacy. The enemy also seeks opportunities to exert this legitimacy by filling societal roles that U.S. forces or host-nation leaders have failed to address. As enemies and adversaries pursue this strategy, they often align with criminal organizations to undermine and attack existing government institutions. The resulting corruption, acceptance of illicit activities, and paralysis undermine political reform and stability efforts and prevent information gathering. The enemy promotes weaknesses within political institutions by disrupting or influencing elections at the local, provincial, and national level by conducting attacks on voting sites, intimidating election officials, manipulating political districts, and by backing corrupt officials. Additionally, the enemy may attempt to assassinate, abduct, or extort key civic, ethnic, or military leaders to undermine security and good governance, degrade friendly forces' morale, garner media attention to gain support and sway populace opinion, raise funds, and attract new recruits. Weak government institutions allow the enemy and other in state stakeholders the freedom and ability to divert state resources without repercussions from law enforcement and rule of law.

3-38. The enemy's political subversion campaign seeks to exploit existing social and political weaknesses. Degrading public opinion of U.S. and host-nation efforts, disrupting U.S. and local force's abilities to provide essential services and security, and alienating the populace from supporting friendly forces are efforts within this campaign. Like the physical capabilities of the enemy, the BCT commander must recognize and counter these efforts to maintain the initiative. The commander must visualize the threat in its political context to understand the dynamics existing within the area of operations and to determine tactical objectives that lead to the achievement of sustainable political outcomes consistent with U.S. vital interests. Understanding the political dynamics of a conflict, enables the commander to reassure and protect indigenous populations while simultaneously identifying, disrupting, and isolating the enemy to defeat the enemy.

SECTION IV – COUNTERING ADAPTATIONS AND RETAINING THE INITIATIVE

3-39. Countering enemy adaptations and retaining the initiative in armed conflicts requires the BCT commander and staff to understand the threat and the operational environment specific to its area of operations. Accurately depicting how an enemy employs forces requires an understanding of the enemy's organization, the enemy's capabilities, and the employment of enemy forces in the past. Overcoming increasingly sophisticated area denial actions and capabilities requires an effective information collection effort (see chapter 5) to develop the enemy situation within the BCT's area of operations.

3-40. In armed conflict, the commander's understanding is not limited to enemy organizations and their capabilities. This understanding includes ethnic groups, political factions, tribes or clans, religious sects, or ideological movements and their agendas. Identifying and distinguishing these groups and the associated dynamics is extremely difficult and requires deliberate information collection and analytical efforts through every phase of the operation. Using the mission variable of civil considerations and its subordinate characteristics identified by the mnemonic ASCOPE, the BCT staff has a standardized baseline for analysis to generate understanding. This baseline is augmented by analyses conducted by organic and attached forces such as social-cultural analysis, target audience analyses, intelligence analyses, population, and area studies.

3-41. While in contact with the enemy and in close proximity to the population, the BCT fights for information to understand and develop the situation. Complementary and integrated information collection capabilities (reconnaissance, surveillance, security operations, and intelligence operations) assist the commander in identifying opportunities to seize, retain, and exploit the initiative and dominate in increasingly challenging and complex environments.

This page intentionally left blank.

Chapter 4

Mission Command

Mission command is the Army's approach to command and control that empowers subordinate decision making and decentralized execution appropriate to the situation. (ADP 6-0). The brigade combat team (BCT) commander uses mission command, with its emphasis on seizing, retaining, and exploiting operational initiative, through mission orders. *Mission orders* are directives that emphasize to subordinates the results to be attained, not how they are to achieve them (ADP 6-0). Disciplined initiative—as it relates to mission command describes individual initiative.

Mission command requires the BCT commander to convey a clear commander's intent and concept of operations. These become essential in operations where multiple operational and mission variables interact with the lethal application of ground combat power. Such dynamic interaction often compels subordinate commanders to make difficult decisions in unforeseen circumstances. Based on a specific idea of how to accomplish the mission, commander and staff refine the concept of operations during planning and adjust the concept of operations throughout the operation as subordinates develop the situation or conditions change. Often, subordinates acting on the higher commander's intent develop the situation in ways that exploit unforeseen opportunities.

The commander uses the mission command approach to command and control to exploit and enhance uniquely human skills. The commander, supported by the staff, combines the art and science of command and control to understand situations, make decisions, direct actions, and lead forces toward mission accomplishment.

This chapter addresses the fundamentals of mission command, to include the principles of mission command, command presence, and the Army's approach to command and control. It addresses the command and control warfighting function and the exercise of command and control.

SECTION I – FUNDAMENTALS OF MISSION COMMAND

4-1.   Understanding the fundamentals of mission command as the Army's approach to command and control is essential to the effective conduct of operations. Military operations are human endeavors conducted in complex and ever-changing operational environments. The BCT commander's ability to visualize relationships among opposing human wills is essential to understanding the fundamental nature of operations. To account for the uncertain nature of operations, mission command (as opposed to detailed command) tends to be decentralized and flexible. This uncertain nature requires an environment of mutual trust and shared understanding among the commander, subordinates, and partners. This section focuses on the fundamentals of mission command and using mission orders to ensure disciplined initiative within the BCT commander's intent, enabling subordinate commanders and leaders to synchronize and converge all elements of combat power. (See ADP 6-0 for additional information.)

PRINCIPLES OF MISSION COMMAND

4-2.   The BCT commander focuses the order on the purpose of the operation through mission orders. Mission orders allow the commander's subordinates the greatest possible flexibility to accomplish assigned tasks. Mission command is enabled by the principles of—

- Competence.
- Mutual trust.
- Shared understanding.
- Commander's intent.
- Mission orders.
- Disciplined initiative.
- Risk acceptance.

Note. (See ADP 6-0 for a detailed discussion of the principles of mission command.)

COMMAND PRESENCE

4-3.   Command presence requires the BCT commander to lead from a position that allows timely decisions based on an operational environment assessment of the operational environment and application of judgment. The commander may find it necessary to locate forward of the main command post (CP). For example, the commander may position with the main effort to gain understanding, prioritize resources, influence others, and mitigate risk. To do this, the commander must understand how the principles of mission command guide and help combine the art of command and the science of control.

4-4.   The Armored Raid on Baghdad in 2003 offers an example of how the mission command approach to command and control enabled 2^d Brigade, 3^d Infantry Division (ID) (Mechanized) to seize, retain, and exploit the initiative in an uncertain environment. The vignette below demonstrates how the seven principles of mission command guided the brigade commander during the operation. It also describes how the commander used the principles of mission command to combine the art and science of command and control to understand situations, make decisions, direct actions, and lead forces toward mission accomplishment.

Armored Raid on Baghdad, 5 April 2003

On 5 April 2003, COL David Perkins' 2d Brigade, 3d Infantry Division (Mechanized), conducted a raid into western Baghdad as part of the division's advance on Baghdad after a two-week march of over 700 kilometers from Kuwait. As part of the advance, 3d Infantry Division (Mechanized) created a partial cordon around the Iraqi capital. The raid, ordered by 3d Infantry Division commander MG Buford Blount and V Corps commander LTG William S. Wallace, was conducted as a battalion sized reconnaissance in force into western Baghdad to determine the composition and strength of Iraqi forces defending the capital.

Staging out of Objective (OBJ) SAINTS, the column of M1A1 Abrams tanks and M2 Bradley Fighting Vehicles from LTC Eric Schwartz's Task Force 1st Battalion, 64th Armor Regiment (Task Force 1-64 AR), would advance north on Highway 8 (the main north-south expressway west of the Tigris River) into western Baghdad. The column then turned west to link with troops at the airport. Since the enemy situation was unclear, the operation required initiative and flexibility from the officers, noncommissioned officers, and Soldiers executing the operation. Wallace judged that such a bold plan was a reasonable risk. The raid was the first armored foray into a major city since World War II. Perkins' concept for the raid, called a "Thunder Run" by the tankers, was for an advance up Highway 8 that "...create[d] as much confusion... inside the city [as possible]." In mitigating the inherent risk of the operation, the 2^d Brigade commander considered "...that my Soldiers or my units [could] react to chaos much better than the enemy [could]." Perkins' specific *guidance to Schwartz was to "conduct a movement to contact north along Highway 8 to determine the enemy's disposition, strength, and will to fight."*

Schwartz praised the straightforward commander's intent and purpose. "The planning was simple," he explained, "The Thunder Run mission was the simplest of all tasks that we were given. There was no maneuver required. It was simply battle orders followed by battle drills." Based on Perkins' intent to maintain tempo, Schwartz chose to leave all lightly armored wheeled vehicles at SAINTS. Departing at 0630 on 5 April, Schwartz's command included 731 men, 30 M1A1 Abrams tanks, 14 Bradley Infantry fighting vehicles, 14 engineer vehicles, and other tracked support vehicles.

Within minutes of moving north of SAINTS, the Americans came under sporadic small arms, mortar and rocket-propelled grenade fire from Iraqi irregular forces firing from hastily prepared positions adjacent to the highway. Within an hour, the small arms fire and rocket-propelled grenade volleys had turned the operation into something akin to running a gauntlet of fire, but they did little to slow the column. The plan prohibited slowing the advance for specific targets, which were passed instead to follow-on vehicles by radio. However, this concept jettisoned temporarily when, 6 kilometers from the line of departure, a rocket propelled grenade round fired from an overpass exploded in the engine compartment of SSG Jason Diaz's C Company tank, immobilizing it. As Diaz's crew struggled to put out the growing fire and rig the tank for recovery, other Abrams and Bradley vehicles formed a defensive perimeter. Using coaxial machine gun fire and main gun rounds, the column repulsed several dismounted attacks and approaches by suicide vehicles. Several Americans were wounded. Since Perkins' order emphasized momentum, LTC Schwartz decided after a half hour delay, to renew the northerly advance and destroy Diaz's tank with incendiaries to keep it out of enemy hands. With the spearhead about halfway to the airport, Iraqi small arms fire fatally wounded SSG Stevon Booker, an A Company tank commander, while a nearby Bradley was disabled by rocket propelled grenade fire that also wounded the driver. In this case, the delay was short, with the wounded men placed in other vehicles and the Bradley rigged for towing. Soon the column was back on the move.

Schwartz's force turned in the direction of the airport at the intersection of Highway 8 and the Qaddissiyah Expressway, the main east-west thoroughfare between the airport west of the city and downtown Baghdad. Hundreds of paramilitary fighters and military personnel continued to fire on the column from all directions, only to fall victim to the Americans' overwhelming firepower. After a total travel time of two hours and 20 minutes, the column arrived at the airport.

COL Perkins concluded that the reconnaissance in force had completely surprised the enemy. "[The Iraqis] thought that they could bloody our nose enough on the outside of the city ... that we just would not push through block by block," Perkins explained. The raid had cost five casualties (one killed and four wounded), one Abrams tank destroyed, and one Bradley heavily damaged. Iraqi losses were estimated to be at least 1000 fighters killed, one T-72 tank, and 30 to 40 BMPs (Boyevaya Mashina Pekhotys) destroyed, and the elimination of a large number of light vehicles and countless roadside bunkers.

The operation demonstrated that United States armored forces could penetrate Baghdad at will, while suffering minimal casualties. The operation provided excellent indicators of enemy tactics, strength, and fighting positions. LTG Wallace and MG Blount praised the 5 April "Thunder Run." They envisioned the operation as a prelude to additional armored missions in and around the city that would disrupt the Baghdad defenses with the ultimate goal of regime collapse. Using the lessons learned on 5 April, Perkins launched a second, larger operation on 7 April, which resulted in the occupation of downtown Baghdad and the final fall of the Baathist government.

Donald P. Wright

ILLUSTRATIONS OF THE SEVEN PRINCIPLES OF MISSION COMMAND

4-5.   The application of the seven principles of mission command, combined with COL Perkins' use of the art of command and science of control helped reduce uncertainty during the planning, preparation, and execution of the 5 April 2003 movement to contact through Baghdad. Soldiers easily understood the mission and intent, which were simple and clear. COL Perkins' command presence forward set a positive example for Task Force 1-64 Armor Regiment and allowed him to assess the situation and apply judgment.

4-6.   The science of control was illustrated by the actions of the 2/3 ID. The 2/3 ID main CP facilitated mission accomplishment by coordinating with the 3^d ID main CP and with 1/3 ID, synchronizing and integrating actions, informing COL Perkins, and providing control for the 2/3 ID units in OBJ SAINTS. The paragraphs below described the seven principles of mission command illustrated in the vignette above.

COMPETENCE

4-7.   The 2^d Brigade was a regular Army unit, which had stabilized both command tours and personnel assignments during its overseas tour. While the campaign was only two weeks old, the brigade had been in Kuwait for over six months prior to that and had trained intensively. By 5 April, two continuous weeks of combat experience augmented the training. COL Perkins' command presence, two-way counseling sessions, professional development sessions, and continuous assessments naturally enhanced the building of a competent team. When leaders are open to candid subordinate feedback as well, such sessions and assessments will yield greater results in building the team's competency during the planning, preparation, and execution of any operation, not just during the after-action review.

MUTUAL TRUST

4-8.   COL Perkins trusted his commanders and Soldiers because of their high level of training and their proven ethical and effective performance in combat. When leaders give clear expectations to subordinates through two-way counseling sessions, professional development sessions, and continuous assessments, it engenders trust.

SHARED UNDERSTANDING

4-9.   The corps, division, and brigade commanders clearly conveyed their intents, objectives, and key tasks to subordinate commanders. The long train up for the campaign in Kuwait and the previous two weeks of operations facilitated shared understanding. Additionally, the raid was essentially a battle drill, which Task Force 1-64 Armor Regiment had executed many times before, both in training and in combat. When leaders and subordinates share personal experiences during counseling, professional development sessions, or after-action reviews, all parties develop a shared understanding. Through shared personal experiences, leaders and subordinates will better understand how each other think, gaining keen insight to how commanders and staff analyze and solve problems. With this level of shared understanding, units can achieve greater levels of synchronization and efficiency at a quicker pace during high tempo environments.

COMMANDER'S INTENT

4-10. Both LTG Wallace and MG Blount provided clear and concise commanders' intents for the 5 April mission. Their intent was to conduct a raid into Baghdad in an armored column to test the Iraqi military's urban defenses, collect information, and pressure the regime. COL Perkins added his own emphasis to maintain momentum throughout the movement and to create as much confusion among enemy elements as possible.

4-11. More than just stating an operations purpose, key tasks, and the desired outcome, commander's intent clearly articulates what criteria or metrics the commander will use define success for terrain, civilians (if applicable), friendly forces, and the enemy. Usually no more than three to five sentences, it gives the reason and broad purpose beyond the mission statement in a way subordinates two echelons down can easily remember. An example of the desired outcome might be, "All friendly forces are north of X River at

80-percent combat power, with minimal loss of civilian structures, no loss of civilian life, and all enemy forces are unable to reach bridge Y."

MISSION ORDERS

4-12. When COL Perkins issued his order for the reconnaissance in force, he directed Task Force 1-64 Armor Regiment to attack up Highway 8 all the way to the Baghdad Airport to collect information about the composition and disposition of the Iraqi forces that were defending the city. He provided clear intent, objectives, and graphics and allowed LTC Schwartz to execute. The directive was unambiguous: maintain momentum, handover targets to trailing armored vehicles, and avoid becoming tied down into a pitched battle. The directive also maximized individual initiative.

4-13. In conjunction with graphical control measures, mission orders provide clear doctrinal language familiar to all. Whether assigning doctrinal tasks, allocating resources, or issuing broad guidance, mission orders tell subordinates what results to attain, without articulating exactly how to achieve such results. Without mission orders, there exists too much room for ambiguity, misinterpretation of discussions, and desynchronization of an operation, especially one with multiple unified action partners.

EXERCISE DISCIPLINED INITIATIVE

4-14. Commanders at all levels had confidence that their subordinates could do the job with minimal direction because of the experience level of the unit, shared understanding, and mutual trust. This prevented the column from bogging down at several points during the operation. When enemy disabled SSG Diaz's tank, for example, LTC Schwartz and COL Perkins knew the crew had done whatever they could to save the tank and accepted its destruction and abandonment.

4-15. Leaders must continue to develop and encourage initiative in their subordinates. During assessments, or counseling and professional development sessions, affirming and rewarding observed instances of disciplined initiative will empower subordinates in the organization. Commanders should consider highlighting in a positive light disciplined initiative that was not completely successful. The enemy may have chosen a different course of action (COA), or more information may have become available after a subordinate made a decision, and the subordinate could easily become discouraged to exercise further disciplined initiative.

RISK ACCEPTANCE

4-16. COL Perkins used armored vehicles to execute the 5 April Thunder Run. The brigade's vulnerable wheeled vehicles remained at OBJ SAINTS. This deprived the task force of certain logistical and sustainment functions during the course of the operation. However, the raid's short duration mitigated the risk.

4-17. COL Perkins used his knowledge of the art of command to position himself and his intelligence staff officer (S-2) in an M113 into the lead task force formation to build understanding and enable timely decisions. This position allowed him to assess the situation, apply judgment, and prioritize resources to accomplish the mission. In addition, COL Perkins' command presence forward gave him the ability to influence Task Force 1-64 Armor Regiment through personal example and guidance.

4-18. Commanders define those areas where they are willing to accept risk and where they are not. Commanders establish this in their commander's intent for specific missions, often by phase, and based on experience with subordinates and staffs during previous missions, and results from previous training events and exercises. When faced with subordinates new to their command, a new or ill-structured problem, commanders thoroughly explain these areas of accepted risk and give examples of what they consider acceptable risk. Commanders at each echelon carefully determine risks, analyze and minimize as many hazards as possible, and then accept risk to accomplish the mission.

COMMAND AND CONTROL

4-19. *Command and control* is the exercise of authority and direction by a properly designated commander over assigned and attached forces in the accomplishment of mission (JP 1). Command and control is fundamental to the art and science of warfare. No single activity in operations is more important than command and control. Command and control by itself will not secure an objective, destroy an enemy target, or deliver supplies. Yet none of these activities could be coordinated towards a common objective, or synchronized to achieve maximum effect, without effective command and control. It is through command and control that the countless activities a military force must perform gain purpose and direction. The goal of command and control is mission accomplishment.

COMMAND

4-20. *Command* is the authority that a commander in the armed forces lawfully exercises over subordinates by virtue of rank or assignment (JP 1). Command includes the authority and responsibility for effectively using available resources and for planning the employment of, organizing, directing, coordinating, and controlling military forces for the accomplishment of assigned missions. It also includes responsibility for health, welfare, morale, and discipline of assigned personnel.

4-21. As an art, command requires the use of judgment. Commanders constantly use their judgment for such things as delegating authority, making decisions, determining the appropriate degree of control, and allocating resources. Although certain facts like troop-to-task ratios may influence a commander, they do not account for the human aspects of command. A commander's character, competence, and commitment, influenced by their experience, training, and education influence their decision-making (see ADP 1 and ADP 6-22). Proficiency in the art of command stems from years of schooling, self-development, and operational and training experiences.

4-22. Command is a human skill sharpened by experience, study, and observation. Commanding at any level is more than simply leading Soldiers and units and making decisions. Commanders use their authority with firmness and care. Commanders strive to understand all aspects of their operational environment. Effective commanders create a positive command climate that instills a sense of mutual trust throughout the command. They use their judgment to assess situations, draw feasible conclusions, and make decisions. Commanders guide operations without stifling individual initiative. (See ADP 6-0 for additional information.) The key elements of command are—

- Authority.
- Responsibility.
- Decision-making.
- Leadership.

CONTROL

4-23. *Control* is the regulation of forces and warfighting functions to accomplish the mission in accordance with the commander's intent (ADP 6-0). Aided by staffs, commanders exercise control over assigned forces in their area of operations. Staffs coordinate, synchronize, and integrate actions, inform the commander, and exercise control for the commander. Control permits commanders to adjust operations to account for changing circumstances and direct the changes necessary to address the new situation. Commanders impose enough control to mass the effect of combat power at the decisive point in time while allowing subordinates the maximum freedom of action to accomplish assigned tasks.

4-24. The science of control supports the art of command. In contrast to the art of command, the science of control is based on objectivity, facts, empirical methods, and analysis. Commanders and staffs use the science of control to overcome the physical and procedural constraints under which units operate. Units are bound by such factors as movement rates, fuel consumption, weapons effects, rules of engagement, and legal considerations. Commanders and staffs strive to understand aspects of operations they can analyze and measure, such as the physical capabilities and limitations of friendly and enemy organizations. Control requires a realistic appreciation for time and distance factors, including the time required to initiate certain actions. The commander's command and control system, especially the staff, assists the commander with control (see section III). However, the commander remains the central figure.

4-25. Commanders use control to direct and coordinate the actions of subordinate forces. They communicate information and receive feedback from subordinates to achieve greater shared understanding of the situation. This allows commanders to update their visualization with respect to the current situation, the end state or their operational approach, and adjust operations to reflect those changes. (See ADP 6-0 for additional information.) The elements of control are—

- Direction.
- Feedback.
- Information.
- Communication.

4-26. Control measures provide control without requiring detailed explanations. Control measures help commanders' direct actions by establishing responsibilities and limits that prevent subordinate unit actions from impeding one another. They foster coordination and cooperation between forces without unnecessarily restricting freedom of action. Good control measures foster freedom of action, decision-making, and individual initiative.

4-27. Commanders use the minimum number of control measures necessary to control their forces. Commanders tailor their use of control measures to conform to the higher commander's intent. They also consider the mission, terrain, and amount of authority delegated to subordinates. Effectively employing control measures requires commanders and staffs to understand their purposes and ramifications, including the permissions or limitations imposed on subordinates' freedom of action and initiative. Each measure should have a specific purpose: mass the effects of combat power, synchronize subordinate forces' operations, or minimize the possibility of fratricide, civilian casualties, and unintended excessive collateral damage. (See chapter 6 section IV and chapter 7 section IV, respectively, for a detailed discussion and examples of offensive and defensive control measures.)

SECTION II – COMMAND AND CONTROL WARFIGHTING FUNCTION

4-28. Command and control—as a warfighting function—assists the BCT commander in the exercise of authority and direction over assigned and attached forces in the accomplishment of a mission. By itself, the command and control warfighting function cannot achieve objectives or accomplish missions. Mission accomplishment requires a common understanding of the principles of mission command and the integration and convergence of combat power, while emphasizing command and control that empowers subordinate decision-making and decentralized execution appropriate to the situation. The command and control warfighting function provides purpose and direction to the other warfighting functions. The command and control warfighting function (depicted in figure 4-1 on page 4-8) consists of the command and control warfighting function tasks and the command and control system. This section focuses on the related tasks and a system that enable the commander to synchronize and converge all elements of combat power. In addition to the major activities of the operations process—the Army's framework for exercising command and control—it addresses the integrating processes used by the commander and staff to synchronize specific functions throughout the operations process. (See ADP 6-0 for additional information.)

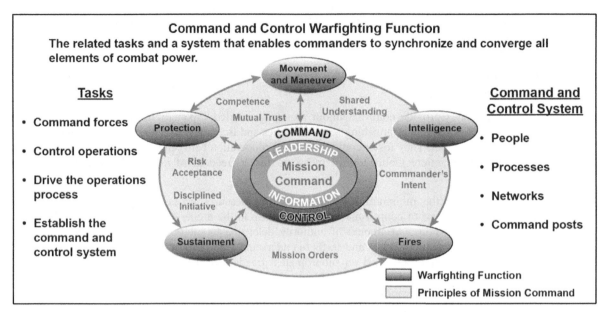

Figure 4-1. Combat power model

COMMAND AND CONTROL WARFIGHTING FUNCTION TASKS

4-29. The command and control warfighting function tasks focus on integrating the activities of the other elements of combat power to accomplish missions. Commanders, assisted by their staffs, integrate numerous processes and activities within their headquarters and across the force through the mission command warfighting function. These tasks are—

- Command forces (described in paragraphs 4-20 to 4-22).
- Control operations (described in paragraphs 4-23 to 4-27).
- Drive the operations process (described in paragraphs 4-43 to 4-137).
- Establish the command and control system (described in paragraphs 4-30 to 4-42).

COMMAND AND CONTROL SYSTEM

4-30. The BCT commander cannot exercise command and control alone. The *command and control system*—the arrangement of people, processes, networks, and command posts that enable commanders to conduct operations (ADP 6-0)—enables the commander's ability to lead the staff and provide direction and motivation to subordinate commanders and Soldiers. The command and control system supports the commander's decision-making, disseminates the commander's decisions to subordinates, and facilitates controlling forces. The commander employs the command and control system to enable the people and formations conducting operations to work towards a common purpose. All equipment and procedures exist to achieve this end. The commander organizes a command and control system to—

- Support decision-making.
- Collect, create, and maintain relevant information and prepare products to support the commander's and leaders' understanding and visualization.
- Prepare and communicate directives.

4-31. To provide these three overlapping functions, the commander must effectively locate, design, and organize the four components of a command and control system. The four components are—

- People.
- Processes.
- Networks.
- CPs.

4-32. At every echelon of command, the most important of these components is people. As the commander's command and control system begins with people, the commander bases command and control systems on human characteristics more than on equipment and procedures. Trained people are essential to an effective command and control system; the best technology cannot support command and control without them. (See ADP 6-0 for additional information.)

PEOPLE

4-33. The Army's approach to mission command is built upon the bedrock of the Army Profession—trust. Mutual trust between commanders, leaders, and Soldiers in cohesive units enables command and control to thrive in the ambiguity and chaos of a complex world. Trusted Army professionals are essential to an effective command and control system.

4-34. Soldiers and leaders exercise disciplined initiative and accomplish assigned missions according to the commander's intent, not technology. Therefore, the BCT commander bases command and control systems on human skills, knowledge, and abilities more than on equipment and procedures. Trained Soldiers and leaders form the basis of an effective command and control system; the commander must not underestimate the importance of providing training.

4-35. Key people within the BCT dedicated to command and control include seconds in command, command sergeants major, and staffs. The second in command is the commander's principal assistant. The command sergeant major is the senior noncommissioned officer of the command. The staff supports the commander with understanding situations, decision-making, and implementing decisions throughout the operations process. The commander systematically arranges the staff as part of the command and control system to perform the following three functions:

- Supporting the commander.
- Assisting subordinate units.
- Informing units and organizations outside the headquarters.

PROCESSES

4-36. The BCT commander establishes and uses systematic processes and procedures to organize the activities within the headquarters and throughout the force. Processes are a series of actions directed to an end state, such as the military decision-making process (MDMP). Procedures are standard, detailed steps, often used by the BCT staff, which describes how to perform specific tasks to achieve the desired end state, such as standard operating procedures (SOPs). Processes and procedures increase organizational competence by improving the staff's efficiency or by increasing the tempo.

4-37. The MDMP provides the commander, staffs, and subordinate commanders an orderly method for planning. SOPs often provide detailed unit instructions on how to configure common operational picture (COP) displays. Adhering to processes and procedures minimizes confusion, misunderstanding, and hesitation as the commander makes frequent, rapid decisions to meet operational requirements.

NETWORKS

4-38. The network connects people and allows sharing of resources and information. The network enables the execution of command and control and supports operations through wide dissemination of data and relevant information. The Army's network is the Department of Defense information network-Army (DODIN-A). The *Department of Defense information network-Army* is an Army-operated enclave of the Department of Defense information network that encompasses all Army information capabilities that collect, process, store, display, disseminate, and protect information worldwide (ATP 6-02.71).

4-39. As networks may degrade during operations, the commander must develop methods and measures to mitigate the impact of degraded networks. The commander may mitigate the impact of degraded networks through exploiting the potential of technology or through establishing trust, creating shared understanding, or providing a clear intent using mission orders.

COMMAND POSTS

4-40. Effective command and control requires continuous, and often immediate, close coordination, synchronization and information-sharing across the staff. To promote this, commanders organize their staffs and other components of the command and control system into CPs to assist them in effectively conducting operations.

4-41. CPs are facilities that include personnel, processes and procedures, and networks that assist commanders in command and control. Commanders employ CPs to help control operations through continuity, planning, coordination, and synchronizing of the warfighting functions. Commanders organize their CPs flexibly to meet changing situations and requirements of different operations.

4-42. CP functions directly relate to assisting commanders in understanding, visualizing, describing, directing, leading, and assessing operations. Different types of CPs, such as the main CP or the tactical CP, have specific functions by design. (CPs are discussed later in this chapter and in FM 6-0.)

OPERATIONS PROCESS

4-43. The Army's framework for exercising command and control is the operations process. The BCT commander, assisted by the staff, uses the operations process to drive the conceptual and detailed planning necessary to understand, visualize, and describe the operational environment and the operations end state; make and articulate decisions; and direct, lead, and assess military operations as shown in figure 4-2.

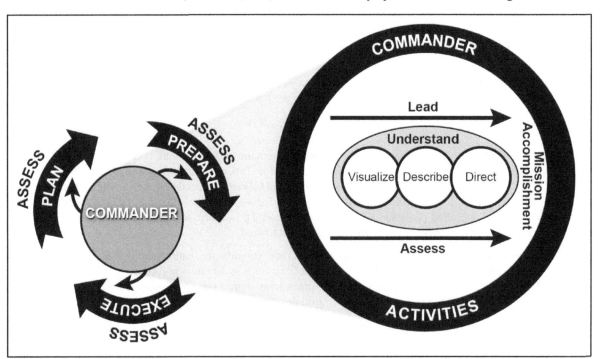

Figure 4-2. The operations process

4-44. The operations process, while simple in concept (plan, prepare, execute, and assess), is dynamic in execution. The BCT commander and staff use the operations process to integrate numerous tasks executed throughout the headquarters and with subordinate units. The commander organizes and trains the staff and subordinates as an integrated team to simultaneously plan, prepare, execute, and assess operations. In addition to the principles of mission command, the commander and staff consider the following principles for the effective employment of the operations process:

- Commanders drive the operation process.
- Build and maintain situational understanding.
- Apply critical and creative thinking.

4-45. The activities of the operations process are not discrete; they overlap and recur as circumstances demand. Planning starts an iteration of the operations process. Upon completion of the initial order, planning continues as leaders revise the plan based on changing circumstances. Preparing begins during planning and continues through execution. Execution puts a plan into action by applying combat power to seize, retain, and exploit the initiative and consolidate gains. (See ADP 5-0 for additional information.)

PLAN

4-46. *Planning* is the art and science of understanding a situation, envisioning a desired future, and determining effective ways to bring that future about (ADP 5-0). Planning consists of two separate but interrelated components, a conceptual component and a detailed component. Successful planning requires the integration of both these components. The BCT commander and subordinate commanders ethically, effectively, and efficiently employ three methodologies for planning operations: the Army design methodology, the MDMP (brigade and battalion/squadron echelons), and troop leading procedures (company and troop echelons and below). Commanders determine how much of each methodology to use based on the scope of the problem, their familiarity with it, and the time available. Planning helps the BCT commander create and communicate a common vision between the staff, subordinate commanders, their staffs, and unified action partners. Planning results in an order that synchronizes the action of forces in time, space, purpose, and resources to achieve objectives and accomplish missions.

> *Note.* The rapid decision-making and synchronization process is a decision-making and planning technique that commanders and staffs commonly use during execution when available planning time is limited. (See paragraph 4-118.)

Army Planning Methodologies

4-47. The *Army design methodology* is a methodology for applying critical and creative thinking to understand, visualize, and describe problems and approaches to solving them (ADP 5-0). The Army design methodology is particularly useful as an aid to conceptual thinking about unfamiliar problems and to gain a greater understanding of the operational environment. To produce executable plans, the commander integrates this methodology with the detailed planning typically associated with the military decision-making process (MDMP). (See ATP 5-0.1 for additional information.)

Conceptual and Detailed Planning

4-48. The BCT commander and staff conduct conceptual and detailed planning to facilitate the activities of the operations process. The commander personally leads the conceptual component of planning. While the commander is engaged in parts of detailed planning, the commander often leaves the specifics to the staff. Conceptual planning provides the basis for all subsequent planning. For example, the commander's intent (see paragraph 4-66) and operational approach provide the framework for the entire plan. This framework leads to a concept of operations (see paragraph 4-69) and associated schemes of support, such as schemes of intelligence, movement and maneuver, fires, sustainment, and protection. In turn, the schemes of support lead to the specifics of execution, including tasks to subordinate units and detailed annexes to the operation order. However, the dynamic does not operate in only one direction, conceptual planning must respond to detailed constraints.

Operational Approach

4-49. Army design methodology is a methodology for applying critical and creative thinking to understand, visualize, and describe problems and approaches to solving them. The BCT commander and staff conduct conceptual planning (Army design methodology, see ATP 5-0.1) to understand, visualize, and describe the operational environment and the *operational approach*—a broad description of the mission, operational concepts, tasks, and actions required to accomplish the mission (JP 5-0)—to the problem. The commander and staff use the Army design methodology, operational variables, and mission variables to analyze an operational environment in support of the operations process. When developing an operational approach, the commander considers methods to employ a combination of defeat mechanisms and stability mechanisms.

Analysis determines the appropriate combination of decisive action (offense, defense, stability) for the operation and conceptual planning activities lead to prioritization of defeat and stability mechanisms.

Defeat Mechanism

4-50. Defeat (see paragraph 2-99) has a temporal component and is seldom permanent. A *defeat mechanism* is a method through which friendly forces accomplish their mission against enemy opposition (ADP 3-0). All echelons within the BCT use a combination of four defeat mechanisms: destroy (see paragraph 2-99), dislocate, disintegrate, and isolate relating to offensive and defensive operations. Commanders consider more than one defeat mechanism simultaneously to produce complementary and reinforcing effects not attainable with a single mechanism. Used individually, a defeat mechanism achieves results relative to how much effort is expended. Using defeat mechanisms in combination creates enemy dilemmas that magnify their effects significantly. Commanders describe defeat mechanisms by the three types of effects they produce:

- Physical effects are those things that are material.
- Temporal effects are those that occur at a specific point in time.
- Cognitive effects those that pertain to or affect the mind.

4-51. The commander conceptualizes an operational approach to attain the end state by formulating the most effective, efficient way to apply defeat mechanisms. For example:

- Physically defeating an enemy deprives enemy forces of the ability to achieve those aims.
- Temporally defeating an enemy anticipates enemy reactions and counters them before they can become effective.
- Cognitively defeating an enemy disrupts decision-making and deprives that enemy of the will to fight.

4-52. *Dislocate* is to employ forces to obtain significant positional advantage, rendering the enemy's dispositions less valuable, perhaps even irrelevant (ADP 3-0). For example, the commander can achieve dislocation by placing forces in locations where the enemy does not expect them.

4-53. *Disintegrate* means to disrupt the enemy's command and control system, degrading its ability to conduct operations while leading to a rapid collapse of the enemy's capabilities or will to fight (ADP 3-0). For example, the commander can achieve disintegration by specifically targeting an enemy's command structure and communications systems.

4-54. *Isolate* means to separate a force from its sources of support in order to reduce its effectiveness and increase its vulnerability to defeat (ADP 3-0). For example, as isolation can encompass multiple domains and can have both physical and psychological effects detrimental to accomplishing a mission, the commander can isolate a force in the electromagnetic spectrum (EMS) thus exacerbate the effects of physical isolation by reducing its situational awareness.

Stability Mechanism

4-55. The BCT's mission, in addition to defeating an enemy, may require performing stability mechanisms related to stability operations, security, and consolidating gains in an area of operations. A *stability mechanism* is the primary method through which friendly forces affect civilians in order to attain conditions that support establishing a lasting, stable peace (ADP 3-0). (See paragraphs 8-23 and 8-24 for a detailed discussion on the four stability mechanisms: compel, control, influence, and support.)

Conceptual Planning Outputs and Activities

4-56. Outputs of conceptual planning include a problem statement, draft mission statement, draft commander's intent, a broad concept sketch, initial decision points, commander's critical information requirements (CCIRs), and initial planning guidance. Conceptual planning activities should include initial framing of branches and sequels to the plan and the ideal end state or posture.

Military Decision-making Process

4-57. The BCT staff uses the outputs of conceptual planning to begin detailed planning (MDMP at the BCT level). The *military decision-making process* is an iterative planning methodology to understand the situation and mission, develop a course of action, and produce an operation plan or order (ADP 5-0). The result of detailed planning is a synchronized plan that provides mission type orders for the staff and subordinate units (see figure 4-3 on page 4-14).

Step 1: Receipt of Mission

Key Inputs	Process		Key outputs
• Higher headquarters' plan and order or a new mission anticipated by the commander	• Alert the staff and other key participants • Gather the tools • Update running estimates	• Conduct initial assessment • Issue the Commander's initial guidance • Issue the initial warning order	• Commander's initial guidance • Initial allocation of time

Warning Order

Step 2: Mission Analysis

Key Inputs	Process		Key outputs
• Commander's initial guidance • Higher headquarters' plan or order • Higher headquarters' intelligence and knowledge products • Knowledge products from other organizations • Army design methodology products	• Analyze the higher headquarters' plan or order • Perform initial IPB • Determine specified, implied, and essential tasks • Review available assets and identify resource shortfalls • Determine constraints • Identify critical facts and develop assumptions • Begin risk management • Develop initial CCIRs (PIR and FFIR) and EEFIs • Develop the initial information collection plan	• Update plan for the use of available time • Develop initial themes and messages • Develop a proposed problem statement • Develop a proposed mission statement • Present the mission analysis briefing • Develop and issue initial commander's intent • Develop and issue initial planning guidance • Develop COA evaluation criteria • Issue a warning order	• Problem statement • Mission statement • Initial commander's intent • Initial planning guidance • Initial CCIRs (PIR and FFIR) and EEFIs • Updated IPB and running estimates • Assumptions • Evaluation criteria for COAs

Warning Order

Step 3: Course of Action (COA) Development

Key Inputs	Process		Key outputs

• Mission statement
• Inital commander's intent, planning guidance, CCIRs (PIR and FFIRs), and EEFIs.
• Updated IPB and running estimates
	• Assess relative combat power
	• Generate options
	• Array forces
	• Develop a broad concept
	• Assign headquarters
		• Develop COA statements and sketches
		• Conduct COA briefing
		• Select or modify COAs for continued analysis
			• COA statements and sketches
			  - Tentative task organization
			  - Broad concept of operations
			• Revised planning guidance
			• Updated assumptions

Step 4: COA Analysis (War Game)

Key Inputs	Process		Key outputs
• Updated running estimates • Revised planning guidance • COA statements and sketches • Updated assumptions	• Gather the tools • List all friendly forces • List assumptions • List known critical events and decision points • Select the war-gaming method	• Select a technique to record and display results • War-game the operation and assess the results • Conduct a war-game briefing (optional)	• Refined COAs • Potential decision points • War-game results • Initial assessment measures • Updated assumptions

LEGEND

CCIR	COMMANDER'S CRITICAL INFORMATION REQUIREMENT	IPB	INTELLIGENCE PREPARATION OF THE BATTLEFIELD
EEFI	ESSENTIAL ELEMENT OF FRIENDLY INFORMATION	PIR	PRIORITY INTELLIGENCE REQUIREMENT
FFIR	FRIENDLY FORCES INFORMATION REQUIREMENTS		

Figure 4-3. Military decision-making process overview

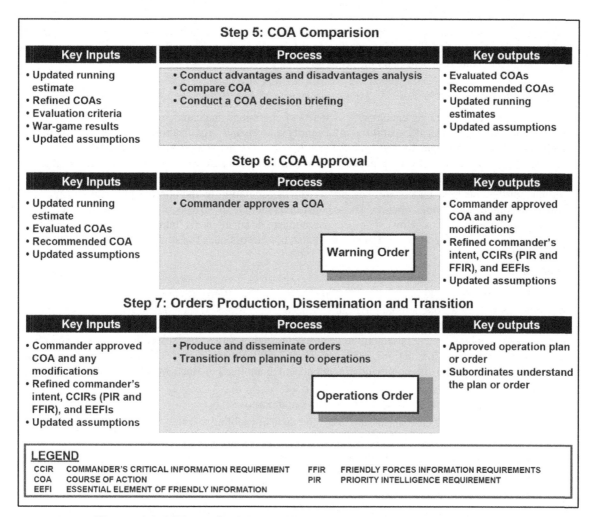

Figure 4-3. Military decision-making process overview (continued)

4-58. Depending on the situation's complexity, the BCT commander can initiate the Army design methodology before or in parallel with the MDMP. If the problem is hard to identify or the operation's end state is unclear, the commander may initiate Army design methodology before engaging in detailed planning. Army design methodology can assist the commander and staff in understanding the operational environment, framing the problem, and considering an operational approach to solve or manage the problem. The understanding and products resulting from Army design methodology guide more detailed planning during the MDMP.

4-59. When staff members use the Army design methodology and MDMP in parallel, the BCT commander may direct some staff members to conduct mission analysis while engaging others in Army design methodology activities before COA development. Results of both mission analysis and Army design methodology inform the commander in development of the commander's intent and planning guidance. In time constrained conditions, or when the problem is not complex, the commander may conduct the MDMP without incorporating formal Army design methodology efforts. During execution, the commander can use Army design methodology to help refine understanding and visualization as well as assessing and adjusting the plan as required. (See FM 6-0 for additional information.)

The Science and Art of Planning

4-60. Planning is both a science and an art. Many aspects of military operations, such as movement rates, fuel consumption, and weapons effects, are quantifiable. They are part of the science of planning. The combination of forces, choice of tactics, and arrangement of activities belong to the art of planning. Soldiers often gain knowledge of the science of planning through institutional training and study. They gain understanding of the art of planning primarily through operational training and experience. Effective planners are grounded in both the science and the art of planning.

Science of Planning

4-61. The science of planning encompasses aspects of operations that can be measured and analyzed. These aspects include the physical capabilities of friendly and enemy organizations. The science of planning includes a realistic appreciation for time-distance factors: an understanding of how long it takes to initiate certain actions, the techniques and procedures used to accomplish planning tasks, and the terms and graphics that compose the language of military operations. The mastery of the science of planning is necessary for military professionals to understand key aspects of the operation to include the physical and procedural constraints under which units operate. These constraints include the effects of terrain, weather, and time on friendly and enemy forces. However—because combat is an intensely human activity—the solution to problems cannot be reduced to a formula. This realization necessitates the study of the art of planning.

Art of Planning

4-62. The art of planning requires understanding the dynamic relationships among friendly forces, the threat, and other aspects of an operational environment during operations. It includes making decisions based on skilled judgment acquired from experience, training, study, imagination, and critical and creative thinking. Commanders apply judgment based on their knowledge and experience to select the right time and place to act, assign tasks, prioritize actions, accept risk, and allocate resources. The art of planning involves developing plans within the commander's intent and planning guidance by choosing from interrelated options, including—

- Arrangement of activities in time, space, and purpose.
- Assignment of tactical mission tasks and tactical enabling tasks.
- Task organization of available forces and resource allocation.
- Choice and arrangement of control measures.
- Tempo.
- The risk the commander is willing to take.

4-63. These interrelated options define a starting point from which planners create distinct solutions to particular problems. Each solution involves a range of options. Each balance competing demands and requires judgment. The variables of mission, enemy, terrain and weather, troops and support available, time available, civil considerations (METT-TC) always combine to form a different set of circumstances. There are no checklists that adequately apply to every situation.

Key Components of a Plan

4-64. The mission statement, commander's intent, and concept of operations are key components of a plan that serve as the framework for an operation. The BCT commander ensures the mission and end state nest with those of their higher headquarters. The commander's intent focuses on the end state; and the concept of operations focuses on the method or sequence of actions by which the force will achieve the end state. Within the concept of operations, the commander may establish *objectives*—a location used to orient operations, phase operations, facilitate changes of direction, and provide for unity of effort (ADP 3-90)—as intermediate goals toward achieving the operation's end state. When developing tasks for subordinate units, the commander ensures that the purpose of each task nests with the accomplishment of another task, with the achievement of an objective, or directly to the attainment of an end state condition. Additional components to the plan include the BCT's task organization, tasks to subordinate units, coordinating instructions, risk acceptance, and control measures.

Mission Statement

4-65. The *mission* is the task, together with the purpose, that clearly indicates the action to be taken and the reason therefore (JP 3-0). The BCT commander analyzes a mission as the commander's intent two echelons above, specified tasks, and implied tasks. The commander considers the mission of adjacent units to understand how they contribute to the decisive operation of their higher headquarters. The analysis results yield the essential tasks that, with the purpose of the operation, clearly specify the action required. This analysis produces the *mission statement*—a short sentence or paragraph that describes the organization's essential task(s), purpose, and action containing the elements of who, what, when, where, and why (JP 5-0), but seldom specifies how.

Commander's Intent

4-66. The *commander's intent* is a clear and concise expression of the purpose of the operation and the desired military end state that supports mission command, provides focus to the staff, and helps subordinate and supporting commanders act to achieve the commander's desired results without further orders, even when the operation does not unfold as planned (JP 3-0). It is critical that staff planners receive the commander's intent as soon as possible after receiving the mission. The commander's intent succinctly describes what constitutes success for the operation. The commander conveys intent in a format determined most suitable to the situation.

4-67. The commander's intent may include the operation's purpose, key tasks, and the conditions that define the end state. When describing the purpose of the operation, the commander's intent does not restate the "why" of the mission statement. Rather, it describes the broader purpose of the unit's operation in relationship to the higher commander's intent and concept of operations. *Key tasks* are those significant activities the force must perform as a whole to achieve the desired end state (ADP 6-0). Key tasks are not specified tasks for any subordinate unit; however, they may be sources of implied tasks. During execution—when significant opportunities present themselves or the concept of operations no longer fits the situation—subordinates use key tasks to keep their efforts focused on achieving the desired end state. *End state* is the set of required conditions that defines achievement of the commander's objectives (JP 3-0).

4-68. The commander's intent links the mission, concept of operations, and tasks to subordinate units. A clear commander's intent facilitates a shared understanding and focuses on the overall conditions that represent mission accomplishment. During execution, the commander's intent spurs disciplined initiative. The commander's intent must be understood two echelons down.

Concept of Operations

4-69. The *concept of operations* is a statement that directs the manner in which subordinate units cooperate to accomplish the mission and establishes the sequence of actions the force will use to achieve the end state (ADP 5-0). The BCT concept of operations expands on the commander's intent by describing how the commander wants the force to accomplish the mission. The concept of operations states the principal tasks required, the responsible subordinate unit, and how the principal tasks complement one another.

4-70. The BCT commander and staff use four components of the operational framework to help conceptualize and describe the concept of operations in time, space, purpose, and resources. First, the commander is assigned an area of operations for the conduct of operations. Second, the commander can designate deep, close, rear, and support areas to describe the physical arrangement of forces in time and space. Third, within this area, the commander conducts decisive, shaping, and sustaining operations to articulate the operation in terms of purpose. In the fourth and final component, the commander designates the main and supporting efforts to designate the shifting prioritization of resources. (See chapter 2 for a detailed discussion on the operational framework.)

> *Note.* The BCT does not conduct operationally significant consolidate gains activities unless tasked to do so, usually within a division consolidation area.

Task Organization

4-71. *Task organization* is a temporary grouping of forces designed to accomplish a particular mission (ADP 5-0). The BCT commander establishes command and support relationships to task organize the force. Command relationships define command responsibility and authority. Support relationships define the purpose, scope, and effect desired when one capability supports another. Establishing clear command and support relationships is fundamental to organizing any operation. The commander designates command and support relationships to weight the decisive operation or main effort and support the concept of operations. The commander considers two organizational principles when task organizing forces; maintain cohesive mission teams and do not exceed subordinates' span of control capabilities.

Tasks to Subordinate Units

4-72. The BCT commander and staff assign tasks to subordinate units. The assignment of a task includes not only the task (what), but also the unit (who), place (where), time (when), and purpose (why). A *task* is a clearly defined action or activity specifically assigned to an individual or organization that must be done as it is imposed by an appropriate authority (JP 1). Tasks are specific activities that direct friendly action and contribute to mission accomplishment and other requirements. The purpose of each task should nest with completing another task, achieving an objective, or attaining an end state condition.

Coordinating Instructions

4-73. Coordinating instructions pertain to the BCT as a whole. Examples include CCIRs, essential element of friendly information (EEFI), fire support coordination measures and airspace coordinating measures, rules of engagement, risk reduction control measures, personnel recovery coordination measures, and the time the operation order becomes effective or the condition of the BCT when the operation order becomes effective.

Risk Acceptance

4-74. The BCT commander uses judgment when identifying risk by deciding how much risk to accept and by mitigating risk where possible. The commander accepts risk to create opportunities and reduces risk with foresight and careful planning. Consideration of risk begins during planning as the commander, with the support of the staff, complete a risk assessment for each COA and proposes control measures. They collaborate and integrate input from higher and subordinate commanders and staffs, and unified action partners. They determine how to manage identified risks. This includes delegating management of certain risks to subordinate commanders who will develop appropriate mitigation measures. Commanders then allocate the resources they deem appropriate to mitigate risks. Subordinates require commanders to underwrite their own risk acceptance. (See paragraph 4-186 for a detailed discussion of risk management [RM].)

Control Measures

4-75. A control measure is a means of regulating forces or warfighting functions. Control measures can be permissive (which allows something to happen) or restrictive (which limits how something is done) to prevent units from impeding one another and to impose necessary coordination. Some control measures are graphic. A graphic control measure is a symbol used on maps and displays to regulate forces and warfighting functions. (See ADP 1-02 for illustrations of graphic control measures and rules for their use.) Tailored to the higher commander's intent, the BCT commander assigns subordinate units' missions and imposes control measures necessary to synchronize and maintain control over the operation.

4-76. The BCT commander or staff assigns graphical control measures such as boundaries and procedural control measures such as target engagement priorities. The commander and staff must understand their purposes and ramifications, including the permissions or limitations imposed on subordinates' freedom of action and initiative, to employ control measures effectively. Each control measure should have a specific purpose and provide the flexibility needed to respond to changes in the situation. The commander uses graphical and written control measures to assign responsibilities, coordinate maneuver, and control the airspace user. The BCT operations staff officer (S-3) nests all coordination measures to include movement and maneuver control measures, fire support coordination measures, and airspace coordinating measures to clarify responsibilities and tasks to reduce risk and facilitate effective military operations. (See FM 3-90-1 and ADP 1-02 for additional information.)

Flexibility and Adaptability

4-77. The mission command approach provides flexibility and adaptability, allowing subordinates to recognize and respond effectively to emerging conditions and to correct for the effects of fog and friction. Control informed by a mission command approach provides information that allows commanders to base their decisions and actions on the results of friendly and opponent actions, rather than rigid adherence to the plan. Commanders seek to build flexibility and adaptability into their plans.

Support to Flexibility and Adaptability

4-78. Control allows organizations to respond to change, whether due to opponent or friendly actions, or environmental conditions. Control supports flexibility and adaptability in two ways. First, it identifies the need to change the plan. It does this through anticipating or forecasting possible opponent actions and by identifying unexpected variances—opportunities or threats—from the plan. This occurs throughout the operations process. Second, control helps commanders develop and implement options to respond to these changes in a timely manner. Flexibility and adaptability provided by the appropriate level of control reduces an opposing force's available options while maintaining or expanding friendly options. Effective control provides for timely action before opposing forces can accomplish their objectives, allowing for the modification of plans as the situation changes.

4-79. Instead of rigidly adhering to the plan, control focuses on information about emerging conditions. The mission command approach to control provides flexibility by—

- Allowing friendly forces to change their tasks, their task organization, or their plan.
- Producing information about options to respond to changing conditions.
- Communicating the commander's decisions quickly and accurately.
- Providing for rapid reframing when the plan changes during execution.
- Allowing collaborative planning to respond to the progress of operations.

Decisions Points

4-80. A *decision point* is a point in space and time when the commander or staff anticipates making a key decision concerning a specific course of action (JP 5-0). Decision points may be associated with the CCIRs, the friendly force, and the status of ongoing operations that describe what information the commander needs to make the anticipated decision. A decision point requires a decision by the commander. It does not dictate what the decision is, only that the commander must make one, and when and where it should be made to maximally impact friendly or enemy COAs or the accomplishment of stability operations tasks. Planners record decision points on a decision support template (DST), decision support matrix (DSM), and execution matrix.

4-81. A *decision support template* is a combined intelligence and operations graphic based on the results of wargaming that depicts decision points, timelines associated with movement of forces and the flow of the operation, and other key items of information required to execute a specific friendly course of action (JP 2-01.3). The DST is refined as planning progresses and during execution. Part of the DST is the DSM.

4-82. The *decision support matrix* is a written record of a war-gamed course of action that describes decision points and associated actions at those decision points (ADP 5-0). It lists decision points, locations of decision points, criteria to be evaluated at decision points, actions that occur at decision points, and the units responsible to act on the decision points. The DSM describes where and when a decision must be made if a specific action occurs. It ties decision points to named area of interest (NAI), target area of interest (TAI), CCIRs, collection assets, and potential friendly response options. The DSM is refined as planning progresses and during execution:

- *Named area of interest*—a geospatial area or systems node or link against which information that will satisfy a specific information requirement can be collected, usually to capture indications of adversary courses of action (JP 2-01.3).
- *Target area of interest*—the geographical area where high-value targets can be acquired and engaged by friendly forces (JP 2-01.3).

4-83. An *execution matrix* is a visual representation of subordinate tasks in relationship to each other over time (ADP 5-0). An execution matrix could be for the entire force, such as an air assault execution matrix, or it may be specific to a warfighting function, such as a fire support execution matrix. The current operations integration cell uses the execution matrix to determine which friendly actions to expect forces to execute in the near term or, in conjunction with the DSM, which execution decisions to make.

Commander's Information Collection Effort

4-84. The BCT's information collection effort answers CCIRs (specifically, priority intelligence requirements and friendly force information requirements) and EEFI. The following key doctrinal terms and definitions are used throughout this and other chapters. See referenced publications for additional information:

- *Commander's critical information requirements*—an information requirement identified by the commander as being critical to facilitating timely decision making (JP 3-0).
- *Friendly force information requirement*—information the commander and staff need to understand the status of friendly and supporting capabilities (JP 3-0).
- *Priority intelligence requirement*—an intelligence requirement that the commander and staff need to understand the threat and other aspects of the operational environment (JP 2-01).
- *Essential element of friendly information*—a critical aspect of a friendly operation that, if known by a threat, would subsequently compromise, lead to failure, or limit success of the operation and therefore should be protected from enemy detection (ADP 6-0).

Flexible Plans

4-85. Flexible plans help the BCT adapt quickly to changing circumstances during operations. Flexible plans provide options to the commander for addressing new or unforeseen circumstances during execution. Ultimately, flexibility enables the commander to mitigate risk and develop options to where best to engage the enemy. For example, the decision on where to fight the enemy is based on the commander and staff's clear understanding of the effects of the terrain, the enemy situation, and what the enemy is expected to do. The commander and staff select the most advantageous location to fight the engagement and then determine other possible locations where the engagement may occur based on a slower- or faster-than-expected enemy advance or the enemy's use of an unlikely avenue of approach (see figure 4-4). The commander identifies these areas as objectives, intermediate objectives, or engagement areas (EAs). Example EA options include—

- Option EA Rain. Enemy lead elements cross phase line (PL) Nita, maneuver battalion engages enemy in EA Rain (lead battalion elements vicinity PL Sally).
- Option EA Hail. Enemy lead elements move east through NAI 3, maneuver battalion engages enemy in EA Hail (lead battalion elements vicinity PL Tracy).
- Option EA Snow. Enemy lead elements move east through NAI 4, maneuver battalion engages enemy in EA Snow (lead battalion elements vicinity PL Tracy).
- Option EA Sleet. Enemy lead elements cross PL Sue, maneuver battalion engages enemy in EA Sleet (lead battalion elements vicinity PL Nita).

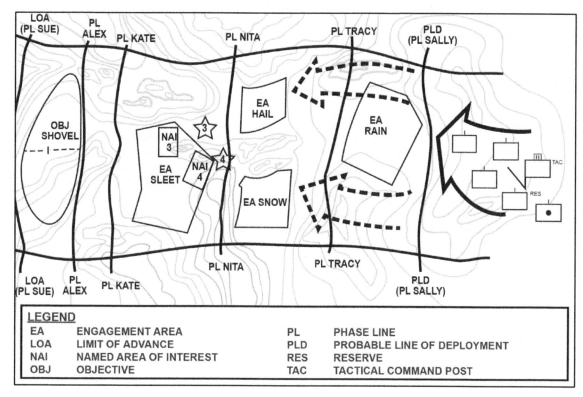

Figure 4-4. Decision points (planning options)

4-86. In coordination with the BCT, the battalion commander and staff in figure 4-4 develop control measures to help coordinate actions throughout its assigned area of operation. The commander, primarily assisted by the S-3 and S-2, develops decision points for the commitment of the battalion to each location based on relative locations and rates of movement of the battalion and the enemy. In this example, the S-2 selected NAIs to identify the enemy's rate and direction of movement to support the commander's decision of where to fight the engagement.

4-87. The commander and staff determine where and under what conditions the reserve force is likely to be employed in order to position it effectively. The commander provides specific planning guidance to the reserve to include priority for planning. The reserve force commander analyzes assigned planning priorities, conducts the coordination with units that will be affected by maneuver and commitment, and provides information to the commander and staff on routes and employment times to designated critical points on the battlefield. The reserve commander should also expect to receive specific decision points and triggers for employment on each contingency. This guidance allows the reserve commander to conduct quality rehearsals and to anticipate commitment as the commander monitors the fight.

Situational Understanding

4-88. Success in operations demands timely and effective decisions based on applying judgment to available information and knowledge. Throughout the conduct of operations, the BCT commander (supported by the staff and subordinate commanders and in coordination with unified action partners) seeks to build and maintain situational understanding. *Situational understanding*—the product of applying analysis and judgment to relevant information to determine the relationships among the operational and mission variables (ADP 6-0)—is used to facilitate decision-making. The BCT staff uses knowledge management and information management to extract knowledge from the vast amount of available information. This enables the staff to provide knowledge to the commander as recommendations and running estimates to help the commander build and maintain situational understanding.

4-89. The BCT commander strives to create shared understanding of the operational environment, the operation's purpose, the problem, and approaches to solving the problem form the basis for unity of effort and trust. Decentralized actions can perform in the context of shared understanding as if they were centrally coordinated. Knowledge management helps create shared understanding through the alignment of people, processes, and tools within the BCT's organizational structure and culture to increase collaboration and interaction. This results in better decisions and enables improved flexibility, adaptability, integration, and synchronization to achieve a position of relative advantage. Knowledge management facilitates situational understanding and acts as a catalyst for enhanced shared understanding.

4-90. Knowledge management and information management assist the commander with progressively adding meaning at each level of processing and analyzing to help build and maintain situational understanding. Knowledge management and information management are interrelated activities that support the commander's decision-making. Four levels of meaning, from the lowest level to the highest level, include data, information, knowledge, and understanding. At the lowest level, processing transforms data into information. Analysis then refines information into knowledge. The BCT commander and staff then apply judgment to transform knowledge into understanding. Commanders and staffs continue a progressive development of learning, as organizations and individuals assign meaning and value at each level. (See figure 4-5.)

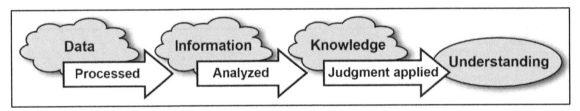

Figure 4-5. Achieving understanding

4-91. In typical organizations, data often flows to CPs from subordinate units. In the context of decision-making, *data* consists of unprocessed observations detected by a collector of any kind (human, mechanical, or electronic) (ADP 6-0). Subordinate units push data to inform higher headquarters of events that facilitate situational understanding. Data can be quantified, stored, and organized in files and databases; however, data only becomes useful when processed into information.

4-92. *Information management* is the science of using procedures and information systems to collect, process, store, display, disseminate, and protect data, information, and knowledge products (ADP 6-0). In the context of decision-making, *information* is data that has been organized and processed in order to provide context for further analysis (ADP 6-0). Information management supports, underpins, and enables knowledge management. Information management and knowledge management link to facilitate understanding and decision-making. Information management is a technical discipline that involves the planning, storage, manipulating, and controlling of information throughout its life cycle in support of the commander and staff. Information management provides a structure so commanders and staffs can process and communicate relevant information and make decisions. The signal staff officer (see paragraph 4-199) of the BCT enables knowledge management by providing network architecture and the technological tools necessary to support content management and knowledge sharing.

4-93. *Knowledge management* is the process of enabling knowledge flow to enhance shared understanding, learning, and decision-making (ADP 6-0). In the context of decision-making, *knowledge* is information that has been analyzed and evaluated for operational implications (ADP 6-0). Knowledge flow refers to the ease of movement of knowledge within and among organizations. Knowledge must flow to be useful. Effective and efficient use of knowledge in conducting operations and supporting organizational learning are essential functions of knowledge management. The BCT executive officer (XO) is the senior knowledge management officer in the BCT and advises the commander on knowledge management policy. The XO is responsible for directing the activities of each staff section and subordinate unit to capture and disseminate organizational knowledge. When staffed, a knowledge management officer (see paragraph 4-217 and FM 6-0), working through the XO, is responsible for developing the knowledge management plan that integrates and synchronizes knowledge and information management within the BCT.

4-94. The staff task of 'conduct knowledge management and information management' is essential to the command and control warfighting function and entails the continuous application of the knowledge management process of assess, design, develop, pilot, and implement activities designed to capture and distribute knowledge throughout the organization. The assessment step begins with determining what information leaders need to make decisions, and how the unit provides information for those leaders. Design is identifying tailored frameworks for knowledge management products or services that effectively and efficiently answer information requirements and meet the objectives established in the assessment. Develop is the step that actually builds the solution derived from the assessment and design steps. Pilot is the phase that deploys the knowledge management solution and tests and validates it with the unit. Implement is the phase that executes the validated knowledge management plan and integrates it into the unit information systems. (See ATP 6-01.1 for additional information.)

Note. The design step of the knowledge management process differs from and should not be confused with Army design methodology. (See paragraph 4-47 for information on Army design methodology.)

4-95. The knowledge management process, used throughout the operations process, puts the knowledge management plan into practice. Example activities involved in the conduct of knowledge management and information management will involve assessments and preparation activities, and reporting, refinement of communications, and collaborative processes. Assessments are critical to the conduct of knowledge management and information management providing feedback to the organization on what is effective. Preparation activities help the commander and staff, and subordinates understand the situation and their roles in upcoming operations. Based on this improved situational understanding, the commander refines the plan, as required, before execution with reporting, refinement of communications, and collaborative processes enabling mission execution.

4-96. Understanding is judgment applied to knowledge in the context of a particular situation. In the context of decision-making, *understanding* is knowledge that has been synthesized and had judgment applied to comprehend the situation's inner relationships, enable decision-making, and drive action (ADP 6-0). Understanding is knowing enough about the situation to change it by applying action. Judgment is based on experience, expertise, and intuition. Ideally, true understanding should be the basis for decisions. However, uncertainty and time preclude achieving perfect understanding before deciding and acting. (See FM 6-0 and ATP 6-01.1 for additional activities involved in the conduct of knowledge and information management.)

PREPARE

4-97. *Preparation* is those activities performed by units and Soldiers to improve their ability to execute an operation (ADP 5-0). The MDMP drives preparation. Since time is a factor in all operations, the BCT commander and staff, as stewards of the Army Profession, conduct a time analysis early in the planning process. This analysis helps them determine what actions they need to take and when to begin those actions to ensure forces are ready and in position before execution. The plan may require the commander to direct subordinates to start necessary movements; conduct task organization changes; begin reconnaissance, surveillance, security operations, and intelligence operations (see chapter 5); and execute other preparation activities before completing the plan. Conduct rehearsal activities are highlighted in paragraphs 4-99 to 4-104.

Preparation Activities

4-98. Mission success depends as much on preparation as on planning. Subordinate and supporting leaders and units of the BCT need enough time to understand plans well enough to execute them and develop their plans and preparations for the operation. After they fully comprehend the plan, subordinate leaders rehearse key portions of the plan and ensure Soldiers position themselves and their equipment to execute the operation. The BCT conducts the activities listed in table 4-1 on page 4-24 to help ensure the force is protected and prepared for execution. (See ADP 5-0 for additional information.)

Table 4-1. Preparation activities

• Coordinate and establish liaison	• Initiate sustainment preparation
• Initiate information collection	• Initiate network preparations
• Initiate security operations	• Manage terrain
• Initiate troop movement	• Prepare terrain
• Complete task organization	• Conduct confirmation briefs
• Integrate new units and Soldiers	• Conduct rehearsals
• Train	• Conduct plans to operations transitions
• Conduct pre-operations checks and inspections	• Revise and refine the plan
	• Supervise

Conduct Rehearsals

4-99. The BCT conducts rehearsals to prepare for upcoming operations. A *rehearsal* is a session in which the commander and staff or unit practices expected actions to improve performance during execution (ADP 5-0). Four primary types of rehearsals are the backbrief, combined arms rehearsal, support rehearsal, and battle drill or SOP rehearsal. Methods for conducting rehearsals are limited only by the commander's imagination and available resources. The BCT commander uses rehearsals as a tool to ensure the staff and subordinates understand the concept of operations and commander's intent. The extent of rehearsals depends on available time. In cases of short-notice requirements, a detailed rehearsal may not be possible.

4-100. The BCT commander often issues orders to subordinates verbally in situations requiring quick reactions. At battalion and higher levels, written fragmentary orders confirm verbal orders to ensure synchronization, integration, and notification of all parts of the force. If time permits, leaders verify that subordinates understand critical tasks. Methods for doing this include the *backbrief*—a briefing by subordinates to the commander to review how subordinates intend to accomplish their mission (FM 6-0) and *confirmation brief*—a briefing subordinate leaders give to the higher commander immediately after the operation order is given to confirm understanding (ADP 5-0). It is their understanding of the commander's intent, their specific tasks, and the relationship between their mission and the other units' mission in the operation. Commanders conduct backbriefs and confirmation briefs between themselves and within staff elements to ensure mutual understanding.

4-101. A mission command or command and control rehearsal ensures that all subordinate elements main CP and tactical CP locations, jump timelines, and battle handovers are synchronized with the maneuver plan. Additionally, the mission command or command and control rehearsal ensures that the locations of the commanders, succession of command, primary, alternate, contingency, and emergency (known as PACE) communication plans, and priorities of signal maintenance and signal support are understood.

4-102. Support rehearsals help synchronize each warfighting function with the BCT's overall operation. Throughout preparation, the BCT conducts support rehearsals within the framework of a single or limited number of warfighting functions. These rehearsals typically involve coordination and procedure drills for sustainment, aviation, fires, engineer support, casualty evacuation, and medical evacuation. Support rehearsals and combined arms rehearsals complement preparations for the operation. Units may conduct rehearsals separately and then combine them into full dress rehearsals. Although these rehearsals differ slightly by warfighting function, they achieve the same result.

4-103. A *battle drill*—rehearsed and well understood actions made in response to common battlefield occurrences (ADP 3-90)—or SOP rehearsal ensures that all participants understand a technique or a specific set of procedures. A battle drill is a collective action rapidly executed without applying a deliberate decision-making process. All echelons use these rehearsal types; however, they are most common for platoons, squads, sections, and teams. Units conduct rehearsals throughout preparation; rehearsals are not limited to published battle drills. Battle drills require a "go" order instead of a plan. All echelons can rehearse such actions as a CP shift change, an obstacle breach lane marking SOP, or a refuel on the move (known as ROM) site operation.

4-104. Subordinate units conduct rehearsals after they complete their plans and issue orders, if possible. Rehearsals allow subordinate leaders and Soldiers to practice synchronizing operations at times and places critical to mission accomplishment. Effective rehearsals throughout the BCT imprint a mental picture of the sequence of the operation's key actions and improve mutual understanding and coordination of subordinate and supporting leaders and units. Four common rehearsals at the BCT level, although not inclusive, are the reconnaissance and security rehearsal, the fire support rehearsal, the sustainment rehearsal and the combined arms rehearsal addressed in paragraphs 4-105 through 4-116. (See FM 6-0 for additional information.)

Reconnaissance and Security Rehearsal

4-105. The BCT conducts reconnaissance and security rehearsals to ensure that the correct information is gathered; and that units and Soldiers gathering the required information have a sound plan for insertion and extraction. Usually, the BCT commander, XO, S-2, S-3, fire support coordinator (field artillery battalion commander) and fire support officer, brigade assistant engineer, Cavalry squadron commander, other subordinate maneuver commanders as required, and military intelligence company commander attend the rehearsal. Other BCT staff cells and elements should have a representative attending (for example, signal, sustainment, information operations, protection, aviation, psychological operations (PSYOP), cyberspace operations, and civil affairs operations).

4-106. The reconnaissance and security rehearsal should last no more than one hour. The documents needed to run the reconnaissance and security rehearsal includes the information collection matrix, the information collection overlay, the reconnaissance and security overlay, and the enemy situation template and event template and its associated matrix (see chapter 5). Rehearse the most important NAI first, then those that answer the BCT commander's priority intelligence requirements. Continue to rehearse subsequent NAIs as time permits. Each participating commander confirms the purpose (such as priority intelligence requirements) and location (such as an NAI) for each of the collection assets. Commanders also confirm to whom the information is reported and the means of communicating that information.

Note. Due to the inherent risk associated with surveillance and reconnaissance teams/units must rehearse withdrawal under fire and "in extremis" extraction to include supporting aviation (both lift and attack assets).

Fire Support Rehearsal

4-107. The BCT fire support rehearsal is crucial to mission accomplishment because it ensures that *fires*—the use of weapons systems or other action to create a specific lethal or nonlethal effects on a target (JP 3-09)—synchronize with the scheme of maneuver. The fire support rehearsal focuses on maximizing the ability of fire support systems to support the maneuver plan and achieve the commander's intent.

4-108. The fire support rehearsal (including any augmenting fire support from the division artillery or a field artillery brigade) may be used to prepare for a combined arms rehearsal or it may be used after a combined arms rehearsal to refine and reinforce key fire support tasks. If the fire support rehearsal is held first, changes from the combined arms rehearsal may require a second fire support rehearsal. If a combined arms rehearsal is not conducted, a fire support rehearsal may serve as the primary preparation for execution of the fire support plan. The unit may conduct the field artillery tactical rehearsal either before or after the fire support rehearsal. The field artillery technical rehearsal is always held last after the target refinement cutoff time.

4-109. The BCT commander, XO, S-3, and subordinate units attend the fire support rehearsal. The BCT staff officers attending include the air liaison officer, assistant brigade engineer (known as ABE), chemical, biological, radiological, and nuclear (CBRN) officer, air and missile defense officer, and brigade aviation officer. Subordinate units often bring personnel that include the S-3, the fire support officer, scout, and mortar platoon leaders. Representatives of reinforcing fire support units should participate when possible. The BCT field artillery battalion commander assisted by the BCT fire support officer usually supervises the rehearsal for the BCT commander.

4-110. The fire support rehearsal should last no more than 90 minutes and should ensure the synchronization of the fire support effort with the maneuver plan. The maneuver plan includes ensuring observers are in the proper location at the proper time to observe planned targets, commonly known as the BCT commander's observation plan.

4-111. Time is inevitably short, so the rehearsal focuses on the critical portions of the plan, to include preplanned fires, and to ensure Soldiers correctly integrate and synchronize within the operational framework. Additionally, the fire support rehearsal should address action during degraded or intermittent communications to ensure interoperability to preserve the effectiveness of the force and maintain the initiative over the enemy. The critical document supporting the fire support rehearsal is the fire support execution matrix, which includes all fire support tasks. To conduct the fire support rehearsal, the BCT follows the same procedures outlined in the combined arms rehearsal sequence of events.

Sustainment Rehearsal

4-112. The BCT sustainment rehearsal ensures the synchronization of sustainment efforts before, during, and after combat operations. The sustainment rehearsal validates the who, what, when, where, and how of support. The sustainment rehearsal usually occurs after the combined arms and fire support rehearsals, which should not last more than 90 minutes.

4-113. The brigade support battalion (BSB) commander hosts the rehearsal for the BCT commander. The support operations officer facilitates the rehearsal to ensure rehearsal of critical sustainment events. BCT attendees include the BCT XO, personnel staff officer (S-1), surgeon, chaplain, S-2 representatives, S-3 representatives, logistics staff officer (S-4) representatives, and signal staff officer (S-6) representatives. Subordinate unit representatives include the BSB commander, the support operations officer, the brigade support medical company (known as BSMC) commander, and each maneuver battalion XO, S-1, S-4, and medical platoon leader, as well as the forward support company (FSC) commander, mobility warrant officer, and distribution company commander. The primary document used at the sustainment rehearsal is the sustainment synchronization matrix. (See chapter 9 for additional information.)

Combined Arms Rehearsal

4-114. The combined arms rehearsal ensures that subordinate plans synchronize with those of other units, and that subordinate commanders understand the intent of the higher headquarters. Usually, the BCT commander, XO, primary staff, and subordinate battalion commanders and their S-3s attend the rehearsal. Based upon the type of operation, the commander can modify the audience, such as the commander of the BCT reserve and attachments to the BCT. If invited, flank units and the higher headquarters may attend the combined arms rehearsal if time and distances permit. The combined arms rehearsal is a critical opportunity for enablers to synchronize activities in support of the BCT.

4-115. The execution matrix, DST, and operation order outline the rehearsal agenda. These tools, especially the execution matrix, drive and focus the rehearsal. The commander and staff use them to control the operation's execution. Any templates, matrixes, or tools developed within each of the warfighting functions should tie directly to the supported unit's execution matrix and DST. Examples include an intelligence synchronization matrix or fires execution matrix.

4-116. The combined arms rehearsal should last no more than two hours; however, the combined arms rehearsal is METT-TC dependent, so if the time allotted is insufficient to rehearse the entire operation, the staff must give priority to those critical events that demand a rehearsal. The staff rehearses the most important events first and continues to rehearse subsequent events as time permits. Rehearsals that integrate airspace use facilitate synchronization of operations and validate airspace user priorities and requirements. All combined arms rehearsal participants arrive at the rehearsal prepared to talk their portion of the operation.

EXECUTE

4-117. *Execution* is the act of putting a plan into action by applying combat power to accomplish the mission and adjusting operations based on changes in the situation (ADP 5-0). The BCT commander positions where best to exercise command and control by applying military expertise ensuring the ethical application of force during execution. This may be forward of the main or tactical command post (TAC [graphic]) to provide command presence, sense the mood of the unit, and to make personal observations. Forward presence of the commander also serves to inspire Soldiers by personal example of the commander's commitment to the mission. A position forward of the CPs and near the main effort or decisive operation facilitates an assessment of the situation and timely decision-making. Staffs synchronize actions, coordinate actions, inform the commander, and provide control to support the commander's ability to assess, use professional judgment, and make decisions.

4-118.   The rapid decision-making and synchronization process is a technique used during execution. While the MDMP seeks the optimal solution, the rapid decision-making and synchronization process seeks a timely and effective solution within the commander's intent, mission, and concept of operations. While identified here with a specific name and method, the commander and staff develop this capability through training and practice. When using this technique, the following considerations apply:

- Rapid is often more important than process.
- Much of it may be mental rather than written.
- It should become a battle drill for the current operations cell, and when established, the plans cell.

4-119.   Using the rapid decision-making and synchronization process lets leaders avoid the time-consuming requirements of developing decision criteria and comparing COAs. As operational and mission variables change during execution, this often invalidates or weakens COAs and decision criteria before leaders can make a decision. Under the rapid decision-making and synchronization process, leaders combine their experience and intuition to quickly reach situational understanding. Based on this, they develop and refine workable COAs.

4-120.   The rapid decision-making and synchronization process facilitates continuously integrating and synchronizing the warfighting functions to address ever-changing situations. This process meets the following criteria for making effective decisions during execution:

- It is comprehensive, integrating all warfighting functions. It is not limited to any one-warfighting function.
- It ensures all actions support the decisive operation by relating them to the commander's intent and concept of operations.
- It allows rapid changes to the order or mission.
- It is continuous, allowing commanders to react immediately to opportunities and threats.

4-121.   The rapid decision-making and synchronization process is based on an existing order and the commander's priorities as expressed in the order. The most important of these control measures are the commander's intent, concept of operations, and CCIRs. The rapid decision-making and synchronization process includes five steps (see figure 4-6). The first two may be performed in any order, including concurrently. The last three are performed interactively until commanders identify an acceptable COA. (See FM 6-0 for additional information.)

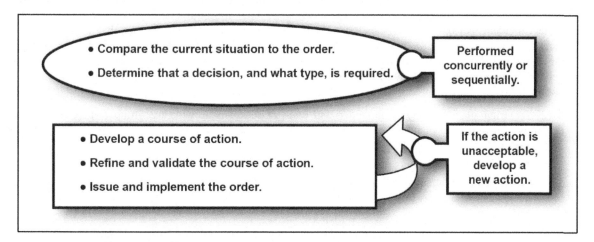

Figure 4-6. Rapid decision-making and synchronization process

Assess

4-122. *Assessment* is the determination of the progress toward accomplishing a task, creating a condition, or achieving an objective (JP 3-0). Assessment is continuous; it precedes and guides every operations process activity and concludes each operation or phase of an operation. The BCT commander and staff conduct assessments by monitoring the current situation to collect information; evaluating progress towards attaining end state conditions, achieving objectives, and performing tasks; and recommending or directing action to modify or improve the existing COA. The commander establishes priorities for assessment in planning guidance, CCIRs, EEFI, and decision points. By prioritizing the effort, the commander avoids excessive analyses when assessing operations.

4-123. Throughout the operations process, the BCT commander's personal assessment is integrated with those of the staff, subordinate commanders, and other unified action partners. Primary tools for assessing progress of the operation include the operation order, the COP, personal observations, running estimates, and the assessment plan. The latter includes measures of effectiveness (MOEs), measures of performance (MOPs), and reframing criteria. The commander's visualization forms the basis for the commander's personal assessment of progress. Running estimates provide information, conclusions, and recommendations from the perspective of each staff section. (See ADP 5-0 for the assessment process during the operations process.)

Assessment Plan Development

4-124. Critical to the assessment process (see paragraph 4-128) is developing an assessment plan. The BCT uses an assessment working group (when established) to develop assessment plans when appropriate. A *working group* is a grouping of predetermined staff representatives who meet to provide analysis, coordinate, and provide recommendations for a particular purpose or function (FM 6-0). A critical element of the commander's planning guidance is determining which assessment plan to develop. An assessment plan focused on attainment of end state conditions often works well. It is also possible, and may be desirable, to develop an entire formal assessment plan for an intermediate objective, a named operation subordinate to the base operation plan, or a named operation focused solely on a single line of operations or geographic area. The time, resources, and added complexity involved in generating an assessment plan strictly limit the number of such efforts.

4-125. The BCT commander and staff integrate and develop an assessment plan within the MDMP. As the commander and staff begin mission analysis, they also need to determine how to measure progress towards the operation's end state.

4-126. In order to measure progress towards an objective, criteria are established which provide measurable points from which analysis can be conducted. The staff then determines the method to collect on these criteria. The criteria or individual criterion may require specific means of measurement. These types of measurement fall under the categories of quantitative measurement and qualitative measurement. Quantitative measurement is observations that are based on measurements and numbers. For example, the number of violent incidents per day in a city block. Qualitative measurements are observations that are not based on measurements and numbers. They typically involve using the senses, knowledge, or insights. For example, the reason why there has been an increase or decrease in violent incidents in a city block. In regard to human behavior, analysis of quantitative measurements should be tempered with qualitative analysis whenever possible.

4-127. During planning, the commander and staff or assessment working group (when established) develops an assessment plan using six steps. FM 6-0 provides a detailed discussion on each step during assessment plan development. The six steps are—

- Step 1—Gather tools and assessment data.
- Step 2—Understand current and desired conditions.
- Step 3—Develop an assessment framework.
- Step 4—Develop the collection plan.
- Step 5—Assign responsibilities for conducting analysis and generating recommendations.
- Step 6—Identify feedback mechanisms.

Assessment Process

4-128. Once the commander and staff or assessment working group (when established) develops the assessment plan, it applies the assessment process of monitor, evaluate, and recommend or direct continuously throughout preparation and execution. (See ADP 5-0 for the assessment process during the operations process in detail.) Broadly, assessment consists of, but is not limited to, the following activities—

Monitoring

4-129. *Monitoring* is continuous observation of those conditions relevant to the current operation (ADP 5-0). Monitoring within the assessment process allows staffs to collect relevant information, specifically that information about the current situation that can be compared to the forecasted situation described in the commander's intent and concept of operations. Progress cannot be judged, nor effective decisions made, without an accurate understanding of the current situation.

4-130. Staff elements record relevant information in running estimates. Staff elements maintain a continuous assessment of current operations to determine if they are proceeding according to the commander's intent, mission, and concept of operations. In their running estimates, staff elements use this new information and these updated facts and assumptions as the basis for evaluation.

Evaluating

4-131. The staff analyzes relevant information collected through monitoring to evaluate the operation's progress toward attaining end state conditions, achieving objectives, and performing tasks. *Evaluating* is using indicators to judge progress toward desired conditions and determining why the current degree of progress exists (ADP 5-0). Evaluation is at the heart of the assessment process where most of the analysis occurs. Evaluation helps the commander determine what is working, what is not working, and insights into how to better accomplish the mission.

4-132. Criteria in the form of MOE and MOP aid in evaluating progress. MOEs help determine if a task is achieving its intended results. MOPs help determine if a task is completed properly. MOEs and MOPs are simply criteria—they do not represent the assessment itself. MOEs and MOPs require relevant information as indicators for evaluation.

4-133. A *measure of effectiveness* is an indicator used to measure a current system state, with change indicated by comparing multiple observations over time (JP 5-0). MOEs help measure changes in conditions, both positive and negative. MOEs are commonly found and tracked in formal assessment plans. MOEs help to answer the question, "Are we doing the right things?"

4-134. A *measure of performance* is an indicator used to assess a friendly action that is tied to measuring task accomplishment (JP 5-0). MOPs help answer questions such as "Was the action taken?" or "Were the tasks completed to standard?" A MOP confirms or denies that a task has been properly performed. MOPs are commonly found and tracked at all echelons in execution matrixes. MOPs are also commonly used to evaluate training. MOPs help to answer the question "Are we doing things, right?" There is no direct hierarchical relationship among MOPs to MOEs. MOPs do not feed MOEs or combine in any way to produce MOEs—MOPs simply measure the performance of a task.

4-135. An *indicator*, in the context of assessment, is a specific piece of information that infers the condition, state, or existence of something, and provides a reliable means to ascertain performance or effectiveness (JP 5-0). Indicators take the form of reports from subordinates, surveys and polls, and information requirements. Indicators help to answer the question "What is the current status of this MOE or MOP?" A single indicator can inform multiple MOPs and MOEs.

Recommending or Directing Action

4-136. Monitoring and evaluating are critical activities; however, assessment is incomplete without recommending or directing action. Assessment may diagnose problems, but unless it results in recommended adjustments, its use to the commander is limited.

4-137. When developing recommendations, staffs draw from many sources and consider their recommendations within the larger context of the operation. While several ways to improve a particular aspect of the operation might exist, some recommendations could impact other aspects of the operation. As with all recommendations, staffs should address any future implications.

PARALLEL, COLLABORATIVE, AND DISTRIBUTED PLANNING

4-138.   Whether planning deliberately or rapidly, all planning requires the skillful use of available time to optimize planning and preparation throughout the BCT. Taking more time to plan often results in greater synchronization; however, any delay in execution risks yielding the initiative-with more time to prepare and act-to the enemy. When allocating planning time to subordinate unit commanders, the BCT commander must ensure subordinates have enough time to plan and prepare their own actions before execution. Both parallel, collaborative, and distributed planning help optimize available planning time. Parallel planning allows each echelon to make maximum use of time available. Collaborative planning is the real-time interaction of commanders and staffs. Distributed planning allows the commander and staff members to execute planning from different locations.

PARALLEL PLANNING

4-139.   *Parallel planning* is two or more echelons planning for the same operations nearly simultaneously facilitated by the use of warning orders by the higher headquarters (ADP 5-0). Parallel planning requires significant interaction between echelons. Parallel planning can happen only when higher headquarters produces timely warning orders and shares information with subordinate headquarters as it becomes available (see figure 4-7).

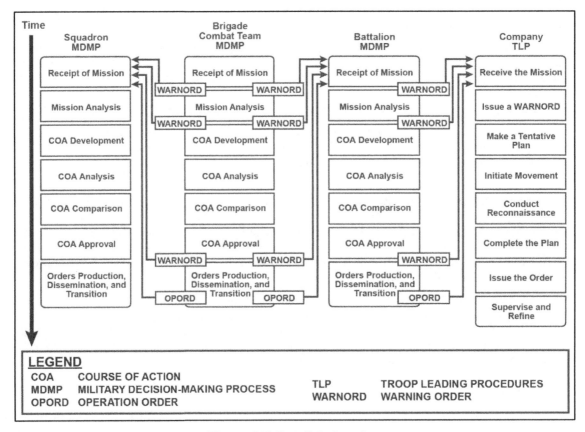

Figure 4-7. Parallel planning

COLLABORATIVE PLANNING

4-140. *Collaborative planning* is two or more echelons planning together in real time, sharing information, perceptions, and ideas to develop their respective plans simultaneously (ADP 5-0). Collaborative planning is the real-time interaction among commanders and staffs at two or more echelons developing plans for a single operation. It must be used judiciously.

4-141. Collaborative planning is most appropriate when time is scarce, and a limited number of options are being considered. It is particularly useful when the commander and staff can benefit from the input of subordinate commanders and staffs.

4-142. Collaborative planning is not appropriate when the staff is working a large number of COAs or branches and sequels, many of which will be discarded. In this case, involving subordinates wastes precious time working options that are later discarded. Collaborative planning also is often not appropriate during ongoing operations in which extended planning sessions take commanders and staffs away from conducting current operations.

4-143. As a rule of thumb, if the commander is directly involved in time-sensitive planning, some level of collaborative planning probably is needed. The commander, not the staff, must make the decision to conduct collaborative planning. Only the commander can commit subordinate commanders to using their time for collaborative planning.

DISTRIBUTED PLANNING

4-144. Digital communications and information systems enable members of the same staff to execute the MDMP without being collocated. Distributed planning saves time and increases the accuracy of available information in that it allows for the rapid transmission of voice and data information, which can be used by staffs over a wide geographical area. (See ATP 3-12.3 for information on appropriate electromagnetic protection [EP] active and passive measures.)

INTEGRATING PROCESSES

4-145. Throughout the operations process, the BCT commander and staff integrate warfighting functions to synchronize the force according to the commander's intent and concept of operations. The commander and staff use integrating processes (see ADP 5-0) to synchronize specific functions throughout the operations process in addition to the major activities. The integrating processes are intelligence preparation of the battlefield (IPB) (see ATP 2-01.3), information collection (see FM 3-55), targeting (see ATP 3-60), RM (see ATP 5-19), and knowledge management (see ATP 6-01.1).

INTELLIGENCE PREPARATION OF THE BATTLEFIELD

4-146. *Intelligence preparation of the battlefield* is the systematic process of analyzing the mission variables of enemy, terrain, weather, and civil considerations in an area of interest to determine their effect on operations (ATP 2-01.3). Led by the BCT S-2, the entire staff participates in the IPB to develop and sustain an understanding of the threat, terrain and weather, and civil considerations. IPB helps identify options available to friendly and threat forces.

4-147. During planning, the commander focuses activities on understanding, visualizing, and describing, while directing and assessing. The IPB is one of the processes the commander uses to aid in planning (see ATP 2-01.3). The IPB consists of four steps. Each step is performed or assessed and refined to ensure that IPB products remain complete and relevant. Figure 4-8 on page 4-32 shows the relationship between IPB and the steps of the MDMP along with key inputs and outputs during the process. The four IPB steps are—

- Define the operational environment.
- Describe environmental effects on operations.
- Evaluate the threat.
- Determine threat COAs.

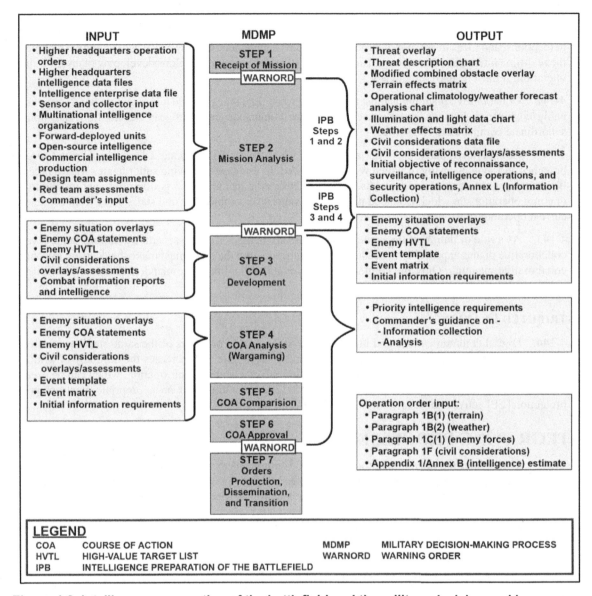

Figure 4-8. Intelligence preparation of the battlefield and the military decision-making process

4-148. IPB supports all activities of the operations process and identifies gaps in current intelligence. IPB results in intelligence products that are used during the MDMP to assist in developing friendly COAs and decision points for the commander. IPB products help the commander and staff, and subordinate commanders and leaders understand the threat, physical environment, and civil considerations throughout the operations process. Additionally, the conclusions reached and the products (which are included in the intelligence estimate) developed during IPB are critical to planning information collection and targeting operations. (See ATP 2-01.3 for additional information.) IPB products include—

● Threat situation templates with associated COA statements and high-value target (HVT) lists.

● Event templates and associated event matrices.

● Modified combined obstacle overlays, terrain effects matrices, and terrain assessments.

● Weather effects work aids—weather forecast charts, weather effects matrices, light and illumination tables, and weather estimates.

● Civil considerations overlays and assessments.

INFORMATION COLLECTION

4-149. *Information collection* is an activity that synchronizes and integrates the planning and employment of sensors and assets as well as the processing, exploitation, and dissemination systems in direct support of current and future operations (FM 3-55). Information collection integrates intelligence and operations staff functions, specifically information collection capabilities, focused on answering information requirements. FM 3-55 describes an information collection capability as any human or automated sensor, asset, or processing, exploitation, and dissemination capabilities that can be directed to collect information that enables better decision-making, expands understanding of the operational environment, and supports warfighting functions in decisive action. Key aspects of information collection, addressed below, influence how the BCT operates as a ground force in close and continuous contact with the environment, including the enemy, terrain and weather, and civil considerations.

Commander and Staff Input

4-150. The BCT commander is the most important participant in collection management. The initial commander's intent, planning guidance, and CCIRs form the foundation of the information collection plan and the basis for assessing its execution. The brigade operations staff officer (S-3) is responsible for the information collection plan and is the tasking authority for collection assets within the BCT. It is important that the entire staff (led by the S-3) collaborate closely to ensure information collection activities are fully synchronized and integrated into the overall operation order.

4-151. Effective planning requirements and assessing collection focuses information collection activities on obtaining the information required by commanders and staffs to influence decisions and operations. Planning requirements and assessing collection—

- Includes commander and staff efforts to synchronize and integrate information collection tasks throughout the operations process.
- Supports the commander's situational understanding and visualization of the operation by—
  - Identifying information gaps.
  - Coordinating assets and resources against requirements for information to fill these gaps.
  - Assessing the collected information and intelligence to inform the commander's decisions.
- Supports the staff during all operations process activities, and integrating processes, for example, during IPB and the MDMP, as well as the RM, targeting, and operations and intelligence processes (see chapter 3).

4-152. During planning and preparation, the operations and intelligence staffs, or the operations and intelligence working group (if formed), work to develop the information collection plan and the staff products required to execute it. During execution, they oversee execution of the plan, keeping the staff products current and using them to keep information collection efforts synchronized with the overall operation. The staff updates planning requirements as operations unfold and modify the plan as necessary to satisfy new information requirements that emerge. (See ATP 2-01 for additional information.)

> *Note.* Depending on the availability of personnel, the BCT commander may designate an operations and intelligence-working group. The S-3 and intelligence staff officer (S-2) direct and manage the efforts of this working group to achieve a fully synchronized and integrated information collection plan.

Develop Understanding

4-153. The integration and synchronization of knowledge and information facilitates the BCT commander's situational understanding for any problem set and the staff's shared understanding. Knowledge is the precursor to effective action (especially within large-scale combat operations) across physical domains and the dimensions of the information environment. Acquiring information about an operational environment requires aggressive and continuous information collection operations. The BCT commander uses information collection to plan, organize, and execute shaping operations that answer the CCIRs no matter what element of decisive action currently dominates. (See figure 4-9.)

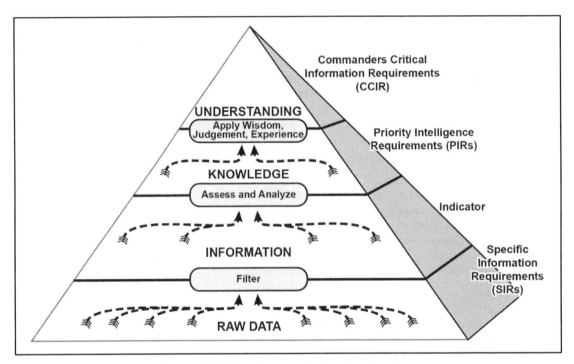

Figure 4-9. Development of understanding

Develop and Manage Requirements

4-154.　Developing requirements is the continuous process of identifying, prioritizing, and refining gaps in data, relevant information, and knowledge concerning relevant aspects of the operational environment. The BCT staff must resolve these gaps in order for the commander to achieve situational understanding. Constant collaboration among all staff sections helps in redefining information requirements as the situation develops. Requirements are generally captured as information requirements; the two types of information requirements are CCIRs and EEFI (see figure 5-5 on page 5-17). Identifying information requirements assists the BCT commander and staff in filtering available information by defining what is important to mission accomplishment.

4-155.　For requirements management, there are two types of requirements that result from planning requirements and assessing collection: priority intelligence requirements that are part of the CCIRs, and information requirements. (See figure 4-10.) Priority intelligence requirements and information requirements may focus on threat units or on capabilities the threat requires to complete missions and tasks. Each requirement is further refined into discrete pieces of information that together answer that requirement. These pieces are referred to as indicators and specific information requirements. Use the indicators and to develop the information collection plan. (See ATP 2-01 and ATP 2-19.4 for additional information.)

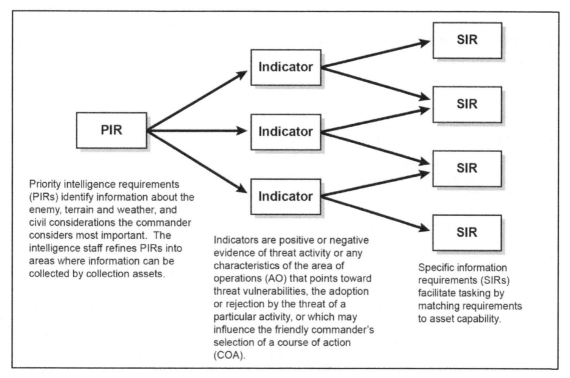

Figure 4-10. Relationship between priority intelligence requirements, indicators, and specific information requirements

Information Collection Planning

4-156. The information collection plan sets information collection in motion. The primary information collection planning objective is tasking subordinate units (all possible assets for example—military intelligence, maneuver, military police, fires, signal, engineer, sustainment, and aviation) to cover NAIs important to the BCT mission or directed by higher headquarters. (See ATP 2-19.4, appendix C and ATP 2-01 for specific information on the information collection plan.) The information collection plan is based on—

- The commander's initial information collection guidance.
- Key information gaps identified by the staff during mission analysis.
- The enemy situation template, event template, and event matrix developed during IPB.

Information Collection Tasks

4-157. During planning, information collection tasks are specified or implied. Subordinate units plan the use of available information collection assets to satisfy BCT taskings as well as their own requirements. Units strive to complete their plans quickly, so these assets have time to prepare and execute the plan. Collection involves the acquisition of information and the provision of this information to processing elements and consists of the following tasks:

- Plan requirements and assess collection.
- Task and direct collection.
- Execute collection.

4-158. Information collection requires a continuous, collaborative, and parallel planning process involving the BCT, its higher headquarters, and subordinate battalion staffs. Subordinate battalion and separate company plans are consolidated and included in the BCT information collection plan. The commander at each echelon must be intimately involved in the information collection planning process and must quickly and clearly articulate priority intelligence requirements to the staff.

4-159. Plan requirements and assess collection is a commander-driven, coordinated staff effort led by BCT S-2, in coordination with the BCT S-3. The continuous functions of planning requirements and assessing collection identify the best way to satisfy a requirement. These functions are not necessarily sequential.

4-160. The BCT S-3 (based on recommendations from the staff, specifically the S-2) tasks, directs, and, when necessary, retasks the information collection assets. Tasking and directing of limited information collection assets are vital to their control and effective use. The staff accomplishes tasking information collection by issuing warning orders, fragmentary orders, and operation orders. The staff accomplishes directing information collection assets by continuously monitoring the operation. The staff conducts retasking to refine, update, or create new requirements.

4-161. Executing collection focuses on requirements tied to the execution of tactical missions (normally reconnaissance, surveillance, security operations, and intelligence operations). Information acquired during collection activities about the threat and the area of interest is provided to intelligence processing and exploitation elements. Typically, collection activities begin soon after receipt of mission and continue throughout preparation for and execution of the operation. They do not cease at the conclusion of the mission but continue as required. This allows the commander to focus combat power, execute current operations, and prepare for future operations simultaneously.

Planning Requirements and Assessing Collection Functions

4-162. Collection management is the task of analyzing requirements, evaluating available assets (internal and external), recommending taskings to the operations staff for information collection assets, submitting requests for information for adjacent and higher collection support, and assessing the effectiveness of the information collection plan. The continuous functions of collection management identify the best way to satisfy the requirements of the supported commander and staff. These functions are not necessarily sequential. Collection management inherently requires an understanding of the relative priority of incoming requests for collection and processing, exploitation, and dissemination. Additionally, collection management includes the staff vetting requirements against current intelligence holdings to ensure resources are not wasted collecting information that is already available. (See figure 4-11.)

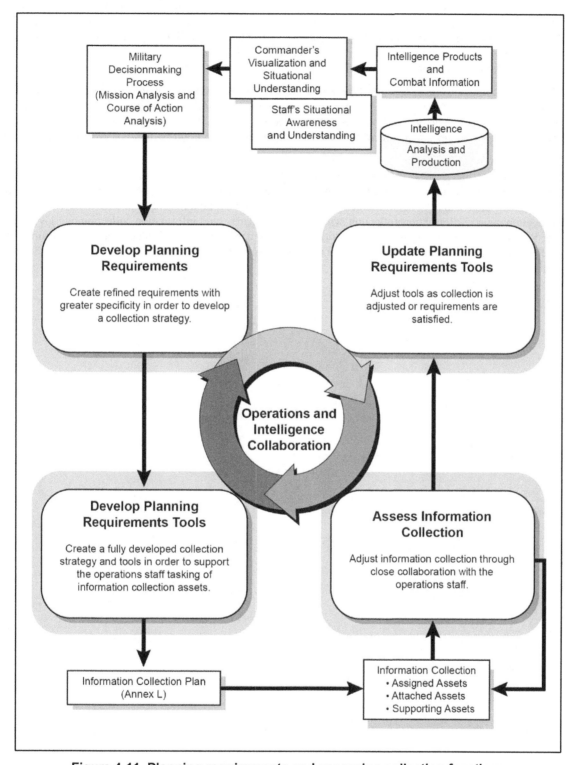

Figure 4-11. Planning requirements and assessing collection functions

4-163. After receiving inputs from the BCT commander and staff—intent, planning guidance, and requirements—the intelligence staff, in close coordination with the operations staff, performs the planning requirements and assessing collection functions. (See ATP 2-01, chapter 2 for a detailed discussion of the planning requirements and assessing collection functions.) The planning requirements and assessing collection functions are the basis for creating an information collection plan that synchronizes activities of the information collection effort to enable the commander's visualization and situational understanding. The intelligence staff, in coordination with the operations staff, monitors available collection assets and assesses their ability to provide the required information. They also recommend adjustments to new requirements or locations of information collection assets, if required.

4-164. The initial information collection plan is crucial to begin or adjust the information collection effort. The initial information collection plan sets reconnaissance, surveillance, security operations, and intelligence operations in motion, and may be issued as part of a warning order or fragmentary order. As more information becomes available, the initial information collection plan is updated and issued as a part of the operation order.

4-165. During this process, it is important for the S-3 and S-2 to collaborate closely in order to ensure information collection activities are fully synchronized and integrated into the BCT's, to include higher and subordinate unit, plan(s). The intelligence staff creates initial planning requirements tools (information collection matrix, information collection synchronization matrix, and information collection overlay) before the S-3 can develop the completed information collection plan. Figure 4-12 is an example of an information collection overlay used in the development of the information collection plan. (See ATP 2-19.4 and ATP 2-01 for additional information.)

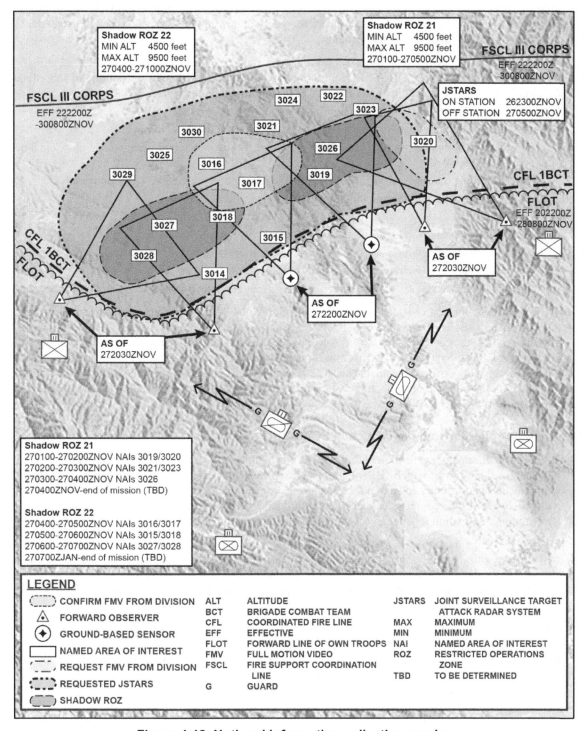

Figure 4-12. Notional information collection overlay

Collector and Processor Availability, Capability, and Limitation

4-166. The BCT intelligence cell must know collectors and processors available within the BCT and echelons above and below. The cell must know how to access those assets and resources to support the collection plan. Theater and joint echelons apportion information collection resources to subordinate echelons. Corps and divisions allocate support and intelligence capabilities to the BCT. The intelligence cell must understand the system of apportionment and allocation to determine what is available and what can be requested by analyzing the higher headquarters order, reviewing the various scheduling or tracking mechanisms, and collaborating across echelons.

4-167. Soldiers within the intelligence cell must know and address the practical capabilities and limitations of BCT (and available external to the BCT) information collection assets and the capability of subordinates to provide information. Capability and limitation issues (although not inclusive) include—

- Range (duration and distance) to provide require target coverage.
- Day and night effectiveness through available optic and thermal crossover.
- Technical characteristics to see/operate through fog or other obscurant/hostile electromagnetic warfare (EW).
- Reporting timeliness regarding established reporting criteria for each collection asset.
- Geolocation accuracy (reliability and precision).
- Durability to move across restricted terrain, launch in high winds or limited visibility.
- Ability to obtain and report required enemy activity.
- Sustainability requirements for extended duration operations (fuel capacity/maintenance issues).
- Vulnerability to enemy in route and in the target area.
- Performance history to meet the commander's requirements (responsiveness/reliability/accuracy).

4-168. The intelligence cell must know the collection capabilities requiring confirmation, especially if targeting is an issue. For example, target selection standards may require reliance on sensors capable of providing targeting accuracy, such as the Advanced Synthetic Aperture Radar System, Joint Surveillance Target Attack Radar System (JSTARS), or an unmanned aircraft system (UAS). If experience shows that the collection capability is often unavailable because of local weather patterns, the intelligence cell considers this in evaluating the information collection asset's performance history, perhaps leading to the selection of an alternate information collection asset.

Information Collection and the Targeting Process

4-169. The information collection plan guides reconnaissance and security forces to answer the CCIRs, to include those high-payoff targets (HPTs) designated as priority intelligence requirements. Effective planning requirements and assessing collection focuses information collection activities on obtaining the information required by the BCT commander and staff to influence targeting decisions and the scheme of fires. Determining information requirements is necessary for the early identification of information gaps.

4-170. The targeting process is comprised of four basic steps: decide, detect, deliver, and assess. The decide step sets priorities for information collection and scheme of fires during detect and deliver steps (see chapter 4). The decide step draws heavily on the commander's intent and concept of operations and a detailed IPB with continuous assessment. Targeting, nested within the operations process, is an effective method for matching friendly force capabilities against enemy targets.

4-171. Information collection priorities must be set for each phase or critical event of an operation. Priorities depicted during targeting value analysis using visual products and matrixes communicate the importance of specific targets to the enemy's COA and those targets that, if destroyed, would contribute favorably to the friendly COA. (See FM 3-09 and ATP 2-19.4 for additional information.)

TARGETING

4-172. *Targeting* is the process of selecting and prioritizing targets and matching the appropriate response to them, considering operational requirements and capabilities (JP 3-0). Targeting personnel within the BCT identify critical target subsets that, when successfully acquired and attacked, significantly diminish enemy capabilities. The commander synchronizes combat power to attack and eliminate critical target(s) using the most effective system in the right time and place.

4-173. Targeting is a complex and multidiscipline effort that requires coordinated interaction among many command and staff elements within and external to the BCT. The functional and integrating cell members (see paragraph 4-235) within the BCT necessary for effective collaboration are represented in the targeting working group. Close coordination among all cells is crucial for a successful targeting effort. Sensors and collection capabilities under the control of external agencies must be closely coordinated and carefully integrated into the execution of attacks especially those involving rapidly moving, fleeting, or dangerous targets. In addition, the appropriate means and munitions must attack the vulnerabilities of different types of targets.

Note. The targeting of fires in decisive action requires the judicious use of lethal force balanced with restraint, tempered by professional judgment. The BCT commander works with BCT fire support coordinator and planners, taking into consideration the civilian populace, noncombatants, friendly forces, and collateral damage when planning fire support. As with the BCT commander, they have the legal and moral obligation to challenge a proposed fire mission if they believe it will violate the Law of War or the moral principles of the Army Ethic. Together, the BCT commander and the fire support coordinator and planners, must plan ahead and have the foresight to mitigate and reduce the risk of unintended effects such as excessive collateral damage and negative psychological impacts on the civilian populace, which create or reinforce instability in the area of operations. Improper planning could lead to severe consequences that adversely affect efforts to gain or maintain legitimacy and impede the attainment of both short term and long-term goals for the BCT commander. To mitigate this risk, they plan and prepare fire support coordination measures to minimize noncombatant casualties and excessive collateral damage.

Commander's Targeting Guidance

4-174. The commander's targeting guidance must be articulated clearly and simply to enhance understanding. Targeting guidance must focus on essential threat capabilities and functions that could interfere with the achievement of the BCT's objectives. The commander's targeting guidance describes the desired effects to be generated by fires, physical attack, cyberspace electromagnetic activities (CEMA), and other information-related capabilities against threat operations. Targeting enables the commander through various lethal and nonlethal capabilities and restrictions and constraints (fire control measures) the ability to produce the desired effects. Capabilities associated with one desired effect may also contribute to other desired effects. For example, delay can result from disrupting, diverting, or destroying enemy capabilities or targets. (See ATP 3-60 for a complete listing of desired effects.)

4-175. The commander can direct a variety of nonlethal actions or effects separately or in conjunction with lethal actions or effects. These actions or effects are framed by the disciplined, ethical application of force. Commanders and subordinate leaders, in formulating plans and orders, consider choices of nonlethal versus lethal means for executing operations in accomplishment of the mission by exercising restraint. Commanders and subordinate leaders using lethal force, exercise restraint tempered by professional judgment when conducting operations.

4-176. The commander provides restrictions as part of their targeting guidance. Targeting restrictions fall into two categories—the no-strike list and the restricted target list.

4-177. The no-strike list consists of objects or entities protected by—
- Law of war.
- International laws.
- Rules of engagement.
- Other considerations.

4-178. A restricted target list is a valid target with specific restrictions such as—
- Limit collateral damage.
- Preserve select ammo for final protective fires.
- Do not strike during daytime.
- Strike only with a certain weapon.
- Proximity to protected facilities and locations.

4-179. The targeting process supports the commander's decision-making with a comprehensive, iterative, and logical methodology for employing the ways and means to create desired effects that support achievement of objectives. Once actions are taken against targets, the commander and staff assess the effectiveness of the actions. If there is no evidence that the desired effects were created, reengagement of the target may be necessary, or another method selected to create the desired effects.

Targeting Categories

4-180. The targeting process can be generally grouped into two categories: deliberate and dynamic. Deliberate targeting prosecutes planned targets. These targets are known to exist in the area of operations and have actions scheduled against them. Examples range from targets on target lists in the applicable plan or order, targets detected in sufficient time to place in the joint air tasking cycle, mission-type orders, or fire support plans. *Dynamic targeting* is targeting that prosecutes targets identified too late, or not selected for action in time to be included in deliberate targeting (JP 3-60). Dynamic targeting (see ATP 3-60.1) prosecutes targets of opportunity and changes to planned targets or objectives. Targets of opportunity are targets identified too late, or not selected for action in time, to be included in deliberate targeting. Targets engaged as part of dynamic targeting are previously unanticipated, unplanned, or newly detected.

4-181. The two types of planned targets are scheduled and on-call:
- Scheduled targets exist in the area of operation and are located in sufficient time so that fires or other actions upon them are identified for engagement at a specific, planned time.
- On-call targets have actions planned, but not for a specific delivery time. The commander expects to locate these targets in sufficient time to execute planned actions.

4-182. The two types of targets of opportunity are unplanned and unanticipated:
- Unplanned targets are known to exist in the area of operations, but no action has been planned against them. The target may not have been detected or located in sufficient time to meet planning deadlines. Alternatively, the target may have been located, but not previously considered of sufficient importance to engage.
- Unanticipated targets are unknown or not expected to exist in the area of operation.

Targeting Methodology

4-183. Targeting methodology is an integral part of the MDMP. Targeting begins with the receipt of the mission and continues through operations process's execution and assessment phases. Like the MDMP, targeting is a commander-driven process. As the MDMP is conducted, targeting becomes more focused based on the commander's guidance and intent. Figure 4-13 illustrates the relationship between the targeting methodology (decide, detect, deliver, and assess) and the MDMP along with products generated during targeting. (See ATP 3-60 for additional information.)

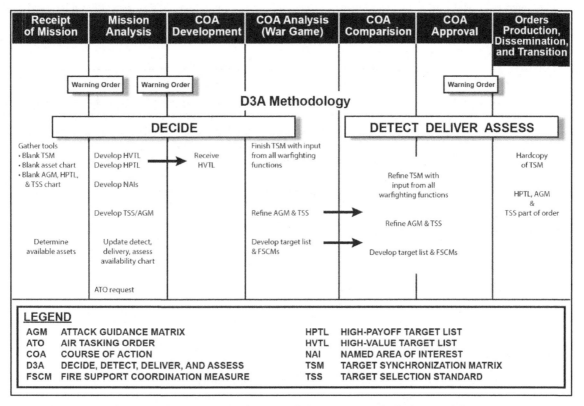

Figure 4-13. Targeting methodology and the military decision-making process

Targeting Working Group

4-184. The BCT targeting working group is used as a vehicle to focus the targeting process within a specified time. The commander is responsible for the targeting effort, with the intelligence, operations, and fire support staff officers forming the core of the targeting working group within the BCT staff. The targeting working group can vary in make-up and size as determined by the commander and SOPs of the BCT. The decide, detect, deliver, and assess process assists the targeting working group determine requirements for combat assessment to assess targeting and attack effectiveness. The targeting working group has three primary functions in assisting the commander:

- Helps in synchronizing operations.
- Recommends targets to acquire and engage. The team also recommends the most efficient and available assets to detect and engage these targets.
- Identifies the level of combat assessment required. Combat assessment can provide crucial and timely information to allow analysis of the success of the plan or to initiate revision of the plan.

4-185. The targeting effort is continuous at all levels of command. Continuity is achieved through parallel planning by targeting working groups from corps through battalion task force. Targeting is not just a wartime function. This process must be exercised before battle if it is to operate effectively. The members of the targeting working group must be familiar with their roles and the roles of the other team members. That familiarity can only be gained through staff training.

RISK MANAGEMENT

4-186. *Risk management* is the process to identify, assess, and control risks and make decisions that balance risk cost with mission benefits (JP 3-0). RM helps organizations and individuals make informed decisions to reduce or offset risk. Using this process increases the force's operational effectiveness and the probability of mission accomplishment. This systematic approach identifies hazards, assesses them, and manages associated risks. RM outlines a disciplined approach to express a risk level in terms readily understood at all echelons. For example, the commander may adjust the level of body armor protection during dismounted movement balancing an increased risk level to individual Soldiers to improve the likelihood of mission accomplishment.

Note. Soldier load is an area of concern for commanders and subordinate leaders. How much is carried, how far, and in what configuration are critical mission considerations. The commander balances the risk to Soldiers from the enemy against the risk to mission accomplishment due to excessive loads and Soldier exhaustion and injury. Soldier load is limited to mission-essential equipment to sustain continuous operations. The commander accepts risks to reduce Soldier load based on a through mission analysis. (See ATP 3-21.20 and ATP 3-21.18 for additional information on Soldier load.)

Principles of Risk Management

4-187. The principles of RM (see ATP 5-19) are—
- Integrate RM into all phases of missions and operations.
- Make risk decisions at the appropriate level.
- Accept no unnecessary risk.
- Apply RM cyclically and continuously.

Five-Step Process

4-188. RM is a cyclical and continuous five-step process to identify and assess hazards; develop, choose, implement, and supervise controls; and evaluate outcomes as conditions change. Except in time-constrained situations, planners complete the process in a deliberate manner-systematically applying all the steps and recording the results. In time constrained conditions, the commander, staff, subordinate leaders, and Soldiers use judgment to apply RM principles and steps. The five steps of RM are—
- Step 1—Identify the hazards.
- Step 2—Assess the hazards.
- Step 3—Develop controls and make risk decisions.
- Step 4—Implement controls.
- Step 5—Supervise and evaluate.

Risk Management and the Military Decision-making Process

4-189. The BCT commander and staff use RM to identify, assess, and control hazards, reducing their effect on operations and readiness. The five steps of RM tend to require emphasis at different times during the MDMP (see table 4-2). While planning doctrine places the beginning of formal RM in mission analysis, the commander and staff can begin identifying hazards upon receipt of the warning order or operation order. For example, when conducting unilateral and partnered operations and training it is important for the commander to assess early in the process the potential risk for an insider attack (see chapter 8 for additional information on insider attacks).

Note. The representation in table 4-2 is not intended to be prescriptive. RM is an adaptable integrating process. The five steps are dynamic and cyclical.

Table 4-2. Risk management and the military decision-making process

Steps in the Military Decision-making Process	Risk Management Steps				
	Identify the Hazards	Assess the Hazards	Develop Controls and Make Risk Decisions	Implement Controls	Supervise and Evaluate
RECEIPT OF MISSION	X				
MISSION ANALYSIS	X	X			
COURSE OF ACTION DEVELOPMENT	X	X	X		
COURSE OF ACTION ANALYSIS	X	X	X		
COURSE OF ACTION COMPARISON			X		
COURSE OF ACTION APPROVAL			X		
ORDERS, PRODUCTION, DISSEMINATION, AND TRANSITION	X	X	X	X	X

Knowledge Management

4-190. Knowledge management facilitates the transfer of knowledge among commanders, staffs, and forces to build and maintain situational understanding. Knowledge management helps get the right information to the right person at the right time to facilitate decision-making. Knowledge management uses a five-step process to create a shared understanding. (See paragraphs 4-88 to 4-96 for a further discussion.) The steps of knowledge management include—

- Assess.
- Design.
- Develop.
- Pilot.
- Implement.

SECTION III – THE EXERCISE OF COMMAND AND CONTROL

4-191. The BCT commander organizes the headquarters into CPs and by staff sections, cells, elements, and teams to assist in the exercise of command and control. This section addresses BCT staff organization, CP organization and operation, cells, staff elements and teams, and staff processes and procedures.

STAFF ORGANIZATION

4-192. The BCT staff supports the commander, assists subordinate units, and informs units and organizations outside the headquarters. The staff supports the BCT commander's understanding, making and implementing decisions, controlling operations, and assessing progress. The staff makes recommendations and prepares plans and orders for the commander. The staff establishes and maintains a high degree of coordination and cooperation with staffs of higher, lower, supporting, supported, and adjacent units. The staff does this by actively collaborating and communicating with commanders and staffs of subordinate and other units to solve problems ethically, effectively, and efficiently to accomplish the mission—consistent with the moral principles of the Army Ethic (see ADP 1). The staff keeps civilian organizations informed with relevant information according to their security classification as well as their need to know. (See figure 4-14 on page 4-46.) The basic BCT staff structure includes an XO and various staff sections. A *staff section* is a grouping of staff members by area of expertise under a coordinating, special, or personal staff officer (FM 6-0).

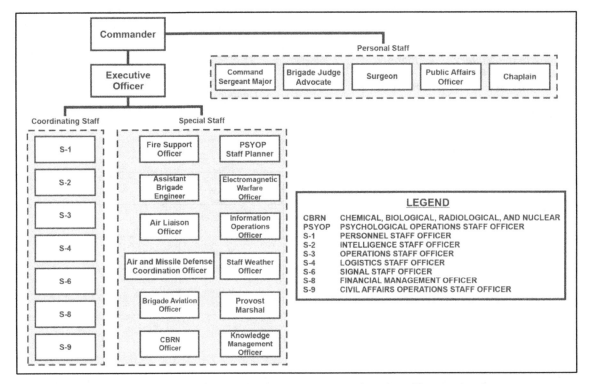

Figure 4-14. Brigade combat team command and staff organization

EXECUTIVE OFFICER

4-193. The XO is the commander's principal assistant and directs staff tasks, manages and oversees staff coordination, and special staff officers. The commander normally delegates executive management authority to the XO. The XO provides oversight of sustainment planning (see chapter 9) and operations for the BCT commander. As the key staff integrator, the XO frees the commander from routine details of staff operations and the management of the headquarters and ensures efficient and prompt staff actions. The XO may be second in command. The XO leans heavily on recommendations of the staff in each member's area of expertise, while at the same time balancing competing priorities for the overall situation. This delicate balance assists the commander as the XO anticipates the decision-making cycle and commander's intent, helping the commander to make informed decisions, while empowering the staff and guiding them. For example, the S-6 may recommend an optimal location for the BCT's main CP based on atmospherics, but the S-2 knows how easily the adversary will identify such a location. The XO might direct them to evaluate the advantages and disadvantages to such a location, along with two additional possible locations, determining which best meets the commander's intent, or presenting the accumulated analysis with a recommendation to the commander. (See FM 6-0 for additional information.)

COORDINATING STAFF OFFICERS

4-194. Coordinating staff officers are the commander's principal staff assistants. Coordinating staff functionalities are organized and described in the paragraphs 4-195 through 4-201.

Personnel Staff Officer

4-195. The S-1 is the principal staff officer for all matters concerning human resources support (military and civilian). (See chapter 9.) Specific responsibilities include manning, personnel services, personnel support, and headquarters management. The S-1 has coordinating staff responsibility for the civilian personnel officer and the equal opportunity advisor and prepares a portion of Annex F (Sustainment) to the operation order. When planning an operation, the S-1 provides accurate information regarding replacement personnel and ensures the rest of the staff includes all attachments or other assets into any plan or operation. For example, have staff sections' adjustments for personnel and manning of vehicles accounted for leave and other absences, or has the staff erroneously assumed 100-percent personnel? (See FM 1-0 and ATP 1-0.1 for additional information.)

Intelligence Staff Officer

4-196.   The S-2 is chief of the intelligence warfighting function. The intelligence staff officer is the principal staff officer responsible for providing intelligence to support current operations and plans. The S-2 gives the commander and the S-3 the initial information collection plan, which facilitates integration of reconnaissance, surveillance, intelligence operations, and security operations. The S-2 helps the S-3 to develop the initial information collection plan. The S-2 is responsible for the preparation of Annex B (Intelligence) and assists the S-3 in the preparation of Annex L (Information Collection). The S-2 sees into the mind of the enemy commander and forces. The S-2 helps the staff, XO, and commander see multiple COAs, to determine why the enemy might choose one or the other. If the S-2 and S-3 synchronize well, the information collection plan will solve gaps in intelligence, further shaping the planned friendly COA, such as the enemy's ability and intent to use a certain piece of terrain, and will they use that piece of terrain. Identifying particular tactics, the enemy will use based upon their disposition and friendly actions. (See FM 2-0 for additional information.)

Operations Staff Officer

4-197.   The S-3 is responsible for coordinating the activities of the movement and maneuver warfighting function. The S-3 is the primary staff officer for integrating and synchronizing the operation as a whole for the commander. The S-3 integrates information collection assets during plans and operations. The S-3 synchronizes information collection with the overall operation throughout the operations process (with the rest of the staff). The S-3 develops plans and orders and determines potential branches and sequels. The S-3 coordinates and synchronizes warfighting functions in all plans and orders. Additionally, the S-3 is responsible for and prepares Annex L (Information Collection) and Annex V (Interagency Coordination). The S-3 prepares Annex A (Task Organization), Annex C (Operations), and Annex M (Assessment) to the operation order. In conjunction with the knowledge management officer, the S-3 prepares Annex R (Reports) and Annex Z (Distribution). The S-3 ensures proper dissemination of the plan, as one unsynchronized staff section or functional cell can quickly unravel accomplishment of the mission and commander's intent. The S-3 examines closely and identifies the best spots on the battlefield for the commander's locations to lead the fight. (The S-3 generally collocates with the commander.) In addition, the S-3 identifies key intersections in canalizing terrain, and timed events for key transitions in phases or decision points. (See FM 6-0 for additional information.)

Logistics Staff Officer

4-198.   The S-4 is the principal staff officer for sustainment planning and operations, supply, maintenance, transportation, services, field services, distribution, and operational contract support (see chapter 9). The S-4 prepares Annex F (Sustainment), Annex P (Host-Nation Support) and Annex W (Operational Contract Support) to the operation order. The S-4 works closely with the BSB support operations officer to ensure successful planning and execution of the sustainment plan. The S-4 identifies requirements for external and higher echelon support requirements. The S-4's concept of support for logistics encompasses multiple considerations. For example, identifying locations for key facilities and services, determining how long it will take assets to reach each node (time-distance analysis), and by what routes, determining what capacity each node can provide, and anticipating times units will require more logistics and when delivery will be a higher risk. (See FM 6-0 for additional information.)

Signal Staff Officer

4-199.   The S-6 is the principal staff officer who advises the commander on all matters related to communications in the brigade. The signal staff officer provides network transport and information services, network sustainment, conducts Department of Defense information network (DODIN) operations, conducts information dissemination management and content staging to enable knowledge management, manages the

BCT's portion of the network and combat net radios assets in the area of operations, and performs spectrum management operations. The S-6 prepares Annex H (Signal) and participates in preparation of Appendix 12 (CEMA) to Annex C (Operations) to the operation order, with input from the S-2 and in coordination with the S-3. Organizing what systems, the unit uses, prioritizing them for various communications is essential as well, not only for redundancies sake, but also for the quick and efficient flow of information. The S-6 is critical to ensure planning includes all considerations for maintaining communications throughout an operation. For example, ensuring the planning accounts for terrain and enemy disruption where certain communications systems will be degraded or inoperative. Including appropriate security measures for signal teams. (See FM 6-02, ATP 6-02.70, and ATP 6-02.71 for additional information.)

Financial Management Staff Officer

4-200. The brigade financial management staff officer (S-8) is the principal staff officer singularly responsible for all financial management (see chapter 9) within the BCT. The S-8 is the focal point in planning financial management support that allows the BCT to accomplish its mission. The S-8 prepares a portion of Annex F (Sustainment). (See FM 1-06 for additional information.)

Civil Affairs Operations Staff Officer

4-201. The brigade civil affairs operations staff officer (S-9) is the dedicated principal staff officer position responsible for all matters concerning civil affairs. The S-9 evaluates civil considerations during mission analysis, recommends the establishment of the civil-military operations center in conjunction with the supporting civil affairs unit commander, and prepares the groundwork for transitioning the area of operations from military to civilian control. The S-9 advises the commander on the military's effect on civilians in the area of operations relative to the complex relationship of these people with the terrain and institutions over time. The S-9 is responsible for enhancing the relationship between Army forces and the civil authorities and people in the area of operations. A supportive civilian population can provide freedom of maneuver, resources, and information that facilitate friendly operations and preserve combat power and lethality, by mitigating the effects of the civil considerations on combat operations. The S-9 is required at all echelons from a BCT through theater Army and in special operations forces formations at battalion and group. The S-9 prepares Annex K (Civil Affairs Operations) to the operation order or operation plan.

PERSONAL STAFF OFFICERS

4-202. The personal staff officers work under the immediate control of, and have direct access to, the BCT commander. They advise the commander, provide input to orders and plans, and interface and coordinate with entities external to the BCT headquarters. Examples of personal staff officers to the BCT commander include the command sergeant major, the brigade judge advocate, the surgeon, the public affairs officer, and the chaplain. Personal staff responsibilities are described below.

Command Sergeant Major

4-203. The command sergeant major is the senior noncommissioned officer within the BCT who advises the commander on issues related to the enlisted ranks. The command sergeant major carries out policies and enforces standards for the performance, training, and conduct of enlisted Soldiers. In operations, a commander employs the command sergeant major throughout the area of operations to extend command influence, assess the morale of the force, and assist during critical events.

Brigade Judge Advocate

4-204. The brigade judge advocate is the senior legal advisor to the BCT commander. The brigade judge advocate advises the commander and staff on operational law, military justice, administrative law, fiscal law, and other areas of the law as required and ensures the delivery of legal services to the brigade across the core legal functions of the Judge Advocate General's Corps. The brigade judge advocate prepares a portion of Annex C (Operations) and Annex F (Sustainment) to the operation order. (See AR 27-1 and FM 1-04 for additional information.)

Surgeon

4-205.   The surgeon is responsible for coordinating Army Health System (AHS) support (see chapter 9) and operations within the command. The surgeon provides and oversees medical care to Soldiers, civilians, and detainees. (See ATP 4-02.3.) The surgeon prepares a portion of Annex E (Protection) and Annex F (Sustainment) of the operation order. When preparing a portion of these annexes, the surgeon additionally provides the BCT S-4 with recommendations on the BCT plan, and subordinate unit plans (with battalion medical officers) for air and ground medical evacuations, medical facility locations, and other considerations regarding heath service support. (See FM 4-02 and ATP 4-02.55 for additional information.)

Public Affairs Officer

4-206.   The public affairs officer develops strategies, leads, and supervises the conduct of public information, community engagements, and command information. The public affairs officer's principal role is to provide advice and counsel to the commander and the staff on how affected external and internal publics will accept and understand the BCT's operations. The BCT public affairs officer understands and coordinates the flow of information to Soldiers, the Army community, and the public and prepares Annex J (Public Affairs) to the operation order. (See FM 3-61 for additional information.)

Chaplain

4-207.   The chaplain is responsible for religious support operations; advises the commander and staff on religion, morale, moral, and ethical issues, within both the command and area of operations. Chaplains and religious affairs specialists are assigned at brigade and battalion echelons. (See chapter 9.) The chaplain prepares a portion of Annex F (Sustainment) to the operation order. (See FM 1-05 for additional information.)

SPECIAL STAFF OFFICERS

4-208.   Every staff organization has special staff officers. The number of special staff officers and their responsibilities vary with authorizations, the desires of the commander, and the size of the command. Special staff officers, common to the BCT, include the fire support officer, the ABE, the air liaison officer, the air and missile defense coordination officer, the brigade aviation officer, the CBRN officer, the PSYOP staff planner, the knowledge management officer, the electromagnetic warfare officer (EWO), the information operations officer, the staff weather officer, and the provost marshal. Paragraphs 3-187 through 3-199 describe the responsibilities for each. (See FM 6-0 for additional information.)

Fire Support Officer

4-209.   The fire support officer serves as the special staff officer for fires and integrates fires into the scheme of maneuver for the commander. The fire support officer leads the targeting process and fire support planning for the delivery of fires to include preparation fires, harassing fires, interdiction fires, suppressive fires, destruction fires, and deception fires. The fire support officer leads the fire support cell and prepares Annex D (Fires) of the operation order. The fires support officer also coordinates with the EWO and the air liaison officer. The BCT S-3 coordinates this position. Key to success is continual communication between the fire support officer and maneuver commander. The BCT commander provides a clear intent for fires, to include guidance for selection of high-payoff targets (HPTs), priority of fires, any special fires or munitions, and recommended fire control measures, such as no fire areas and final protective fires. The fire support officer identifies targeting capabilities, all fire support assets available, nominating HPTs and evaluating them for the commander. (See ADP 3-19 for additional information.)

Note. The BCT's organic field artillery battalion commander, as the fire support coordinator, is the BCT commander's primary advisor for the planning, coordination, and integration of field artillery and fire support to execute assigned tasks. (See FM 3-09 for additional information.)

Assistant Brigade Engineer

4-210.   The ABE is the senior engineer on staff responsible for coordinating engineer support to combined arms operations. The ABE integrates specified and implied engineer tasks into the maneuver force plan. The ABE ensures that mission planning, preparation, execution, and assessment activities integrate supporting engineer units. The ABE oversees any contract construction activity planning, preparation, and execution in support of the S-4 contracting support plan. The ABE prepares Annex G (Engineer) to the operation order. (See FM 3-34 for additional information.)

4-211.   The ABE, with the BCT S-2, oversees a geospatial engineering team (located in the S-2 section) that performs the analysis, management, and dissemination of geospatial data and products in support of BCT planning, preparation, execution, and assessment. This team maintains the BCT COP on the BCT server and provides updates to the brigade portion of the theater geospatial database. The team primarily supports the S-2 and S-3 sections, and as directed other staff sections and subordinate units of the BCT. The team works to fuse intelligence and geospatial information into a COP for the commander. This BCT level team is too small to provide continuous support, but it forms improvised geospatial intelligence cells as necessary to support operations. The geospatial engineering team requires access to the classified tactical local area network and Secret Internet Protocol Router Network to update and disseminate geospatial information and products. The ABE also supports with integrating geospatial products into the planning process by coordinating with the geospatial team. (See ATP 3-34.80 for additional information.)

> *Note*. The brigade engineer battalion (BEB) commander is the senior engineer in the BCT and advises the BCT commander on how best to employ combat, general, and geospatial engineering capabilities to conduct combined arms integration in support of decisive action. (See ATP 3-34.22 for additional information.)

Air Liaison Officer

4-212.   The air liaison officer is the senior United States Air Force officer with each tactical air control party (TACP). The air liaison officer plans close air support, in direct support of the BCT commander, in accordance with the joint force air component commander's guidance and intent. The air liaison officer is responsible for coordinating aerospace assets and operations such as close air support, air interdiction, air reconnaissance, airlift, and joint suppression of enemy air defenses. At battalion or squadron level, the senior member of the TACP is called a battalion air liaison officer—a specially trained and experienced noncommissioned officer or officer. (See JP 3-09.3, FM 3-52, and ATP 3-52.1 for additional information.)

Air and Missile Defense Coordination Officer

4-213.   The air and missile defense coordination officer leads the air defense airspace management (ADAM) cell, responsible for planning, coordinating, integrating, and controlling air defense and airspace management for the BCT. This includes providing the capability to integrate command and control systems to provide the brigade aviation element (BAE) with the COP, developing air defense plans, air defense artillery task organization, scheme of air defense operations, surveillance, and reconnaissance planning. In addition, the air and missile defense coordination officer integrates and coordinates tasks between the BCT and any augmented air and missile defense assets and units not directly task organized to BCT subordinate units. The coordination officer within the ADAM element prepares a portion of Annex D (Fires) to the operation order. (See ATP 3-01.50, FM 3-52, and ATP 3-52.1 for additional information.)

Brigade Aviation Officer

4-214.   The brigade aviation officer leads the BAE and assists the BCT S-3 with the planning and synchronization of Army aviation and other airspace users to support the BCT commander's ground scheme of maneuver. The brigade aviation officer standardizes unmanned aircraft systems (UASs) employment for the BCT, advises and plans the use of reconnaissance, surveillance, attack, air assault, air movement, sustainment, and medical evacuation. As the BCT's aviation subject matter expert, the brigade aviation officer is responsible for advising the BCT commander and staff on the status and availability of aviation assets, their capabilities and limitations. The brigade aviation officer recommends priorities for and allocations of Army aviation assets, coordinates the employment of those air assets, and assists with the synchronization of airspace coordinating measures with fire support coordination measures and movement and maneuver control measures within the BCT area of operations. The brigade aviation officer participates in the operations process, targeting and the development of the BCT unit airspace plan. The brigade aviation officer helps prepare portions of Annex C (Operations) to include Appendix 10 (Airspace Control) along with portions of Annex D (Fires) of the operation order. (See FM 3-52 and ATP 3-52.1 for additional information.)

Chemical, Biological, Radiological, and Nuclear Officer

4-215. The CBRN officer is the principle advisor to the commander on CBRN hazard awareness and understanding and is responsible for CBRN operations and CBRN asset use. The CBRN officer leads the CBRN working group. When established, the CBRN working group includes members from the protection working group, subordinate commands, host-nation agencies, and other unified action partners (see paragraph 7-82). The CBRN officer prepares a portion of Annex E (Protection) and a portion of Annex C (Operations) of the operation order. (See ATP 3-11.36 for additional information.)

Psychological Operations Staff Planner

4-216. The PSYOP staff planner, a noncommissioned officer authorized at the BCT level, is responsible for synchronizing and coordinating information activities with information-related capabilities. If no PSYOP noncommissioned officer is assigned, the commander of an attached PSYOP unit may assume those responsibilities. The PSYOP staff planner prepares Appendix 13 (PSYOP) and a portion of Appendix 14 (Military Deception) and Appendix 15 (Information Operations) to Annex C (Operations) to the operation order. (See FM 3-53 for additional information.)

Knowledge Management Officer

4-217. Working through the BCT XO, the knowledge management officer is responsible for developing the knowledge management plan that integrates and synchronizes knowledge and information management. (The BCT XO is responsible for the organization's knowledge management program.) The knowledge management officer synchronizes knowledge and information management to facilitate the BCT commander's situational understanding for any problem set and to provide the staff shared understanding. The knowledge management officer accomplishes this by using the tools, processes, and people available to facilitate an environment of shared understanding. When required, the knowledge management officer is responsible for Annex Q (Knowledge Management) to the operation order. (See FM 6-0 and ATP 6-01.1 for additional information.)

Electromagnetic Warfare Officer

4-218. The EWO serves as the BCT commander's designated staff officer for the planning, integration, synchronization, and assessment of electromagnetic warfare (EW) (see ATP 3-12.3), to include CEMA (see FM 3-12). The EWO coordinates through other staff members to integrate EW or/and CEMA into the commander's concept of operations. The EWO prepares Appendix 12 (CEMA) to Annex C (Operations) to the operation order and contributes to any section that has a CEMA subparagraph such as Annex N (Space Operations) in the operation order (see FM 3-14). As the cyberspace planner, the EWO is responsible for understanding policies relating to cyberspace operations, EW, and spectrum management operations to provide accurate information to the commander for proper planning, coordination, and synchronization of cyberspace operations, EW, and spectrum management operations into all operations (see paragraph 4-335).

Information Operations Officer

4-219. The information operations officer, authorized at the BCT level, is responsible for synchronizing and deconflicting information-related capabilities employed in support of BCT operations. An *information-related capability* is a tool, technique, or activity employed within a dimension of the information environment that can be used to create effects and operationally desired conditions (JP 3-13). Led by the information operations officer, the BCT staff synchronizes capabilities that communicate information to audiences and affect information content and flow of enemy or adversary decision-making while protecting friendly information flow. The information operations staff planner prepares appendix 15 and a portion of appendixes 12, 13, and 14 to Annex C (Operations) to the operation order. (See FM 3-13 for additional information.)

Staff Weather Officer

4-220. The staff weather officer is a U.S. Air Force officer or noncommissioned officer who coordinates operational weather support and weather service matters through the S-2 and other staff members. The staff weather officer collects environmental information and uses this information to produce and disseminate an environmental running estimate, mission execution forecast, and watches warnings and advisories. The staff weather officer integrates weather effects into planning and execution and responds to weather requests for information. The staff weather officer prepares Tab B (Weather) to Appendix 1 (Intelligence Estimate) to Annex B (Intelligence) to the operation order. (See FM 2-0 for additional information.)

Provost Marshal

4-221. The provost marshal is responsible for planning, coordinating, requesting, and employing all assigned or attached military police assets. Usually, the provost marshal is the senior military police officer in the command. The provost marshal augments the staff with a small planning cell that works within the S-3 typically. The provost marshal prepares a portion of Annex C (Operations) and a portion of Annex E (Protection) to the operation order. (See FM 3-39 for additional information.)

AUGMENTATION

4-222. Often, the BCT and its subordinate units receive support from (or attached) augmentation teams to assist in the exercise of command and control. Commanders within the BCT integrate this support or attached teams or detachments into their CPs and operations. For example, a division may receive an Army space support team when deployed. An Army space support team within a division can provide the BCT with space related planning that may directly affect BCT operations. Critical space related information provided to BCT operations includes navigation accuracy forecasts for planning and conducting mission and maneuver operations in support of fires and targeting effects. Space operations identify deliberate enemy interference activities such as attempts to jam friendly communications systems and navigation warfare that directly impacts targeting and maneuver forces. The BCT commander may request staff augmentation. Augmentation teams include but are not limited to—

- Army space support team. (See FM 3-14.)
- Army cyberspace operations support team. (See FM 3-12.)
- Civil affairs company. (See FM 3-57.)
- Combat camera team. (See FM 3-61.)
- Legal support teams. (See FM 1-04.)
- Mobile public affairs detachment. (See FM 3-61.)
- Military history detachment. (See ATP 1-20.)
- PSYOP units. (See FM 3-53.)
- Army information operations field support team. (See FM 3-13.)
- Individual augmentation by specialty (assessment or economic development).

COMMAND POST ORGANIZATION AND OPERATIONS

4-223. A *command post* is a unit headquarters where the commander and staff perform their activities (FM 6-0). The BCT commander balances the need to create a capable CP organization(s) to support the capacity to plan, prepare, execute, and continuously assess operations with the resulting diversion of capabilities to fight the enemy due to the size of the CP itself. Larger CPs ease face-to-face coordination: however, they are vulnerable to multiple acquisitions and means of attack. Smaller CPs can be hidden and protected more easily, but they may not exercise the degree of command and control necessary to control all BCT subordinate units. Striking the right balance provides a responsive yet agile organization. This section provides guidelines for CP organization and operations to include the importance of establishing running estimates, SOPs and a battle rhythm. In addition, this section considers various factors that degrade the efficiency of command and control systems within organizations and considerations for digital and analog command and control systems techniques. (See FM 6-0 and ATP 6-0.5 for additional information.)

COMMON COMMAND POST CONSIDERATIONS

4-224. The BCT commander organizes CPs by staff sections or staff cells. Organizing the staff among CPs, and into cells within CPs, expands the commander's ability to exercise command and control and makes the system survivable. The commander assigns functions and tasks to each CP. The commander determines the sequence, timing of the deployment or movement, initial locations, and exact organization of CPs.

4-225. CP survivability is vital to the success of the BCT mission. CPs often gain survivability at the price of effectiveness. When concentrated, the enemy can easily acquire and target most CPs. However, when elements of a CP disperse, they often have difficulty maintaining a coordinated staff effort. When developing CP SOPs and organizing the headquarters into CPs for operations, the BCT commander uses dispersion, size, redundancy, and mobility to increase survivability.

4-226. Echelons within the BCT man, equip, and organize CPs to control operations for extended periods. CP personnel maintain communication with all subordinate units and higher and adjacent units. The commander arranges CP personnel and equipment to facilitate internal coordination, information-sharing, and rapid decision-making. The BCT commander and staff use SOPs, battle rhythms, and meetings to assist with CP operations. The commander positions CPs within areas of operation to maintain flexibility, redundancy, survivability, and mobility. Activities common in all CPs include but are not limited to—

- Maintaining running estimates.
- Controlling operations.
- Assessing operations.
- Developing and disseminating orders.
- Coordinating with higher, lower, and adjacent units.
- Conducting knowledge management and information management.
- Conducting DODIN operations.
- Providing a facility for the commander to control operations, issue orders, and conduct rehearsals.
- Maintaining the COP.
- Performing CP administration (includes sleep plans, security, and feeding schedules).
- Supporting the commander's decision-making process.

COMMAND POST CONFIGURATION

4-227. The BCT design, combined with robust communications, gives the commander two CPs, the main CP and the TAC, and a command group. The BCT commander may designate the main CP of a subordinate battalion, normally the BEB, field artillery battalion, or BSB as the BCT alternate CP. Either, the BSB or the BEB main CP may be assigned responsibility for the brigade support area (BSA) (see chapter 9).

Main Command Post

4-228. The *main command post* is a facility containing the majority of the staff designed to control current operations, conduct detailed analysis, and plan future operations (FM 6-0). The main CP (graphically depicted as the MAIN) is the BCT's principal CP. The main CP includes representatives of all staff sections and a full suite of information systems to plan, prepare, execute, and assess operations. The main CP is larger, more staffing and less mobile than the TAC. Normally, the BCT XO leads and supervises the staff of the main CP. Functions of the main CP include the following:

- Planning current operations including branches and sequels.
- Developing contingency plans from identified branches to the plan.
- Developing plans from information from higher headquarters.
- Developing plans from sequels identified during the planning process.

- Controlling and synchronizing current operations.
- Synchronizing all aspects of the operational framework (see ADP 3-0) such as—
  - Area of operations.
  - Deep, close, rear, and support areas.
  - Decisive, shaping, and sustaining operations.
  - Main and supporting efforts.
- Monitoring and assessing current operations for their impact on future operations.
- Coordinating fires and effects.
- Synchronizing information-related capabilities; capabilities complemented by capabilities such as—
  - Operations security.
  - Information assurance.
  - Counterdeception.
  - Physical security.
  - Electromagnetic support.
  - EP.
- Coordinating CEMA including—
  - EW operations.
  - Cyberspace operations.
  - Spectrum management operations.
- Planning for future operations.
- Employing information collection.
- Anticipating and monitoring the commander's decision points and critical information requirements.
- Coordinating with higher headquarters and adjacent or lateral units.
- Informing higher headquarters and units of ongoing missions.
- Supporting the commander's situational understanding through information and knowledge management.
- Defense Information Systems Network services and DODIN operations. (See ATP 6-02.71.)
- Planning, monitoring, and integrating airspace users.
- Synchronizing sustainment including—
  - COP across all echelons of support.
  - Synchronization with the operations process; plan, prepare, execute, and assess.
  - Alignment with military actions in time and space, prioritization, and purpose.
  - Material readiness reports of combat power platforms.
  - Coordination of echelons above brigade sustainment support.
- Developing and implementing—
  - Safety and occupational health. (See AR 385-10.)
  - RM. (See ATP 5-19.)
  - Accident prevention requirements, policies, and measures.
- Coordinating air-ground operations.
- Coordinating personnel recovery operations. (See FM 3-50.)

4-229. Positioning the main CP includes the following considerations:
- Where the enemy can least affect main CP operations.
- Where the main CP can achieve the best communications (digital and voice).
- Where the main CP can control operations best.

Note. In contiguous areas of operation, the BCT main CP locates behind the battalion TAC and main CP, the BCT TAC, and out of enemy medium artillery range, if practical. In noncontiguous areas of operation, the BCT main CP usually locates within a subordinate battalion's area of operations.

Tactical Command Post

4-230.   A *tactical command post* is a facility containing a tailored portion of a unit headquarters designed to control portions of an operation for a limited time (FM 6-0). The BCT commander employs the TAC as an extension of the main CP to help control the execution of an operation or task. The BCT commander can employ the TAC to direct the operations of units close to each other when direct command is necessary. The commander can use the TAC to control a special task force or to control complex tasks such as reception, staging, onward movement, and integration. When the TAC is not used, the staff assigned to it reinforces the main CP. BCT SOPs should address procedures to detach the TAC from the main CP.

4-231.   The TAC is fully mobile and is usually located near the decisive point of the operation. The *decisive point* is a geographic place, specific key event, critical factor, or function that, when acted upon, allows commanders to gain a marked advantage over an adversary or contribute materially to achieving success (JP 5-0). As a rule, the post includes the personnel and equipment essential to the tasks assigned; however, sometimes the TAC requires augmentation for security. The TAC relies on the main CP for planning, detailed analysis, and coordination. Usually the BCT S-3 leads the tactical CP. TAC functions include the following when employed:

- Control current operations.
- Provide information to the COP.
- Assess the progress of operations.
- Assess the progress of higher and adjacent units.
- Perform short-range planning.
- Provide input to targeting and future operations planning.
- Provide a facility for the commander to control operations, issue orders, and conduct rehearsals.

4-232.   Airborne command and control support often require independent operations by aircrews and aircraft under operational control (OPCON) to commanders and staffs down to the BCT and battalion level. An Army airborne command and control platform provides the maneuver commander with a highly mobile, self-contained, and reliable airborne digital CP. The CP is equipped with the command and control systems needed to operate with joint forces and components, multinational forces, and U.S. Government agencies and departments. The airborne command and control platform allows the commander and staff to maintain voice and digital connectivity with required elements, roughly replicating the systems and capabilities of a digitized maneuver BCT commander's TAC. The commander and staff can perform all command and control and coordination functions from the airborne platform. The airborne platform provides tactical internet access to manipulate, store, manage, and analyze data, information, intelligence, mission plans, and mission progress. The size and functions required of an airborne TAC is mission dependent and within the capabilities and limitations of the aircraft. Ideally, as a minimum the S-3, S-2, fire support officer, and air liaison officer accompany the commander. (See FM 3-99 for information on airborne TAC operations.)

Command Group

4-233.   While not part of the BCT's table of organization and equipment, the commander can establish the command group (see ATP 6-0.5) from the main or TAC (personnel and equipment). The command group, led by the BCT commander, consists of whomever the commander designates. The command group can include the command sergeant major and representatives from the S-2, S-3, and the fire support cell. The command group gives the commander the mobility and protection to move throughout the area of operations and to observe and direct BCT operations from forward positions. Normally, the command group is task organized with a security element whenever it departs the main or TACs. For example, a maneuver platoon from one of the BCTs maneuver battalions may be tasked to provide that element. The commander positions the command group near the most critical event, usually with or near the main effort or decisive operation. The BCT XO may establish a second command group when required.

Early-Entry Command Post

4-234. While not a separate section of the unit's table of organization and equipment, the commander can establish an early-entry CP to assist in controlling operations during the deployment phase of an operation. An *early-entry command post* is a lead element of a headquarters designed to control operations until the remaining portions of the headquarters are deployed and operational (FM 6-0). The early-entry CP normally consists of personnel and equipment from the TAC with additional intelligence analysts, planners, and other staff officers from the main CP based on the situation. The early-entry CP performs the functions of the main and TACs until they are deployed and operational. The BCT XO or S-3 normally leads the early entry CP, when established.

> *Note.* (See FM 3-99, for information on airborne assault and air assault CP organization and operation.)

COMMAND POST CELLS

4-235. The situation determines CP cell organization. A *command post cell* is a grouping of personnel and equipment organized by warfighting function or by planning horizon to facilitate the exercise of mission command (FM 6-0). Staff elements, consisting of personnel and equipment from staff sections, form CP cells. Typically, a BCT organizes into two types of CP cells: integrating cells (current operations and plans) and functional cells (intelligence, movement and maneuver, fire support, protection, sustainment). Integrating and functional cells provide staff expertise, communications, and information systems that work in concert to aid the commander in planning and controlling operations. (See figure 4-15.)

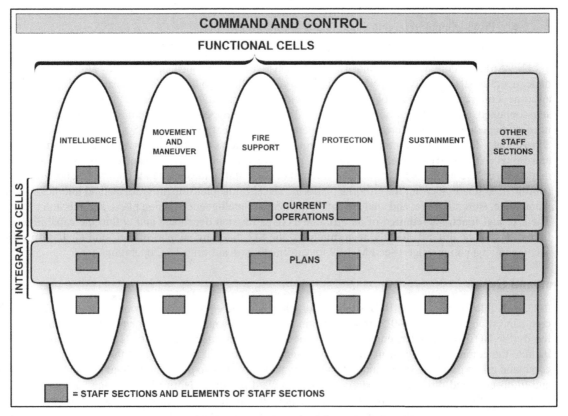

Figure 4-15. Integrating and functional cells

Integrating Cells

4-236. Cross functional by design, integrating cells coordinate and synchronize forces and warfighting functions within a specified planning horizon. A *planning horizon* is a point in time that commanders use to focus the organization's planning efforts to shape future events (ADP 5-0). The three planning horizons are short, mid, and long. These planning horizons correspond to the integrating cells within a headquarters, which are the current operations cell, future operations cell (typically division and above level, see FM 6-0 for discussion), and the plans cell. The BCT has a current operations cell and a small, dedicated planning cell. The BCT is not resourced for a future operations cell. Planning horizons are situation dependent; they can range from hours and days to weeks and months. As a rule, the higher the echelon, the more distant the planning horizon with which it is concerned.

Current Operations Cell

4-237. The current operations cell is the focal point for all operational matters. The cell oversees execution of the current operation. The current operations cell assesses the current situation while regulating forces and warfighting functions according to the commander's intent and concept of operations.

4-238. The current operations cell displays the COP and conducts shift change, battle updates, and other briefings as required. The cell provides information on the status of operations to all staff members and to higher, lower, and adjacent units. The movement and maneuver cell forms the core of the current operations cell. Typically, a BCT designates a chief of operations to lead the current operations cell from the main CP. The current operations cell has representatives from all staff sections, who are either permanent or on-call as well as attached or supporting subordinate units, special operations forces, unified action partners, and liaison officers.

Plans Cell

4-239. The plans cell is responsible for planning operations for the mid- to long-range planning horizons. The plans cell develops plans, orders, branches, and sequels using the MDMP to prepare for operations beyond the scope of the current order. The plans cell oversees military deception planning.

4-240. The plans cell consists of a core group of planners and analysts led by the plans officer. All staff sections assist as required. While the BCT has a small, dedicated plans element, the majority of its staff sections balance their efforts between the current operations and plans cells. Upon completion of the initial operation order, the plans cell normally develops plans for the next operation or the next phase of the current operation. In addition, the plans cell develops solutions to complex problems resulting in orders, policies, and other coordinating or directive products such as memorandums of agreement. In some situations, planning teams form to solve specific problems, such as redeployment within the theater of operations. These planning teams dissolve when planning is complete.

Functional Cells

4-241. Functional cells coordinate and synchronize forces and activities by warfighting function. The functional cells within a CP are movement and maneuver, fire support, intelligence, protection, and sustainment.

Movement and Maneuver Cell

4-242. The movement and maneuver cell coordinates activities and systems that move forces to achieve an advantageous position in relation to the enemy. Activities include tasks that employ forces in combination with direct fire or fire potential (maneuver), force projection (movement) related to gaining a positional advantage over an enemy, and mobility and countermobility. Elements of the operation, airspace management, aviation, and engineer staff sections form this cell. The S-3 leads this cell. Staff elements in the movement and maneuver cell form the core of the current operations cell, also. Additional staff officers and elements residing in the movement and maneuver cell may include information operations officer, PSYOP officer, EWO, and brigade judge advocate.

Fire Support Cell

4-243.   *Fire support* are fires that directly support land, maritime, amphibious, space, cyberspace, and special operations forces to engage enemy forces, combat formations, and facilities in pursuit of tactical and operational objectives (JP 3-09). The fire support cell and its elements integrate the fires warfighting function within the BCT. The BCT fire support officer leads this cell. Soldiers who have expertise integral to the fires warfighting function staff the fire support cell. The cell has resources to plan for future operations from the main CP and to support current operations from the TAC when deployed. Additionally, the cell has the limited capability to provide coverage to the command group when deployed.

4-244.   The fire support cell plans, prepares, executes, and assesses fires. The cell synchronizes the effects of fires with other elements of combat power to accomplish the commander's intent. During the targeting process, the fire support cell develops *high-payoff targets*—a target whose loss to the enemy will significantly contribute to the success of the friendly course of action (JP 3-60)—and, prioritizes targets for attack. *High-value targets*—a target the enemy commander requires for the successful completion of the mission (JP 3-60)—are developed such as enemy artillery formations, reserves, and command and control. The cell matches a wide range of targeting and delivering systems and integrates air defense and airspace management. The fire support cell coordinates with the joint air-ground integration center (JAGIC), in the division's current operations integrating cell, for the execution of fires in support of current and future operations (see ATP 3-91.1). The JAGIC ensures the fire support cell has current fire support coordination measures and airspace coordinating measures and that all BCT fires are executed within BCT airspace parameters. The JAGIC may also execute fires, through the BCT fire support cell, in specific situations.

4-245.   The ADAM element and the BAE collocate within the fire support cell (see ATP 3-01.50 and FM 3-04, respectively). The ADAM/BAE cell composed of Army air and missile defense and aviation staff supports the BCT commander and staff by providing situational understanding of the airspace and early warning via connectivity with airspace users and with multinational partner's sensors and command networks. The ADAM/BAE cell coordinates closely with the BCT TACP to identify close air support airspace requirements and facilitate air-ground integration. The cell coordinates airspace and aviation support issues with other BCT cells, participates directly in the targeting process, airspace management, air defense, and may be a part of most working groups and meetings. The ADAM/BAE is responsible for integrating airspace requirements in the BCT unit airspace plan and submits airspace requirements to the division airspace element.

4-246.   The TACP is the principal air liaison unit collocated with the fire support cell in the main CP. Selected portions of the cell can deploy with the TAC when used. The air liaison officer is the senior TACP member attached to the BCT who functions as the primary advisor to the BCT commander on air operations. The TACP has two primary missions: advise the BCT commander and staff on the capabilities and limitations of air operations and provide the primary terminal attack control of close air support. TACPs may employ joint terminal attack controllers (JTACs) at any echelon. The JTAC is a qualified and certified Service member, who directs the action of combat aircraft engaged in close air support and other air operations. The JTAC provides the ground commander recommendations on the use of close air support and its integration with ground maneuver. (See JP 3-09.3 and ATP 3-04.1 for additional information.)

4-247.   The brigade judge advocate participates in the planning and targeting processes. Additionally, the trial counsel assists the brigade judge advocate on national law matters and is a standing member of working groups, targeting boards, and the fire support cell. The brigade legal section's inclusion in planning, on-boarding, and working groups helps the legal section to have a full awareness of all the issues. The legal section should advise the command about matters such as rules of engagement. The legal section also should review any output for legal sufficiency and provide responsive advice for proposed follow-on operations. (See FM 1-04 for additional information.)

4-248.   The EWO leads the EW and the CEMA working groups. The determination of which working group is appropriate is situation dependent based on which portion of the information environment desired effects occur. The EWO plans, coordinates, assesses, and supports the execution of EW and other CEMA, supports the BCT S-2 during IPB and the fire support officer to ensure electromagnetic attack fires are prioritized and integrated with all other effects. The EWO plans, assesses, and implements friendly electronic security measures, serves as EW subject matter expert on existing EW rules of engagement, and maintains a current assessment of available EW resources. (See FM 3-12 and ATP 3-12.3 for additional information.) Tasks of the BCT CEMA working group include—

- Develop and integrate cyberspace and EW actions into operation plans and exercises.
- Support CEMA policies.
- Plan, prepare, execute, and assess cyberspace and EW operations.
- Integrate intelligence preparation of the battlefield into the operations process.
- Identify and coordinate intelligence support requirements for BCT and subordinate units' cyberspace and EW operations.
- Assess offensive and defensive requirements for cyberspace and EW operations.
- Maintain current assessment of cyberspace and EW resources available to the unit.
- Nominate and submit approved targets within cyberspace to division.
- Prioritize BCT targets within cyberspace and the EMS.
- Plan, coordinate, and assess friendly CEMA.
- Implement friendly electronic and network security measures (for example, EMS mitigation and network protection).
- Ensure cyberspace and EW operations actions comply with applicable policy and laws.
- Identify civilian and commercial cyberspace and EMS-related capacity and infrastructure within the area of operations.

Note. Within the BCT, subordinate maneuver battalions and squadron rely on the BCT for core services, network accessibility, and network defense. The battalion/squadron S-6 performs the planning and operations associated with the main and tactical command posts, including establishing connectivity with adjacent, subordinate, and higher elements. Currently, battalions/squadron do not have organic capabilities to plan and integrate all aspects of cyberspace operations. They do have capabilities to support cybersecurity policies and request for information regarding cyberspace and the EMS. Companies/troops rely on their battalion/squadron for network service, access, and network defense. The company/troop performs the planning and operations associated with the command post, including establishing connectivity with adjacent, subordinate, and higher elements. Commanders at this echelon are responsible for applicable cybersecurity measures.

4-249. The Army and Air Force can augment the main CP's fire support cell as the mission variables of METT-TC dictate. Various capabilities such as PSYOP, civil affairs, and cyberspace operations can augment the cell as needed. Additional functions within the fire support cell include—

- Targeting working group. (See ATP 3-60.)
- Preparing fires portion of operation order including scheme of fires. (See FM 3-09.)
- Managing changes to fire support coordination measures. (See ATP 3-09.32.)
- Coordinating clearance for attacks against targets (clearance of fires). (See FM 3-09.)
- Preparing products for targeting working group and targeting board. (See ATP 3-60.)
- Implementing, updating, managing, and disseminating all targeting guidance in the Advanced Field Artillery Tactical Data Systems. (See FM 3-09.)
- Recommending radar employment and functional dissemination of rocket, artillery, and mortar warnings. (See ATP 3-01.60.)

Intelligence Cell

4-250. Intelligence core competencies within the intelligence cell are intelligence synchronization, intelligence operations, intelligence processing, exploitation, and dissemination and intelligence analysis. *Processing, exploitation, and dissemination* is the execution of the related functions that converts and refines collected data into usable information, distributes the information for further analysis, and, when appropriate, provides combat information to commanders and staffs (ADP 2-0). Processing, exploitation, and dissemination conducted by intelligence personnel or units are called intelligence processing, exploitation, and dissemination. *Intelligence synchronization* is the art of integrating information collection; intelligence processing, exploitation, and dissemination; and intelligence analysis with operations to effectively and efficiently fight for intelligence in support of decision making (ADP 2-0). Intelligence operations (see chapter 5) are the tasks undertaken by military intelligence units and Soldiers to obtain information to satisfy validated requirements. *Intelligence analysis* is the process by which collected information is evaluated and integrated with existing information to facilitate intelligence production (ADP 2-0). Intelligence core competencies are the basic activities and tasks used to describe and drive the intelligence warfighting function and leverage national to tactical intelligence. (See ADP 2-0 for additional information.)

4-251.   The BCT intelligence officer leads the intelligence cell. The BCT intelligence staff section is the core around which the intelligence officer forms the BCT intelligence cell along with designated Soldiers from the BCT military intelligence company and an assigned U.S. Air Force weather team. Higher headquarters may augment this cell with additional capabilities to meet mission requirements. The BCT intelligence cell requests, receives, and analyzes information from all sources to produce and distribute intelligence products. Although there are intelligence staff elements in other CP cells, most of the intelligence staff section resides in the intelligence cell. (See FM 2-0 for additional information.) The BCT intelligence cell performs the following functions:

Facilitate Commander's Visualization and Understanding

4-252.   The BCT intelligence cell facilitates the commander's visualization and understanding of the threat, terrain and weather, and civil considerations as well as other relevant aspects of the operational environment within the BCT area of interest. The intelligence cell provides information and intelligence to support the commander's visualization and understanding (see ADP 5-0). The cell performs IPB (see ATP 2-01.3), indications and warning (see FM 2-0), and situation development tasks (see FM 2-0) to provide information and intelligence.

Support Targeting and Protection

4-253.   The intelligence cell provides the commander and staff with information and intelligence to target threat forces, organizations, units, and systems through lethal and nonlethal effects. The BCT intelligence cell conducts tasks to deny or degrade the threat's effort to access and gain intelligence about friendly forces. The intelligence cell develops target systems, locates targets, and performs battle damage assessment to support targeting (see ATP 3-60). The intelligence cell performs counterintelligence by reporting the capabilities and limitations of threat intelligence services to the commander. (See ATP 2-22.2-1 and ATP 2-22.2-2 for additional information.)

Assisting in Information Collection Planning

4-254.   The BCT intelligence cell integrates military intelligence collection assets so the commander can gain situational understanding to produce intelligence. Information collection (see chapter 5) is the activity within the BCT that synchronizes and integrates the planning and employment of sensors and assets as well as intelligence processing, exploitation, and dissemination capabilities in direct support of current and future operations. The intelligence cell identifies, prioritizes, and validates information collection tasks. The information collection plan is developed and synchronized with the concept of operations. The BCT intelligence cell performs the collection management tasks to support information collection planning. (See ADP 2-0, FM 2-0, and ATP 2-01 for additional information.)

Produce Intelligence Products

4-255.   Intelligence informs the commander and staff of where and when to look. Reconnaissance, surveillance, security operations, and intelligence operations are the collection means (see chapter 5). The collection means range from national and joint collection capabilities to individual Soldier observations and reports. The result or product is intelligence that supports the commander's decision-making. (See ATP 2-01 for additional information.)

Disiminating and Integrating Intelligence

4-256.   The cell uses various command and control networks to disseminate and integrate within the BCT area of operations. The cell uses verbal reports, documents, textual reports, graphic products, softcopy products, and automated databases to disseminate intelligence. The commander and staff integrate the intelligence to assist them in maintaining situational awareness. Establishing communications networks and knowledge and information management procedures accomplishes this function.

Protection Cell

4-257.   The protection cell synchronizes, integrates and organizes protection capabilities and resources to preserve combat power and identify and prevent or mitigate the effects of threats and hazards. Protection is not a linear activity—planning, preparing, executing, and assessing protection is a continuous and enduring activity. Protection integrates all protection capabilities, to include those of united action partners, to safeguard the force, personnel (combatants and noncombatants), systems, and physical assets of the United States and its mission partners. Primary protection tasks and systems include coordinate air and missile defense, personnel recovery, explosive ordnance disposal, antiterrorism, survivability, force health protection, CBRN operations, detention operations, RM, physical security, police operations, and populace and resources control. Additional protection tasks to protect the force, critical assets, and information, and to preserves combat power include area and local security activities, operations security, and cyberspace and EW operations.

4-258.   The S-3 supervises the protection cell within the BCT. Protection synchronization, integration, and organization in the BCT may require the commander to designate a staff lead as the protection officer. The protection officer understands how threats, hazards, vulnerability, and criticality assessments are used to prioritize and determine which assets should be protected given no constraints and which assets can be protected with available resources. There are seldom sufficient resources to simultaneously provide all assets the same level of protection. For this reason, the commander makes decisions on acceptable risks and provides guidance to the staff so that they can employ protection capabilities based on protection priorities. Commanders place high priority sites on the critical asset list. With finite resources, those sites that require additional force protection, security, and survivability are prioritized on the defended asset list. Managed resources include horizontal engineer platoons, Infantry platoons, and air defense artillery platoons.

4-259.   Working groups established within the protection cell may include CEMA (see FM 3-12), CBRN (see ADP 3-37), antiterrorism (see ATP 3-37.2), and personnel recovery (see FM 3-50). For example, protection requires the integration and coordination of tasks to defend the network, as well as protect individuals and platforms. Thus, the S-3 designates and relies on the CEMA working group, when established. The EWO, with representation from the S-2, S-6 and other staff elements, leads the CEMA working group to achieve the level of protection required. In all cases, protection officers and coordinators work with higher and lower echelons to nest protection activities with complementary and reinforcing capabilities. (See ADP 3-37 for additional information.)

Sustainment Cell

4-260.   The sustainment cell coordinates activities and systems that provide support and services to ensure freedom of action and to prolong endurance. The sustainment cell includes tasks associated with logistics, financial management, personnel services, and health service support. The following staff section elements work in the sustainment cell, logistics, human resources, and the surgeon. The BCT sustainment cell may collocate with the BSB within the brigade support area. The BCT S-4 leads this cell. (See chapter 9 for additional information.)

STAFF PROCESSES AND PROCEDURES

4-261.   A BCT must man, equip, and organize CPs to control operations for extended periods. The BCT commander, assisted by the staff, arranges CP personnel and equipment to facilitate 24-hour operations, internal coordination, information-sharing, and rapid decision-making. The commander ensures procedures to execute the operations process within the headquarters enable mission command. The staff uses SOPs, battle rhythm, meetings, running estimates, information, battle captains, and command and control systems and techniques to assist in CP operations.

Standard Operating Procedures

4-262. SOPs that assist with effective command and control serve two purposes. Internal SOPs standardize each CP's internal operations and administration. External SOPs, developed for the entire force, standardize interactions among CPs and between subordinate units and CPs. CPs are organized to permit continuous and rapid execution of operations. SOPs for each CP should be established, known to all, and rehearsed. These SOPs should include at a minimum the following:

- Organization and setup.
- Plans for teardown and displacement.
- Eating and sleeping plans.
- Shift manning, shift changes and operation guidelines.
- Physical security plans.
- Priorities of work.
- Loading plans and checklists.
- Orders production.
- Techniques for monitoring enemy and friendly situations.
- Posting of map boards.
- Maintenance of journals and logs.

4-263. Effective SOPs require all Soldiers to know their provisions and to train to their standards. (See ADRP 1-03 for additional information on tasks for CP operations.) Critical BCT SOPs include tactical SOPs (see ATP 3-90.90), targeting SOPs (see ATP 3-60), and CP battle drill SOPs (see ADRP 1-03).

Battle Rhythm

4-264. *Battle rhythm* is a deliberate daily cycle of command, staff, and unit activities intended to synchronize current and future operations (FM 6-0). Within the operations process, the BCT commander and staff integrates and synchronizes numerous activities, meetings, and reports with their headquarters and higher headquarters and with subordinate units. The BCT's battle rhythm sequences the actions and events within a headquarters that are regulated by the flow and sharing of information that supports decision-making. The establishment of a battle rhythm is always dependent on the availability of time (including speed of decision-making, speed of action, and operational tempo). For example, time considerations (especially within large-scale combat operations) across physical domains and the dimensions of the information environment help the commander determine the correct battle rhythm in regards to the type and number of activities and events relevant to operational tempo. An effective battle rhythm—

- Establishes a routine (dependent upon the availability of time) for staff interaction and coordination.
- Facilitates interaction (dependent upon mission constraints) among the commander, staff, and subordinate units.
- Facilitates (dependent upon planning horizon) staff planning and the commander's decision-making.

4-265. As a practical matter, a BCT's battle rhythm consists of a series of meetings, report requirements, and other activities synchronized by time and purpose. These activities may be daily, weekly, monthly, or quarterly depending on the planning horizon.

4-266. The BCT commander adjusts the unit's battle rhythm as operations progress. For example, early in the operation, a commander may require a commander's update every several hours. As the situation changes, the commander may require only a daily commander's update. Some factors that help determine a unit's battle rhythm include the staff's proficiency, higher headquarters' battle rhythm, and current mission. The BCT commander and XO consider the following when developing the unit's battle rhythm:

- Higher headquarters' battle rhythm and report requirements.

- Duration and intensity of the operation.
- Planning requirements of the integrating cells (current operations and plans).

Meetings

4-267. Meetings (including working groups and boards) take up a large amount of a BCT's battle rhythm. Meetings are gatherings to present and exchange information, solve problems, coordinate action, and make decisions. Meetings may involve the staff; the commander and staff; or the commander, subordinate commanders, staff, and other partners. Who attends a meeting depends on the issue. The BCT commander establishes meetings to integrate the staff and enhance planning and decision-making within the headquarters. Two critical meetings that happen as a part of the BCT battle rhythm are the operations update and assessment briefing and the operations synchronization meeting.

Operation Update and Assessment Briefing

4-268. An operation update and assessment briefing may occur daily or anytime the commander calls for one. The content is similar to the shift change briefing but has a different audience. The staff presents the briefing to the commander and subordinate commanders. The briefing provides all key personnel with common situational awareness. Often the commander requires this briefing shortly before an operation begins to summarize changes made during preparation, including changes resulting from information collection efforts.

4-269. Staff sections present their running estimates during the briefing. Subordinate commanders brief their current situation and planned activities. Rarely do all members conduct this briefing face-to-face. All CPs and subordinate commanders participate using available communications, including radio, conference calls, and video teleconference. The briefing follows a sequence and format specified by SOPs that keeps transmissions short, ensures completeness, and eases note taking. The briefing normally has a format similar to a shift-change briefing. However, this briefing omits CP administrative information and includes presentations by subordinate commanders in an established sequence.

Operations Synchronization Meeting

4-270. The key event in the battle rhythm is the operations synchronization meeting, which supports the current operation. The primary purpose of the meeting is to synchronize all warfighting functions and other activities in the short-term planning horizon. The meeting ensures that all staff members have a common understanding of current operations including upcoming and projected actions at decision points.

4-271. The operations synchronization meeting does not replace the shift change briefing or operation update and assessment briefing. The S-3 or XO chairs the meeting. Representatives of each CP cell and separate staff section attend the meeting. The operations synchronization meeting includes a fragmentary order addressing any required changes to maintain synchronization of current operations, and any updated planning guidance for upcoming working groups and boards. All warfighting functions are synchronized, and appropriate fragmentary orders are issued to subordinates based on the commander's intent for current operations.

Running Estimate

4-272. Effective plans and successful executions hinge on accurate and current running estimates. A *running estimate* is the continuous assessment of the current situation used to determine if the current operation is proceeding according to the commander's intent and if planned future operations are supportable (ADP 5-0). Failure to maintain accurate running estimates may cause errors or omissions resulting in flawed plans or bad decisions during execution.

4-273. Running estimates are principal knowledge management tools the BCT commander and staff use throughout the operations process. In their running estimates, the commander and each staff section member continuously consider the effect of new information and update the following:

- Facts.
- Assumptions.

- Friendly force status.
- Enemy activities and capabilities.
- Civil considerations.
- Conclusions and recommendations.

4-274. Running estimates always include recommendations for anticipated decisions. During planning, the BCT commander uses these recommendations to select feasible, acceptable, and suitable COAs for further analysis. The commander uses recommendations from running estimates in decision-making during preparation and execution.

4-275. The BCT staff maintains formal running estimates while the commander's estimate is a mental process directly tied to the commander's vision. The commander integrates personal knowledge of the situation with analysis of the operational and mission variables, with subordinate commanders and other organizations assessments, and with relevant details gained from running estimates. The BCT commander uses a running estimate to crosscheck and supplement the staff's running estimates. A running estimate format is included in FM 6-0.

Information

4-276. CPs within the BCT monitor communications nets, receive reports, and process information to satisfy commander needs or critical information requirements. This information is maintained, in addition to digital systems, on maps, charts, and logs. Each section or cell maintains daily journals to log messages and radio traffic. CPs maintain information as easily understood map graphics and charts. Status charts can be combined with situation maps to give the commander and staff friendly and enemy situation snapshots for the planning process. This information is updated continuously. For simplicity, all map boards should be the same size and scale and overlay mounting holes should be standard on all map boards. This allows easy transfer of overlays from one board to another. The following procedures for posting friendly and enemy information on the map aid the commander and staff in following the flow of battle:

- Friendly and enemy unit symbols are displayed on clear acetate placed on the operations overlay. These symbols can be marked with regular stick cellophane tape or with marking pen.
- Units normally keep track of subordinate units, two levels down. This may be difficult during offensive operations. It may be necessary to track locations of immediate subordinate units instead.

Battle Captain

4-277. Battle captain positions are habitually filled, and found in most, if not all CPs. They coordinate the day-to-day staff activities, in effect acting as an assistant XO, and provide continuity for the staff's actions. The battle captain's informal role is to plan, coordinate, supervise, and maintain communication flow throughout the CP to ensure the successful accomplishment of all assigned missions. The battle captain assists the commander, XO, and S-3 by being the focal point in the CP for communications, coordination, and knowledge and information management. The battle captain is also the CP officer in charge in the absence of the commander, XO, and S-3. To function effectively, the battle captain must have a working knowledge of all elements in the CP, understand unit SOPs, and ensure CP personnel use them. The battle captain must know the current plan and task organization of the unit and understand the commander's intent.

4-278. Battle captains integrate into the decision-making process and know why certain key decisions were made. Battle captains must know the technical aspects of the battle plan and understand the time-space relationship to execute any specific support task. Battle captains must understand and enforce the battle rhythm, the standard events or actions that happen during a normal 24-hour period and ensure that the CP staff is effective throughout the period. Understanding their assigned authorities, battle captains use judgment to adjust activities and events to accomplish the mission across different shifts, varying tactical circumstances, and changes in the CP location. Battle captains have the overall responsibility for the smooth functioning of the facility and its staff elements. This range of responsibility includes—

- Maintaining continuous operations (while static and mobile).
- Tracking the current situation.
- Ensuring communications are maintained and all messages and reports are logged.

- Assisting the XO in ensuring a smooth and continuous information flow.
- Processing essential data to ensure tactical and logistical information is gathered and provided to staff members on a regular basis.
- Tracking CCIRs and providing recommendations.
- Approving fabrication and propagation of manual unit icons.
- Sending reports to higher and ensuring relevant information passes to subordinate units.
- Monitoring security within and around the CP.
- Organizing the CP to displace rapidly.
- Conducting battle drills and enforcing the SOP.

4-279. The battle captain ensures all staff elements in the CP understand their actions in accordance with the SOP and operation order, and provides coordination for message flow, staff briefings, updates to charts, and other coordinated staff actions. As the focal point in the CP, the battle captain processes essential information from incoming data, assesses it, ensures dissemination, and makes recommendations to the commander, XO, and S-3. The battle captain ensures the consistency, accuracy, and timeliness of information leaving the CP, including preparing and issuing fragmentary orders and warning orders. The battle captain monitors and enforces the updating of charts and status boards necessary for battle management and ensures this posted information is timely, accurate, and accessible.

Digital and Analog Command and Control Systems and Techniques

4-280. Digital command and control systems within a CP bring a dramatic increase in the level of informational dominance units may achieve. Techniques for digital procedures and for integrating analog and digital units contribute to battlefield lethality and tempo, and the ability to maintain information dominance. These techniques can significantly speed the process of creating and disseminating orders, allow for extensive collection of information, and increase the speed and fidelity of coordination and synchronization of battlefield activities. At the same time, achieving the potential of these systems requires extensive training, a high level of technical proficiency by both operators and supervisors, and the disciplined use of detailed SOPs. Communications planning and execution to support the digital systems is significantly more demanding and arduous than is required for units primarily relying on Combat Net Radio communication and Joint Capabilities Release (known as JCR)/Joint Battle Command-Platform (known as JBC-P). (See ATP 6-02.53 for additional information on CP tactical radio operations.)

4-281. Whether to use Combat Net Radio or digital means for communication is a function of the situation and SOPs. Some general considerations can help guide the understanding of when to use which mechanism at what time. Frequency modulation communication is normally the initial method of communications when elements are in contact. Before and following an engagement, the staff and commanders use digital systems for disseminating orders and graphics and conducting routine reporting. During operations, however, the staff uses a combination of systems to report and coordinate with higher and adjacent units.

4-282. The BCT staff must remain sensitive to the difficulty and danger of using digital systems when moving or in contact. The staff should not expect digital reports from subordinate units under such conditions. Other general guidelines include the following:

- Initial contact at any echelon within the BCT should be reported on frequency modulation voice; digital enemy spot reports should follow as soon as possible to generate the enemy COP.
- Elements moving about the battlefield (not in CPs) use frequency modulation voice unless they can stop and generate a digital message or report.
- Emergency logistical requests, especially casualty evacuation (see ATP 4-25.13) requests, should be initiated on frequency modulation voice with a follow-up digital report, if possible.
- Combat elements moving or in contact should transmit enemy spot reports on frequency modulation voice; their higher headquarters should convert frequency modulation reports into digital spot reports to generate the COP.
- Calls for fire on targets of opportunity should be sent on frequency modulation voice; fire support teams submit digitally to advanced field artillery tactical data system.
- Plan calls for fire digitally and execute them by voice with digital back up.

- Routine logistical reports and requests are sent digitally.
- Routine reports from subordinates to the BCT before and following combat are sent digitally.
- Orders, plans, and graphics should be done face-to-face, if possible. If these products are digitally transmitted, they should be followed by frequency modulation voice call to alert recipients that critical information is being sent. The transmitting element should request a verbal acknowledgement of both receipt and understanding of the transmitted information by an appropriate Soldier, who usually is not the computer operator.
- Obstacle and CBRN 1 reports should be sent initially by voice followed by digital reports to generate a geo-referenced message portraying the obstacle or contaminated area across the network.

Friendly Common Operational Picture

4-283. The creation of the friendly COP is extensively automated, requiring minimal manipulation by CPs or platform operators. Each platform creates and transmits its own position location and receives the friendly locations, displayed as icons, of all the friendly elements in that platform's wide area network. This does not necessarily mean that all friendly units in the general vicinity of that platform are displayed because some elements may not be in that platform's network. For example, a combat vehicle in a BCT probably will not have information on an artillery unit in another division operating nearby because the two are in different networks. The COP provided by the JCR/JBC-P (a situational awareness and command and control system) transmitted to CPs across the BCT utilizes a separate satellite-based network to provide position location information. The BCT S-6 ensures the proper alignment and interoperability between the command and control applications and the networks the BCT utilizes. (See ATP 6-02.53 for additional information on tactical radio networks.)

4-284. Commanders must recognize limitations in the creation of the friendly COP which results from vehicles or units that are not equipped with the JCR. The following are two aspects to consider:

- Not all units will be equipped with all command and control system components, particularly multination partners and organizations. It is likely analog units or organizations—those operating in an area of operations without compatible digital-based command and control systems—will enter the BCT's area of operations.
- Most dismounted Soldiers will not be equipped with a digital device that transmits information.

4-285. The following are ways to overcome these limitations:

- A digitally equipped element tracks the location of specified dismounts and manually generates and maintains an associated friendly icon.
- The main CP tracks analog units operating within the area and generates associated friendly icons. The main CP must keep the analog equipped unit informed of other friendly units' locations and activities.
- A digitally equipped platform acts as a liaison or escort for analog units moving or operating in the same area. Battalion and squadron, and higher elements must be informed of the association of the liaison officer icon with the analog unit.
- Do not use friendly positional information to clear fires because not all elements will be visible. Friendly positional information can be used to deny fires and can aid in the clearance process, but it cannot be the sole source for clearance of fires. This holds true for all Army command and control systems.

Enemy Common Operational Picture

4-286. The most difficult and critical aspect of creating the COP is creating the picture of the enemy. The enemy COP is the result of multiple inputs (for example, frequency modulation spot reports, UAS and Joint Surveillance Target Attack Radar System [JSTARS] reports, reports from JCR-equipped platforms in subordinate units, electronic or signals intelligence feeds) and inputs from BCT information collection efforts through the BCT S-2. Enemy information generation is a complex process requiring automated intelligence all source inputs and detailed analysis from within the BCT.

4-287. Generation of the enemy COP occurs at all echelons. At BCT level and below, the primary mechanism for generating information is the JBC-P. When an observer acquires an enemy element, they create and transmit a spot report, which automatically generates an enemy icon that appears throughout the network. Only those in the address group to whom the report was sent receive the text of the report, but all platforms in the network can see the icon. As the enemy moves or its strength changes, the observer must update this icon. If the observer must move, the observer ideally passes responsibility for the icon to another observer. If multiple observers see the same enemy element and create multiple reports, the S-2 or some other element that has the capability must eliminate the redundant icons.

4-288. Unit SOPs must clearly establish who has the ability, authority, and responsibility to create and input enemy icons. Without the establishment of these procedures, it is highly probable that the enemy COP will not be accurate.

4-289. JBC-P spot reports must include the BCT S-2 in the address group for the data to be routed through the CP server into the Distributed Common Ground System-Army (DCGS-A) to feed the larger intelligence picture. Frequency modulation reports received at a CP can be inputted manually into the DCGS-A database by the S-2 section. JBC-P and frequency modulation voice reports are the primary source of enemy information in the BCT's area of operations.

4-290. Fusion of all the intelligence feeds normally occurs at BCT and higher levels. The BCT S-2 routinely (every 30 minutes to every hour) sends the updated enemy picture to subordinate units down to platform level. Since the fused DCGS-A database focuses on the deep areas of the battlefield; its timeliness may vary. Subordinate elements of the BCT normally use only the JBC-P generated COP. Companies and troops stay focused entirely on the JCR/JBC-P generated COP. Battalion and squadron leaders and staffs refer occasionally to the JCR/JBC-P generated intelligence picture to keep track of enemy forces they might encounter in the near future, but not yet in the close or deep area.

4-291. Automation and displays contribute enormously to the ability to disseminate information and display it in a manner that aids comprehension. However, information generation must be rapid for it to be useful. Information also must be accompanied by analysis; pictures alone cannot convey all that is required, nor will they be interpreted the same by all viewers. The BCT S-2 and section must be particularly careful about spending too much time operating a DCGS-A terminal while neglecting the analysis of activities for the BCT and subordinate commanders and other staff members. (See ATP 6-02.53 and ATP 6-02.60 for additional information.) The success of the BCT's intelligence effort depends primarily on the ability of staffs to—

- Analyze enemy activities effectively.
- Develop and continuously refine effective IPB.
- Create effective collection requirements management.
- Execute effective collection operations management.

Graphics and Orders

4-292. The advent of digitization does not mean that acetate and maps have no use and will disappear, at least not in the near future. Maps remain the best tools when maneuvering and fighting on the battlefield, and for controlling and tracking operations over a large area. The combination of a map with digital information and terrain database is ideal; both are required and extensively used.

4-293. Army command and control system components support the creation and transmission of operation orders. BCT staff sections normally develop their portions of orders and send them to the S-3 where they are merged into a single document. The S-3 deconflicts, integrates, and synchronizes all elements of the order. Once the order is complete, it is transmitted to subordinate, higher, and adjacent units. The tactical internet does not possess high transmission rates and therefore orders and graphics should be concise to reduce transmission times. Orders transmitted directly to JCR/JBC-P-equipped systems within the battalion or squadron must meet the size constraints of the order formats in the JCR/JBC-P. Graphics and overlays are constructed with the same considerations for clarity and size.

4-294. Digital graphics must interface and be transmittable. The interface and commonality of graphics will continue to evolve technologically and will require further software corrections. The following guidelines apply when creating graphics:

- Create control measures based on readily identifiable terrain, especially if analog units are part of the task organization.
- Boundaries are important, especially when multiple units must operate in close proximity or when it becomes necessary to coordinate fires or movement of other units.
- Intent graphics that lack the specificity of detailed control measures are an excellent tool for use with warning and fragmentary orders and when doing parallel planning. Follow them with appropriately detailed graphics, as required.
- Use standardized colors to differentiate units. This is articulated in the tactical SOP. For example, BCT graphics may be in black, battalion A in purple, battalion B in magenta, and battalion C in brown and so forth. This adds considerable clarity for the viewer. Subordinate company and team colors are then specified.
- Use traditional doctrinal colors for other graphics (green for obstacles, yellow for contaminated areas, and so forth).

4-295. In order to accelerate transmission times when creating overlays, use multiple smaller overlays instead of a single large overlay. System operators can open the overlays they need, displaying them simultaneously. The technique also helps operators in reducing screen clutter. The S-3 should create the initial graphic control measures on a single overlay and distribute it to the staff. The overlay is labeled as the operations overlay with the appropriate order number. Staff elements should construct their appropriate graphic overlays using the operations overlay as a background but without duplicating the operations overlay. This avoids unnecessary duplication and increase in file size and maintains standardization and accuracy. Each staff section labels its overlay appropriately with the type of overlay and order number. Before overlays are transmitted to subordinate, higher, and adjacent units, the senior battle captain or the XO checks them for accuracy and labeling. Hard copy (traditional acetate) overlays are required for the CPs and any analog units. Transmit graphics for on-order missions or branch options to the plan before the operation as time permits. If time is short, transmit them with warning orders.

Digital Standard Operating Procedures

4-296. The BCT SOP should contain standards for digital operations, in addition to analog operations. Most of digital operating procedures are established at the BCT level with the battalion SOP complying and adding detail when required. One of the critical requirements when task organized to another unit is to receive and disseminate that units SOP.

4-297. To create a common picture, JCR/JBC-P must have the same information filter settings. This is particularly important for the enemy COP so that as icons go stale, they purge at the same time on all platforms. Standard filter settings should be established in unit SOPs and be the same throughout the BCT. For enemy offensive operations, the filter setting times should be short; for enemy defensive operations, the setting times should be longer, reflecting the more static nature of the enemy picture.

4-298. The standardization of friendly and enemy situational filter settings is of great importance in maintaining a COP. JCR/JBC-P provides three methods for updating individual vehicle locations: time, distance, and manual. When the system is operational, it automatically updates friendly icons using time, distance traveled, or both, based on the platform's friendly situational filter settings. The unit should standardize filter settings across the force based on both the mission and the function of the platform or vehicle. Use shorter refresh rates for combat vehicles and vehicles that frequently move and longer refresh rates for static vehicles such as CPs. Tailoring the frequency of these automatic updates reduces the load on the tactical internet, freeing more capacity for other types of traffic.

4-299. The BCT node is probably the most effective place to standardize the situational filter settings using the BCT tactical SOP. There are no set rules for what these settings should be. The commander must establish them based on the unit's experience using JCR/JBC-P and the capacity of the tactical internet. Subordinate units should use the capability to update a vehicle's position manually only when a platform's system is not fully functional, and it has lost the ability to maintain its position automatically.

Reporting and Tracking of Battles

4-300. Having all platforms and units on the battlefield send spot reports digitally may result in mass confusion. However, to eliminate confusion, there should be one designated individual within the unit authorized to initiate digital spot reports. While the designated individual will be somewhat removed from the fight, that individual can assist those who execute the direct firefight by filtering multiple reports of the same event.

4-301. Another technique, used at company level, to eliminate duplicate reporting problems is to limit the creation of enemy icons through digital spot reports to reconnaissance and security elements and the company leadership (commander, XO, or first sergeant) or other designated individual. Others report to their higher headquarters, which creates and manages the icon. At company level, the XO, first sergeant, or CP personnel become the primary digital reporters. These assignments cannot be completely restrictive. Unit SOPs and command guidance must allow for and encourage Soldiers who observe the enemy and know they are the sole observer (because there is no corresponding enemy icon displayed in the situational COP) to create a digital spot report. BCT and subordinate unit SOPs should define the schedule for report submissions, the message group for the reports, and the medium (digital system or verbal) used.

4-302. Battle tracking is the process of monitoring designated elements of the COP tied to the commander's criteria for success. Battle tracking requires special attention from all staff officers, and normally done digitally and manually with situation maps and boards. The XO and S-3 must continue to monitor the progress of the operation and recommend changes as required.

4-303. The BCT XO establishes a schedule for routine systems updates. For example, the S-2 section should continuously update the DCGS-A database and should transmit the latest COP to the network every 30 minutes during operations if the commander, S-3, or reconnaissance and security elements need it. Staff sections should print critical displays on an established schedule. These printed snapshots of the COP are used for continuity of battle tracking in the event of system failures and can contribute to after-action reviews and unit historical records.

4-304. SOPs define the technical process for creating, collating, and transmitting orders and overlays, both analog and digital and in degraded environments. For interoperability and clarity, BCT SOPs should define the naming convention and filing system for all reports, orders, and message traffic. This significantly reduces time and frustration associated with lost files or changes in system operators or the environment. Information systems will inevitably migrate to a web-based capability. This allows information in a database to be accessed by users as needed or when they are able to retrieve it. For example, the S-2 may transmit an intelligence summary to all subordinates. Inevitably, some will lose the file or not receive it. The S-2 can simultaneously post that same summary to an established homepage so users can access it as required. If this technique is used, the following are a few things to consider—

- Posting a document to a homepage does not constitute communications. The right people are alerted when the document is available.
- Keep documents concise and simple. Elaborate digital slide presentations take longer to transmit, causing delays in the tactical internet.
- The amount of information entered in a database and personnel who have access is carefully controlled, both to maintain security and to keep from overloading the tactical internet.
- Assign responsibility to personnel who are authorized to input and delete both friendly and enemy unit icon information.

4-305. In combat against a peer enemy, the probability of losing digital connectivity for short and extended periods if high. The BCT commander and staff must be able to transition to pure analog means of battle tracking immediately upon loss of digital capability and degraded capability. Procedures for integrating digital and analog units and operations within degraded environments are essential and should consider the following:

- Frequency modulation and joint network node/CP node are the primary communications mediums with the analog unit.
- Hard copy orders and graphics are required.

- Graphic control measures require a level of detail necessary to support operations of a unit without situational information. This requires more control measures tied to identifiable terrain, especially during operations within degraded environments.
- Liaison teams are critical in both digital and analog situations through direct liaison with the partner unit(s).
- The staff must recognize that integrating an analog unit into a digital unit requires retention of most of the analog control techniques. In essence, both digital and analog control systems must be in operation, with particular attention paid to keeping the analog unit apprised of all pertinent information that flows digitally.
- The staff establishes redundant communication, especially when the BCT shares its area of operations with other entities that have cultural differences and lack of or degraded communications.

Considerations Concerning the Degradation of BCT Command and Control Systems

4-306. As the staff supports the commander in the exercise of command and control, assists subordinate units, and informs units and organizations outside the BCT, a broad array of actors and activities challenge the BCT's freedom of action in cyberspace and space. Enemies and adversaries utilize cyberspace and space to degrade the BCT's capability to communicate and operate command and control systems. The ability for adversaries and enemies to operate in the cyberspace and space domains increases the need for the BCT to maintain the capability to conduct offensive and defensive cyberspace operations (see paragraph 4-329) to affect the operational environment and to protect friendly command and control systems. For example, enemy global positioning satellite jamming capabilities could render precision fires and blue force tracker inaccurate.

4-307. During the operations process, the BCT and its subordinate units prepare for degraded command and control systems and reduced access to cyberspace and space capabilities. Considerations concerning the BCT in denying and degrading adversary and enemy use of cyberspace and the EMS and other effects that degrade friendly command and control systems include—

- Enemy capabilities (cyberspace, space, EW) to degrade, planned and targeted against command and control systems. Enemy efforts include jamming, spoofing, intercepting, hacking, and direction finding (leads to targeting).
- Friendly effects that degrade include—
  - Lack of familiarity with Army command and control systems.
  - Lack of protection or countermeasures at BCT and below echelons.
  - Lack of understanding of threat capabilities and doctrine for employment.
  - Terrain and weather, and other environmental variables.

4-308. Key indicators that BCT command and control systems are being degraded include—

- Reliable voice communications are degraded.
- Increased latency for data transmissions.
- Frequent and accurate targeting by threat lethal and nonlethal effects.
- Increased pings/network intrusions.
- Inconsistent digital COP, for example spoofing.
- Inaccurate Global Positioning System (GPS) data/no satellite lock and inconsistency between inertial navigation aids and GPS-enabled systems.

4-309. BCT efforts to counter the effects of degraded mission command systems include—

- Train to recognize indicators that it is happening.
- Develop contingency plans and rehearse implementation during the planning process and preparations.
- Maintain analog COP at all echelons.
- Train to operate from the commander's intent, and analog graphics and synchronization matrixes.
- Keep plans as simple as possible so that they are less susceptible to friction.

4-310. BCT efforts to prevent degraded command and control systems include—

- Minimize length of frequency modulation transmissions.
- Use terrain to mask transmission signatures.
- Employment of directional antennas.
- Require physical presence of leaders at briefings, for example distribute information via analog means in person.
- Use of camouflage and deception in all domains.
- Use of communications windows to reduce transmissions.
- Employment of encryption/cypher techniques.

AIR GROUND OPERATIONS

4-311. Air and ground forces must integrate effectively and properly plan, coordinate, synchronize, and conduct operations with a combination of lethal joint fires and nonlethal actions to minimize the potential for fratricide, allied casualties, noncombatant casualties, and un-intended excessive collateral damage. *Integration*, the arrangement of military forces and their actions to create a force that operates by engaging as a whole (JP 1), maximizes combat power through synergy of both forces. The integration of air operations into the ground commander's scheme of maneuver may also require integration of other Services or multinational partners. Integration continues through planning, preparation, execution, and assessment. The BCT commander and staff must consider the following framework fundamentals to ensure effective integration of air and ground maneuver forces:

- Understanding capabilities and limitations of each force.
- SOPs.
- Habitual relationships.
- Regular training events.
- Airspace management.
- Maximizing and concentrating effects of available assets.
- Employment methods.
- Synchronization.

AIRSPACE MANAGEMENT

4-312. *Airspace management* is the coordination, integration, and regulation of the use of airspace of defined dimensions (JP 3-52). Airspace management is essential to integrate all airspace uses (manned and unmanned aircraft and indirect fires). Properly developed airspace coordinating measures facilitate reconnaissance and security operations and the BCT's employment of aerial and surface-based fires simultaneously as well as unmanned assets to maintain surveillance. Airspace management includes identifying airspace users' requirements and processing airspace coordinating measures requirements to satisfy the synchronization of operational timelines and events. It is critical to process airspace coordinating measures requirements per higher headquarters battle rhythm to meet joint force suspense to get airspace requirements integrated into the theater airspace control system and published on the airspace control order. (See FM 3-52 and ATP 3-52.1 for additional information.)

DIVISION JOINT AIR-GROUND INTEGRATION CENTER

4-313. The BCT commander uses the division JAGIC to ensure continuous collaboration with unified action partners to integrate fires and to use airspace effectively. The BCT fire support cell sends requests for division level Army and joint fires to the JAGIC in the current operations integrating cell of the division. Upon receipt of the request for fire or joint tactical air strike request (DD Form 1972 *[Joint Tactical Air Strike Request]),* the JAGIC develops targeting solutions and coordinates airspace requirements. Additionally, the JAGIC conducts collateral damage estimation and reviews available ground and air component fires capabilities to determine the most effective attack method. (See ATP 3-91.1 for additional information on the JAGIC.)

AIR DEFENSE AIRSPACE MANAGEMENT AND BRIGADE AVIATION ELEMENT

4-314. The ADAM element and BAE, located within the BCT fire support cell, provides the BCT commander and staff with the aerial component of the COP. These elements coordinate airspace management requirements with higher headquarters and enable air and missile defense and aviation considerations throughout the operations process. By providing the BCT commander and staff with near real time situational awareness of the airspace dimension, these elements allow the commander to optimize the air battle and airspace management at all levels.

4-315. The ADAM element integrates within the BCT's fire support cell and always deploys with the BCT. Upon mission notification, the ADAM element conducts an assessment to determine if air and missile defense augmentation from the division air and missile defense battalion is required. The element conducts continuous planning and coordination proportionate with the augmented sensors deployed within the brigade's area of operations. The ADAM element and tailored air and missile defense augmentation force provide the active air defense within the BCT's area of operations. (See ATP 3-01.50 for additional information.)

4-316. The BAE plans and coordinates the incorporation of Army aviation into the ground commander's scheme of maneuver and synchronizes aviation operations and airspace coordinating measures. The element provides employment advice and initial planning for aviation missions to include employment of UASs, airspace planning and coordination, and synchronization with other air liaison officers and the fire support coordinator. The BAE coordinates directly with the supporting combat aviation brigade or aviation task force. The combat aviation brigade commander exercises an informal oversight role for the brigade aviation officer and the BAE. The combat aviation brigade commander interfaces with the supported BCT commander to ensure the BAE is manned properly to meet the BCT commander's intent. (See FM 3-04 and ATP 3-04.1 for additional information.)

COMBAT AVIATION BRIGADE

4-317. The combat aviation brigade is a modular and tailorable force organized and equipped to integrate and synchronize operations of multiple aviation battalions. The combat aviation brigade can operate as a maneuver headquarters and can employ subordinate battalions and other augmenting forces in deliberate and hasty operations. The combat aviation brigade headquarters provides tailored support to adjacent supported maneuver commanders at the BCT level and below when employed in this role. While a BAE works directly for the BCT commander as a permanent member of the BCT staff, aviation liaison teams represent the supporting aviation task force at designated maneuver headquarters for the duration of a specific operation. If collocated with a BAE, the liaison team normally works directly with the brigade aviation officer as a functioning addition to the BAE staff section. Effective employment of liaison officers is imperative for coordination and synchronization. Often aviation liaison teams coordinate with the BAE and proceed to a supported ground maneuver battalion or squadron location.

4-318. Air-ground integration is merging air and ground operations into one fight. The goal is to apply aviation capabilities according to the BCT commander's intent. Ideally, integration begins early in the planning process with the BAE's involvement. The BAE advises the BCT commander on aviation capabilities and on how to best use aviation to support mission objectives. The employment of aviation assets is dependent upon providing the supporting aviation units with a task and purpose, integrating them into the BCT commander's scheme of maneuver. This integration allows the aviation commander and staff to identify the best available platform(s) for the mission, to identify the proper utilization of aviation assets, and to increase the BCT's maneuver capabilities, as well as the commander's ability to conduct command and control on the move. BCT planners, down through the supporting aviation unit to the individual aircrews, should consider these imperatives as elements of air-ground operations. A failure to properly consider these imperatives can result in the lack of synchronization, wasted combat power, the loss of friendly forces by enemy actions, or fratricide. (See FM 3-04 and ATP 3-04.1 for additional information.)

4-319. Combat aviation brigade attacks may be in close proximity or in direct support of ground maneuver forces or the attacks may be against enemy forces not in direct contact with friendly ground forces (interdiction). Army aviation attacks are coordinated attacks by Army attack reconnaissance aircraft (manned and unmanned) against targets that are in close proximity to friendly forces. Army aviation attacks are not synonymous with close air support flown by joint aircraft. Detailed integration with ground forces is required due to the close proximity of friendly forces. *Interdiction* is an action to divert, disrupt, delay, or destroy the enemy's military surface capability before it can be used effectively against friendly forces, or to otherwise achieve objectives (JP 3-03). Interdiction is at such a distance from friendly forces that detailed integration with ground forces is not required.

4-320.   Air-ground operations include the movement of maneuver forces. An *air assault* is the movement of friendly assault forces by rotary-wing or tiltrotor aircraft to engage and destroy enemy forces or to seize and hold key terrain (JP 3-18). Air assaults use the firepower, mobility, protection, and total integration of aviation assets in their air and ground roles to attain the advantage of surprise. Air assaults allow friendly forces to strike over extended distances and terrain to attack the enemy when and where it is most vulnerable. By their very nature, air assaults are high risk, high payoff operations that are resource intensive and require extensive planning and preparation to be successful. (See FM 3-99 for additional information.)

4-321.   Army *air movements* are operations involving the use of utility and cargo rotary-wing assets for other than air assaults (FM 3-90-2). Air movements are a viable means of transport and distribution to support maneuver and sustainment conducted to reposition units, personnel, supplies, equipment, and other critical combat elements. In addition, to airdrop and air landing, these operations include external carry by sling-load. Army rotary wing aircraft conduct airdrop and air landing movement as well as sling-load operations. Sling operations are unique to helicopters with external cargo hooks. The utility and cargo helicopters of the combat aviation brigade supplement ground transportation to help sustain continuous operations. The aviation unit performs air movements on a direct support or general support basis with utility and cargo aircraft. The same general planning considerations that apply to air assaults apply to air movements. (See FM 3-04 for additional information.)

4-322.   The combat aviation brigade has an organic air ambulance medical company, also referred to as the medical company (air ambulance), found in the general support aviation battalion. The air ambulance medical company has a company headquarters and four forward support medical evacuation platoons or forward support medical evacuation teams. Air ambulance aircraft are equipped with medical personnel and equipment enabling the provision of en route care of patients. Air ambulance medical company assets can collocate with AHS support organizations, the aviation task force, the supported BCT, or higher to provide air ambulance support throughout the area of operations. (See ATP 4-02.3 and FM 3-04 for additional information.)

INTELLIGENCE SUPPORT TEAMS

4-323.   The military intelligence company within the BCT distributes intelligence support teams regardless of which element of decisive action (offense, defense, or stability) currently dominates. Dependent on the situation, these teams can be employed down to maneuver company level. The intelligence support teams' mission is to provide basic analytic support, develop basic-level intelligence products, serve as a conduit for effective intelligence communications, and when resourced, manage some information collection programs. Some of those information collection programs include friendly force debriefings, basic document and media exploitation, and biometric and forensic collections (when properly equipped and trained).

4-324.   The BCT can employ anywhere from two intelligence analysts, for example to a maneuver company, or a large team of intelligence analysts as an intelligence support team to support, based on the situation, a maneuver battalion or squadron, BEB, field artillery battalion, BSB, or to further augment the BCT intelligence cell or brigade intelligence support element. A supported maneuver unit or element may subsequently augment the intelligence analysts with nonmilitary intelligence Soldiers to form a larger intelligence support team. When this occurs, it is critical that the appropriate S-2 section thoroughly train all nonmilitary intelligence personnel on intelligence support team activities.

4-325.   The BCT S-3 and S-2 work together with the battalion and squadron S-3s and S-2s to determine the intelligence support teams' task organization, based on the mission variables of METT-TC, using standard command and support relationships as part of the overall BCT intelligence architecture. Planning considerations for the intelligence support team includes the supported unit's—

- Commander's guidance.
- Decisive and shaping operations and main and supporting efforts.

- Specific tasks and the requirement for quick analysis at the point of action or to help manage a unit's information collection effort.
- Ability to provide transportation and logistical support.
- Communications capacity for the intelligence support teams.
- Use of a specific intelligence support team to support or train with a specific unit.

CYBERSPACE ELECTROMAGNETIC ACTIVITIES

4-326. Cyberspace electromagnetic activities is the process of planning, integrating, and synchronizing cyberspace and EW operations. Incorporating CEMA throughout all phases of an operation is key to obtaining and maintaining freedom of maneuver in cyberspace and the EMS while denying the same to enemies and adversaries. CEMA synchronizes capabilities across domains and warfighting functions and maximizes complementary effects in and through cyberspace and the EMS. Intelligence, signal, information operations, cyberspace, space, protection, and fires operations are critical to planning, synchronizing, and executing cyberspace and EW operations. CEMA optimizes cyberspace and EW effects when integrated throughout operations. This section provides an understanding of cyberspace and EW operations, and the roles, relationships, responsibilities, and capabilities within the BCT CEMA cell. (See FM 3-12 and ATP 3-12.3 for additional information.)

> *Note.* In alignment with JP 3-85, this publication introduces the following doctrinal terms: electromagnetic warfare, electromagnetic attack, electromagnetic protection, and electromagnetic support, replacing electronic warfare, electronic attack, electronic protection, and electronic warfare support, respectively. JP 3-85 replaced JP 3-13.1.

CYBERSPACE AND ELECTROMAGNETIC SPECTRUM

4-327. *Cyberspace* is a global domain within the information environment consisting of the interdependent networks of information technology infrastructures and resident data, including the Internet, telecommunications networks, computer systems, and embedded processors and controllers (JP 3-12). Friendly, enemy, adversary, and host-nation networks, communications systems, computers, cellular phone systems, social media, and technical infrastructures are all part of cyberspace. Cyberspace can be described in three layers: physical network layer (geographic land, air, maritime, or space where elements or networks reside), logical network layer (components of the network related to each other in a way abstracted from the network), and cyber-persona layer (digital representations of individuals or entities in cyberspace).

4-328. The *electromagnetic spectrum* is the range of frequencies of electromagnetic radiation from zero to infinity. It is divided into 26 alphabetically designated bands (JP 3-85). Superiority in cyberspace and the EMS to support BCT operations results from effectively synchronizing DODIN operations, offensive cyberspace operations, defensive cyberspace operations, electromagnetic attack, EP, electromagnetic support, and spectrum management operations. Through CEMA, the BCT plans, integrates, and synchronizes these missions, supports and enables the command and control system, and provides an interrelated capability for information and intelligence operations.

> *Note.* The Army plans, integrates, and synchronizes cyberspace operations through CEMA as a continual and unified effort. The continuous planning, integration, and synchronization of cyberspace and EW operations, enabled by spectrum management operations, can produce singular, reinforcing, and complementary effects. Though the employment of cyberspace operations and EW differ because cyberspace operates on wired networks, both operate using the EMS.

CYBERSPACE OPERATIONS

4-329. *Cyberspace operations* are the employment of cyberspace capabilities where the primary purpose is to achieve objectives in or through cyberspace (JP 3-0). Cyberspace operations range from defensive to offensive. These operations establish and maintain secure communications, detect and deter threats in cyberspace to the DODIN, analyze incidents when they occur, react to incidents, and then recover and adapt while supporting Army and joint forces from strategic to tactical levels while simultaneously denying enemy and adversary effective use of cyberspace and the EMS. The Army's contribution to the DODIN is the technical network that encompasses the Army information management systems and information systems that collect, process, store, display, disseminate, and protect information worldwide. Army cyberspace operations provide support to, and receive support from, joint cyberspace operations. The close coordination and mutual support with joint cyberspace operations provides Army commanders and staffs enhanced capabilities for operations.

4-330. A cyberspace capability is a device, computer program, or technique, including any combination of software, firmware, or hardware, designed to create an effect in or through cyberspace. The analysis of mission variables specific to cyberspace operations enables the BCT to integrate and synchronize cyberspace capabilities to support the operation. For cyberspace operations, mission variables provide an integrating framework upon which critical questions can be asked and answered throughout the operations process. The questions may be specific to either the wired portion of cyberspace, the EMS, or both. For example—

- Where can we integrate elements of cyberspace operations to support the BCT's mission?
- What enemy vulnerabilities can be exploited by cyberspace capabilities?
- What are the opportunities and risks associated with the employment of cyberspace operations capabilities when terrain and weather may cause adverse impacts on supporting information technology infrastructures?
- What resources are available (internal and external) to integrate, synchronize, and execute cyberspace operations?
- How can we synchronize offensive and defensive cyberspace operations and related desired effects with the scheme of maneuver within the time available for planning and execution?
- How can we employ cyberspace operations without negative impacts on noncombatants?
- How can an Army cyberspace operations support team support cyberspace operations in the BCT area of operations?

4-331. Cyberspace operations provide a means by which BCT forces can achieve periods or instances of cyberspace superiority to create effects to support the commander's objectives. The employment of cyberspace capabilities tailored to create specific effects is planned, prepared, and executed using existing processes and procedures. However, there are additional processes and procedures that account for the unique nature of cyberspace and the conduct of cyberspace operations to support BCT operations. The BCT commander and staff, along with subordinate commanders and staffs, apply additional measures for determining where, when, and how to use cyberspace effects.

4-332. Commanders and staffs at each echelon will coordinate and collaborate regardless of whether the cyberspace operation is directed from higher headquarters or requested from subordinate units. The BCT's intelligence process, informed by the division and corps intelligence process, provides the necessary analysis and products from which targets are vetted and validated and aimpoints are derived. Because of the BCT IPB, informed by echelons above brigade IPB and joint intelligence preparation of the operational environment, network topologies are developed for enemy, adversary, and host-nation technical networks. In the context of cyberspace and the EMS, network topology are overlays that graphically depict how information flows and resides within the operational area and how the network transports data in and out of the area of interest.

4-333. As part of CEMA, staffs at each echelon perform a key role in target network node analysis. As effects are determined for target and critical network nodes, staffs prepare, submit, and track the cyber effects request format (known as CERF) or electromagnetic attack request format (see paragraph 4-336). The CERF is the format forces use to request effects in and through cyberspace. Effects in cyberspace can support operations in any domain. Execution orders provide authorization to execute cyberspace effects. Support in response to CERFs may be from joint cyberspace forces such as the combat mission teams, from other joint or service capabilities, or from service retained cyberspace forces.

4-334. The CERF will elevate above the corps echelon and integrate into the joint targeting cycle for follow-on processing and approval. The joint task force, combatant command, and U.S. Cyber Command staff play a key role in processing the CERF and coordinating follow-on cyberspace capabilities. (See FM 3-12 for additional information.)

> *Note.* The distinctions between cyberspace and EW capabilities allow each to operate separately and support operations distinctly. However, this also necessitates synchronizing efforts to avoid unintended interference. Any operational requirement specific to electronic transfer of information through the wired portion of cyberspace must use a cyberspace capability for affect. If the portion of cyberspace uses only the EMS as a transport method, then it is an EW capability that can affect it. Any operational requirement to affect an EMS capability not connected to cyberspace must use an EW capability.

ELECTROMAGNETIC WARFARE

4-335. *Electromagnetic warfare* is military action involving the use of electromagnetic and directed energy to control the electromagnetic spectrum or to attack the enemy (JP 3-85). The commander integrates EW capabilities into operations through CEMA. EW capabilities are applied from the air, land, sea, space, and cyberspace by manned, unmanned, attended, or unattended systems. EW capabilities assist the commander in shaping the operational environment to gain an advantage. For example, EW may be used to set favorable conditions for cyberspace operations by stimulating networked sensors, denying wireless networks, or other related actions. Operations in cyberspace and the EMS depend on EW activities maintaining freedom of action in both. (See ATP 3-12.3.) EW consists of three functions, electromagnetic attack, EP, and electromagnetic support.

Electromagnetic Attack

4-336. Electromagnetic attack involves the use of electromagnetic energy, directed energy, or antiradiation weapons to attack personnel, facilities, or equipment with the intent of degrading, neutralizing, or destroying enemy combat capability. Electromagnetic attack is a form of fires. The electronic attack request format and the electromagnetic attack 5-line briefing format are used to request specific electromagnetic attack support and on-call electromagnetic attack support. The staff requests electromagnetic attack effects via normal request processes and provides specific effects requests using the electronic attack request format. The electronic attack request format normally accompanies DD Form 1972. For more information on DD Form 1972 see ATP 3-09.32, figure 1. (See FM 3-12, table D-1 for an example of the electronic attack request format and table D-2 for an example of the electronic attack 5-line briefing.)

> *Note.* Once FM 3-12 is revised, the electromagnetic attack request format and electromagnetic attack 5-line briefing format will replace the electronic attack request format and electronic attack 5-line briefing format, respectively.

4-337. Electromagnetic attack includes—
- Actions taken to prevent or reduce an enemy's effective use of the EMS.
- Employment of weapons that use either electromagnetic or directed energy as their primary destructive mechanism.
- Offensive and defensive activities, including countermeasures.

4-338. Examples of offensive electromagnetic attack include—
- Jamming enemy radar or electronic command and control systems.
- Using antiradiation missiles to suppress enemy air defenses. (Antiradiation weapons use radiated energy emitted from a target, as the mechanism for guidance onto the target.)
- Using electronic deception to confuse enemy intelligence, surveillance, reconnaissance, and acquisition systems.
- Using directed-energy weapons to disable an enemy's equipment or capability.

4-339. Defensive electromagnetic attack uses the EMS to protect personnel, facilities, capabilities, and equipment. Examples include self-protection and other protection measures such as the use of expendables (flares and active decoys), jammers, towed decoys, directed-energy infrared countermeasures, and counter radio-controlled improvised explosive device (IED) systems.

Electromagnetic Protection

4-340. EP involves the actions taken to protect personnel, facilities, and equipment from any effects of friendly or enemy use of the EMS that degrade, neutralize, or destroy friendly combat capability. For example, EP includes actions taken by the commander to ensure friendly use of the EMS, such as frequency agility in a radio or variable pulse repetition frequency in radar. The commander avoids confusing EP with self-protection. Both defensive electromagnetic attack and EP protect personnel, facilities, capabilities, and equipment. However, EP protects from the effects of electromagnetic attack (friendly and enemy) and electromagnetic interference, while defensive electromagnetic attack primarily protects against lethal attacks by denying enemy use of the EMS to guide or trigger weapons.

Commander's Electromagnetic Protection Responsibilities

4-341. EP is a command responsibility. The more emphasis the commander places on EP, the greater the benefits, in terms of casualty reduction and combat survivability, in a hostile environment or degraded information environment. The commander ensures support and consolidation areas on and practices sound EP techniques and procedures. The commander continually measures the effectiveness of EP techniques and procedures used within the BCT throughout the operations process. Commander EP responsibilities are—

- Review all information on jamming and deception reports, and assess the effectiveness of defensive EP.
- Ensure the BCT S-6 and S-2, in coordination with the EWO, report and properly analyze all encounters of electromagnetic interference, deception, and jamming.
- Analyze the impact of enemy efforts to disrupt or destroy friendly communications systems on friendly operation plans.
- Ensure the BCT staff exercises communications security (see ATP 6-02.75) techniques daily. Subordinate units should—
  - Change network call signs and frequencies often (in accordance with the signal operating instructions).
  - Use approved encryption systems, codes, and authentication systems.
  - Control emissions.
  - Make EP equipment requirements known through quick reaction capabilities designed to expedite procedure for solving, research, development, procurement, testing, evaluation, installations modification, and logistics problems as they pertain to EW.
  - Ensure quick repair of radios with mechanical or electrical faults; this is one way to reduce radio-distinguishing characteristics.
  - Practice network discipline.

Staff Electromagnetic Protection Responsibilities

4-342. The BCT staff assists the commander in accomplishing EP requirements. Specifically, the staff responds immediately to the commander and subordinate units. The staff—

- Keeps the commander informed.
- Reduces the time to control, integrate, and coordinate operations.
- Reduces the chance for error.

4-343. The BCT staff provides information, furnishes estimates, and provides recommendations to the commander. Specific staff officer responsibilities include the—

- S-2. Advise the commander of enemy capabilities that could be used to deny the unit effective use of the EMS. Keep the commander informed of the BCT's signal security posture.

- S-3. Exercise staff responsibility for EP. Include electromagnetic support and electromagnetic attack considerations throughout the operations process and evaluate EP techniques and procedures employed. Ensure EP training is included in all unit-training programs, the MDMP, and troop leading procedures during operations.
- S-6. Exercise staff responsibility for signal security and support EP. The S-6 in coordination with the EWO—
  - Prepares and conducts the unit EP training program.
  - Ensures alternate means of communications for those systems most vulnerable to enemy jamming.
  - Ensures distribution of available communications security equipment to those systems most vulnerable to enemy information gathering activities.
  - Ensures measures are taken to protect critical friendly frequencies from intentional and unintentional electromagnetic interference.

Signal Security

4-344. EP and signal security are closely related; they are defensive arts based on the same principle. If adversaries and enemies do not have access to the EEFI, they are much less effective. The BCT's goal of practicing sound EP techniques is to ensure the continued effective use of the EMS. The BCT's goal of signal security is to ensure the enemy cannot exploit the friendly use of the EMS for communication. Signal security techniques are designed to give the commander confidence in the security of BCT transmissions. Signal security and EP are planned by the BCT based on the enemy's ability to gather intelligence and degrade friendly communications systems. (See ATP 6-02.53 for additional information.)

Communications Planning Considerations

4-345. The BCT staff, specifically the S-6 in coordination with the S-2, S-3, and EWO, assesses threats to friendly communications during the communications planning process. Planning counters the enemy's attempts to take advantage of the vulnerabilities of friendly communications systems. Ultimately, the commander, subordinate commanders, and staff planners and radio and network operators are responsible for the security and continued operation of all command and control systems.

4-346. When conducting communications planning, the S-6 uses spectrum management tools to assist in EMS planning and to define and support requirements. The S-6 coordinates all frequency use before any emitter is activated to mitigate or eliminate electromagnetic interference or other negligible effects and considers the following when conducting EMS management planning:

- Transmitter and receiver locations.
- Antenna technical parameters and characteristics.
- Number of frequencies desired and separation requirements.
- Nature of the operation (fixed, mobile land, mobile aeronautical, and over water or maritime).
- Physical effects of the operational environment (ground and soil type, humidity, and topology).
- All EMS-dependent equipment to be employed to include emitters, sensors, and unmanned aerial sensors.
- Start and end dates for use.

4-347. The PACE plan is a communication plan that exists for a specific mission or task, not a specific unit, as the plan considers both intra- and inter-unit sharing of information. A CP establishes a PACE plan with each unit it is required to maintain communications with during a mission. The PACE plan designates the order in which an element will move through available communications systems until contact can be established with the desired distant element. The S-6 develops a PACE plan for each phase of an operation to ensure that the commander can maintain command and control of the formation. The plan reflects the training, equipment status, and true capabilities of the formation. The BCT S-6 evaluates its communication requirements with the subordinate units and their S-6 to develop an effective plan. Upon receipt of an order, the S-6 evaluates the PACE plan for two key elements as follows:

- Does the BCT have the assets to execute the plan?

- How can subordinate units' nest with the plan when they develop their own plan?

4-348. Accurate PACE plans are crucial to the commander's situational awareness. A subordinate unit (considerations include those for host nation or multinational forces) that is untrained on a particular communication system or lacks all of the subcomponents to make the system mission capable, does not ensure continuity of command and control by including the communication system in the PACE plan. The commander's ability to exercise command and control during an operation can suffer due to communications systems that are in transit or otherwise unavailable. If the BCT or a subordinate unit does not have four viable methods of communications, it is appropriate to issue a PACE plan that may only have two or three systems listed. If the unit cannot execute the full PACE plan to its higher command, it must inform the issuing headquarters with an assessment of shortfalls, gaps, and possible mitigations as part of the mission analysis process during the MDMP. During COA development, the S-6 nests the subordinate unit's plan with the BCT's plan whenever practical. This aids in maintaining continuity of effort. (See FM 6-02 for additional information.)

Terrain Analysis

4-349. The BCT S-6 analyzes the terrain and determines the method(s) to make the geometry of the operations work support the commander's plan. Adhering rigidly to standard CP deployment makes it easier for the enemy to use the direction finder and aim jamming equipment. Deploying units and communications systems perpendicular to the forward line of own troops enhance the enemy's ability to intercept communication by aiming transmissions in the enemy's direction. When possible, install terrestrial line-of-sight communications parallel to the forward line of own troops. This supports keeping the primary strength of U.S. transmissions in friendly terrain.

4-350. Single-channel tactical satellite systems reduce friendly CP vulnerability to enemy direction efforts. Tactical satellite communication (see ATP 6-02.54) systems are relieved of this constraint because of their inherent resistance to enemy direction finder efforts. When possible, utilize terrain features to mask friendly communication from enemy positions. This may require moving headquarters elements farther forward and using more jump or TACs to ensure the commander can continue to direct units effectively.

4-351. Location of CPs requires carefully planning as CP locations generally determine antenna locations. The proper installation and positioning of antennas around CPs is critical. Disperse and position antennas and emitters at the maximum remote distance and terrain dependent from the CP to ensure that not all of a unit's transmissions are coming from one central location system design.

4-352. Establish alternate routes of communication when designing communications systems. This involves establishing sufficient communications paths to ensure that the loss of one or more routes does not seriously degrade the overall system. The commander establishes the priorities of critical communications links. Provide high priority links with the greatest number of alternate routes. Alternate routes enable friendly units to continue to communicate despite the enemy's efforts to deny them the use of their communications systems. Alternate routes can also be used to transmit false messages and orders on the route that is experiencing electromagnetic interference, while they transmit actual messages and orders through another route or means. A positive benefit of continuing to operate in a degraded system is that the problematic degraded system causes the enemy to waste assets used to impair friendly communication elsewhere. Three routing concepts, or some permutation of them, can be used in communications as follows:

- Straight-line system. Provides no alternate routes of communications.
- Circular system. Provides one alternate route of communications.
- Grid system. Provides as many alternate routes of communications as can be practically planned.

4-353. Avoid establishing a pattern of communication. Enemy intelligence analysts may be able to extract information from the pattern, and the text, of friendly transmissions. If easily identifiable patterns of friendly communication are established, the enemy can gain valuable information.

4-354. The number of friendly transmissions tends to increase or decrease according to the type of tactical operation being executed. Execute this deceptive communication traffic by using false peaks, or traffic leveling. Utilize false peaks to prevent the enemy from connecting an increase of communications with a tactical operation. Transmission increases on a random schedule create false peaks. Tactically accomplish traffic leveling by designing messages to transmit when there is a decrease in transmission traffic. Traffic leveling keeps the transmission traffic constant. Coordinate messages transmitted for traffic leveling or false peaks to avoid operations security violations, electromagnetic interference, and confusion among friendly equipment operators.

4-355.   During operations, dismounted tactical unit area coverage and distance extension is a major concern to the commander. Communications inside buildings or over urban terrain is a challenge. For these conditions, the multiband inter/intra team radio system provides a "back-to-back" (two radios) retransmission (known as RETRANS) capability for communications security and plain text modes. Beside two radios, the only hardware required for RETRANS is a small cable kit and some electronic filters. When configured for RETRANS operations, a true digital repeater form. Since the radios repeat the transmitted digits and since the radios do not have to have any communications security keys loaded in them, the radios do not degrade signal quality. (See ATP 6-02.72 for additional information.)

4-356.   Automated Communications Engineering Software equipment and subsequent signal operating instructions development resolve many problems concerning communications patterns; they allow users to change frequencies often, and at random. This is an important aspect of confusing enemy traffic analysts. Enemy traffic analysts are confused when frequencies, network call signs, locations, and operators are often changed. Communications procedures require flexibility to avoid establishing communications patterns. (See ATP 6-02.53 for additional information.)

Control of Electromagnetic Emissions

4-357.   The control of electromagnetic emissions is essential to successful defense against the enemy's attempts to destroy or disrupt the BCT's communications. Emission control is the selective and controlled use of electromagnetic, acoustic, or other emitters to optimize command and control capabilities while minimizing, for operations security. When operating radios, the BCT exercises emission control at all times within all echelons and only transmit when needed to accomplish the mission. Enemy intelligence analysts look for patterns they can turn into usable information. Inactive friendly transmitters do not provide the enemy with useable intelligence. Emission control can be total; for example, the commander may direct radio silence whenever desired. *Radio silence* is the status on a radio network in which all stations are directed to continuously monitor without transmitting, except under established criteria (ATP 6-02.53).

4-358.   Radio operators keep transmissions to a minimum (20 seconds absolute maximum, 15 seconds maximum preferred) and transmit only mission-critical information. Good emission control makes the use of communications equipment appear random and is therefore consistent with good EP practices. This technique alone will not eliminate the enemy's ability to find a friendly transmitter; but when combined with other EP techniques, it makes locating a transmitter more difficult.

Replacement and Concealment

4-359.   Replacement involves establishing alternate routes and means of doing what the commander requires. Frequency modulation voice communications are the most critical communications used by the commander during enemy engagements and require reserving critical systems for critical operations. The enemy should not have access to information about friendly critical systems until the information is useless.

4-360.   The BCT utilizes alternate means of communication before enemy engagements. This ensures the enemy cannot establish a database to destroy primary means of communication. If the primary means degrades, replace primary systems with alternate means of communication. Replacements require preplanning and careful coordination; if not, compromise of the alternate means of communication occurs and is no longer useful as the primary means of communication. Users of communications equipment require knowledge of how and when to use the primary and alternate means of communication. This planning and knowledge ensure the most efficient use of communications systems.

4-361.   The BCT commander and subordinate commanders ensure effective employment of all communications equipment, despite the enemy's concerted efforts to degrade friendly communication to the enemy's tactical advantage. Operation plans should include provisions to conceal communications personnel, equipment, and transmissions. As it is difficult to conceal most communications systems, installing antennas as low as possible on the backside of terrain features and behind manmade obstacles help conceal communications equipment while still permitting communication.

Training and Procedures Countering Enemy Electromagnetic Attack

4-362.   EP includes the application of training and procedures for countering enemy electromagnetic attack. Once the threat and vulnerability of friendly electronic equipment to enemy electromagnetic attack are identified, the commander takes appropriate actions to safeguard friendly combat capability from exploitation and attack. EP measures minimize the enemy's ability to conduct electromagnetic support and electromagnetic attack operations successfully against the BCT. To protect friendly combat capabilities, units—

- Regularly brief friendly force personnel on the EW threat.
- Ensure that they safeguard electronic system capabilities during exercises, workups, and predeployment training.
- Coordinate and deconflict EMS usage.
- Provide training during routine home station planning and training activities on appropriate EP active and passive measures under normal conditions, conditions of threat electromagnetic attack, or otherwise degraded networks and systems.

Electromagnetic Support

4-363.   Electromagnetic support is a division of EW involving actions tasked by or under direct control of the BCT commander to search for, intercept, identify, and locate or localize sources of intentional and unintentional radiated electromagnetic energy for the purpose of immediate threat recognition, targeting, planning, and conduct of future operations. Electromagnetic support assists the BCT in identifying the electromagnetic vulnerability of an enemy or adversary's electronic equipment and systems. The commander takes advantage of these vulnerabilities through EW operations.

4-364.   Electromagnetic support systems are a source of information for immediate decisions involving electromagnetic attack, EP, avoidance, targeting, and other tactical employment of forces. Electromagnetic support systems collect data and produce information to—

- Corroborate other sources of information or intelligence.
- Conduct or direct electromagnetic attack operations.
- Initiate self-protection measures.
- Task weapons systems.
- Support EP efforts.
- Create or update EW databases.
- Support information-related capabilities.

CYBERSPACE ELECTROMAGNETIC ACTIVITIES CELL

4-365.   The CEMA cell, within the BCT staff, synchronizes cyberspace and EW operations for effective collaboration across staff elements. It includes the EWO (who is also the cyberspace planner), the spectrum manager, the EW technician, and EW noncommissioned officers. The section participates in the planning and targeting process, leads the CEMA working group to support the MDMP, as the cyberspace planner requests nonorganic resource effects. CEMA effects must be briefed and rehearsed, to include a shared understanding of CEMA capabilities, collaborating CEMA targets across all warfighting functions (while validating them for the targeting process), and integrating CEMA into schemes of maneuver and phases of the operation. Listed below in paragraphs 4-366 through 4-368 is each member of the CEMA section with key duties and responsibilities (for a full list, see FM 3-12).

4-366.   The EWO cyberspace planner plans, integrates, synchronizes, and assesses cyberspace and EW operations as the commanders designated staff officer. The EWO cyberspace planner—

- In coordination with the appropriate legal support, advises the commander on effects in cyberspace (including associated rules of engagement, impacts, and constraints).
- Nominates offensive cyberspace operations and EW targets for approval from the fire support coordinator and commander.

- Advises the commander on how cyberspace and EW effects can affect the operational environment.
- Provides recommendations on CCIRs.
- Assists the S-2 during IPB.
- Provides information requirements to support planning, integration, and synchronization of cyberspace and EW operations.

4-367.   The EW technician or noncommissioned officer plans, coordinates, and supports EW as part of CEMA. EW personnel—

- Provide information collection requirements to the S-2 to support the assessment, planning, preparation, and execution of EW.
- Support the fire support coordinator to ensure the integration of electromagnetic attack with all other effects.
- Plan and coordinate EW operations across functional and integrating cells.
- Maintain a current assessment of available EW resources.

4-368.   The CEMA section spectrum manager's role is to plan and synchronize EP, integrating and synchronizing operational spectrum considerations across cyberspace and EW operations, collaborating with the S-6. The CEMA spectrum manager—

- Leads, develops, and synchronizes the EW-EP plan by assessing EA effects on friendly force emitters.
- Mitigates harmful impact of EA on friendly forces through coordination with higher and subordinate units.
- Synchronizes cyberspace operations to protect radio frequency enabled transport layers.
- Collaborates with staff, subordinate, and senior organizations to identify unit emitters for inclusion on the joint restricted frequency list.

Chapter 5

Reconnaissance and Security

Reconnaissance and security is essential to all operations. Brigade combat teams (BCTs) develop and sustain situational understanding to defeat the enemy. Reconnaissance and security forces within the BCT provide flexibility, adaptability, and depth to the maneuver commander's plan by synchronizing and integrating combined arms teams based on a relevant understanding of the situation. BCT commanders understand the tactical, human, and political environment, visualize operations, develop the situation, and identify or create options to seize, retain, and exploit the initiative through reconnaissance and security. Reconnaissance and security forces protect the force being protected from surprise, reduce the unknowns in any situation, and answer the commander's critical information requirements (CCIRs) to enable the commander to make decisions, and direct forces to achieve the mission.

SECTION I – RECONNAISSANCE AND SECURITY FORCES

5-1.   Reconnaissance and security forces, through effective information collection, help develop and sustain the BCT's understanding of the operational environment within its area of operation to defeat adaptive and determined enemies and set conditions to consolidate tactical gains. Forces conducting security operations normally orient on the force or facility being protected, while forces conducting reconnaissance normally orient on the enemy and terrain. This section addresses the planning and preparation for, and the employment of reconnaissance and security forces within the BCT. It addresses the commander's guidance, information collection, air-ground operations, and sustainment for reconnaissance and security.

RECONNAISSANCE AND SECURITY OPERATIONS

5-2.   BCTs conduct combined arms reconnaissance and security operations utilizing their Cavalry squadron and organic maneuver battalions. By employing reconnaissance and security forces, in the context of the mission variables of mission, enemy, terrain and weather, troops and support available, time available, civil considerations (METT-TC), the BCT commander can fight, collect, and exploit information and develop the situation against a broad range of threats. The resulting tactical effects of these combined arms provide the BCT commander with tactical depth, freedom to maneuver, and flexibility. As the eyes and ears of the BCT commander, reconnaissance and security forces can also better enable decision-making by confirming or denying the CCIRs, as well as identify and develop opportunities to seize, retain, and exploit the initiative and consolidate gains. Reconnaissance and security operations enable the BCT commander to—

- Understand the tactical, human, and political dynamics within an area of operations.
- Visualize operations in the context of mission variables.
- Achieve tactical depth.
- Develop the situation through action in close contact with enemy and civilian populations.
- Execute decisive operations with higher degrees of flexibility, adaptability, synchronization, and integration.
- Identify or create options to seize, retain, and exploit the initiative and consolidate gains.

5-3.   The BCT commander and maneuver battalion commanders use reconnaissance and security forces, the BCT's Cavalry squadron and the maneuver battalions' scout platoon, respectively, to develop the situation under conditions of uncertainty in close contact with the enemy and civilian populations. Additional collection enablers internal to the BCT (although not inclusive) can include maneuver battalions, engineers, the target acquisition radar platoon, and the tactical unmanned aircraft system (known as TUAS) platoon; and signals intelligence, geospatial intelligence, and human intelligence (HUMINT) assets from the military intelligence company. Reconnaissance and security forces enable the BCT's fight to gain a position of relative advantage over the enemy and to strike the enemy in a time, manner, and place where the enemy is not prepared. The BCT commander then prevents the enemy's recovery by rapidly following up with a series of actions that destroy enemy capabilities, seize decisive terrain, protect populations and critical infrastructure, and degrade the integrity of the enemy force, and then defeat or destroy the force before the enemy can recover.

5-4.   Reconnaissance and security operations are essential in providing the BCT commander with the freedom of action required to conduct decisive action (offensive and defensive operations and stability operations tasks). Knowing when, where, and how to conduct decisive action, as well as protecting fleeting opportunities to do so, is a result of effective reconnaissance and security operations. Additionally, BCT reconnaissance and security forces accomplish a secondary mission to defeat enemy reconnaissance and surveillance efforts through counterreconnaissance. *Counterreconnaissance* is a tactical mission task that encompasses all measures taken by a commander to counter enemy reconnaissance and surveillance efforts. Counterreconnaissance is not a distinct mission, but a component of all forms of security operations (FM 3-90-1). Counterreconnaissance prevents hostile observation of a force or area and is an element of most local security measures. Counterreconnaissance involves both active and passive elements and includes combat action to destroy or repel enemy reconnaissance units and surveillance assets.

5-5.   During decisive action, reconnaissance and security forces must provide information for the BCT to develop an accurate understanding of the tactical situation. Effective reconnaissance and security operations assist the BCT to ease transitions and mitigate information gaps between units. In other words, if the BCT is to conduct operations characterized by flexibility, lethality, adaptability, depth, and synchronization, then the BCT commander must have the combat information on the enemy, the terrain, and indigenous populations to do so. With this information, the commander can maneuver to positions of relative advantage, and apply effective firepower against enemies to accomplish the mission. Effective reconnaissance and security operations allow the commander to direct friendly strengths against enemy weaknesses, while simultaneously protecting friendly forces, infrastructure, and populations. In the end, reconnaissance and security operations allows the commander to confirm information requirements, identify or create options, and employ the most appropriate forms of maneuver to defeat enemy forces.

5-6.   The BCT commander and staff identify information gaps during the military decision-making process (MDMP) and continuously assess, adapt, add, and delete requirements throughout the operation. During the process, the BCT staff identifies specified, implied, and essential tasks necessary for mission success during mission analysis, while reviewing available assets and when identifying resource and information shortfalls. During mission analysis, the staff identifies certain critical facts and assumptions that aid in the development of initial CCIRs. The CCIRs include priority intelligence requirements and friendly force information requirements. CCIRs and essential element of friendly information (EEFI) facilitate timely decision-making during the intelligence preparation of the battlefield (IPB) process and the MDMP, as well as the targeting, risk management (RM), and operations and intelligence processes.

5-7.   Priority intelligence requirements are information requirements necessary to understand an adversary or enemy and the operational environment. They identify information about the threat, terrain, weather, and civil considerations that the commander considers most important. Priority intelligence requirements have an impact upon future decisions. Friendly force information requirements identify information about friendly forces and supporting capabilities and information that affects future courses of action (COAs) and decisions from a friendly perspective. The BCT staff assigns tasks to prioritize, manage, and develop collection of information requirements based upon identified information requirements leading to future decisions. As the staff identifies requirements necessary for successful execution, they recommend and assign tasks for reconnaissance forces so the commander can make decisions and capitalize on opportunities.

5-8.   Surveillance and intelligence operations (two of four primary means of information collection) conducted to satisfy validated information requirements (normally specified in the information collection plan [see paragraph 4-156]) enable reconnaissance and security efforts within the BCT. These requirements, assigned to surveillance and intelligence operations collection assets, drive intelligence production to support the commander's situational awareness and understanding. Surveillance assets (see paragraph 5-64) assigned or distributed to the BCT, for example, can monitor and collect information on geographic areas well beyond the BCT area of operations for early warning of threat actions. Intelligence operations (see paragraph 5-90), tasks undertaken by military intelligence units and Soldiers (within and external to the BCT), collect information about the intent, activities, and capabilities of threats and about relevant aspects of the operational environment to support the BCT commander's decision-making.

COMMANDER'S RECONNAISSANCE AND SECURITY GUIDANCE

5-9.   The BCT commander's reconnaissance guidance and security guidance gives a clear understanding of the reconnaissance and security organization's task, purpose, and end state, specifically the BCT's Cavalry squadron. Reconnaissance guidance and security guidance explains tempo, the level of detail, and covertness required, the reconnaissance objective, and guidelines for engagement, disengagement, bypass criteria, and displacement criteria. The commander develops reconnaissance guidance and security guidance based on the BCT mission, commander's intent, timeline, and enemy to satisfy information requirements and identify opportunities to seize, retain, and exploit the initiative and consolidate gains. The BCT commander specifies different reconnaissance guidance and security guidance for each phase of an operation and adjusts the components of the guidance when appropriate. The commander's reconnaissance guidance and security guidance consists of the following components:

- Tempo, level of detail, and covertness required.
- Focus (reconnaissance objective and security objective).
- Engagement, disengagement, and bypass criteria.
- Displacement criteria.

TEMPO, LEVEL OF DETAIL, COVERTNESS REQUIRED

5-10. Tempo, the level of detail, and covertness required of the Cavalry organization to accomplish reconnaissance or security operations are described in four ways: rapid, deliberate, stealthy, and forceful. (See figure 5-1 on page 5-4.) *Tempo* is the relative speed and rhythm of military operations over time with respect to the enemy (ADP 3-0). Rapid and deliberate are levels of detail that are mutually exclusive in all cases, as one cannot be rapid and deliberate at the same time. However, Cavalry organizations can oscillate between the two from phase to phase or even within sub phases of an operation. Stealthy and forceful indicate mutually exclusive levels of covertness. Commanders choose the appropriate type of reconnaissance or security operations, balanced with the mission variables of METT-TC, to complete the mission.

5-11. Rapid action dictates that the level of detail for reconnaissance and security operations is limited to a prescribed list of critical tasks or priority intelligence requirements. Rapid action is appropriate when time is of the essence and only a limited number of critical tasks or information requirements are necessary to accomplish the mission.

5-12. Deliberate action implies that the organization must accomplish all critical tasks to ensure mission success. Deliberate action allows the organization more time to answer all information requirements. Detailed and thorough reconnaissance and security operations require time intensive, comprehensive, and meticulous mounted and dismounted efforts to observe reconnaissance objectives and develop the situation.

5-13. Stealthy action emphasizes avoiding detection and engagement dictated by restrictive engagement criteria. Stealthy reconnaissance and security operations typically take more time than aggressive reconnaissance and security operations. Stealthy reconnaissance or security operations utilize dismounted scouts to take maximum advantage of cover and concealment to reduce signatures that lead to compromise. The BCT commander uses stealthy reconnaissance or security operations when time is available, detailed reconnaissance and stealth is required, enemy forces are likely to be in a specific area, when dismounted scouts encounter danger areas, and when restrictive terrain limits effectiveness of mounted reconnaissance or security operations.

5-14. Forceful action develops the situation by employing reconnaissance and security forces, technical means, and direct and indirect-fire systems that can move rapidly to develop the situation. Forceful reconnaissance and security operations require firepower, aggressive exploitation of action on contact, security, and training to survive and accomplish the mission. Forceful reconnaissance and security operations are appropriate when time is limited, detailed reconnaissance is not required, terrain is open, environmental conditions allow for mounted movement, and when dismounted movement cannot complete the mission within existing time constraints. Forceful reconnaissance and security operations do not preclude the judicious use of dismounted movement to reduce risk as long as the organization maintains the tempo of the operation.

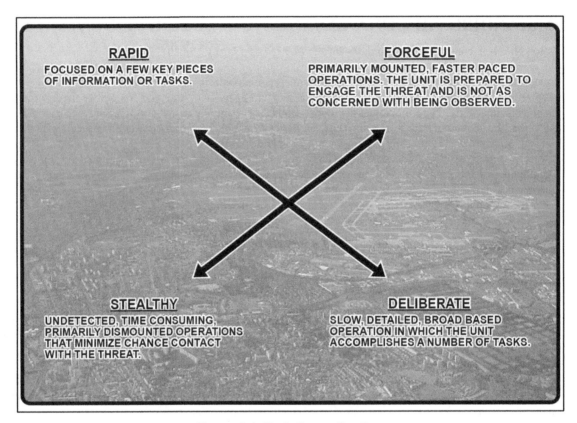

Figure 5-1. Variations of action

FOCUS

5-15. Focus for reconnaissance and security serves as a guide to indicate the tasks associated with the type of reconnaissance and security operation conducted although not a set checklist. Focus does not limit the reconnaissance and security forces' ability to collect on and report all information but instead allows commanders to prioritize tasks to accomplish, and the assets used to accomplish them. Commanders help refine the focus of reconnaissance and security by describing the reconnaissance objective and security objective.

Reconnaissance Objective

5-16. *Reconnaissance objective* is a terrain feature, geographic area, enemy force, adversary, or other mission or operational variable about which the commander wants to obtain additional information (ADP 3-90). The reconnaissance objective clarifies the intent of the reconnaissance effort by stating the most important result of the reconnaissance effort. A reconnaissance objective focuses the Cavalry (or other maneuver organization) organization's area of emphasis. Four categories form the area of emphasis—threat, infrastructure, terrain and weather effects, and civil considerations. The commander often assigns more than one category to Cavalry units even though the commander recognizes that a broad focus in multiple areas dilutes the Cavalry organization's ability to collect information. Narrowing the scope of operations helps to focus the Cavalry organization to acquire the information to develop the situation for future operations.

5-17. Threat focused reconnaissance prescribes the identification of the enemy's locations, composition, disposition, and strength within an assigned area of operations. Infrastructure dictates gathering information pertinent to the understating of the operational environment. Terrain and weather effects confirm step two (describe environmental effects on operations) of the IPB process and is accomplished by analyzing and determining the influences that the five military aspects of terrain and the military aspects of weather will have on future operations. The five military aspects of terrain are observation and fields of fire, avenues of approach, key terrain, obstacles, and cover and concealment (OAKOC). The military aspects of weather include visibility, wind, precipitation, cloud cover, temperature, humidity, and atmospheric pressure (as required). Civil considerations reflect the influence of manmade infrastructure, civilian institutions, and attitudes and activities of the civilian leaders, populations, and organizations within the operational environment on the conduct of military operations. The commander and staff analyze civil considerations in terms of the following characteristics: areas, structures, capabilities, organizations, people, and events, expressed in the memory aid (ASCOPE). (See ATP 2-01.3 and ATP 3-34.80 for additional information.)

Security Objective

5-18. Similar to a reconnaissance objective, the security objective clarifies the intent of the security effort by stating the most important result of the security effort. A security objective focuses the Cavalry (or other maneuver organization) organization's area of emphasis. Three categories form the area of emphasis—the protected force, activity, and facility. As in reconnaissance, the commander often assigns more than one category to Cavalry units even though the commander recognizes that a broad focus in multiple areas dilutes the Cavalry organization's ability to conduct the security mission(s). The BCT commander narrows the scope of the security mission(s) by providing clear security guidance that offers freedom of action and direction to focus the Cavalry organization's area of emphasis. Focusing the Cavalry organization's area of emphasis helps to ensure it can accomplish stated objectives within the required timeframe. The commander does this by providing a clear understanding of the Cavalry organization's task, purpose, and end state, and the protection requirements of the security mission. The ultimate goal of any security mission is to provide early and accurate warning of enemy operations, to provide the force being protected with time and maneuver space within which to react to the enemy, and to develop the situation to allow the commander to effectively use the protected force.

ENGAGEMENT, DISENGAGEMENT, AND BYPASS CRITERIA

5-19. During reconnaissance and security operations, engagement, disengagement, and bypass criteria prescribe events and conditions that require initiation of engagement with the enemy, disengagement from enemy contact, or bypassing the enemy. Engagement, disengagement, and bypass criteria outlines parameters for Cavalry units to engage the enemy with direct or indirect fire based on the level of threat, levels of risk, required levels of covertness, and preservation of the force.

5-20. *Engagement criteria* are protocols that specify those circumstances for initiating engagement with an enemy force (FM 3-90-1). Regardless of engagement criteria, it is not enough to state in the operation order that engagement criterion is either restrictive or permissive; the operation order must describe conditions relative to the enemy situation to ensure complete understanding.

5-21. *Disengage* is a tactical mission task where a commander has the unit break contact with the enemy to allow the conduct of another mission or to avoid decisive engagement (FM 3-90-1). Disengagement criteria describe the events and conditions that necessitate disengaging from enemy contact or temporarily breaking enemy contact to preserve the force. Compromised Cavalry units or scouts who find themselves in a position of disadvantage provide no information or security value and should temporarily break contact to re-establish observation as soon as the tactical situation permits. As with engagement criteria, specific conditions are described that require disengagement.

5-22. *Bypass criteria* are measures established by higher echelon headquarters that specify the conditions and size under which enemy units and contact may be avoided (ADP 3-90). Bypass criteria describes the events and conditions that necessitate maneuver around an obstacle, position, or enemy force to maintain the momentum of the operation. Bypass criteria describes the conditions that necessitate maneuver so as not to decisively engage or fall below a certain combat strength when deliberately avoiding combat with an enemy force.

DISPLACEMENT CRITERIA

5-23. Displacement criteria define triggers for planned withdrawals, passage of lines, or reconnaissance handovers (battle handover for security operations) between units. As with engagement, disengagement, and bypass criteria, the conditions and parameters set in displacement criteria integrate the BCT commander's intent with tactical feasibility. Conditions and parameters are event-driven, time-driven, or enemy-driven. Displacement criteria conditions and parameters are key rehearsal events due to the criticality of identifying the triggers to anticipate during the BCT's mission. An example of event-driven conditions and parameters are associated priority intelligence requirements being met, enemy contact not expected in the area, and observed named area of interest (NAI) or avenue of approach denied to the enemy. Time-driven conditions and parameters ensure the time triggers are met (for example, latest time information is of value). An observation post compromised by threat or local civilian contact is a threat-driven condition. Failure to dictate conditions and parameters of displacement, nested within the higher scheme of maneuver, results in mission failure.

COMBINED ARMS, AIR-GROUND RECONNAISSANCE AND SECURITY

5-24. The commander uses information and intelligence from combined arms, air-ground reconnaissance (and when available air and ground surveillance and military intelligence assets internal and external to the BCT) and security to reduce uncertainty and facilitate rapid decision-making. Reconnaissance collects information so the commander can understand the situation, visualize the battlefield, and shape decisions. Security protects the force, provide reaction time, and maneuver space to enable decisions and prudent use of combat power. The commander uses air-ground reconnaissance and security to answer priority intelligence requirements to fill information gaps, mitigate risk, prioritize tasks, and allocate resources. Lastly, air-ground reconnaissance and security create advantageous conditions for future operations that seize, retain, and exploit the initiative.

5-25. Army attack reconnaissance aircraft, both manned and unmanned, provide direct fire, observation, and rapid movement during reconnaissance and security and counterreconnaissance. Army attack reconnaissance units conduct Army aviation attacks to destroy high-value targets (HVTs) and HPTs within a target area of interest (TAI). Army aviation attack reconnaissance aircraft can provide additional observation to assist reconnaissance and security forces, specifically the Cavalry squadron in maintaining contact. Utility and cargo helicopters support reconnaissance and security operations through air movements, (including casualty evacuation and emergency resupply operations) depending on the enemy's air defense threat.

> *Note.* The same general planning considerations that apply to air assaults apply to air movements. (See FM 3-99 for additional information.)

5-26. Air-ground operations require detailed planning of synchronized timelines, aviation task and purpose, and airspace management (see ATP 3-04.1). Shared graphics ensure common operational language, reduce fratricide risk, reduce the chance of an accidental compromise of a ground unit, and increase the effectiveness of mixing collection sources. Development of detailed mission statements for the supporting aviation is essential for aviation commanders and staffs to employ the right platforms and munitions. Understanding the threat and the commander's intent and desired effects drives the aviation units' task organization of air elements and selection of weapon systems. Aircraft fuel consumption rates, forward arming and refueling, and fighter management can limit aircraft availability.

RECONNAISSANCE AND SECURITY FORCE SUSTAINMENT

5-27. Sustainment for reconnaissance and security forces requires deliberate planning. Logistics units supporting reconnaissance and security operations must contend with long lines of communication, dispersed forces, poor trafficability, and contested terrain. Planners must consider protection requirements to protect sustainment units against bypassed enemy forces and the effects of extended lines of communications. Reconnaissance and security force sustainment must be rehearsed and war gamed.

5-28. Reconnaissance and security forces often require a basic load in excess of the typical three days of supply configuration due to mission requirements. Supplies can be pre-positioned in collocated trains with a maneuver battalion's echelon support. In restricted terrain, the most important commodities are likely class I (subsistence-priority to water) and class III (petroleum, oils, and lubricants [POL]) and depending on the enemy situation and terrain class V (ammunition). Possible examples of restricted terrain requiring an increased basic load include moderate-to-steep slopes or moderately-to-densely spaced obstacles, swamps, and rugged terrain and operation in urban terrain.

5-29. Forces conducting reconnaissance generally have a greater requirement for class III and class V for indirect-fire assets and antiarmor systems. Similar to offensive operations, reconnaissance requires refuel on the move (known as ROM). Security forces have a greater reliance on class V and reduced requirements for class III during security operations. Reconnaissance and security forces generally do not have large barrier class IV (construction and barrier materials) requirements. Possible exceptions for security forces are during the execution of long-term guard missions or during a defensive cover.

5-30. When units task organize, particularly from outside the BCT, planners must incorporate and rehearse supporting logistics assets. The nature of reconnaissance and security operations stresses medical evacuation and requires wargaming and close coordination with external assets. Casualty evacuation (see ATP 4-25.13) planning and requirements for reconnaissance and security forces focuses on ground movement assets and must balance with survivability and stealth. Planners plan for and utilize aviation casualty backhaul as aircraft become available. (See chapter 9 for additional information.)

SECTION II – RECONNAISSANCE

5-31. *Reconnaissance* is a mission undertaken to obtain, by visual observation or other detection methods, information about the activities and resources of an enemy or adversary, or to secure data concerning the meteorological, hydrographic, or geographic characteristics of a particular area (JP 2-0). Reconnaissance employs many tactics, techniques, and procedures throughout the course of an operation, one of which may include an extended period of surveillance.

PURPOSE OF RECONNAISSANCE

5-32. The purpose of reconnaissance is to gather information so the commander can create plans, make decisions, and issue orders. The BCT commander's focus for reconnaissance usually falls in three general areas: CCIRs, targeting, and voids in information. The BCT staff, primarily the S-2 in coordination with the S-3, identifies gaps in available intelligence based on the initial IPB (see ATP 2-19.4, appendix G) and the situationally dependent CCIRs. The BCT's reconnaissance effort and the IPB process are interactive and iterative, each feeding the other. For example, the IPB process helps determine factors that affect reconnaissance during collection, such as—

- Avenues of approach that support friendly movement and exploit enemy weaknesses.
- Key terrain, choke points, obstacles, and hazard areas.
- Enemy positions, especially flanks that can be exploited.
- Observation points.

5-33. Conversely, reconnaissance drives the refinement of IPB results (see paragraph 4-157) as appropriate by confirming or denying priority intelligence requirements that support tentative plans. The results of the IPB process contribute to the BCT's security by developing products that help the commander protect subordinate forces, including identification of key terrain features, manmade and natural obstacles, trafficability and cross-country mobility analysis, line-of-sight overlays, and situation templates. For example, line-of-sight overlays help protect the force. If an enemy cannot observe the friendly force, the enemy cannot engage the friendly force with direct fire weapons. Situation templates also help protect the force. If a commander knows how fast an enemy force can respond to the unit's offensive actions, unit operations can be sequenced, so they occur at times and places where the enemy cannot respond effectively. This occurs through determining enemy artillery range fans, movement times between enemy reserve assembly area locations and advancing friendly forces, and other related intelligence items.

5-34. The commander, with the support of the staff, employs the appropriate combinations of mounted, dismounted, and aerial manned (and unmanned) reconnaissance (and surveillance) to obtain the information required to answer the CCIRs (priority intelligence requirements and friendly force information requirements) and to support the targeting process. At the same time, reconnaissance forces must be prepared to conduct counterreconnaissance (see paragraph 5-4) and continuously develop detailed information on both the enemy and terrain. Reconnaissance forces fight for information as in a reconnaissance in force (see paragraph 5-52) designed to discover or test the enemy's strength, dispositions, and reactions.

5-35. Surveillance complements and informs reconnaissance as reconnaissance complements and informs surveillance by cueing (see paragraph 5-64) the commitment of capabilities against specific locations or specially targeted enemy units. Throughout planning and preparation, the BCT S-3 and S-2 integrate actions within the BCT's overall information collection plan (see paragraph 4-146) and other higher and lateral information collection efforts to ensure that each asset is used effectively. The S-2 develops an initial synchronization plan to acquire information to help answer priority intelligence requirements based on the available reconnaissance and surveillance assets supporting the intelligence scheme of support. The S-3, in coordination with the commander, assigns specific intelligence acquisition tasks to specific units for action.

5-36. The S-2 and S-3, in coordination with the rest of the staff, develop a synchronized and integrated information collection plan that satisfies the commander's maneuver, targeting, and information requirements. As stated earlier, the S-3 is overall responsible for the information collection plan. The S-3 is also responsible for ground and air reconnaissance assets, which includes engineers, chemical, biological, radiological, and nuclear (CBRN), artillery, and Army attack reconnaissance aircraft. The S-2's primary responsibility is to integrate ground surveillance systems and special electronics mission aircraft. The brigade civil affairs operations staff officer's (S-9) primary responsibility is planning civil reconnaissance and the integration of civil information into the common operational picture (COP). The commander's requirements, dictated by the mission variables of METT-TC, commonly include—

- Locations, composition, equipment, strengths, and weaknesses of the enemy force, to include high-priority targets and enemy reconnaissance, security, and surveillance capabilities.
- Locations of obstacles, prepared fighting positions, enemy engineer units, earth moving equipment, breaching assets, and barrier material.
- Probable locations of enemy reconnaissance objectives.
- Locations of possible enemy assembly areas.
- Locations of enemy indirect-fire weapon systems and units.
- Locations of gaps, assailable flanks, and other enemy weaknesses.
- Locations of areas for friendly and enemy air assault and parachute assault operations.
- Locations of enemy air defense gun and missile units and air defense radars.
- Locations of enemy EW units.
- Effects of weather and terrain on current and projected operations.
- ASCOPE related information about civilians located within the unit's area of operations.
- Likely withdrawal routes for enemy forces.
- Anticipated timetable schedules for the enemy's most likely COA and other probable COAs.
- Locations of enemy command and control centers, intelligence nodes, reconnaissance, security, and surveillance systems, and the frequencies used by the information systems linking these systems.
- Locations of enemy sustainment assets.

5-37. When reconnaissance forces (and surveillance assets) cannot answer the commander's information requirements. The commander's options include—

- The S-2 sending a request for information to higher and adjacent units.
- The commander committing additional resources.
- The commander deciding to execute task with the current information.

5-38. CBRN reconnaissance and surveillance is the detection, identification, reporting, and marking of CBRN hazards. CBRN reconnaissance consists of search, survey, surveillance, and sampling operations. Due to limited availability and number of the CBRN reconnaissance vehicles within the BCT, the commander considers alternate means of conducting CBRN reconnaissance such as reconnaissance elements, engineers, and maneuver units. (See ATP 3-11.37.) At a minimum, the commander and staff consider the following actions when planning and preparing for CBRN reconnaissance and surveillance:

- Use the IPB process to orient on CBRN enemy NAI.
- Pre-position reconnaissance and surveillance assets to support requirements.
- Establish command and support relationships.
- Assess the time and distance factors for the conduct of CBRN reconnaissance and surveillance.
- Report all information rapidly and accurately to higher.
- Plan for resupply activities to sustain CBRN reconnaissance and surveillance operations.
- Determine possible locations for post-mission decontamination.
- Plan for fire support requirements.
- Plan fratricide prevention measures.
- Establish medical evacuation procedures.
- Identify CBRN warning and reporting system procedures and frequencies.

RECONNAISSANCE OPERATIONS

5-39. Reconnaissance operations validate the IPB process by confirming or denying natural and manmade obstacles, trafficability of routes, viability and utility of key terrain, and enemy composition, disposition, and strength. As mission analysis identifies information gaps, the BCT commander and staff develop information requirements to fill those gaps. During the operations process, information requirements develop into priority intelligence requirements, which further develop tasks that, when executed, answer priority intelligence requirements. The commander and staff continuously reevaluate information gaps and refocus the reconnaissance effort with the seven reconnaissance fundamentals. The commander establishes priorities for assessment in planning guidance, CCIRs (priority intelligence requirements and friendly force information requirements), EEFI, and decision points. The commander utilizes one of the five types of reconnaissance as they collect and assess information. (See FM 3-90-2 for additional information.)

RECONNAISSANCE FUNDAMENTALS

5-40. Reconnaissance fundamentals, discussed in paragraphs 5-41 to 5-46, remind planners and practitioners of the inherent characteristics required to execute successful reconnaissance. Failure to understand the following seven fundamentals results in incomplete reconnaissance and missed opportunities.

Ensure Continuous Reconnaissance

5-41. The BCT conducts reconnaissance before, during, and after all operations. Before an operation, reconnaissance fills gaps in information about the enemy, the terrain, and civil considerations. During an operation, reconnaissance provides the BCT commander with updated information that verifies the enemy's composition, dispositions, and intentions as the battle progresses. After an operation, reconnaissance forces maintain contact with the enemy to determine the enemy's next move and collect information, including terrain and civil considerations, necessary for planning subsequent operations. When current operational information is adequate, reconnaissance forces gather information for branches and sequels to current plans. As operations transition from a focus on one element of operations to another, the nature of priority intelligence requirements and information requirements change. Reconnaissance over extended distances and time may require pacing reconnaissance assets (surveillance assets can enable this effort) to maintain the effort, or rotating units to maintain continuous coverage. The human and technical assets used in the reconnaissance effort must be allowed time for rest, resupply, troop leading procedures, and preventive maintenance checks and services. The commander must determine not only where, but also when, the maximum reconnaissance effort is required and pace the commitment of available reconnaissance assets to ensure adequate assets are available at those critical times and places.

Do Not Keep Reconnaissance Assets in Reserve

5-42. Never keep reconnaissance assets in reserve. The BCT commander commits reconnaissance forces and assets with specific missions designed to help reduce uncertainty through the collection of information related to priority intelligence requirements and information requirements. Although noncontiguous operations may necessitate orientation of reconnaissance assets in multiple directions, reconnaissance forces maximize all assets at their disposal to information collection focused on the CCIRs. This does not mean that all reconnaissance forces and assets are committed all the time. The BCT commander uses reconnaissance forces and assets based on their capabilities and the mission variables of METT-TC to achieve the maximum coverage needed to answer CCIRs. At times, this requires the commander to withhold or position reconnaissance forces and assets to ensure that they are available at critical times and places.

Orient on the Reconnaissance Objective

5-43. The BCT commander orients reconnaissance assets by identifying a reconnaissance objective in the area of operations. The reconnaissance objective clarifies the intent of the reconnaissance effort by specifying the most important result to obtain from the reconnaissance effort. Every reconnaissance mission specifies a reconnaissance objective. The commander assigns a reconnaissance objective based on priority intelligence requirements resulting from the IPB process and the capabilities and limitations of the reconnaissance force or asset. The reconnaissance objective can be information about a specific geographical location, such as the cross-country trafficability of a specific area, a specific enemy or adversary activity to be confirmed or denied, or a specific enemy or adversary unit to be located and tracked. When the reconnaissance force does not have enough time to complete all the tasks associated with a specific type of reconnaissance, it uses the reconnaissance objective to guide it in setting priorities. The commander may need to provide additional detailed instructions beyond the reconnaissance objective, such as the specific tasks and their priorities. The commander issues additional guidance to the reconnaissance force or specifies these instructions in tasks to subordinates in a warning order, fragmentary order, or the operation order.

Report Information Rapidly and Accurately

5-44. Reconnaissance assets acquire and report accurate and timely information on the enemy, terrain, and civil considerations of the area over which the commander conducts operations. As information may quickly lose its value over time, the BCT commander must have accurate reports quickly to make informed decisions as to where to concentrate combat power. Rapid reporting allows the staff maximum time to analyze information and make timely recommendations to the commander. Information requirements, tied to decision points, define a latest time information is of value date-time group. Reconnaissance forces report exactly what they see and, if appropriate, what they do not see. Seemingly, unimportant information may be extremely important when combined with other information. Reports of no enemy activity are as important as reports of enemy activity. Failing to report tells the commander nothing.

Retain Freedom of Maneuver

5-45. Reconnaissance forces must maintain battlefield mobility, as fixed reconnaissance forces are ineffective. Reconnaissance forces must have clear engagement criteria that support the BCT commander's intent. They must employ proper movement and reconnaissance techniques, use overwatching fires, and follow standard operating procedures (SOPs). Initiative and knowledge of both the terrain and the enemy reduce the likelihood of decisive engagement and help maintain freedom of movement. Before initial contact, the reconnaissance force adopts a movement technique designed to gain contact with the smallest friendly element possible. This movement technique provides the reconnaissance force with the maximum opportunity for maneuver and enables the force to avoid having the entire reconnaissance force decisively engaged. The IPB is used to identify anticipated areas of contact. Indirect fires to provide suppression, obscuration, and to destroy point targets is a method reconnaissance forces use to retain freedom of maneuver.

Gain and Maintain Enemy Contact

5-46. Once reconnaissance forces gain contact with the enemy, it maintains that contact unless the commander directing the reconnaissance orders a change of mission, disengagement or displacement criteria is met, when the force conducts a reconnaissance handover, or the survival of the unit is at risk. Contact can range from surveillance to close combat. Surveillance, combined with stealth, is often sufficient to maintain contact and can limit exposure of reconnaissance assets. Units conducting reconnaissance avoid combat unless it is necessary to gain essential information, in which case the reconnaissance force uses maneuver (fire and movement) to maintain contact while avoiding decisive engagement. Maintaining contact provides real time information on the enemy's composition, disposition, strength, and actions that allow the staff to analyze and make recommendations to the commander.

Develop the Situation Rapidly

5-47. When reconnaissance forces make contact with an enemy force or obstacle, it must act instinctively to develop the situation and quickly determine the threat it faces. For an enemy force, reconnaissance forces must determine the enemy's composition, disposition, activities, and movements and assess the implications of that information to allow the BCT commander freedom of action. For an obstacle, reconnaissance forces must determine the type and extent of the obstacle and whether fire is covering the obstacle. Obstacles can provide information concerning the enemy force, weapon capabilities, and organization of fires. Reconnaissance forces, in most cases, develop the situation using *actions on contact*—a series of combat actions, often conducted simultaneously, taken on contact with the enemy to develop the situation (ADP 3-90)—in accordance with the commander's plan and intent. Actions on contact are deploy and report, evaluate and develop the situation, choose a COA, execute selected COA, and recommend a COA to the higher commander.

TYPES OF RECONNAISSANCE OPERATIONS

5-48. The five types of reconnaissance operations, discussed in paragraphs 5-49 through 5-53, are zone reconnaissance, area reconnaissance, route reconnaissance, reconnaissance in force, and special reconnaissance. Each type of reconnaissance operation provides a specific level of detail in information collection specific to the mission, conditions, and end state of the BCT commander. All types of reconnaissance operations satisfy priority intelligence requirements to understand and visualize the environment, develop the situation, create options, and identify opportunities to seize, retain, and exploit the initiative.

Zone Reconnaissance

5-49. *Zone reconnaissance* is a type of reconnaissance operation that involves a directed effort to obtain detailed information on all routes, obstacles, terrain, and enemy forces within a zone defined by boundaries (ADP 3-90). Zone reconnaissance is a deliberate and time-intensive operation that takes more time to conduct than any other type of reconnaissance. The BCT commander assigns a zone reconnaissance when the enemy situation is vague or when information related to terrain, infrastructure, or civil considerations is limited. A zone reconnaissance conducted over an extended distance begins at the line of departure and concludes at a specified limit of advance. The BCT commander specifies information requirements based upon time constraints and commander's intent and relates reconnaissance objectives to follow-on missions. Reconnaissance forces find and report enemy activities within the area of operations for the zone reconnaissance, reconnoiter specific terrain, and report all information in a timely manner.

Area Reconnaissance

5-50. *Area reconnaissance* is a type of reconnaissance operation that focuses on obtaining detailed information about the terrain or enemy activity within a prescribed area (ADP 3-90). The commander assigns an area reconnaissance when information on the enemy situation is limited, when focused reconnaissance in a given area likely yields specific information related to decision points, or when information that is more thorough is required in a designated area. The commander defines the area as an NAI to focus the unit on a relatively small area such as a building, bridge, or key terrain. Area reconnaissance allows for focused reconnaissance over a wide area concentrated in specific locations that answer priority intelligence requirements and develop the situation to provide the commander with options. An area reconnaissance differs from a zone reconnaissance in that the units conducting an area reconnaissance first move to the area in which the reconnaissance will occur. In a zone reconnaissance the units conducting the reconnaissance start from a line of departure.

Route Reconnaissance

5-51. *Route reconnaissance* is type of reconnaissance operation to obtain detailed information of a specified route and all terrain from which the enemy could influence movement along that route (ADP 3-90). A route can be a road, highway, trail, mobility corridor, avenue of approach, or axis of advance. Routes begin at a start point and end at a destination release point. The commander assigns a route reconnaissance either as a discrete mission or as a specified task of a zone or area reconnaissance. Route reconnaissance is not to be confused with route classification, which requires technical measurements and analysis typically performed by mission tailored engineer reconnaissance teams. Typically, a route classification is included as a specified task for the engineer reconnaissance team as part of an assigned route reconnaissance. Reconnaissance forces collect information about roads, bridges, tunnels, fords, waterways, and other natural and manmade terrain features that can affect traffic flow. Route reconnaissance provides the commander with detailed information on the route and terrain that can influence the route to prevent surprise, determine trafficability for follow-on forces, and to confirm or deny running estimates made during the operations process.

Reconnaissance in Force

5-52. *Reconnaissance in force* is a type of reconnaissance operation designed to discover or test the enemy's strength, dispositions, and reactions or to obtain other information (ADP 3-90). A reconnaissance in force is a limited objective operation normally conducted by a battalion-sized or larger task force. The BCT commander assigns a reconnaissance in force when the enemy is operating within an area and the commander cannot obtain adequate intelligence by any other means. Reconnaissance in force is an aggressive reconnaissance, which develops information in contact with the enemy to determine and exploit enemy weaknesses. The commander plans for the extrication of the force or the exploitation of success in advance. For example, the BCT commander positions forces to extricate the reconnaissance in force element if required or to seize on opportunities identified by the element.

Special Reconnaissance

5-53. *Special reconnaissance* is reconnaissance and surveillance actions conducted as a special operation in hostile, denied, or diplomatically and/or politically sensitive environments to collect or verify information of strategic or operational significance, employing military capabilities not normally found in conventional forces (JP 3-05). Special reconnaissance operations support the collection of the joint task force commander's priority intelligence requirements. Special reconnaissance may occur before conventional forces entering a designated area of operations, such as during an airborne or air assault operation, or other anti-assess or area-denial operation (see FM 3-99). A special operations liaison may provide a responsive reporting capability in situations where the special operations task force commander has been requested to provide intelligence information that supports the intelligence requirements of a conventional force commander. The BCT commander and staff must understand when, where, and why the force is conducting special reconnaissance operations to establish unity of purpose. The BCT and the special operations forces element may establish a liaison capacity to understand collection task prioritization, and to understand associated reporting requirements and mechanisms. Reconnaissance forces often may be the first friendly units to encounter special operations forces because of their forward proximity in the BCT's area of operations. Depending on the command relationship, conventional reconnaissance forces may operate in conjunction with special operations forces. (See FM 3-18 for additional information.)

> *Note.* A special operations forces element will not suspend or alter their collection efforts to support another collection plan unless directed to do so by the joint task force commander.

RECONNAISSANCE HANDOVER

5-54. *Reconnaissance handover* is the action that occurs between two elements in order to coordinate the transfer of information and/or responsibility for observation of potential threat contact, or the transfer of an assigned area from one element to another (FM 3-98). Reconnaissance handover facilitates observation or surveillance of enemy contact or an assigned NAI or TAI. Reconnaissance handover is associated with a trigger, coordination point, or phase line (PL) designated as the reconnaissance handover line. (See figure 5-2.) A *reconnaissance handover line* is a designated phase line on the ground where reconnaissance responsibility transitions from one element to another (FM 3-98). The reconnaissance handover line ensures control and chain of custody from the initial force to the force assuming responsibility and control.

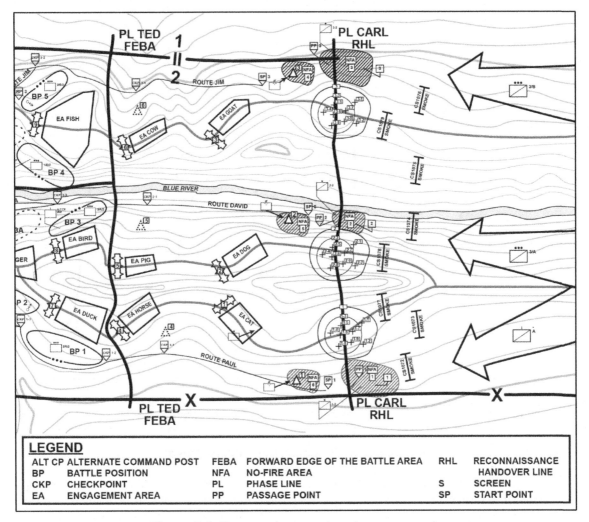

Figure 5-2. Reconnaissance handover, example

5-55. Reconnaissance handover prevents gaps or seams to emerge that the enemy can exploit. Once handover is complete, the reconnaissance force transferring control either passes to the rear through the main body assuming responsibility for the reconnaissance objective as a rearward passage of lines or continues further into the zone to continue the reconnaissance mission. Reconnaissance handover assures that information requirements are transferred between units to maintain initiative, tempo, and to ease transitions. Well-planned and executed reconnaissance handover eases transitions in plans, phases, and priorities of effort and mitigates information gaps between units.

RECONNAISSANCE-PULL VERSES RECONNAISSANCE-PUSH

5-56. *Reconnaissance-pull* is reconnaissance that determines which routes are suitable for maneuver, where the enemy is strong and weak, and where gaps exist, thus pulling the main body toward and along the path of least resistance. This facilitates the commander's initiative and agility (FM 3-90-2). In reconnaissance-pull (see figure 5-3 on page 5-14), the commander uses the products of the IPB process in an interactive and repetitive way. The commander obtains combat information from available reconnaissance assets to determine a preferred COA for the tactical situation presented by the mission variables of METT-TC.

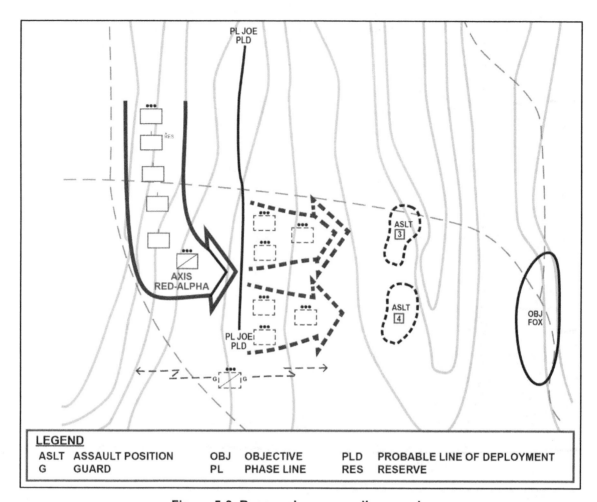

Figure 5-3. Reconnaissance-pull, example

5-57. *Reconnaissance-push* is reconnaissance that refines the common operational picture, enabling the commander to finalize the plan and support shaping and decisive operations. It is normally used once the commander commits to a scheme of maneuver or course of action (FM 3-90-2). In reconnaissance-push (see figure 5-4), the commander uses the products of the IPB process in an interactive way with combat information from reconnaissance assets in support of a COA.

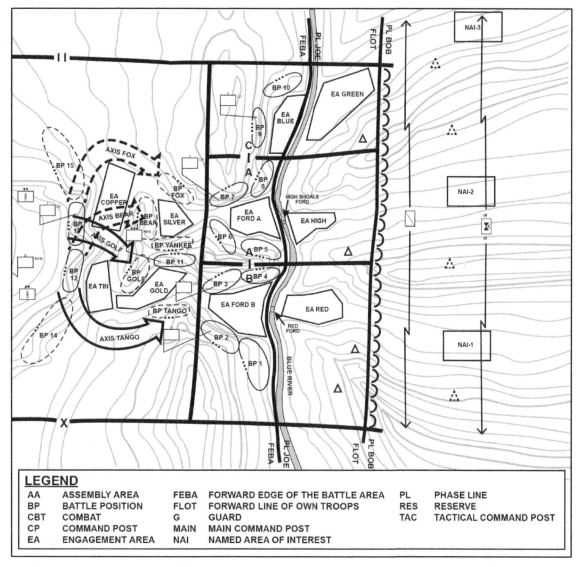

Figure 5-4. Reconnaissance-push, example

5-58. The chief reason for preferring one method to the other is the time available or confidence in the IPB. The time required to develop a COA can give the enemy enough time to recover and prepare so that taking an objective may cause higher casualties than necessary. Commanders balance the time needed to develop a COA with the need to act rapidly and decisively on the battlefield. There is no available model that a commander can use to determine how much enough is; that determination is part of the art of command (the creative and skillful exercise of authority through decision-making and leadership).

SURVEILLANCE

5-59. *Surveillance* is the systematic observation of aerospace, cyberspace, surface, or subsurface areas, places, persons, or things by visual, aural, electronic, photographic, or other means (JP 3-0). Surveillance may be a stand-alone mission or part of a reconnaissance mission (particularly area reconnaissance). Both reconnaissance and surveillance produce raw data and information, some of which may be combat information that meets one or more of the CCIRs or intelligence requirements. A key difference between surveillance missions and reconnaissance is that surveillance is tiered and layered with technical assets and it is passive and continuous. Reconnaissance is active in the collection of information (such as maneuver) and usually includes human participation.

COLLECTION EFFORT

5-60. Surveillance is one of the four tasks of the information collection effort to assist in answering requirements. Although surveillance platforms, devices, and assets change with technology, the general principles behind surveillance do not. When the BCT commander and staff systematically integrate the passive collection of information through surveillance, it enhances reconnaissance and security operations throughout the area of operations. The observation and data gained through surveillance contributes simultaneously to a greater understanding of the adversary or threat, while also increasing the protection of friendly forces, yielding time and space to react to the enemy. Weather, temperature, meteorological data, atmospherics, and other factors can hinder or increase visual effects in surveillance. The commander and staff account for these considerations when planning and integrating assets capable of observation.

COLLECTION INTEGRATION, SYNCHRONIZATION, AND COLLABORATION

5-61. While reconnaissance missions are specifically conducted to obtain information about the threat or the operational environment, surveillance missions consist of the systematic observation of places, persons, or things. To ensure these two types of missions are fully integrated into the overall collection effort, the BCT S-2 and S-3 staffs must continuously collaborate to synchronize the employment of assigned and allocable platforms and sensors against specified collection targets. The S-2, in coordination with the S-3, ensures raw data is routed to the appropriate processing and exploitation system so that it may be converted into useable information and disseminated to the user in a timely fashion. The S-2 is responsible for identifying potential collection targets and prioritizing anticipated collection requirements that are then used to drive surveillance and reconnaissance mission planning. The S-3 deconflicts the physical employment of the various platforms with other operations to be conducted within the land, air, and maritime domains.

5-62. Decision points (see chapter 4) are events or locations where decisions are required during mission execution. Decision points relate information requirements to identified critical events and are linked to NAIs and TAIs. Priorities for the apportionment and allocation of collection capabilities to subordinates are typically based on a decision point and the CCIRs (priority intelligence requirements and friendly force information requirements) and EEFI supporting a decision point (see figure 5-5). The S-3 recommends to the commander the apportionment of platforms to subordinate echelons to inform their planning efforts and in collaboration with the S-2 makes recommendations regarding their allocation during execution. Adaptive collection planning by the S-2 and continuous collaboration between the S-2 and S-3 staffs during development of the information collection plan provides for the effective management and optimal employment of all available platforms, sensors, and associated intelligence processing, exploitation, and dissemination capabilities (see paragraph 4-156).

> *Note.* Processing, exploitation, and dissemination is not exclusive to military intelligence organizations; other branches employ sensor collection capabilities. Intelligence processing, exploitation, and dissemination is the way the intelligence warfighting function processes collected data and information, performs an initial analysis (exploitation), and provides information in a useable form for further analysis or combat information (see chapter 2) to commanders and staffs. (See ADP 2-0 for additional information.)

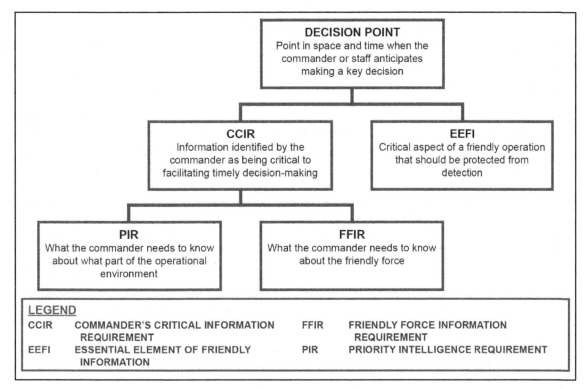

Figure 5-5. Information requirements

5-63. An integrated collection plan that fully optimizes the use of all available United States, unified action partner, and host-nation collection capabilities assets is essential to persistent surveillance. Information from BCT reconnaissance and security forces, to include internal forward and engaged combat forces, can be integrated with intelligence obtained from information collection assets external to the BCT. In many situations, even negative reporting from operational forces may be valuable (for example, a lack of contact with adversary forces may be just as significant as positive contact). Likewise, special operations forces provide a unique manned and unmanned deep look capability, especially useful in areas where other sensors are not available or cannot provide situational awareness. Based on operational requirements, the BCT S-2, in coordination with the division assistant chief of staff, intelligence, identify the priority intelligence requirements and associated reporting criteria to focus special operations forces assets. The continuous real-time monitoring of the status, location, and reporting of intelligence platforms and sensors by the BCT S-2 and higher headquarters assistant chief of staff, intelligence provides real-time cross cueing (see paragraph 5-64) and a basis for re-tasking and time-sensitive decision-making. (See ATP 2-01 for additional information.)

ASSET MANAGEMENT

5-64. When allocating information collection assets, no single asset can answer every intelligence requirement, and there are rarely enough assets to cover every requirement. The BCT staff, and division and battalion staffs, use a mix of reconnaissance management methods, such as cueing, mixing, redundancy, and task organizing, in an attempt to use limited assets most effectively and collect the most critical information with the fewest assets as quickly as possible. While several technical systems can perform reconnaissance, most systems are considered surveillance platforms. Surveillance complements reconnaissance by cueing the commitment of reconnaissance assets against specific locations or specially targeted enemy units.

5-65. *Cueing* is the integration of one or more types of reconnaissance or surveillance systems to provide information that directs follow-on collecting of more detailed information by another system (FM 3-90-2). Cueing helps to focus limited reconnaissance assets, especially limited ground reconnaissance assets, which can rarely examine every part of a large area closely. Electronic, thermal, visual, audio, and other technical assets with wide-area surveillance capabilities, often working from aerial platforms, can quickly determine areas of enemy concentration or areas where there is no enemy presence. These assets may cue ground and air reconnaissance assets to investigate specific areas to confirm and amplify information developed by technical assets. For example, JSTARS and Guardrail-equipped (signals intelligence collection platform) aircraft can cover large areas and cue ground reconnaissance or unmanned aircraft once they identify an enemy force. The BCT commander may dispatch ground reconnaissance or unmanned aircraft (RQ-7 Shadow UAS, brigade echelon asset or RQ-11 Raven UAS, company echelon asset) to verify the information and track the enemy for targeting purposes. Similarly, a ground reconnaissance asset can cue surveillance assets. The commander uses reconnaissance assets based on their capabilities and uses the complementary capabilities of other assets, such as surveillance assets to verify and expand information.

5-66. *Mixing* is using two or more different assets to collect against the same intelligence requirement (FM 3-90-2). Employing a mix of systems not only increases the probability of collection, but also tends to provide information that is more complete. For example, a JSTARS aircraft may detect and locate a moving enemy tactical force, while the division assistant chief of staff, intelligence analysis and control element uses organic and supporting assets to determine its identity, organizational structure, and indications of future plans. When available from echelons above division and corps, a U2 Advanced Synthetic Aperture Radar System provides high-resolution, multimode, long-range, air-to-ground radar that provides operators with critical intelligence. This all weather, day or night capability detects and accurately locates fixed and moving ground targets with precision. Employing a mix of systems is always desirable if the situation and available resources permit. Mixing systems can also help uncover military deception attempts by revealing discrepancies in reports from different collectors.

5-67. *Redundancy* is using two or more like assets to collect against the same intelligence requirement (FM 3-90-2). Based on the priority of the information requirement, the commander must decide which NAI justifies having more than one asset covering it. When more than one asset covers the same NAI, a backup is available if one asset cannot reach the NAI in time, the first asset suffers mechanical failure, or the enemy detects and engages the first asset. Redundancy also improves the chances of information collection.

5-68. To increase the effectiveness and survivability of an information collection asset, the commander may task organize it by placing additional assets under a subordinate unit's control. For example, to conduct an area reconnaissance of possible river crossing sites at extended distances from the division's current location, a ground Cavalry troop of an attached Armored brigade combat team (ABCT) can be task organized with a signal retransmission (known as RETRANS) element, an engineer reconnaissance element, a joint fires observer, and a tank platoon. The engineers provide additional technical information on proposed crossing sites; the signal RETRANS elements allow the Cavalry troop's combat net radios to reach the division main command post (CP). The joint fires observer provides additional observation, lazing, and fire coordination capabilities. The tank platoon provides additional combat capabilities and protection for the Cavalry troop.

Vulnerability

5-69. The BCT staff, led by the S-2 and in coordination with the S-3, evaluates the collector's vulnerability to threat forces, not only in the target area but also along the entire route of travel. It is important to evaluate the threat's ability to locate, identify, and destroy collection assets. For example, a helicopter's capabilities may make it a suitable collection asset; however, its vulnerabilities could make it too risky to use if the enemy possesses surface-to-air missiles. Another consideration is the signature associated with the collection asset. For example, a UAS engine emits an uncommon noise that is distinctly identifiable and may alert the target they are under surveillance.

SECTION III – SECURITY OPERATIONS

5-70. *Security operations* are those operations performed by commanders to provide early and accurate warning of enemy operations, to provide the forces being protected with time and maneuver space within which to react to the enemy and to develop the situation to allow commanders to effectively use their protected forces (ADP 3-90). The main difference between the conduct of security operations and reconnaissance is that the conduct of security operations orients on the protected force or facility, while reconnaissance is enemy and terrain oriented. Security missions protect the BCT from observation, indirect fires, harassment, surprise, and sabotage. At the same time, security forces conducting security operations provide information about the size, composition, location, and movement of enemy forces including information about the terrain and populations within a BCT's area of operations. Effective security operations can also draw enemy forces into exposed positions, trade space for time, allow the BCT to concentrate forces elsewhere, deceive the enemy, attrite enemy forces, and hold, deny, or control key terrain. Security forces must be prepared to destroy enemy reconnaissance efforts and fight for information to seize, retain, or exploit the initiative. (See FM 3-90-2 for additional information.)

FUNDAMENTALS OF SECURITY OPERATIONS

5-71. Five fundamentals of security operations establish the framework for security operations. These fundamentals, discussed below, provide a set of principles that remind planners and practitioners of the inherent characteristics required to execute security operations. These fundamentals include provide early and accurate warning, provide reaction time and maneuver space, orient on the force or facility to be secured, perform continuous reconnaissance, and maintain enemy contact.

PROVIDE EARLY AND ACCURATE WARNING

5-72. The security force detects, observes, and reports threat forces that can influence the protected force. Early detection and warning through rapid reporting enables the BCT commander to make timely and well-informed decisions to apply forces relative to the threat. As a minimum, security forces should operate far enough from the protected force to prevent enemy ground forces from observing or engaging the protected force with direct fires. The BCT commander and staff plan for the positioning of ground security, aerial scouts, and UASs to provide long-range observation of expected enemy avenues of approach. The commander reinforces and integrates them with available intelligence collection systems, such as unattended ground sensors, surveillance systems, and moving target indicators to maximize warning time.

PROVIDE REACTION TIME AND MANEUVER SPACE

5-73. Security forces provide the protected force with enough reaction time and maneuver space to effectively respond to likely enemy actions by operating at a distance from the protected force and by offering resistance (within its capabilities and mission constraints) to enemy forces. Providing the security force with an area of operations that has sufficient depth to operate enhances its ability to provide reaction time and maneuver space to the protected force. The commander determines the amount of time and space required to respond from the information provided by the IPB process and the protected force commander's guidance regarding time the protected force requires to react to enemy COAs based on the mission variables of METT-TC. Reaction time and maneuver space relates to decision points driven by information requirements and indicators given the latest time information is of value parameters to ensure the commander makes decisions that place maximum firepower at the decisive point in a timely manner.

ORIENT ON THE FORCE, AREA, OR FACILITY

5-74. While reconnaissance forces orient on the enemy, security forces orient on the protected force by understanding their scheme of maneuver and follow-on mission. The security force focuses all its actions on protecting and providing early warning operating between the protected force and known or suspected enemy. The security force moves as the protected force moves and orients on its movement. The value of terrain occupied by the security force hinges on the protection it provides to the protected force. In addition to orienting on a force, security operations may orient on an area or facility.

PERFORM CONTINUOUS RECONNAISSANCE

5-75. Reconnaissance fundamentals are implicit in all security operations. Security forces continuously seek the enemy and reconnoiter key terrain. Security forces use continuous reconnaissance to gain and maintain enemy contact, develop the situation, report rapidly and accurately, retain freedom of maneuver to provide early and accurate warning, and provide reaction time and maneuver space to the protected force. Security forces conduct area reconnaissance or zone reconnaissance to detect enemy movement or enemy preparations for action and to learn as much as possible about the terrain with the ultimate goal to determine the enemy's

COA and to assist the protected force in countering it. Terrain information focuses on its possible use by the enemy or the friendly force, either for offense or for defense. Civil consideration is a key focus for information during the elements (offense, defense, and stability) of decisive action. Stationary security forces use combinations of observation posts, aviation, patrols, intelligence collection assets, and battle positions to perform reconnaissance. Moving security forces perform zone, area, or route reconnaissance along with using observation posts and battle positions to detect enemy movements and preparations.

MAINTAIN ENEMY CONTACT

5-76. Once the security force makes enemy contact, it does not break contact unless the main force commander specifically directs it. However, the individual security asset that first makes contact does not have to maintain that contact, if the entire security force maintains contact with the enemy. The security force commander ensures that subordinate security assets hand off contact with the enemy from one security asset to another in this case. The security force must continuously collect information on the enemy's activities to assist the main body in determining potential and actual enemy COAs and to prevent the enemy from surprising the protected force. Depth in space and time enables security forces to maintain continuous visual contact, to use direct and indirect fires, and to maneuver freely.

TYPES OF SECURITY OPERATIONS

5-77. Security operations provide the protected force with varying levels of protection and are dependent upon the size of the unit conducting the security operation. All security operations provide protection and early warning to the protected force. Security operations encompass four types: screen, guard, cover, and area security.

5-78. Security operations conducted in the *security area*—that area occupied by a unit's security elements and includes the areas of influence of those security elements (ADP 3-90)—by one force or a subordinate element of a force that provides security for the larger force are screen, guard, and cover. The screen, guard, and cover security operations, respectively, contain increasing levels of combat power and provide the main body with increasing levels of security. The more combat power in the security force means less combat power for the main body. Normally, the BCT commander designates a security area in which security forces provide the BCT with reaction time and maneuver space to preserve freedom of action. (See FM 3-90-2 for additional information.)

SCREEN

5-79. *Screen* is a type of security operation that primarily provides early warning to the protected force (ADP 3-90). The screen provides the least protection of any security mission; it does not have the combat power to develop the situation. A screen is appropriate to cover gaps between forces, exposed flanks, or the rear of stationary and moving forces. The commander can place a screen in front of a stationary formation when the likelihood of enemy action is small, the expected enemy force is small, or the main body needs only limited time, once it is warned, to react effectively. If a significant enemy force is expected or a significant amount of time and space is needed to provide the required degree of protection, the commander assigns and resources a guard mission instead of a screen.

5-80. A screening force observes, identifies, and reports enemy actions. The unit performing a screen may engage, repel, or destroy an enemy's reconnaissance and surveillance element within its capabilities, augmented by indirect fires, Army aviation attacks, or close air support, but otherwise fights only in self-defense. The screen has the minimum combat power necessary to provide the desired early warning, which allows the commander to retain the bulk of the main body's combat power for commitment at the decisive place and time. The depth of the screen is critical to allow reconnaissance handover of threat contact from one element to another without displacement from established observation posts. Screening forces use depth to delay, impede, and harass the enemy with indirect fires or air support to cause the enemy to deploy early and to prevent the enemy from identifying, penetrating, and exploiting the screen.

5-81. Within an assigned area of operations, a security force normally conducts a screen by establishing a series of observation posts and patrols to ensure adequate observation of designate NAIs and TAI. The commander uses reconnaissance patrols (mounted, dismounted, and aerial), relocates observation posts, and employs technical assets to ensure continuous and overlapping surveillance. The commander also employs terrain data base analytical support systems to ensure the integration of friendly reconnaissance and surveillance assets to provide the necessary coverage. (See FM 3-90-2 for additional information.)

GUARD

5-82. *Guard* is a type of security operation done to protect the main body by fighting to gain time while preventing enemy ground observation of and direct fire against the main body (ADP 3-90). Units performing a guard task cannot operate independently because they rely upon fires and functional and multifunctional support assets of the main body. A guard force differs from a screen in that it routinely engages enemy forces with direct and indirect fires. A screening force primarily uses indirect fires or air support to destroy enemy reconnaissance elements and slow the movement of other enemy forces.

5-83. The BCT commander assigns a guard mission when expecting contact or has an exposed flank that requires greater protection than a screen can provide. The three types of guard operations are advance, flank, and rear guard. The commander can assign a guard mission to protect either a stationary or a moving force. The guard force commander normally conducts the guard mission as an area defense, a delay, a zone reconnaissance, or a movement to contact mission in the security area to provide reaction time and maneuver space to the main body. A guard operates within the range of the main body's fire support weapons, deploying over a narrower front than a comparable sized screening force to permit concentrating combat power. Guards are most effective when air assets are integrated. The commander's intent and end state determine the nature and extent of required augmentation. (See FM 3-90-2 for additional information.)

COVER

5-84. *Cover* is a type of security operation done independent of the main body to protect them by fighting to gain time while preventing enemy ground observation of and direct fire against the main body (ADP 3-90). Security forces protect the main body by fighting to gain time, observe and report information, and prevent enemy ground observation of and direct fire against the main body. In Army doctrine, a *covering force* is a self-contained force capable of operating independently of the main body, unlike a screen or guard force to conduct the cover task (FM 3-90-2). A covering force performs all the tasks of screening and guard forces.

5-85. A division covering force is usually a reinforced BCT that performs reconnaissance or other security missions. The *covering force area* is the area forward of the forward edge of the battle area out to the forward positions initially assigned to the covering force. It is here that the covering force executes assigned tasks (FM 3-90-2). The width of the covering force area is the same as the main body's area of operations. An adequately reinforced combined arms battalion, ABCT or Stryker brigade combat team (SBCT) Cavalry squadron, or SBCT Infantry battalion may perform a covering force mission if the division area of operations is narrow enough. These reinforcements typically revert to their parent organizations on passage of the covering force. BCTs and battalions typically organize a guard force instead of a covering force because their resources are limited.

5-86. A covering force's distance forward of the main body depends on the main body commander's intentions and instructions, reinforcements, the terrain, the enemy location and strength, and the main body and covering force's rate of march. Covering forces often become decisively engaged with enemy forces and therefore, must have substantial combat power to engage the enemy and accomplish the mission. A covering force develops the situation earlier than a screen or a guard force, fights longer and more often and defeats larger enemy forces. (See FM 3-90-2 for additional information.)

AREA SECURITY

5-87. *Area security* is a type of security operation conducted to protect friendly forces, lines of communications, and activities within a specific area (ADP 3-90). Area security operations allow commanders to provide protection to critical assets without a significant diversion of combat power. Protected forces range from echelon headquarters through artillery and echelon reserves to the sustaining base. Protected installations can be part of the sustaining base or they can constitute part of the area's critical infrastructure.

5-88. During the offense, various military organizations may be involved in conducting area security operations in an economy-of-force role to protect lines of communications, convoys, or critical fixed sites and radars. Route security operations are defensive in nature and are terrain oriented. A route security force may prevent an enemy force from impeding, harassing, or destroying lines of communications. Establishing a movement corridor for traffic along a route or portions of a route is an example of route security operations.

5-89. Areas to secure range from specific points, (bridges and defiles) and terrain features (ridgelines and hills), to large civilian population centers and their adjacent areas. Population-centric area security missions are common across the range of military operations. Population-centric area security operations typically combine aspects of the area defense and offensive operations to eliminate the power to produce internal defense threats. (See chapter 8, section IV for a detailed discussion.)

> *Note.* Area security activities take advantage of the local security measures performed by all units (regardless of their location) within an area of operations. For example, the BCT would link all local security activities conducted by its subordinate units to its broader area security activities. (See paragraph 8-126 for further information on establishing local security measures.)

SECTION IV – INTELLIGENCE OPERATIONS

5-90. *Intelligence operations* are the tasks undertaken by military intelligence units through the intelligence disciplines to obtain information to satisfy validated requirements (ADP 2-0). They are also one of the four primary tactical tasks and missions the Army conducts as part of information collection, along with the other three primary tasks: reconnaissance, surveillance, and security operations. Intelligence drives operations and operations support intelligence; this relationship is continuous. The BCT commander and staff need accurate, relevant, and predictive intelligence in order to understand threat centers of gravity, goals and objectives, and COAs. Precise intelligence is also critical to target threat capabilities at the right time and place and to open windows of opportunity across domains during large-scale combat operations. The commander and staff must have detailed knowledge of threat strengths, weaknesses, organization, equipment, and tactics to plan for and execute BCT operations.

EMPLOYMENT OF MILITARY INTELLIGENCE ASSETS

5-91. Successful intelligence operations, like reconnaissance fundamentals and the fundamentals of security, include effective and efficient employment of military intelligence assets, based on the following guidelines:

- Maintain readiness—the training of military intelligence personnel, the maintenance and status of their equipment (by dedicated field service representatives if necessary), and any necessary augmentation from outside personnel, or sustainment related resources for mission success.
- Ensure continuous intelligence operations—before, during, and after execution of all operations; intelligence operations focus on every relevant aspect of the operational environment before execution, constant updates for the commander verifying threat composition, disposition, and intention during execution, and after execution maintaining contact with threat forces to collect necessary information for planning subsequent operations while protecting the BCT.
- Orient on requirements—as the commander prioritizes intelligence operations by providing guidance and intent, military intelligence personnel identify and update priority intelligence requirements, ensuring they tie to the concept of the operation and decision points, focus on most critical needs, and consider the latest time the information is of value. The staff assists the commander in approving requests beyond a BCT's capabilities, as well as seeking higher echelon intelligence operations when needed.
- Provide mixed and redundant coverage—the commander integrates assets to ensure careful employment by layering through cueing (follow-on collection of more detail by another system), redundancy (using two or more like assets against the same requirement), and mixing (using two or more different assets against the same requirement) for maximum efficiency in information collection. This ensures a balance of requirements, available capabilities, and areas to be covered.

- Gain and maintain sensor contact—intelligence operations must gain and maintain sensor contact when it occurs, and the collection asset is able to observe or receive a signal.

- Report information rapidly and accurately—military intelligence staff closely collaborates with the signal staff to ensure communications plans incorporate military intelligence processing, exploitation, and dissemination and collection assets. The staff must test for value all relevant information within the area of interest in the operational environment, while quickly assessing it. Each collection asset must have a primary, alternate, contingency, and emergency (known as PACE) communications plan (see chapter 4).

- Provide early warning—to ascertain threat COA and timing, the commander and staff orient assets to observe the right locations for indicators to yield early timely and complete reporting.

- Retain freedom of movement—collection assets require battlefield mobility for mission accomplishment, refraining from close combat, as decisive engagement reduces or stops collection. Knowledge of terrain, weather, and threat reduce this likelihood, as does IPB to identify likely contact areas.

INTELLIGENCE CAPABILITIES

5-92. The intelligence warfighting function executes the intelligence process by employing intelligence capabilities. All-source intelligence and single-source intelligence are the building blocks by which the intelligence warfighting function facilitates situational understanding and supports decision-making. As the intelligence warfighting function receives information from a variety of capabilities, some of these capabilities are commonly referred to as single-source capabilities. Single-source capabilities are employed through intelligence operations with the other means of information collection (reconnaissance, surveillance, and security operations). Intelligence processing, exploitation, and dissemination capabilities (see paragraph 4-254) process information and prepare it for subsequent analysis. The intelligence produced based on all of those capabilities is called all-source intelligence.

ALL-SOURCE INTELLIGENCE

5-93. Army forces conduct operations based on all-source intelligence assessments and products developed by intelligence staffs. *All-source intelligence* is the integration of intelligence and information from all relevant sources in order to analyze situations or conditions that impact operations (ADP 2-0). In joint doctrine, *all-source intelligence* is intelligence products and/or organizations and activities that incorporate all sources of information in the production of finished intelligence (JP 2-0).

All-Source Analysis

5-94. The fundamentals of all-source intelligence analysis comprise intelligence analysis techniques and the all-source analytical tasks: situation development, generating intelligence knowledge, IPB, and support to targeting and information operations. Within the BCT, the intelligence staff determines the significance and reliability of incoming information, integrates incoming information with current intelligence holdings, and through analysis and evaluation determines changes in threat capabilities, vulnerabilities, and probable COAs. The intelligence staff supports the integrating processes (IPB, targeting, RM, information collection, and knowledge management [see paragraph 4-145]) by providing all-source analysis of threats, terrain and weather, and civil considerations. The intelligence staff uses all-source intelligence to develop the products necessary to aid situational understanding, support the development of plans and orders, and answer information requirements.

All-Source Production

5-95. *Fusion*—consolidating, combining, and correlating information together (ADP 2-0)—facilitates all-source production. Fusion occurs as an iterative activity to refine information as an integral part of all-source analysis. All-source intelligence production is continuous and occurs throughout the intelligence and operations processes. Most of the products from all-source intelligence are initially developed during planning and updated, as needed, throughout preparation and execution based on information gained from continuous assessment.

All-Source and Identity Activities

5-96. Identity activities (as described in joint doctrine) are a collection of functions and actions that appropriately recognize and differentiate one person from another to support decision-making. For example, they may include the collection of identity attributes and physical materials; their processing and exploitation; all-source analytic efforts, production of identity intelligence and criminal intelligence products; and dissemination of those products to inform and assess, and the appropriate action at the point of encounter. These all-source activities result in the discovery of true identities; link identities to events, locations, and networks; and reveal hostile intent. These outputs enable tasks, missions, and actions that span the range of military operations.

SINGLE-SOURCE INTELLIGENCE

5-97. Single-source intelligence includes the joint intelligence disciplines and complementary intelligence capabilities. Intelligence processing, exploitation, and dissemination capabilities are a critical aspect of single-source intelligence activities that ensure the results of collection inform single- and all-source analysis. Military intelligence units can conduct intelligence operations with a single intelligence discipline or complementary intelligence capability or as a multifunction intelligence operation, which combines activities from two or more intelligence disciplines or complementary intelligence capabilities.

5-98. Command and support relationships direct the flow of reported information during intelligence operations along various echelon specific transmission paths or channels. Channels assist in streamlining information dissemination by ensuring the right information passes promptly to the right people. Commanders and staffs normally communicate through three channels: command, staff, and technical. (See ADP 6-0 and ATP 6-02.71 for additional information.)

Intelligence Disciplines and Complementary Intelligence Capabilities

5-99. Intelligence disciplines, supported by military intelligence personnel, include—counterintelligence, geospatial intelligence, HUMINT, measurement and signature intelligence, open-source intelligence, signals intelligence, technical intelligence. Intelligence disciplines support reconnaissance, surveillance, and security operations through which intelligence units and staffs complete tasks in intelligence operations. Additionally, complementary intelligence capabilities such as biometrics-enabled, cyber-enabled, and forensic-enabled intelligence, along with document and media exploitation, ensure the successful accomplishment of intelligence tasks. (See ADP 2-0 for a detailed description of capabilities and disciplines.)

5-100. As the BCT's organic intelligence organization, the military intelligence company (see paragraph 1-24) supports the BCT and its subordinate units through collection and analysis of information and dissemination of intelligence, signals intelligence, and HUMINT. The military intelligence company provides continual input for the BCT commander by enabling the BCT S-2 in maintaining the threat portion of the COP. Military intelligence company elements working in the intelligence cell (see chapter 3) collaborate with the BCT operations staff to integrate information collection tasks and coordinate requirements as directed by the BCT S-3. The military intelligence company commander directs the employment of the company in accordance with missions and guidance from the BCT headquarters. The military intelligence company commander locates where best to exercise command and control of company intelligence assets. The military intelligence company CP is usually co-located with the BCT main CP to facilitate control of company intelligence assets and maximize support to the BCT intelligence cell. The military intelligence company CP includes the company headquarters element and representatives from each platoon. During BCT operations, the analysis platoon normally augments the BCT intelligence cell under control of the BCT S-2. (See FM 2-0 for additional information.)

5-101. Military intelligence units, external to the BCT, conduct both reconnaissance and surveillance missions. They provide electronic intercept, UASs sensor feeds, and HUMINT, counterintelligence, and downlinks from theater of operations and national assets. Theater and national-level reconnaissance and surveillance systems provide broadcast dissemination of information and intelligence and provide near real-time imagery as a part of an integrated intelligence effort. Artillery and air defense target acquisition radars complement military intelligence surveillance systems as a part of that effort. HUMINT collection (see chapter 8 for information on HUMINT collection teams) occurs through face-to-face interrogation of captured enemy soldiers, screening of the civilian population, and debriefing of friendly Soldiers, such as scouts and special operations forces. (See FM 2-0 for additional information.)

Technical Channels

5-102. Technical channels, while not a command or support relationship, often affect intelligence operations. For intelligence operations, technical channels are the transmission paths between paths between intelligence units and sections performing a technical function requiring special expertise. Technical channels control the performance of technical functions. They neither constitute nor bypass command authorities; rather, they serve as the mechanism for ensuring the execution of clearly delineated technical tasks, functions, and capabilities to meet the dynamic requirements of decisive actions. (See FM 2-0 for additional information.)

RECONNAISSANCE AND SECURITY ACROSS ALL DOMAINS

5-103. During reconnaissance and security operations, the BCT commander and staff, as well as subordinate commanders and staffs, consider intelligence operations to collect information across all relevant or necessary domains (air, land, maritime, space, and cyberspace). The BCT commander attempts to leverage advantages in one or multiple domains, such as seizing key terrain or denying enemy freedom of movement and action in the maritime and air domains, disrupting cyber access, or controlling narratives to influence population support. The BCT's higher headquarters (usually division or corps) supplements the information collection capabilities of the military intelligence company in one or multiple domains. Even as aggressive air-ground reconnaissance and security operations remain the key means of information collection, intelligence operations in multiple domains can increase the commander's situational understanding. The aim is to integrate information collection by all means (reconnaissance, surveillance, security operations, and intelligence operations) in all domains throughout the depth of the battlefield to prevent surprise, protect the force, and preclude enemy options.

5-104. Within the space domain, intelligence operations capabilities can provide information collection, early warning, environmental monitoring, and satellite-based communications, positioning, navigation, and timing. Such activities enable freedom of action for operations in all other domains, and operations in other domains can create effects in and through the space domain. The BCT commander considers the use of such capabilities in this domain for reconnaissance and surveillance operations, but also an enemy's use of capabilities. Reconnaissance and security forces must be prepared to operate in a denied, degraded, and disrupted operational environment. For example, in the space domain enemies will deny U.S. and partner space-based intelligence, reconnaissance, Global Positioning System (GPS), and secure satellite communications. Since space operations are inherently joint, the BCT commander and staff must not assume U.S. forces will have unconstrained use of space-based capabilities. (See FM 3-14 for doctrine on Army space operations.)

5-105. In the cyberspace domain, capabilities yield the same advantages to friendly forces and enemies and are therefore key for intelligence operations during reconnaissance and security operations. Cyberspace is highly vulnerable for several reasons, including ease of access, network and software complexity, lack of security considerations in network design and software development, and inappropriate user activity (see paragraph 4-329). Access to cyberspace by an individual or group with a networked device is easy, and an individual with a single device may be able to disable an entire network. An enemy could implant malicious code on the United States and its partners disrupting logistics, communications, reconnaissance, battlefield information systems, and ultimately security measures in other domains. Vulnerabilities in the systems that operate in cyberspace contribute to a continuous obligation to manage risk and protect portions of cyberspace.

5-106. Within the air domain, enemy forces will employ integrated air defense capabilities, particularly man-portable systems that are difficult to detect, or capabilities which are resistant to electronic suppression. UASs, fixed and rotary wing aircraft, in addition to airborne and air assault operations all fall under this domain. From a security standpoint, intelligence operations must consider enemy use of each of these systems during planning in order to protect the force. From a surveillance, reconnaissance, and security standpoint, the BCT commander and staff consider the same assets or capabilities within U.S. and partner forces in order to support intelligence operations.

5-107. The maritime domain, though mostly higher echelon Army and Navy forces, still holds intelligence operations considerations for a BCT. Reconnaissance can occur by assets with capabilities of lift, sustainment, and ship-to-shore fires to support maneuver (for targeting). Ground forces can also provide direct or indirect fires in littoral areas in support of maritime operations. Reconnaissance and surveillance must also occur on similar assets owned by the enemy in this domain, while friendly units conduct security operations to protect against these, while also protecting U.S. and partner maritime domain assets.

5-108. In the land domain, adversaries employ precision and extended range munitions, requiring BCTs to rapidly transition from movement to maneuver, and disperse forces to avoid enemy fire. The enemy will use advanced multifunctional mines to deny friendly freedom of movement and maneuver while protecting its own land force. The enemy will deploy camouflage, deception, and security forces to mitigate collection capabilities. Surveillance and reconnaissance focus upon all of these to allow for successful intelligence operations, while security incorporates such considerations into planning for the safety of information collection forces. BCTs will not only operate in the land domain but will also provide a large amount of information about the land domain to higher echelons.

Chapter 6

Offense

The brigade combat team (BCT) conducts offensive operations to defeat and destroy enemy forces and seize terrain, resources, and population centers. Offensive actions impose the BCT commander's will on the enemy. Offensive actions capitalize on accurate and timely intelligence and other relevant information regarding enemy forces, weather and terrain, and civil considerations. Protection tasks, such as security operations, operations security, and information protection prevent or inhibit the enemy from acquiring accurate information about friendly forces. As the commander maneuvers forces to advantageous positions before contact, contact with enemy forces before the decisive operation is deliberate and designed to shape the optimum situation for the decisive operation. When committed, the decisive operation is a sudden, shattering action that capitalizes on subordinate initiative and a common operational picture (COP). This chapter addresses the characteristics of the offense, common offensive planning considerations, forms of maneuver, offensive operations, and planning considerations when transitioning to other tactical operations.

SECTION I – CHARACTERISTICS OF THE OFFENSE

6-1. Successful offenses share the following characteristics: surprise, concentration, tempo, and audacity. Offensive characteristics, used in concert, create the foundation for an effective offense in any operational environment. The tactical vignette below is an example of this foundation and illustrates why U.S. forces must be able to transition from one type of military action (counterinsurgency) to another (close combat) seamlessly and rapidly.

6-2. Close combat, as experienced by Russian forces in Grozny and U.S. forces in Fallujah illustrate two approaches for conducting the offense in urban terrain. In each case, noncombatants were told to evacuate in advance of the attack, and anyone left was a de facto enemy fighter. These geographically remote cities were, in effect, besieged and then stormed, with attacks supported by massive firepower. The result was high casualties on both sides and rubbled cities. The 2008 battle for Sadr City offers a different approach. The challenges during the battle for Sadr City were in some cases even more formidable than the challenges posed by Grozny and Fallujah.

The Battle for Sadr City

Sadr City is part of Baghdad and has an estimated population of 2.4 million. Forcing noncombatants to evacuate was not an option, there was nowhere for them to go. However, the approach to ridding Sadr City of Jaish al-Mahdi fighters' was quite different from that used in Grozny or Fallujah. The operation in Sadr City focused on enemy fighters and their capabilities, rather than seizing and clearing the city.

The trigger for the battle was Jaish al-Mahdi fighter's response to the Iraqi government's offensive against insurgents in Basra. Jaish al-Mahdi fighters launched their own offensive, overrunning Iraqi government security forces and firing rockets and mortars into the International Zone, also known as the Green Zone. In response, a U.S. Army brigade and Iraqi security forces (army and police), featuring Abrams tanks, Bradley fighting vehicles, and Strykers, along with engineers, civil affairs, and psychological operations personnel and other support troops, attacked Jaish al-Mahdi fighters. The command and control arrangements gave the brigade commander direct access to crucial joint intelligence, surveillance, and reconnaissance assets, and fire support including attack helicopters, fighter aircraft, armed Predator unmanned aircraft systems, and Shadow unmanned aircraft systems. This arrangement gave the brigade commander a short decision response time, rapidly increasing the tempo of attacks to disrupt Jaish al-Mahdi operations.

An early priority was to stop the rocket and mortar attacks on the International Zone. Jaish al-Mahdi fighters could launch these attacks quickly and almost at will. These attacks simply required pulling a vehicle into a firing position, unloading the rocket and its firing rail, firing off the rocket, and driving back to a hide position. U.S. forces quickly realized that the International Zone was at the extreme end of the 107-millimeter (mm) rocket's range. The solution was to force Jaish al-Mahdi fighters out of their firing positions and back into Sadr City. This approach did not stop Jaish al-Mahdi infiltration. The brigade then employed an innovative but straightforward approach: It walled off two neighborhoods south of Sadr City, including the one containing the Jamilla market where Jaish al-Mahdi fighters obtained much of their resources. This inventive plan consisted of T-wall sections, each twelve feet tall and weighing 9000 pounds. The wall became impenetrable; a nearly five-kilometer long barrier that denied Jaish al-Mahdi fighters what had been terrain and avenues of movement crucial to its operations. The fighting was particularly intense and required the brigade commander to commit Abrams tanks and Bradley fighting vehicles to dislodge Jaish al Mahdi fighters and protect the soldiers building the wall. As soon as the wall started to go up, Jaish al Mahdi fighters instantly recognized the threat posed to their operations and launched numerous attacks to stop its construction. The wall, in the words of one U.S. commander, became a terrorist magnet. U.S. forces fought from a position of advantage, massing the effects of combat power to defeat the Jaish al-Mahdi fighters' assaults. While the construction of the T-wall ultimately squelched the rocket attacks by defeating Jaish al-Mahdi fighters, U.S. forces waged an intense and instructive concentration of counterfire operation.

Key to the counterfire operation was giving the brigade commander direct access to joint intelligence, surveillance, and reconnaissance assets so that the commander could identify firing locations almost immediately without having to consult with another headquarters. The commander could also pass intelligence rapidly by using secure communications down to the company level. The commander could attack enemy firing points around the clock with a formidable array of assets, including Apache helicopters, air component fighter aircraft, and armed Predator unmanned aircraft systems. Brigade intelligence analysts honed their techniques over time and learned to follow Jaish al-Mahdi rocket teams to their source rather than attack them immediately. This tactic allowed the U.S. forces to strike ammo dumps and senior leaders at a time or place the enemy did not expect. This tactic had a profound effect, more so than if they had destroyed a vehicle and a few fighters.

The overall results were impressive. In about two months, U.S. and Iraqi forces obliterated Jaish al-Mahdi fighters, killing an estimated 700, won back significant numbers of the population, and re-established control of what had been an insurgent stronghold. U.S. forces killed in action numbered fewer than ten. Furthermore, the Multi-national Division-Baghdad exploited the success of the combat gains in Sadr City with an intensive campaign of providing local security and reconstruction, all complemented by information operations. In addition to the key lessons highlighted above, other key lessons emerged. First, persistent intelligence, surveillance, reconnaissance, security operations, and responsive precision strikes were crucial to success because they were integrated at low levels. Second, ground maneuver forces were essential. Aggressive ground maneuver forced the enemy to react and enabled U.S. forces to seize control of the terrain south of Sadr City and to erect the barrier. Finally, capable indigenous forces were decisive in securing gains. Their presence signaled that Iraqis were in charge, not coalition forces that would leave eventually.

Jerry Salinger

SURPRISE

6-3. As in the vignette above, the BCT commander achieves surprise by striking the enemy at a time or place the enemy does not expect or in a manner that the enemy is unprepared. The commander assesses the enemy's intent to prevent the enemy from gaining situational understanding. The BCT identifies and avoids enemy strengths while attacking enemy weaknesses.

6-4. The BCT strikes the enemy where and when the enemy least expects it through night attacks, infiltrations, or rapid insertion of airborne or air assault forces or in a manner the enemy is unprepared, for example in mass, leading with tanks, or attacking earlier or later than anticipated. Thus, forcing the enemy to deal with multiple forms of contact or to take advantage of their lack of unpreparedness for a specific type of contact at a specific time and place.

6-5. The BCT focuses assigned or attached reconnaissance and security forces and information collection efforts, and exploits echelon above brigade enablers, for example space and cyberspace capabilities, to gain accurate and timely information about the enemy. The BCT then capitalizes on this information by maneuvering forces to advantageous positions on the battlefield and by imposing lethal and nonlethal effects to undermine the integrity of the enemy's tactical plan and thereby manufacture an advantage through the presentation of a tactical problem that the enemy is not prepared to, or limits the enemy's ability to react.

CONCENTRATION

6-6. Concentration, as displayed in the vignette above, is the massing of overwhelming effects of combat power to achieve a single purpose. During the offense, the BCT commander must avoid set patterns or obvious movements that would indicate the timing or direction of the attack. The commander designates, sustains, and shifts the decisive operation or main effort as necessary. The BCT concentrates combat power against the enemy using company level enhanced digital communications and information systems. Simultaneously, the BCT synchronizes information from adjacent units, higher headquarters, and unified action partners. Synchronizing allows the BCT to gain an understanding of the terrain and threat forces in its area of operations and to concentrate assigned or attached and echelon above brigade information collection capabilities to enable the commander's specific information requirements.

6-7. The division commander assists the BCT commander to achieve concentration by task organizing additional resources from augmenting units or forces from within the division. The division commander, through the division artillery headquarters, may provide additional artillery support from the division artillery (when tasked organized) or a field artillery brigade. If lacking external resources, the division commander for example, may direct the organic field artillery battalion of the BCT in reserve to reinforce the fires of the field artillery battalion organic to the BCT conducting the main effort until the reserve is committed. Another example is to direct the division artillery target acquisition platoon radars (when tasked organized) to provide coverage while BCT radars are moving.

TEMPO

6-8. Commanders build the appropriate tempo to provide the necessary momentum for successful attacks that achieve the objective. Controlling or altering tempo, as demonstrated in the vignette, by the commander's direct access to crucial joint intelligence, surveillance, and reconnaissance assets was essential to retaining the initiative and maintaining the rapid tempo of the operation. During the offense, rapid tempo focuses on key pieces of information and terrain at the tactical level. A rapid tempo entails a small number of tasks and allows attackers to penetrate barriers and defenses quickly to destroy enemy forces in-depth before they can react. A rapid tempo allows the BCT to deliver multiple blows in-depth from numerous directions to seize, retain, and exploit the initiative. Blows from multiple directions and multiple domains cause a multiple dimensional and domain dilemma for the enemy.

6-9. Commanders adjust the tempo to achieve synchronization. Speed is preferred to keep the enemy off balance. Establishing the conditions for offensive actions may require slowing the tempo as the pieces are set in place. Once ready, the tempo is increased, and the action takes place rapidly.

6-10. The BCT operations staff officer (S-3), in coordination with the commander, ensures combat operations flow smoothly during phases or transitions (sequels) between offensive and defensive operations. The right mix of forces available to quickly transition combat operations between the offense and defense enables the tempo necessary to maintain momentum.

AUDACITY

6-11. As seen in the vignette a simple but boldly executed plan of action, walling off the two neighborhoods south of Sadr City, demonstrated audacity through action to seize the initiative and press the battle. The offense favors the bold execution of plans. The BCT commander exercises audacity by developing inventive plans that produce decisive results while violently applying combat power. The commander compensates for any lack of information to develop the plan by developing the situation aggressively to seize the initiative, and then continuously engage in combat to exploit opportunities as they arise.

6-12. Audacity is a willingness to take bold risks. The BCT commander displays audacity by accepting risk commensurate with the value of the BCT's objective. The commander must understand when and where to take risks and avoid hesitation when executing the plan.

SECTION II – COMMON OFFENSIVE PLANNING CONSIDERATIONS

6-13. The BCT commander begins with a designated area of operations, identified mission, and assigned forces. The commander develops and issues planning guidance based on visualization relating to the physical means to accomplish the mission. The following paragraphs discuss activities, functions, and specific operational environments as the framework for discussing offensive planning considerations. (See FM 3-90-1 for additional information.)

COMMAND AND CONTROL

6-14. As with all operations, the BCT commander drives the operations process through the activities of understanding, visualizing, describing, directing, leading, and assessing. The BCT commander uses the principles of mission command (specifically mutual trust and shared understanding [see paragraph 4-2]) and inspires the BCT to accomplish the mission. The BCT commander develops shared understanding, by clearly stating the intent, assigns responsibility, delegates authority, and allocates resources to enable subordinates to take disciplined initiative, accept prudent risk, and act on mission orders to achieve success. For example, a movement to contact includes the general plan, direction, objectives, general organization of forces, general guidance of actions on contact, bypass criteria, and other guidance as required. The commander's location is also specified. (See ADP 6-0 and FM 6-0 for additional information.)

OPERATIONAL FRAMEWORK

6-15. The BCT commander and staff use four components of the operational framework to help conceptualize and describe the concept of operations in time, space, purpose, and resources. (See chapter 2.) First, the commander is assigned an area of operations for the conduct of operations. Second, the commander can designate deep, close, rear, and support areas to describe the physical arrangement of forces in time and space. Third, within this area, the commander conducts decisive, shaping, and sustaining operations to articulate the operation in terms of purpose. (See figure 2-4 on page 2-26.) In the fourth and final component, the commander designates the main and supporting efforts to designate the shifting prioritization of resources. (See FM 3-0 for additional information.)

Note. The BCT does not conduct operationally significant consolidate gains activities unless tasked to do so, usually within a division consolidation area.

OPERATIONS PROCESS

6-16. Command and control within the operations process involves a continuous development process of estimates, decisions, assigning tasks and missions, executing tasks and missions, and acquiring feedback. The operations process includes deriving missions, formulating concepts, and communicating the commander's intent successfully. Information products and the interpretations result in decisions and directives. Based on the commander's guidance, the staff recommends—

- Suspected enemy locations and courses of action (COAs).
- Formation and task organization of forces (planned two levels down, tasked one level down).
- Information collection plan (enemy's strength, disposition, and location).
- Decision points:
    - To support changes in the movement formation.
    - To counter/take advantage of enemy action.
    - To commit/not commit additional assets.
- Security plans to protect the main body.
- Priorities of fire.
- Bypass criteria.
- Missions (task and purpose) for subordinate units.
- Control measures.

INFORMATION

6-17. Command and control involves acquiring and displaying information. All units continually acquire information about the mission, enemy, terrain and weather, troops and support available, time available, civil considerations (METT-TC) through a variety of means. Units send and receive information, manage the means of communicating the information, and filter and maintain the information in a form that is convenient to the decision-making process. The commander records decisions as plans and orders that serve as input to the command and control process at the next lower echelon. Feedback from subordinate units provides input to the BCT's command and control process thus contributing to an ongoing process. (See FM 3-90-1 for additional information.)

CONTROL MEASURES

6-18. The BCT commander and staff select control measures, including graphics, to control subordinate units during operations associated with offensive operations (including subordinate tasks and special purpose attacks) and form of maneuver to establish responsibilities and limits that prevent subordinate units' actions from impeding one another. The commander and staff establish control measures that foster coordination and cooperation between forces without unnecessarily restricting freedom of action. Control measures should foster decision-making and subordinate unit initiative. For example, the lateral boundaries of the unit making the decisive operation are narrowly drawn to help establish the overwhelming combat power necessary at the area of penetration.

6-19. The commander assigns, as a minimum, an area of operations to every maneuver unit, a line of departure or line of contact; time of the attack or time of assault; phase lines (PLs); objective; and a limit of advance to control and synchronize the attack. The commander can use a *battle handover line*—a designated phase line where responsibility transitions from the stationary force to the moving force and vice versa (ADP 3-90)—instead of a limit of advance if the commander knows where the likely commitment of a follow-and-assume force will occur. The commander locates the limit of advance beyond the enemy's main defensive position to ensure that suitable terrain in immediate proximity to the objective is denied to the enemy, and that maneuver space is available for consolidation and reorganization in anticipation of an enemy counterattack. If the operation results in opportunities to exploit success and pursue a beaten enemy, the commander adjusts existing boundaries to accommodate the new situation. (See section IV for a discussion of control measures associated with offensive operations.)

AIRSPACE MANAGEMENT

6-20. Airspace management begins with the development of a unit airspace plan. There are two airspace control methods: positive and procedural. The Army primarily conducts procedural control of airspace use. Preplanned airspace coordinating measures that are integrated with each Army echelon's unit airspace plan and approved on the theater airspace control order makes procedural control highly effective. Although echelon above brigade units may exercise positive control methods over small areas for limited periods of time, BCTs are not able to perform positive control of airspace over any of their assigned area of operations without significant augmentation. *Positive control* is a method of airspace control that relies on positive identification, tracking, and direction of aircraft within an airspace, conducted with electronic means by an agency having the authority and responsibility therein (JP 3-52). *Procedural control* is a method of airspace control which relies on a combination of previously agreed and promulgated orders and procedures (JP 3-52). Properly developed airspace coordinating measures facilitate the BCT's employment of aerial and surface-based fires simultaneously. (See JP 3-52, FM 3-52, and ATP 3-52.1 for additional information on airspace control and ATP 3-91.1 for information on the joint air-ground integration center [JAGIC].)

JOINT, INTERORGANIZATIONAL, AND MULTINATIONAL TEAMS

6-21. The operational environment may require the BCT to maintain direct links with joint and multinational forces and U.S. and foreign governmental and nongovernmental organizations involved in the conflict, crisis, or instability. In many situations, such as when an adversary or enemy is primarily employing unconventional activities, the BCT benefits from exploiting the knowledge and capabilities residing within these organizations.

6-22. The BCT headquarters or subordinate elements actively participate in civil military operations and may synchronize their operations with those of different civil military organizations. Unity of effort with these organizations is essential and facilitates best through the exchange of a liaison officer. The fact that the BCT's communications systems may not be compatible with the civil-military organization increases the need for an exchange of knowledgeable liaison officers who are properly equipped to communicate according to the table of organization and equipment.

MOVEMENT AND MANEUVER

6-23. The BCT commander conducts movement and maneuver to avoid enemy strengths and to create opportunities to increase friendly fire effects. The commander makes unexpected maneuvers, rapidly changes the tempo of ongoing operations, avoids observation, and uses deceptive techniques and procedures to surprise the enemy. The commander overwhelms the enemy with one or more unexpected blows before the enemy has time to react in an organized fashion. Attacking the enemy force from an advantageous position in time and space, such as engaging the enemy from a location or at a time when unprepared, facilitates defeating the enemy force.

SEIZE, RETAIN, AND EXPLOIT THE INITIATIVE

6-24. The offense is the most direct means of seizing, retaining, and exploiting the initiative to gain a physical and psychological advantage. In the offense, the decisive operation is a sudden, shattering action directed toward enemy weaknesses and capitalizing on speed, surprise, and shock. If that operation fails to destroy an enemy, operations continue until enemy forces are defeated. The offense compels an enemy to react, creating new or larger weaknesses the attacking force can exploit.

6-25. The commander maneuvers to close with and destroy the enemy by close combat and shock effect. Close combat defeats or destroys enemy forces or seizes and retains ground. Close combat encompasses all actions that place friendly forces in immediate contact with the enemy where the commander uses fire and movement. Swift maneuver against several points supported by precise, concentrated fire can induce paralysis and shock among enemy troops and commanders. The key to success is to strike hard and fast, overwhelm a portion of the enemy force, and then quickly transition to the next objective or phase, thus maintaining the momentum of the attack without reducing the pressure on the enemy.

6-26. During combined arms operations, commanders compel the enemy to respond to friendly action. Such friendly actions nullify the enemy's ability to conduct their synchronized, mutually supporting reactions. The offense involves taking the fight to the enemy and never allowing enemy forces to recover from the initial shock of the attack.

6-27. The commander integrates and synchronizes all available combat power to seize, retain, and exploit the initiative, and to sustain freedom of movement and action. The commander employs joint capabilities across multiple domains when provided, such as close air support and surveillance assets to complement or reinforce BCT capabilities.

6-28. Air-ground operations (see chapter 4) support the commander's objectives. Relationships, common understanding, and mutual trust enhance the planning, coordination, and synchronized employment of ground and air maneuver. Air-ground operations require detailed planning of synchronized timelines, aviation task and purpose, and airspace management. Aircraft are limited in time due to fuel requirements and fighter management of aircrew duty day. The commander and staff use friendly timelines and synchronization matrixes to ensure air assets are at the right place at the right time, that they nest with the ground maneuver plan, and provide the desired effects to support the BCT mission. Aviation commanders and staffs use these timelines to manage aircrew duty day and aircraft readiness.

SCHEME OF MANEUVER

6-29. The scheme of maneuver covers the actions from before line of departure to consolidation and reorganization. The BCT operation order scheme of maneuver paragraph addresses the following:
- Task and purpose of subordinate elements.
- Reconnaissance and surveillance efforts and security operations.
- Actions at known or likely enemy contact locations.
- Scheme of fires.
- Direct fire control measures.
- Fire support coordination measures and airspace coordinating measures.
- Commander's critical information requirements (CCIRs).
- Methods for moving through and crossing dangerous areas.
- Movement formation and known locations where the formation changes.
- Employment of battlefield obscuration.
- Actions and array of forces at the final objective or limit of advance.
- Decision points and criteria for execution of maneuver options (attack, report and bypass, defend and retrograde), that may develop during execution.

6-30. On the objective once seized, assault forces conduct consolidation and reorganization. *Consolidation* is the organizing and strengthening a newly captured position so that it can be used against the enemy (FM 3-90-1). *Reorganization* is all measures taken by the commander to maintain unit combat effectiveness or return it to a specified level of combat capability (FM 3-90-1). During consolidation and reorganization, the BCT forces execute follow-on missions as directed. One mission is to continue the attack against targets of opportunity in the objective area. Whether a raid, attack (as a hasty or deliberate operation), or movement to contact, BCT subordinate units' posture and prepare for continued action and to defeat local counterattacks. The BCT commander may pass follow-on forces through assault forces or have them bypassed by other forces to continue the attack. On the objective, assault forces may be tasked to establish support by fire and attack by fire positions. BCT subordinate commanders or assault force leaders identify initial support by fire and attack by fire positions to reinforce the unit's limit of advance and to support follow-on missions.

MOBILITY

6-31. Engineer priority of support typically is to mobility, although it may rapidly change to countermobility in anticipation of an enemy attack. Engineer reconnaissance teams, from the brigade engineer battalion (see chapter 1, section I), join reconnaissance and security forces (see chapter 5) to reconnoiter obstacles based on an analysis of the mission variables of METT-TC. Planned suppression and obscuration fires support breaching operations. Additional combat engineers, task organized for breaching, reducing obstacles, and making expedient repairs to roads, trails, and ford sites may travel with the advance guard during a movement to contact. The combat engineers' purpose is to ensure that the advance guard and main body during a movement to contact, or an assault force during an attack remain mobile. (See ATP 3-34.22 for additional information.)

6-32. Mobility planning based upon the mission variables of METT-TC includes identifying requirements for military police support and augmentation. Military police contribute to the maneuver and mobility by—

- Preserving the freedom of movement over main supply routes (see ATP 3-39.10).
- Improving the protection of high-risk personnel and facilities during security and mobility (see ATP 3-39.30).
- Providing temporary detention operations (see FM 3-63) for detained individuals.
- Integrating police intelligence through operations (see ATP 3-39.20) to enhance situational understanding, protection, civil control, and law enforcement efforts.
- Providing military working dogs (patrol explosive detection dog) support to route clearance (see ATP 3-39.34).

6-33. The BCT provost marshal is responsible for requesting and coordinating military police assets and activities for the BCT. Mobility planning should integrate the security and mobility support discipline to support the BCT, when required, with a distribution of military police forces throughout the area of operations. (See FM 3-39 for additional information.)

PASSAGE OF LINES

6-34. A *passage of lines* is an operation in which a force moves forward or rearward through another force's combat positions with the intention of moving into or out of contact with the enemy (JP 3-18). Maneuver forces conduct passage of lines when at least one of the mission variables of METT-TC does not permit the bypass of a friendly unit. A passage of lines is a complex operation requiring close supervision and detailed planning, coordination, and synchronization between the commander of the unit conducting the passage and the unit being passed. A passage of lines occurs under two conditions:

- A *forward passage of lines* occurs when a unit passes through another unit's positions while moving toward the enemy (ADP 3-90).
- A *rearward passage of lines* occurs when a unit passes through another unit's positions while moving away from the enemy (ADP 3-90).

6-35. BCT units conduct a passage of lines to sustain the tempo of an offensive operation or to transfer responsibility from one unit to another to maintain the viability of the defense. Units also conduct passage of lines to transition from a delay or security operation by one force to a defense or to free a unit for another mission or task. A passage of lines involves transferring the responsibility for an area of operation between two commanders. That transfer of authority usually occurs when roughly two-thirds of the passing force has moved through the passage point or on meeting specific conditions agreed to by the commanders of the unit conducting the passage and the unit being passed.

MOVEMENT FORMATIONS

6-36. A *movement formation* is an ordered arrangement of forces for a specific purpose and describes the general configuration of a unit on the ground (ADP 3-90). The seven movement formations are column, line, echelon (left or right), box, diamond, wedge, and vee.

6-37. Movement formations are threat- or terrain-based. The BCT may use more than one formation within a given movement, especially if the terrain or enemy situation changes during a movement. For example, a battalion may use the column formation during the passage of lines and then change to another formation such as the wedge. Companies within the battalion formation may conduct movement-using formations different from that of the battalion. For example, one company may be in a wedge, another in an echelon right, and yet another in a column. Other factors, such as the distance of the move or enemy dispositions may prompt the commander to use more than one formation. Distances between units are METT-TC dependent.

6-38. Movement formations allow the unit to move in a posture suited to the commander's intent and mission. The commander considers the advantages and disadvantages of each formation to determine the appropriate formation for a situation. A series of movement formations may be appropriate during the course of an attack. All movement formations use one or more of the three movement techniques, which are traveling, traveling overwatch, and bounding overwatch.

6-39. During operations, the commander designates a movement formation to establish a geographic relationship between units and to posture for an attack. The commander considers probable reactions on enemy contact, indicates the level of security desired, and establishes the preponderant orientation of subordinate weapon systems when directing formations. The commander provides flexibility to subordinate units to shift from one formation to another based on changes to METT-TC. (See FM 3-90-1 for additional information.)

TROOP MOVEMENT

6-40. *Troop movement* is the movement of Soldiers and units from one place to another by any available means (ADP 3-90). The BCT commander must be able to move forces to a position of advantage relative to the enemy. Troop movement places troops and equipment at the destination at the proper time, ready for combat. METT-TC dictates the level of security required and the resulting speed of movement. (See FM 3-90-2 for additional information.) The three types of troop movement are administrative movement, tactical road march, and approach march.

Administrative Movement

6-41. *Administrative movement* is a movement in which troops and vehicles are arranged to expedite their movement and conserve time and energy when no enemy ground interference is anticipated (ADP 3-90). The commander only conducts administrative movements in secure areas. Examples of administrative movements include rail and highway movement in the continental United States. Once units deploy into a theater of war, commanders normally do not employ administrative movements. Since these types of moves are nontactical, the echelon assistant chief of staff, logistics or brigade/battalion logistics staff officer (S-4) usually supervises the movement. (See FM 4-01 for additional information.)

Tactical Road Marches

6-42. A *tactical road march* is a rapid movement used to relocate units within an area of operations to prepare for combat operations (ADP 3-90). The unit maintains security against enemy air attack and prepares to take immediate action against an enemy ambush, although contact with the enemy ground forces is not expected.

6-43. The march column is the organization for a tactical road march. All elements use the same route for a single movement under control of a single commander. The commander organizes a march column into four elements: reconnaissance, quartering party, main body, and trail party.

Approach March

6-44. An *approach march* is the advance of a combat unit when direct contact with the enemy is intended (ADP 3-90). An approach march emphasizes speed over tactical deployment. Commanders employ an approach march when they know the enemy's approximate location, since an approach march allows units to move with greater speed and less physical security or dispersion.

6-45. Units conducting an approach march are task organized before the march begins (for example, in an assembly area) to allow transition to another movement technique without slowing the tempo. Units likely to occupy assembly areas are units preparing to conduct a tactical movement or to move forward to execute a forward passage of lines (additionally includes a unit establishing a tactical reserve, completing a rearward passage of lines, and conducting reconstitution). The approach march terminates at a march objective, such as an attack position, assembly area, or assault position, or an approach march and can be used to transition to an attack. Follow and assume and reserve forces may conduct an approach march forward of a line of departure.

INTELLIGENCE

6-46. The BCT commander considers the entire area of operations, the enemy, and information collection activities (intelligence operations, reconnaissance, security operations, and surveillance) necessary to shape an operational environment and civil conditions (see chapter 5). Intelligence helps commanders visualize the operational environment, organize forces, and control operations to achieve objectives. Intelligence answers specific requirements focused in time and space. Intelligence leaders within the BCT ensure that the intelligence warfighting function operates effectively and efficiently. The intelligence staff officer is the BCT commander's primary advisor on employing information collection assets and driving information collection (see chapter 5).

6-47. Information collection is an activity that synchronizes and integrates the planning and employment of sensors and assets as well as the intelligence processing, exploitation, and dissemination of capabilities in direct support of current and future operation. The information collection plan should be the first consideration for the conduct of an offensive task. The BCT staff must integrate, synchronize, and coordinate the plan among the BCT subordinate units, with the higher echelon assets, and the other elements executing the overall information collection plan.

6-48. Information identified early and incorporated into the information collection plan includes potential enemy missions, COAs, objectives, defensive locations, uses of key terrain, avenues of approach and routes, enemy engagement areas (EAs), population locations and characteristics, and obstacles. Information collection supports situational understanding and intelligence support to targeting and information capabilities across all domains. Information collection efforts result in the timely collection and reporting of relevant and accurate information, which supports intelligence production and the commander's decision points. Information collection can disseminate as combat information, also.

6-49. Commanders use reconnaissance, security operations, surveillance, and intelligence operations to obtain information. All activities that help to develop understanding of the area of operations are information collection activities. Planners must understand all collection assets and resources available to them and the procedures to request or task collection from those assets and resources. For example, effective reconnaissance and security operations in the offense by the Cavalry squadron allows the BCT commander to gain and maintain contact with the enemy and to direct subordinate units into the fight at opportune times and places. Units within the squadron conducting reconnaissance and security orient on reconnaissance and security objectives. They fight for information to provide the commander with the necessary information to keep other BCT maneuver units free from contact as long as possible, so that they can concentrate on conducting the BCT's decisive operation. (See ATP 3-20.96 for additional information on the role of the Cavalry squadron within the intelligence warfighting function.)

6-50. Intelligence operations conducted by the military intelligence company (see chapter 1, section I) supports the BCT and its subordinate commands. The military intelligence company commander validates the ground collection and dissemination plan for military intelligence company organic and assigned collectors in support of the commander's offensive concept of operations. The company's scheme of information collection supports information requirements regarding locations, composition, equipment, strengths, and weaknesses of defending enemy forces. The company, through collection, analysis, and dissemination of intelligence information, supports the BCT and its subordinate units to locate high-payoff targets (HPTs) and enemy reconnaissance, armor, air assault, and air defense and withdrawal routes of enemy forces. The company provides analysis and intelligence synchronization support to the BCT intelligence staff officer (S-2) to locate enemy command and control facilities and electromagnetic warfare (EW) systems, obstacles, security zone, and main defensive area. Prior to and during movement and maneuver to an objective area, the company supports the BCT S-2 with maintaining a timely and accurate picture of the enemy situation to increase the BCT commander's situational understanding and to support the lethal and nonlethal targeting process. (See FM 2-0 and ATP 2-19.4 for additional information.)

FIRES

6-51. The BCT, in coordination with the field artillery headquarters, positions its field artillery batteries to provide continuous indirect fires. Battalions do the same with their heavy mortars. Companies often have their mortars follow behind the forward platoons, so they are prepared to provide immediate indirect fires.

Army attack reconnaissance helicopters and close air support may be available to interdict enemy counterattack forces or to destroy defensive positions.

6-52. The BCT plans for, integrates, coordinates, and synchronizes joint fires capabilities (sensors and weapon systems) into the BCT's scheme of maneuver to achieve synergy and provide redundancy in coverage from a particular asset. By definition, maneuver is the employment of forces in the operational area through movement in combination with fires to achieve a position of advantage in respect to the enemy.

6-53. During the offense, using preparation fires, counterfire, suppression fires, and cyberspace and EW assets provides the BCT commander with numerous options for gaining and maintaining fire superiority. The commander uses long-range artillery systems (cannon and rocket) and air support (rotary- and fixed-wing) to engage the enemy throughout the depth of the enemy's defensive positions. (See FM 3-09 and FM 3-12 for additional information.)

6-54. Fires can be time- or event-driven. The two types of triggers associated with a target are tactical (event-driven) and technical (time-driven). A tactical trigger is the maneuver related event or action that causes the initiation of fires. This event can be friendly, or enemy based and is usually determined during COA development. A technical trigger is the mathematically derived solution for fires based on the tactical trigger to ensure that fires arrive at the correct time and location to achieve the desired effects. Triggers can be marked using techniques similar to those for marking target reference points (TRPs).

6-55. The tactical air control party (TACP) is collocated with the fire support cell at the BCT main command post (CP). Air liaison officers and joint terminal attack controllers (JTACs) make up the TACP. The air liaison officer is the BCT commander's principal air support advisor. The air liaison officer leverages the expertise of the TACP with linkage to the higher (division and corps) echelon to plan, prepare, execute, and assess air support for BCT operations to include the integration of all forms of unified action partner fires. Joint fires observers may assist JTACs with conducting Type 2 or 3 terminal attack control of close air support or with the proper authorization, conduct autonomous terminal guidance operations. (See JP 3-09.3 for additional information.)

6-56. Considerations for supporting the scheme of maneuver during the offense include—
- Weight the main effort.
- Consider positioning fires assets to exploit weapons ranges.
- Prevent untimely displacement when fires are needed the most.
- Provide counterfire.
- Provide early warning and dissemination.
- Provide wide area surveillance.
- Provide fires to protect forces preparing for and assets critical to offensive actions.
- Disrupt enemy counterattacks.
- Plan fires to support breaching operations.
- Plan fires to deny enemy observation or screen friendly movements.
- Allocate responsive fires to support the decisive operation.
- Allocate fires for the neutralization of bypassed enemy combat forces.
- Plan for target acquisition and sensors to provide coverage of named area of interest (NAI), target area of interest (TAI), and critical assets.

6-57. The BCT's brigade aviation element (BAE) and air defense airspace management (ADAM) element, normally located in the fire support cell, are key monitoring and managing assets for the airspace over the area of operations. The ADAM and BAE element must process a unit airspace plan on time and maintain communications with the airspace element at Division JAGIC. The element must execute airspace management procedures per higher headquarters in accordance with Appendix 10 (Airspace Control) to have responsive fires from direct support artillery and mortars, and for the employment of other supporting airspace users (for example, unmanned aircraft system [UAS], close air support, and aviation). The BAE and ADAM element assist the commander to coordinate and employ air and missile defense and aviation assets to support the BCT's scheme of maneuver. (See FM 3-04 and ATP 3-01.50 for additional information.)

SUSTAINMENT

6-58. The BCT commander and staff normally plan for increased sustainment demands during the offense. Sustainment planners synchronize and coordinate with the entire BCT staff to determine the scope of the operation. Sustainment planners develop and continually refine the sustainment concept of support. Coordination between staff planners must be continuous to maintain momentum, freedom of action, prolong endurance, and extended operational reach. The brigade support battalion (BSB) commander anticipates where the greatest need may occur to develop a priority of support that meets the BCT commander's operational plan. Sustainment planners may consider positioning sustainment units in close proximity to operations to reduce critical support response times. Establishment of a forward logistics element (FLE) provides the ability to weight the main effort for the operation by drawing on all sustainment assets across the BCT. The commander and staff may consider alternative methods for delivering sustainment during emergencies. (See chapter 9 for a detailed discussion.)

6-59. Logistics within the BCT is planning and executing the movement and support of forces, synchronized with, and in support of, operations. During the offense, the most important commodities typically are fuel (class III bulk), ammunition (class V), and repair parts (class IX). Movement control is critical to ensuring supply distribution. The concept of support must include a responsive medical evacuation plan (see ATP 4-02.2) and resupply plan. Long lines of communication, dispersed forces, poor trafficability, contested terrain, and congested road networks are factors that impede the transportation system. The BCT commander must consider all of these factors when developing the distribution plan that supports the operational plan. (See FM 4-0 for additional information.)

6-60. Personnel services are sustainment functions that man and fund the force, promote the moral and ethical values, and enable the fighting qualities of the BCT during the conduct of operations. Personnel services staff planning is a continuous process that evaluates current and future operations from the perspective of the personnel services provider. Providers consider how the information being developed impacts personnel services that support each phase of the operation. During the offense, unit casualty reporting and personnel accountability demand will increase along with the demands to accomplish other support tasks due in great part to a higher tempo in maneuver and extended lines of communication. Personnel services complement logistics by planning and coordinating efforts that provide and sustain personnel. Personnel services within the BCT include human resources support, financial management operations, legal support, and religious support. (See FM 1-0 and ATP 1-0.1 for additional information.)

6-61. The burden on health service support (casualty care, medical evacuation, and medical logistics) increases due to the intensity of offensive actions and the increased distances over which support is required as the force advances. The BCT has organic medical resources within maneuver unit headquarters (brigade surgeon's section), the BCT (Role 2 medical company), and subordinate battalions or squadron (medical platoons). The commander reallocates medical resources as the tactical situation changes. Within the echelons above brigade, the medical command (deployment support) (MEDCOM [DS]) or the medical brigade (support) (known as MEDBDE [SPT]) serves as the medical force provider and is responsible for developing medical force packages for augmentation to the BCT as required. Slight differences exist between the medical capabilities or resources of the three BCTs due to differences in types and quantities of vehicles and numbers of personnel assigned; however, the mission remains the same for all health service support units or elements and they execute their mission in a similar fashion. (See ATP 4-02.5, ATP 4-02.2, and FM 4-02 for additional information.)

6-62. BCT planners must consider protection requirements to protect sustainment units against bypassed enemy forces. Planners must also factor time and distance when developing the offensive plan. The BSB and its supporting sustainment units must balance maintaining manageable distances to resupply the maneuver battalions and squadron and receiving resupply from their next higher sustainment echelon. The BSB commander must articulate to the BCT commander any potential sustainment shortfall risks as the BCT's offensive movement extends logistic lines of communication. The BSB receives its resupply from the supporting division sustainment support battalion (known as DSSB) within the division sustainment brigade. During the offense, the BSB must synchronize the operational plan with supporting higher sustainment echelons to ensure that echelons above brigade sustainment support is responsive as the maneuver plan is incorporated. (See chapter 9 for a detailed discussion.)

6-63. Transportation shortfalls can occur during the offense. BCT planners integrate a combination of surface and aerial delivery methods to meet distribution requirements. Distribution managers synchronize the BCT's movement plans and priorities according to the commander's priority of support. Regulating traffic management through movement control is essential to coordinate and direct movements on main supply routes and alternate supply routes. (See chapter 9 for a detailed discussion.)

6-64. The anticipated nature and tempo of the missions during the conduct of offensive operations normally associates with a higher casualty rate and an increase in requirements for medical resources and nonstandard transportation support. Additional combat and operational stress control teams may be required to treat casualties following operations. Higher casualty rates increase the emphasis on personnel accountability, casualty reports, and replacement operations. The offense support plan must incorporate religious support. Religious support through counseling and appropriate worship can help reduce combat and operational stress, increase unit cohesion, and enhance performance.

PROTECTION

6-65. Survivability operations enhance the ability to avoid or withstand hostile actions by altering the physical environment. Conduct of survivability operations in the offense (fighting and protective position development) is minimal for tactical vehicles and weapons systems. The emphasis lies on force mobility. Camouflage and concealment typically play a greater role in survivability during offensive operations than the other survivability operations. Protective positions for artillery, air and missile defense, and logistics positions, however, still may be required in the offense. Stationary CPs, and other facilities for the exercise of command and control, may require protection to lessen their vulnerability. The use of terrain provides a measure of protection during halts in the advance, but subordinate units of the BCT still should develop as many protective positions as necessary for key weapons systems, CPs, and critical supplies based on the threat level and unit vulnerabilities. During the early planning stages, geospatial engineer teams can provide information on soil conditions, vegetative concealment, and terrain masking along march routes to facilitate the force's survivability. (See ATP 3-37.34 for additional information.)

6-66. BCT forces engage in area security operations to protect the forces, installations, routes, areas, and assets across its entire area of operations. Area security normally is an economy-of-force mission, often designed to ensure the continued conduct of sustainment operations and to support decisive and shaping operations by generating and maintaining combat power. Area security operations often focus on an NAIs in an effort to answer CCIRs, aiding in tactical decision-making and confirming or denying threat intentions. In the offense, security forces engaged in area security operations typically organize in a manner that emphasizes their mobility, lethality, and communications capabilities. (See chapter 8, section IV.) As in all operations, the commander has the inherent responsibility to analyze the risks and implement control measures to mitigate them. The BCT commander and staff must understand and factor into their analysis how the execution of the operation could adversely affect Soldiers. Incorporating protection within the risk management (RM) integrating process ensures a thorough analysis of the risk and the implementation of controls to mitigate their effects. RM integration during the activities of the operations process is the primary responsibility of the operations officer and protection officer within the BCT. (See ATP 5-19.)

6-67. Air and missile defense planning in support of the BCT integrates protective systems by using the six employment guidelines, mutual support, overlapping fires, balanced fires, weighted coverage, early engagement, and defense in-depth, and additional considerations necessary to mass and mix air and missile defense capabilities. The BCT's ADAM element is a key monitoring and managing asset for the airspace over the BCT's area of operations. The ADAM element is usually located in the fire support cell with the BAE. The ADAM element assists the commander to employ air defense assets to support the scheme of maneuver. (See ATP 3-01.50 for additional information.)

6-68. The purpose of operations security is to reduce the vulnerability of the BCT from successful enemy exploitation of critical information. Operations security applies to all activities that prepare, sustain, or employ units of the BCT. The operations security process is a systematic method used to identify, control, and protect critical information and subsequently analyze friendly actions associated with the conduct of the offense. Tailored to the operations security process, intelligence preparation of the battlefield (IPB) is a useful methodology for the intelligence section to perform mission analysis on friendly operations. IPB provides insight into potential areas where the adversary could collect information and identify essential element of friendly information (EEFI). (See ADP 3-37 for additional information.)

6-69. Identification of EEFI assists operations security planners to ensure all operations security related critical unclassified information is included in the critical information list. Unlike security programs that seek to protect classified information and controlled unclassified information, operations security is concerned with identifying, controlling, and protecting unclassified information that is associated with specific military operations and activities. The BCT's operations security program and any military deception or survivability efforts should, as a minimum, conceal the location of the friendly objective, the decisive operation, the disposition of forces, and the timing of the offensive operation from the enemy or mislead the enemy regarding this information. (See JP 3-13.3 for additional information.)

6-70. The task provides intelligence support to protection alerts the commander to threats and assists in preserving and protecting the force. Intelligence support to protection includes providing intelligence that supports measures, which the BCT takes to remain viable and functional by protecting the force from the effects of threat activities. Intelligence support to protection includes analyzing the threats, hazards, and other aspects of an operational environment and utilizing the IPB process to describe the operational environment and identify threats and hazards that may influence protection. Intelligence support develops and sustains an understanding of the enemy, terrain and weather, and civil considerations that affect the operational environment. (See FM 2-0 and ATP 2-01.3 for additional information.)

6-71. Information collection can complement or supplement protection tasks (see chapter 5, section I). All-source analysts that the BCT receives depend on information collection assets internal and external to the BCT for accurate and detailed information about threats and relevant aspects of the operational environment. All-source analysts make the most significant contributions when they accurately assess possible threat events and actions. Assessments facilitate the commander's visualization and support decision-making. Intelligence planners use the plan requirements and assess collection task (see ATP 2-01) to answer specific requirements focused in time and space and identifying any threats to mission accomplishment. The intelligence staff of the BCT provides the commander with assessments that consider all aspects of threats, terrain and weather, and civil considerations. The commander should receive an estimate regarding the degree of confidence the intelligence officer places in each analytic assessment using assessments. (See FM 2-0 for additional information.)

6-72. The BCT commander and staff synchronize and integrate the planning and employment of sensors and assets (specifically reconnaissance and security forces) as well as the intelligence processing, exploitation, and dissemination capabilities in direct support of current and future operations. These assets and forces collect, process, and disseminate timely and accurate information to satisfy the CCIRs and other intelligence requirements. (See chapter 5.) When necessary, ground- and space-based reconnaissance and surveillance activities focus on special requirements, such as personnel recovery (see paragraphs 6-86 to 6-90) or ad hoc groupings such as patrols, ground convoys, combat outposts, and human intelligence (HUMINT) teams or civil affairs teams.

6-73. Within the BCT, *physical security* is that part of security concerned with physical measures designed to safeguard personnel; to prevent unauthorized access to equipment, installations, material, and documents; and to safeguard them against espionage, sabotage, damage, and theft (JP 3-0). The BCT employs physical security (see ATP 3-39.32) measures in-depth regardless of which element of decisive action (offense, defense, or stability) currently dominates to protect personnel, information, and critical resources in all locations and situations against various threats through effective security policies and procedures. This total system approach is based on the continuing analysis and employment of protective measures, including physical barriers, clear zones, lighting, access and key control, intrusion detection devices, biometrically-enabled base access systems, CP security, defensive positions both hasty and deliberate, and nonlethal capabilities. The goal of physical security systems is to employ security in-depth to preclude or reduce the potential for sabotage, theft, trespass, terrorism, espionage, or other criminal activity. To achieve this goal, each security system component has a function and related measures that provide an integrated capability for—

- Deterrence to a potential aggressor.
- Detection measures to sense an act of aggression.
- Assessment of an unauthorized intrusion or activity.

- Delay measures to protect assets from actual or perceived intrusion.
- Response measures to assess:
  - Unauthorized acts.
  - Report detailed information.
  - Defeat an aggressor.

6-74. Subordinate units of the BCT may be involved in area security in an economy-of-force role to protect lines of communications, convoys, and critical fixed sites and radars during the conduct of the offense. Units identify antiterrorism measures through mission analysis to counter terrorist tactics. The BCT commander, with the assistance of the antiterrorism officer and staff, assesses the threat, vulnerabilities, and criticality associated with conducting the offense. The BCT's protection cell provides staff oversight and recommends the emplacement of security forces to thwart identified threats and to conduct populace and resource control. The protection cell increases overall protection through implementation of antiterrorism measures to protect the force. Staff members weight the probability of terrorist organizations attacking forces en route to execute offensive operations within the protection cell. Staff members analyze the susceptibility of terrorist attacks on other BCT subordinate units along lines of communications with the reduction of available combat forces other than an economy of force role by security forces. (See ATP 3-37.2 for additional information.)

6-75. Military police support to protection, when requested and received, includes security and mobility support (see ATP 3-39.30), detention (specifically detainee operations [see FM 3-63]), and police operations (see ATP 3-39.10). The security and mobility support discipline, discussed earlier under mobility, provides the BCT with a distribution of military police forces throughout the area of operations. These military police forces support mobility operations, and conduct area security, local security, main supply route regulation enforcement, and populace and resource control. Military police forces, when assigned, patrol aggressively and conduct reconnaissance to protect units, critical facilities, high-risk personnel, and civilian populations within the BCT's area of operations. Military police support planning includes identifying requirements for task organization of military police elements. (See FM 3-39 for additional information.)

6-76. Military police support to the offense includes missions and tasks that support uninterrupted movement, allow maneuver forces to preserve combat power so that it may be applied at decisive points and times, and foster rapid transitions in operations. Military police operations supporting the offense include the simultaneous application of military police capabilities. Military police operations in close support of maneuver forces are the primary focus during offensive operations; however, military police apply all three disciplines simultaneously to some degree. The primary focus is support that enables movement and maneuver, provides detention tasks to support captured or detained individuals, and provides protection.

6-77. Military police operations during the conduct of the offense include early shaping operations to establish conditions for preparing follow-on efforts for civil security and civil control. Military police operations, in concert with other elements, begin the initial efforts to—

- Restore and maintain order in areas passed by maneuver forces.
- Assess the criminal environment and begin the identification of criminal elements.
- Identify and establish rapport with existing host-nation police or friendly security elements.

6-78. Force health protection encompasses measures to promote, improve, or conserve the mental and physical well-being of Soldiers. (See ATP 4-02.8.) Force health protection measures enable a healthy and fit force, prevent injury and illness, protect the force from health hazards, contributes to esprit de corps, resilience, and a professional organizational climate to sustain or create forward momentum or eliminate negative momentum, and include the prevention aspects of—

- Preventive medicine (medical surveillance, occupational and environmental health surveillance).
- Veterinary services (food inspection, animal care missions, prevention of zoonotic disease transmissible to humans).
- Combat and operational stress control.
- Dental services (preventive dentistry).
- Laboratory services (area medical laboratory support).

Note. AR 600-20 addresses the policy and ADP 1 addresses the expectation commanders and staffs will strive to create for a professional command climate within a culture of trust. Prevention and mitigation of "moral injury" is part of leadership and a command responsibility.

6-79. Soldiers must be physically and behaviorally fit; therefore, programs must promote and improve the capacity of personnel to perform military tasks at high levels, under extreme conditions, and for extended periods. Preventive and protective capabilities include physical exercise, nutritional diets, dental hygiene and restorative treatment, combat and operational stress management, rest, recreation, and relaxation geared to individuals and organizations. (See ATP 4-02.3.)

6-80. *Countering weapons of mass destruction* is the efforts against actors of concern to curtail the conceptualization, development, possession, proliferation, use, and effects of weapons of mass destruction, related expertise, materials, technologies, and means of delivery (JP 3-40). At the tactical level combined arms teams conduct specialized activities to understand the environment, threats, and vulnerabilities; control, defeat, disable, and dispose of weapons of mass destruction (WMD); and safeguard the force and manage consequence. (See ATP 3-90.40 for additional information.)

6-81. Countering weapons of mass destruction (CWMD) is described as actions undertaken in a hostile or uncertain environment to systematically locate, characterize, secure, and disable, or destroy WMD programs and related capabilities. Collecting forensic evidence from the WMD program during CWMD is a priority for ascertaining the scope of a WMD program and for follow-on attribution. Nuclear disablement teams (specialized forces) perform site exploitation and disable critical radiological and nuclear infrastructure during CWMD. (See ATP 3-37.11 for additional information.)

6-82. CWMD missions require extensive collaborative planning, coordination, and execution oversight by the BCT commander and staff, and subordinate commanders. CWMD will likely involve teams of experts to include both technical forces (but are not limited to, chemical, biological, radiological, and nuclear [CBRN] reconnaissance teams, hazardous response teams, CBRN dual-purpose teams, and explosive ordnance disposal elements) and specialized forces (but are not limited to, technical escort units, nuclear disablement teams, and chemical analytical remediation activity elements). Associated planning will begin at echelons above the BCT characterized by centralized planning and decentralized execution of CWMD missions to ensure that the right assets are provided. (See FM 3-94, ATP 3-91, and ATP 4-32.2 for additional information.)

6-83. CWMD operations may be lethal or nonlethal as indicators are identified that meet the CCIRs and priority intelligence requirements suggesting that a site contains sensitive information. CWMD operations may develop intelligence that feeds back into the planning process to include the IPB and targeting process. The priority for CWMD activities is to reduce or eliminate the threat. CWMD operations may be conducted under two circumstances—planned and opportunity. While planned operations are preferred, some operations involving WMD sensitive sites may occur because the opportunity presents itself during operations to accomplish another mission. Not every operation requires destruction tasks—tactical isolation or exploitation may be the only elements executed. Nonetheless, the BCT commander and staff, and subordinate commanders always consider each element of CWMD operations (isolation, exploitation, destruction, monitoring, and redirection) and its relevance to the situation. A particular element may be unnecessary, but making that judgment is the appropriate level commander's responsibility. (See ATP 3-11.23 for additional information.)

6-84. An explosive ordnance disposal company, when tasked, provides explosive ordnance disposal, protection planning, and operations support to the BCT and subordinate battalions. The explosive ordnance disposal company supporting the BCT may provide an operations officer and noncommissioned officer to the BCT to provide appropriate explosive ordnance disposal planning and to perform liaison officer duties that include facilitating cooperation and understanding among the BCT commander, staff, subordinate battalions and the squadron, and explosive ordnance disposal battalion and company commanders. The explosive ordnance disposal company coordinates tactical matters to achieve mutual purpose, support, and action. In addition, the company ensures precise understanding of stated or implied coordination measures to achieve synchronized results. (See ATP 4-32, ATP 4-32.1, and ATP 4-32.3 for additional information.)

6-85. Explosive ordnance disposal elements supporting subordinate maneuver units can neutralize hazards from conventional unexploded ordnance, explosives and associated materials, improvised explosive devices (IEDs), booby traps containing both conventional explosives and CBRN explosives that present a threat to those units. These elements may dispose of hazardous foreign or U.S. ammunition, unexploded ordnance, individual mines, booby-trapped mines, and chemical mines. Breaching and clearance of minefields is primarily an engineer responsibility. (See ATP 4-32.2 and ATP 4-32.3 for additional information about unexploded ordnance procedures.)

Note. (See chapter 7 for information on the CBRN environment, CBRN defense measures, and CBRN working group.)

6-86. The BCT commander is committed to the safety and security of the members of and attached to the BCT. The commander emphasizes personnel recovery throughout the operations process to prevent forces or individual Soldiers from becoming isolated, missing, or captured. Individuals or groups become isolated for a variety of reasons, including their own behavior, enemy actions, and interaction with the physical environment. The BCT commander and staff must guard against treating personnel recovery as episodic, must anticipate requirements, and integrate personnel recovery throughout all operations. The commander develops three interrelated categories of information to exercise command and control of personnel recovery: personnel recovery guidance, isolated Soldier guidance, and evasion plan of action.

6-87. The BCT commander and staff must have an understanding of the complex, dynamic relationships among friendly forces and enemies and the other aspects of the operational environment (including the populace). This understanding helps the commander visualize and describe the intent for personnel recovery and to develop focused planning guidance. Effective personnel recovery planning guidance accounts for the operational environment and the execution of operations. Personnel recovery guidance provides a framework for how the BCT and subordinate units synchronize the actions of isolated personnel and the recovery force. As the commander develops personnel recovery guidance for subordinate units, the commander must ensure that subordinates have adequate combat power for personnel recovery. The commander must also define command relationships with the requisite flexibility to plan and execute personnel recovery operations.

6-88. The commander translates personnel recovery guidance into recommendations usually known as isolated Soldier guidance. Isolated Soldier guidance focuses on awareness, accountability, and rapid reporting of isolation incidents. Isolated Soldier guidance anticipates the potential situation. As with personnel recovery guidance, there is no set format. At the BCT level, where there are no dedicated personnel recovery staff officers and noncommissioned officers, the guidance is a part of the general protection guidance. The commander gives guidance for developing isolated Soldier guidance during initial planning and establishes isolated Soldier guidance for operations in any area with a risk of isolation.

6-89. The commander determines if units or individuals require an evasion plan of action. Typically, evasion plans of action contain specific instructions developed for short-term aviation operations (air movements and air assaults) and ad hoc groupings such as combat and reconnaissance patrols, ground convoys, combat outposts, and HUMINT teams or civil affairs teams. These operations and ad hoc groupings develop an evasion plan of action when the risk of isolation is elevated and make modification to the plan when conditions change.

6-90. Personnel recovery guidance, isolated Soldier guidance, and evasion plan of action apply also to civilians and contractors. Because the isolated person may include Department of Defense civilians and contractors authorized to accompany the force, the BCT commander and staff must develop a communications program to inform these individuals. Civilian and contractor members of the organization need the guidance necessary for their safety, especially the isolated Soldier guidance that enables them to contribute to prevention, preparation, and self-recovery if they become isolated. When dealing with local national or third-country national contractors, culture and language complicates this process. (See FM 3-50 for additional information.)

6-91. The conduct of offensive operations often requires the temporary resettlement of dislocated civilians (see FM 3-39 and FM 3-57) and the conduct of detainee operations (see FM 3-63). The BCT can expect to accumulate a sizeable number of dislocated civilians or detainees, all with varying classifications, depending on the situation. The BCT monitors the actual number closely to avoid devoting too many or too few resources to the performance of dislocated civilian operations or detainee operations. The BCT protection cell works with the sustainment cell to ensure resources are available to construct and operate dislocated civilian camps or detention facilities for individuals acquired during the conduct of the mission. Military police and civil affairs organize to establish and support dislocated civilian operations and detainee operations.

SECTION III – FORMS OF MANEUVER

6-92. *Forms of maneuver* are distinct tactical combinations of fire and movement with a unique set of doctrinal characteristics that differ primarily in the relationship between the maneuvering force and the enemy (ADP 3-90). The forms of maneuver are envelopment, turning movement, infiltration, penetration, and frontal assault. Combined arms organizations synchronize the contributions of all units to execute the forms of maneuver to accomplish the mission. The BCT commander generally chooses one form to build a COA. The higher commander rarely specifies the specific offensive form of maneuver; however, commander's guidance and intent, along with the mission, may impose constraints such as time, security, and direction of attack that narrows the form of maneuver to one option. The area of operations' characteristics and the enemy's dispositions also determine the offensive form of maneuver selected. A single operation may contain several forms of maneuver, such as a frontal assault to clear an enemy security zone, followed by a penetration to create a gap in the enemy's defense.

> *Note.* Flank attack is no longer used as a defined Army term or listed as a form of maneuver. Commanders seek to engage an enemy's *assailable flank*—a flank exposed to attack or envelopment (ADP 3-90). An exposed flank usually results from the terrain, the weakness of forces, the technical capability of an opponent, or a gap between adjacent units. See paragraphs 6-93 through 6-95 for additional information.

ENVELOPMENT

6-93. An *envelopment* is a form of maneuver in which an attacking force seeks to avoid the principal enemy defenses by seizing objectives behind those defenses that allow the targeted enemy force to be destroyed in their current positions (FM 3-90-1). At the BCT level, envelopments focus on seizing terrain, destroying specific enemy forces, and interdicting enemy withdrawal routes. The BCT commander's decisive operation focuses on attacking an assailable flank. The operation avoids the enemy's front, which is generally the enemy's strength, where the effects of fires and obstacles are the greatest. The BCT commander prefers to conduct an envelopment instead of a penetration or a frontal assault because the attacking force tends to suffer fewer casualties while having the most opportunities to destroy the enemy. If no assailable flank is available, the attacking force creates one.

6-94. The BCT commander uses boundaries to designate areas of operations for each unit participating in the envelopment. The commander designates PLs, support by fire and attack by fire positions, and contact points. The commander also designates appropriate fire support coordination measures, such as a restricted fire line or boundaries between converging forces, and any other control measures the commander feels are necessary to control the envelopment. The three variations of envelopment are single envelopment, double envelopment, and vertical envelopment.

> *Note.* An encirclement, no longer listed as a type of envelopment, typically results from penetrations and envelopments, or is an extension of exploitation and pursuit operations. See paragraphs 6-99 through 6-101 for information on encirclements.

SINGLE ENVELOPMENT

6-95. *Single envelopment* is a form of maneuver that results from maneuvering around one assailable flank of a designated enemy force (FM 3-90-1). The commander envisioning a single envelopment organizes forces into the enveloping force and the fixing force. (See figure 6-1.) The commander also allocates forces to conduct reconnaissance, security, reserve, and sustaining operations. The enveloping force, conducting the decisive operation, attacks an assailable enemy flank and avoids the enemy's main strength en route to the objective. The fixing force conducts a frontal assault as a shaping operation to fix the enemy in its current positions to prevent its escape and reduce its capability to react against the enveloping force. (See FM 3-90-1 for additional information.)

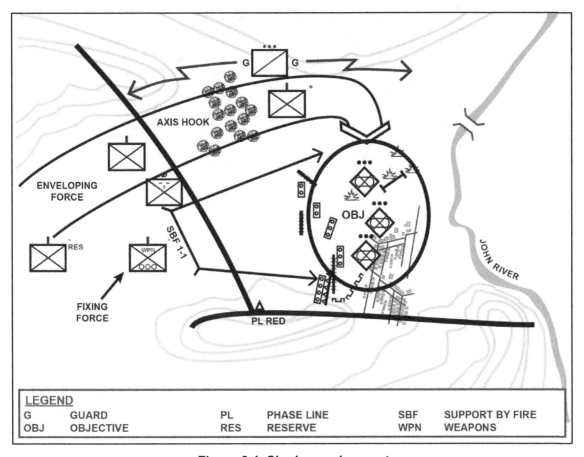

Figure 6-1. Single envelopment

DOUBLE ENVELOPMENT

6-96. *Double envelopment* results from simultaneous maneuvering around both flanks of a designated enemy force (FM 3-90-1). A commander executing a double envelopment organizes friendly forces into two enveloping forces and one fixing force in addition to allocating reconnaissance, security, reserve, and sustaining forces. (See figure 6-2 on page 6-20.) The commander typically designates the more important of the two enveloping forces as the main effort for resources. The enveloping force is the commander's decisive operation if its action accomplishes the mission. Maneuver control graphics, and appropriate fire coordination measures, such as a restrictive fire line or boundary between converging forces and communications are essential to mission accomplishment and preventing fratricide during this envelopment. (See FM 3-90-1 for additional information.)

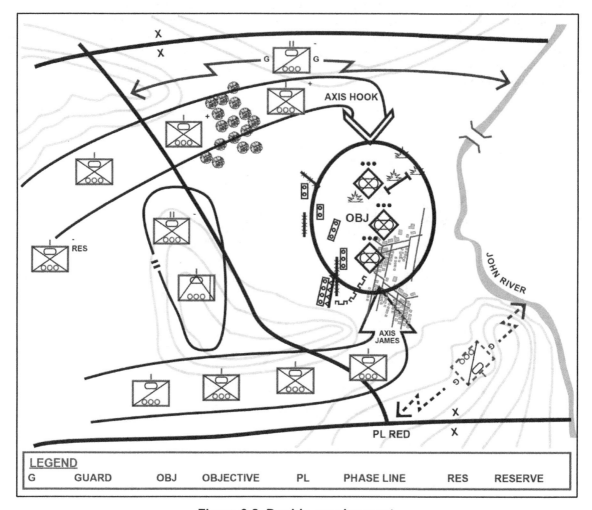

Figure 6-2. Double envelopment

VERTICAL ENVELOPMENT

6-97. Vertical envelopment is a variation of envelopment where airdropped or air-landed troops, attack the rear and flanks of a force, to cut off or encircle that force. For a discussion of airborne and air assault operations, see FM 3-99. A vertical envelopment, airborne assault or air assault (see figure 6-3), allows the commander to threaten the enemy's rear areas causing the enemy to divert combat elements to protect key terrain, vital bases or installations, and lines of communications. An *airborne assault* is the use of airborne forces to parachute into an objective area to attack and eliminate armed resistance and secure designated objectives (JP 3-18). An *air assault* is the movement of friendly assault forces by rotary wing or tiltrotor aircraft to engage and destroy enemy forces or to seize and hold key terrain (JP 3-18). Vertical envelopment allows the commander to—

- Overcome distances quickly, overfly barriers, and bypass enemy defenses.
- Extend the area over which the commander can exert influence.
- Disperse reserve forces widely for survivability reasons while maintaining their capability for effective and rapid response.
- Exploit combat power by increasing tactical mobility.

6-98. Entry operations, airborne operations or air assault operations, occupy advantageous ground to shape the operational area and accelerate the momentum of the engagement. An *airborne operation* is an operation involving the air movement into an objective area of combat forces and their logistic support for execution of a tactical, operational, or strategic mission (JP 3-18). An *air assault operation* is an operation in which assault forces, using the mobility of rotary-wing or tiltrotor aircraft and the total integration of available fires, maneuver under the control of a ground or air maneuver commander to engage enemy forces or to seize and hold key terrain (JP 3-18). An enemy may or may not be in a position to oppose the maneuver. While the commander should attempt to achieve an unopposed landing when conducting a vertical envelopment, the assault force must prepare for the presence of opposition. (See FM 3-99 for additional information.)

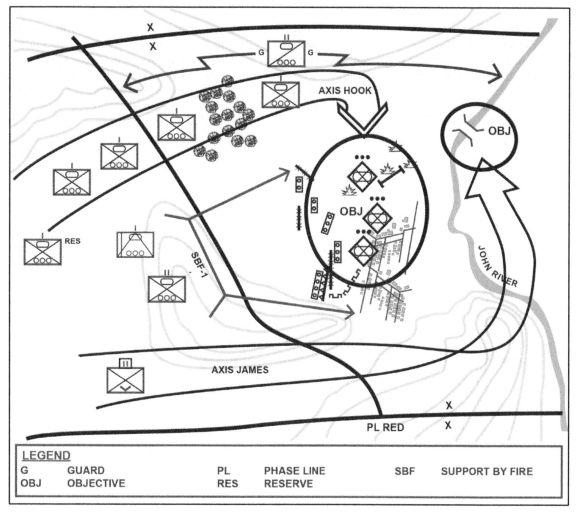

Figure 6-3. Vertical envelopment (example air assault)

ENCIRCLEMENT

6-99. Encirclement is a result of envelopment where a force loses its freedom of maneuver because an opposing force is able to isolate it by controlling all ground lines of communications and reinforcement. The commander conducts offensive encirclements to isolate an enemy force. Typically, encirclements result from penetrations and envelopments, or are extensions of exploitation and pursuit operations. As such, encirclements are not a separate form of offensive operations but an extension of an ongoing operation. Encirclements may be planned sequels or result from exploiting an unforeseen opportunity. Encirclements usually result from the linkup of two encircling arms conducting a double envelopment. However, encirclements can occur in situations where the attacking commander uses a major obstacle, such as a shoreline, as a second encircling force. Although a commander may designate terrain objectives in an encirclement, isolating and defeating enemy forces are the primary goals. Ideally, an encirclement results in the surrender of the encircled force. This minimizes friendly force losses and resource expenditures.

6-100. *Encirclement operations* are operations where one force loses its freedom of maneuver because an opposing force is able to isolate it by controlling all ground lines of communication and reinforcement (ADP 3-90). An encirclement operation usually has at least two phases-the actual encirclement and the action taken against the isolated enemy. The commander considers adjusting subordinate units' task organizations between phases to maximize unit effectiveness in each phase. The first phase is the actual encirclement that isolates the enemy force. The organization of forces for an encirclement is similar to that of a movement to contact or an envelopment. The commander executing an encirclement operation organizes encircling forces into a direct pressure force and one or more encircling arms. (See figure 6-4.)

6-101. The commander organizes an inner encircling arm only if there is no possibility of the encircled forces receiving relief from enemy forces outside the encirclement. The commander organizes both inner and outer encircling arms if there is any danger of an enemy relief force reaching the encircled enemy force. The commander assigns the outer encircling arm a security mission, an offensive mission to drive away any enemy relief force, or a defensive mission to prevent the enemy relief force from making contact with the encircled enemy force. Once the encirclement is complete, these inner or outer encircling arms form a perimeter. (See FM 3-90-2 for additional information.)

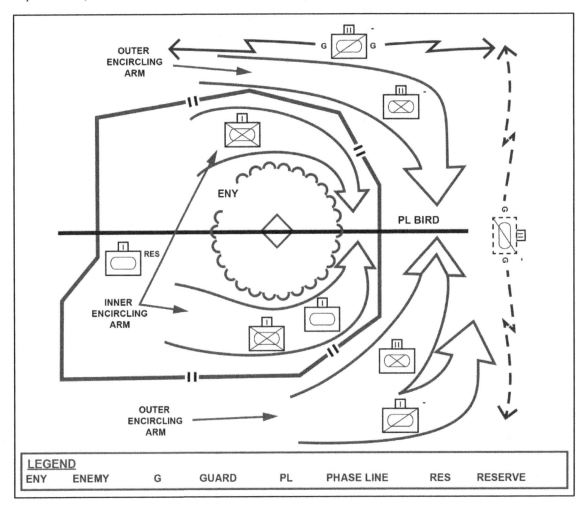

Figure 6-4. Encirclement operations

TURNING MOVEMENT

6-102. A *turning movement* is a form of maneuver in which the attacking force seeks to avoid the enemy's principle defensive positions by seizing objectives behind the enemy's current positions thereby causing the enemy force to move out of their current positions or divert major forces to meet the threat (FM 3-90-1). A turning movement differs from an envelopment in that the turning movement force seeks to make the enemy displace from current locations, whereas an enveloping force seeks to engage the enemy in its current location from an unexpected direction. A turning movement is particularly suited when forces possess a high degree of tactical mobility. Commanders frequently use a turning movement to transition from an attack to an exploitation or pursuit.

6-103.   The BCT commander organizes friendly forces into a turning force, a main body, and a reserve. Either the turning force or the main body can conduct the decisive operation based on the situation. Normally, a turning force conducts the majority of its operations outside of the main body's supporting range and distance; therefore, the turning force must contain sufficient combat power and sustainment capabilities to operate independently of the main body for a specific period. The turning force seizes vital areas to the enemy's rear before the main enemy force can withdraw or receive support or reinforcements. The maneuver of the turning force causes the enemy to leave its position.

6-104.   The commander organizes the main body, so the turning force is successful. The main body conducts operations, such as attacks to divert the enemy's attention away from the area where the turning force maneuvers. The main body can be the decisive or shaping operation. The commander organizes the reserve to exploit success of the turning force or the main body. The reserve also provides the commander with the flexibility to counter unexpected enemy actions.

6-105.   The BCT commander establishes boundaries to designate the area of operations for each force participating in the turning movement. (See figure 6-5.) The commander designates control measures, such as PLs, contact points, objectives, limit of advance, and appropriate fire support coordination measures to synchronize the operation. (See FM 3-90-1 for additional information.)

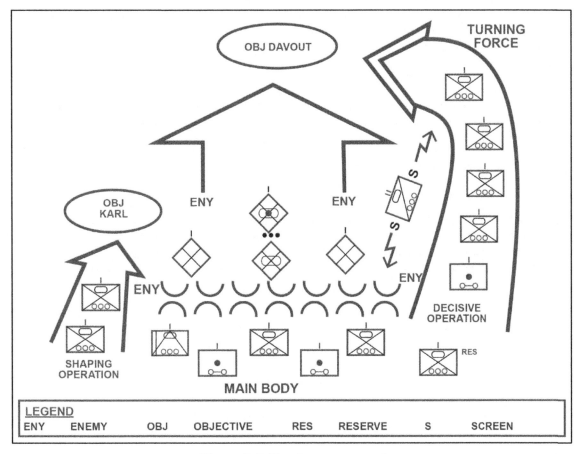

Figure 6-5. Turning movement

INFILTRATION

6-106.   An *infiltration* is a form of maneuver in which an attacking force conducts undetected movement through or into an area occupied by enemy forces to occupy a position of advantage behind those enemy positions while exposing only small elements to enemy defensive fires (FM 3-90-1). Infiltration occurs by land, water, air, or a combination of means. Moving undetected by enemy forces is paramount to success. Moving and assembling forces covertly through enemy positions takes a considerable amount of time. Limits on the size and strength of an infiltrating force mean that the force can rarely defeat an enemy force alone. The commander uses infiltration to—

●   Support other forms of maneuver.

●   Attack lightly defended positions or stronger positions from the flank and rear.

●   Secure key terrain in support of the decisive operation.

●   Disrupt or harass enemy defensive preparations or operations.

●   Relocate maneuver units by moving to battle positions around an EA.

●   Reposition to attack vital facilities or enemy forces from the flank or rear.

6-107.   The infiltrating force's size, strength, and composition are limited usually. The infiltrating unit commander organizes the main body into one or more infiltrating elements. The largest element that is compatible with the requirement for stealth and ease of movement conducts the infiltration. This increases the commander's control, speeds the execution of the infiltration, and provides responsive combat power. The exact size and number of infiltrating elements are situation dependent. The commander with the responsibility for the infiltration considers the following factors when determining how to organize friendly forces. Smaller infiltrating elements are not as easy to detect and can get through smaller defensive gaps. Even the detection of one or two small elements by the enemy does not prevent the unit from accomplishing its mission in most cases. Larger infiltrating elements are easier to detect, and their discovery is more apt to endanger the success of the mission. In addition, larger elements require larger gaps to move through as opposed to smaller elements. A unit with many smaller infiltrating elements requires more time to complete the infiltration and needs more linkup points than a similar size unit that has only a few infiltrating elements. Many infiltrating elements are harder to control than fewer, larger elements. The commander may establish security forces that move ahead of, to the flanks of, and/or to the rear of each infiltrating element's main body to provide early warning, reaction time, and maneuver space. The sizes and orientations of security elements are situation dependent. Each infiltrating element is responsible for its own reconnaissance effort, if required. Sustainment of an infiltrating force normally depends on the force's basic load of supplies and medical and maintenance assets accompanying the infiltrating force. After completing the mission, the commander reopens lines of communication to conduct normal sustainment operations.

6-108.   The commander responsible for the infiltration establishes routes and boundaries to designate the area of operations for the unit(s) conducting the infiltration. (See figure 6-6.) The commander also designates additional control measures as necessary to synchronize the operations of subordinates. Additional control measures include one or more infiltration lanes, a line of departure or points of departure, movement routes, linkup or rally points, objective rally point (known as ORP), assault positions, objectives, and a limit of advance. (See ATP 3-21.10 for a detailed discussion of infiltration missions and special purpose attacks.)

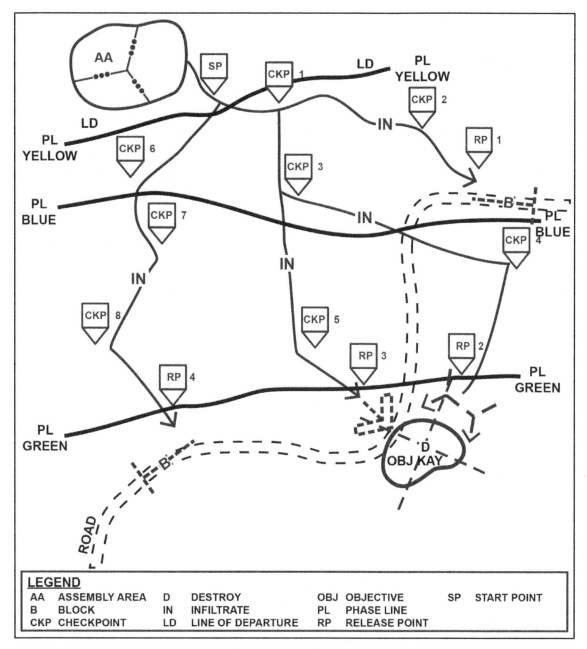

Figure 6-6. Infiltration

6-109. An *infiltration lane* is a control measure that coordinates forward and lateral movement of infiltrating units and fixes fire planning responsibilities (FM 3-90-1). Single or multiple infiltration lanes can be planned. Using a single infiltration lane—facilitates navigation, control, and reassembly, reduces susceptibility to detection, reduces the area requiring detailed intelligence, and increases the time required to move the force through enemy positions. (See figure 6-7 on page 6-26.) Using multiple infiltration lanes—reduces the possibility of compromise, allows more rapid movement, and makes control more challenging.

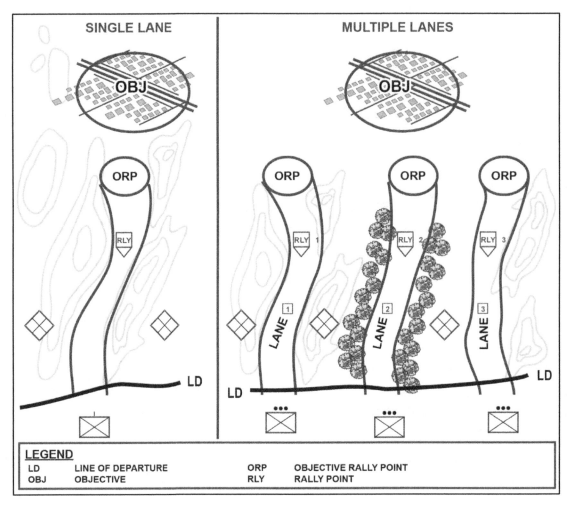

Figure 6-7. Infiltration lane

PENETRATION

6-110. A *penetration* is a form of maneuver in which an attacking force seeks to rupture enemy defenses on a narrow front to disrupt the defensive system (FM 3-90-1). Destroying the continuity of the enemy's defense causes the enemy's isolation and defeat in detail. The penetration extends from the enemy's security zone through the main defensive positions and the rear area. A commander executes a penetration when time pressures do not permit an envelopment, there is no assailable flank, enemy defenses are overextended, and weak spots are detected in the enemy's positions through reconnaissance, surveillance, and security operations.

6-111. Penetrating a well-organized position requires massing overwhelming combat power at the point of penetration and combat superiority to continue the momentum of the attack. The BCT commander designates a breach force, support force, assault force, and a reserve. The commander can designate these elements for each defensive position that requires penetration. The commander assigns additional units follow and support or follow-and-assume missions to ensure rapid exploitation of initial success. The commander designates forces to fix enemy reserves in their current locations and isolate enemy forces within the area selected for penetration. (See figure 6-8.)

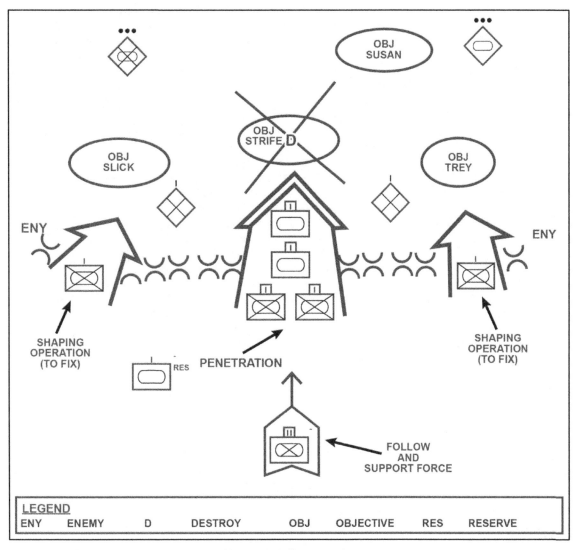

Figure 6-8. Penetration

6-112. The commander assigns, as a minimum, an area of operations to every maneuver unit, a line of departure or a line of contact, a time of the attack or a time of assault, a PL, an objective, and a limit of advance to control and synchronize the attack. (See figure 6-9 on page 6-28.) The commander can use a battle handover line instead of a limit of advance if the commander knows where to commit a follow-and-assume force. The commander designates the limit of advance beyond the enemy's main defensive position. If the operation results in opportunities to exploit success and pursue a beaten enemy, the commander adjusts existing boundaries to accommodate the new situation. (See FM 3-90-1 for additional information.)

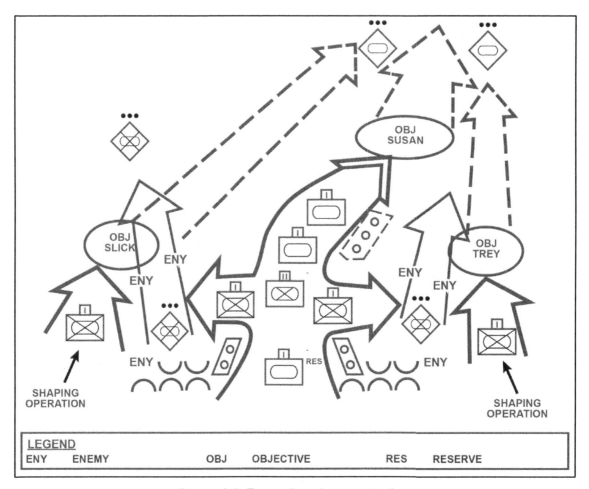

Figure 6-9. Expanding the penetration

FRONTAL ASSAULT

6-113.   A frontal assault is a form of maneuver in which an attacking force seeks to destroy a weaker enemy force or fix a larger enemy force in place over a broad front. The BCT commander uses a frontal assault as a shaping operation in conjunction with other forms of maneuver. The commander employs a frontal assault to clear enemy security forces, overwhelm a depleted enemy during an exploitation or pursuit, and to fix enemy forces in place. The BCT commander conducts a frontal assault when assailable flanks do not exist. While a penetration is a sharp attack designed to rupture the enemy position, the BCT commander designs a frontal assault to maintain continuous pressure along the entire front until either a breach occurs or the attacking forces succeed in forcing back the enemy. Frontal assaults conducted without overwhelming combat power are seldom decisive. (See figure 6-10.)

6-114.   A unit conducting a frontal assault normally has a wider area of operations than a unit conducting a penetration does. A commander conducting a frontal assault may not require any additional control measures beyond those established to control the overall mission. Control measures include an area of operations defined by unit boundaries, and an objective, at a minimum. The commander uses other control measures necessary to control the attack, including attack positions, lines of departure, PLs, assault positions, limits of advance, and direction of attack or axis of advance for every maneuver unit. (See FM 3-90-1 for additional information.)

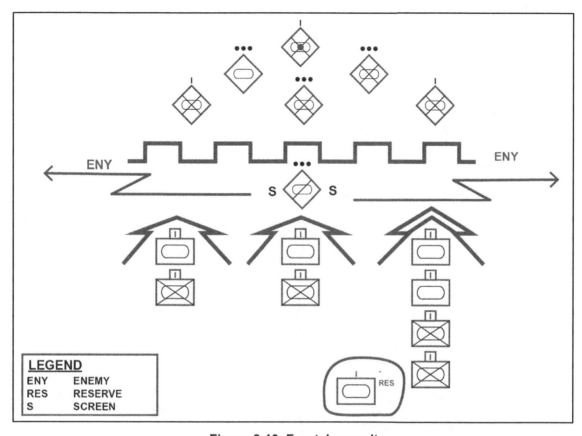

Figure 6-10. Frontal assault

SECTION IV – OFFENSIVE CONTROL MEASURES

6-115. A control measure is a means of regulating forces or warfighting functions. Control measures provide control without requiring detailed explanations. Control measures can be permissive (allowing something to happen) or restrictive (limiting how something is done). Some control measures are graphic. A *graphic control measure* is a symbol used on maps and displays to regulate forces and warfighting functions (ADP 6-0). (See ADP 1-02 for illustrations of graphic control measures and rules for their use.)

COMMON OFFENSIVE CONTROL MEASURES

6-116. Control measures provide the ability to respond to changes in the situation. They allow the attacking commander to concentrate combat power at the decisive point. At a minimum, commanders include an area of operations, defined by unit boundaries, and an objective to control their units and tailor their use to the higher commander's intent. The commander can also use any other control measure necessary to control the operation, including those listed in figure 6-11 on page 6-30, illustrating a BCT's use of the following control measures:

- Assembly area.
- Assault positions.
- Attack positions.
- Axis of advance.
- Battle handover line.
- Boundaries.
- Contact point.
- Coordinated fire line.

- Direction of attack or axis of advance for every maneuver unit.
- Free-fire area.
- Limit of advance.
- Line of contact.
- Line of departure.
- No-fire area.
- PL.
- Probable line of deployment.
- Restrictive fire line.
- Support by fire.

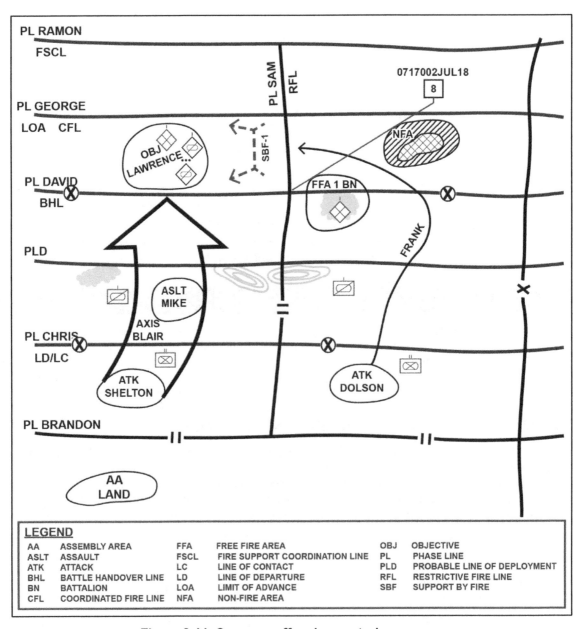

Figure 6-11. Common offensive control measures

EMPLOYING CONTROL MEASURES

6-117. Effectively employing control measures requires a BCT commander and staff to understand their purposes and ramifications, including the permissions or limitations imposed on subordinates' freedom of action and initiative. Each measure should have a specific purpose. Commanders use graphic control measures to assign responsibilities, coordinate fires and maneuver, and airspace management to assist the division's control of airspace. Well-planned fire control measures permit the proper distribution of fires and prevent multiple weapons from firing upon prominent targets while less prominent targets escape destruction. The BCT uses both fire support coordination measures and direct fire control measures.

6-118. Commanders maintain tight control over operations conducted under limited-visibility conditions to prevent fratricide, noncombatant casualties, and excessive or unintended collateral damage due to a loss of situational awareness by small units. These conditions require commanders to impose additional control measures beyond those used in daylight.

EMPLOYMENT DURING OFFENSIVE OPERATIONS

6-119. When executing a movement to contact, commanders start movement from line of departure as specified in the operation order, and control movement using PLs, contact points, and checkpoints as required. They control the depth of the movement to contact using a limit of advance or forward boundary, placed on suitable terrain for the force to establish a hasty defense where it will not be at risk to enemy-controlled terrain features. Additionally, commanders can designate boundaries and a series of PLs that successively become the new rear boundary of the forward security element. The same applies to the main body and rear security element, to delineate limits of responsibility.

6-120. In an attack, before the line of departure or line of contact, a commander may designate assembly areas and attack positions where the unit prepares for offensive actions or waits for the establishment of the required conditions to initiate the attack. Depending on conditions and risk, a commander may use an axis of advance, a direction of attack, or point of departure to further control maneuver forces. Between the probable line of deployment and the objective, a commander may designate a final coordination line, assault positions, attack by fire and support by fire positions, or a limit of advance beyond the objective if the commander does not wish to conduct exploitation or pursuit operations.

6-121. For an exploitation, a commander will likely use more permissive fire support coordination measures. Commanders use targets and checkpoints as required. Moving the coordinated fire line as the force advances is especially important, as is placement of the forward boundary. Placing the forward boundary too deep could limit higher echelon forces and effects, which would normally assist the BCT.

6-122. Pursuit control measures include an area of operations for each maneuver unit, PLs to designate forward and rearward boundaries for the direct-pressure force, and often a route, axis of advance, or an area of operations for the encirclement force to move parallel in order to get ahead of the fleeing enemy. Commanders establish a boundary or restrictive fire line between the force conducting the encirclement and the force exerting the direct pressure before the encircling force reaches its objective. Commanders may also establish a free-fire area, or a no-fire area.

ATTACK AND ASSAULT POSITIONS

6-123. An *attack position* is the last position an attacking force occupies or passes through before crossing the line of departure (ADP 3-90). An attack position facilitates the deployment and last-minute coordination of the attacking force before it crosses the line of departure. *Assault position* is a covered and concealed position short of the objective from which final preparations are made to assault the objective (ADP 3-90). Such final preparations can involve tactical considerations, such as a short halt to coordinate the final assault, reorganize to adjust to combat losses, or make necessary adjustments in the attacking force's dispositions. Ideally assaulting units do not stop in a planned assault position unless necessary.

Note. A rally point and an ORP differ from an assault position, in that, an assault point is a position short of an objective from which final preparations are made before the immediate assault of an objection. A *rally point* is an easily identifiable point on the ground at which units can reassemble and reorganize if they become dispersed (ATP 3-21.20). Forces conducting a patrol, or an infiltration commonly use this control measure. The *objective rally point* is an easily identifiable point where all elements of the infiltrating unit assemble and prepare to attack the objective (ADP 3-90). An ORP is typically near the infiltrating unit's objective; however, there is no standard distance from the objective to the ORP. It should be far enough away from the objective so that the enemy will not detect the infiltrating unit's attack preparations. An assault position, when designated, is generally positioned in the last covered and concealed position prior to the objective and after leaving the ORP, when an ORP is used. Infiltrating units move to an ORP to consolidate their combat power, refine the plan, and conduct any last-minute coordination before to continuing the mission. The unit then conducts those tasks needed to accomplish its mission, which could be an attack, raid, ambush, seizing key terrain, capturing prisoners, or collecting specific combat information.

NO-FIRE AREA, FREE-FIRE AREA, AND RESTRICTIVE FIRE LINE

6-124. A *no-fire area* is an area designated by the appropriate commander into which fires or their effects are prohibited (JP 3-09.3). Its purpose is to identify locations and facilities required for consolidation of gains. A *free-fire area* is a specific region into which any weapon system may fire without additional coordination with the establishing headquarters (JP 3-09). Its purpose is to enclose a bypassed or encircled enemy. A *restrictive fire line* is a specific boundary established between converging, friendly surface forces that prohibits fires or their effects from crossing (JP 3-09). Its purpose is to prevent interference between converging friendly forces, such as what occurs during a linkup operation.

LINE OF CONTACT AND LINE OF DEPARTURE

6-125. The *line of contact* is a general trace delineating the location where friendly and enemy forces are engaged (ADP 3-90). In the offense, a PL as a line of contact is often combined with the line of departure. In the defense, a line of contact is often synonymous with the forward line of troops. A *line of departure* is in land warfare, a line designated to coordinate the departure of attack elements (JP 3-31). Its purpose is to coordinate the advance of the attacking force, so its elements strike the enemy in the order and at the time desired.

PROBABLE LINE OF DEPLOYMENT AND LIMIT OF ADVANCE

6-126. A *probable line of deployment* is a phase line that designates the location where the commander intends to deploy the unit into assault formation before beginning the assault (ADP 3-90). Usually a linear terrain features perpendicular to the direction of attack, it is used primarily at the battalion level and below when the unit does not cross the line of departure in its assault formation. A *limit of advance* is a phase line used to control forward progress of the attack (ADP 3-90). The attacking unit does not advance any of its elements or assets beyond the limit of advance, but the attacking unit can maneuver its security forces to that limit. Commanders use a limit of advance to prevent overextending the attacking force and reduce the possibility of fratricide and friendly fire incidents by fires supporting the attack.

SECTION V – OFFENSIVE OPERATIONS

6-127. The BCT conducts *offensive operations* to defeat or destroy enemy forces and gain control of terrain, resources, and population centers (ADP 3-0). Offensive operations are movement to contact, attack, exploitation, and pursuit. BCTs conduct operations according to the capabilities and limitations inherent in their organizational structure and the knowledge of the situation (enemy) or the advantage relationship to enemy (see figure 6-12).

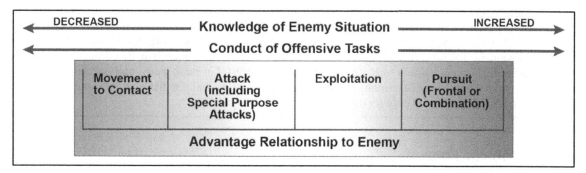

Figure 6-12. Knowledge of situation (enemy) and advantage relationship to enemy

MOVEMENT TO CONTACT

6-128.   A *movement to contact* is a type offensive operation designed to develop the situation and establish or regain contact (ADP 3-90). A movement to contact employs purposeful and aggressive movement, decentralized control, and the hasty deployment of combined arms formations from the march to create favorable conditions for subsequent tactical actions. Close air support, air interdiction, and counterair operations are essential to the success of large-scale movements to contact. Local air superiority or, as a minimum, air parity is vital to the operation's success. The fundamentals of a movement to contact are—

- Focus all efforts on finding the enemy.
- Make initial contact with the smallest force possible, consistent with protecting the force.
- Make initial contact with small, mobile, self-contained forces to avoid decisive engagement of the main body on ground chosen by the enemy. (This allows the commander maximum flexibility to develop the situation.)
- Task organize the force and use movement formations to deploy and attack rapidly in any direction.
- Keep subordinate forces within supporting distances to facilitate a flexible response.
- Maintain contact regardless of the COA adopted once contact is gained.

6-129.   A movement to contact may result in a meeting engagement. A *meeting engagement* is a combat action that occurs when a moving force, incompletely deployed for battle, engages an enemy at an unexpected time and place (ADP 3-90). In a meeting engagement, the force that reacts first to the unexpected contact generally gains an advantage over its enemy. However, a meeting engagement may also occur when the opponents are aware of each other and both decide to attack immediately to obtain a tactical advantage or seize key or decisive terrain. A meeting engagement may also occur when one force attempts to deploy into a hasty defense while the other force attacks before its opponent can organize an effective defense. Acquisition systems may discover the enemy before the security force can gain contact. No matter how the force makes contact, seizing the initiative is the overriding imperative. Prompt execution of battle drills at platoon level and below, and standard actions on contact for larger units, can give that initiative to the friendly force.

6-130.   The BCT commander considers requirements for maneuver (fire and movement) upon contact. The commander develops decision points to support changes in the force's movement formation or a change from an approach march to a movement formation. Using both human and technical means to validate decision points, the commander must determine the acceptable degree of risk, based on the mission. The commander's confidence in the products of the IPB process and the acceptable risk determines the unit's movement formation and scheme of maneuver. In a high-risk environment, it is usually better to increase the distance between forward elements and the main body than to slow the speed of advance. Once the commander makes contact with the enemy, the commander has five options: attack, bypass, defend, delay, or withdraw (see paragraph 6-152). Search and attack and cordon and search are subordinate tasks of movement to contact.

Note. Figures 6-13 through 6-16 on pages 6-36 through 6-42 introduce a fictional organization of forces scenario as a discussion vehicle for illustrating one of many ways that a Stryker brigade combat team (SBCT) can conduct a movement to contact. These figures illustrate example movement formations and movement techniques that maneuver battalions/squadrons and subordinate companies/troops use when part of the BCT's main body and security forces. Illustrated movement formations and movement techniques, and the distances between units are notional, they are used only for discussion purposes.

ORGANIZATION OF FORCES

6-131.   The BCT commander organizes friendly forces into security forces and a main body in a movement to contact. (See figure 6-13.) A maneuver battalion organizes its forces the same as the BCT when conducting a movement to contact independently. When the battalion moves as part of the BCT, a maneuver battalion moves as part of the main body or with the requisite attachments, which may be part of the security force. The Cavalry squadron normally moves as part of the security force.

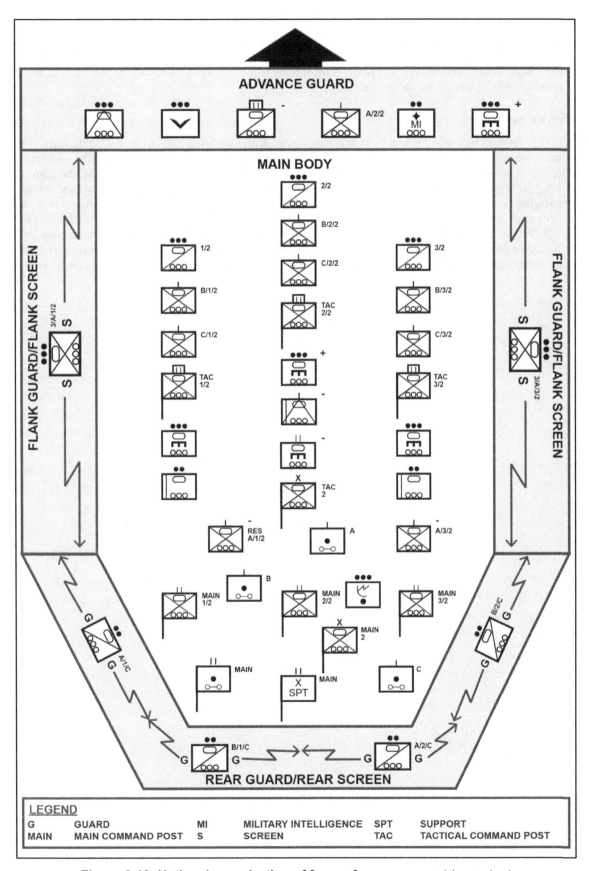

Figure 6-13. Notional organization of forces for a movement to contact

Security Forces

6-132.   The security forces for a BCT conducting a movement to contact, normally consists of the advance guard and, if required, flank and rear security forces. The advance guard has sufficient forces to protect the main body from surprise attack. The positioning of flank and rear forces depends on the proximity of friendly units to the flank or rear and to the enemy.

Advance Guard

6-133.   An advance guard is a task organized combined arms unit that precedes the main body and provides early warning, reaction time, and maneuver space (see figure 6-14). The BCT commander organizes an advance guard to lead the BCT with or without a covering force from a higher echelon. When a covering force from a higher echelon is employed forward of the BCT, the advance guard maintains contact with the covering force. The advance guard requires antiarmor and engineer support and remains within range of the main body's indirect-fire systems. The advance guard reduces obstacles to create passage lanes, repairs roads and bridges, and locates bypasses. For obstacles not covered by fire, the advance guard can either seek a bypass or create the required number of lanes to support its maneuver or the maneuver of a supported unit's maneuver. For obstacles covered by fire, the unit can either seek a bypass or conduct a breaching operation.

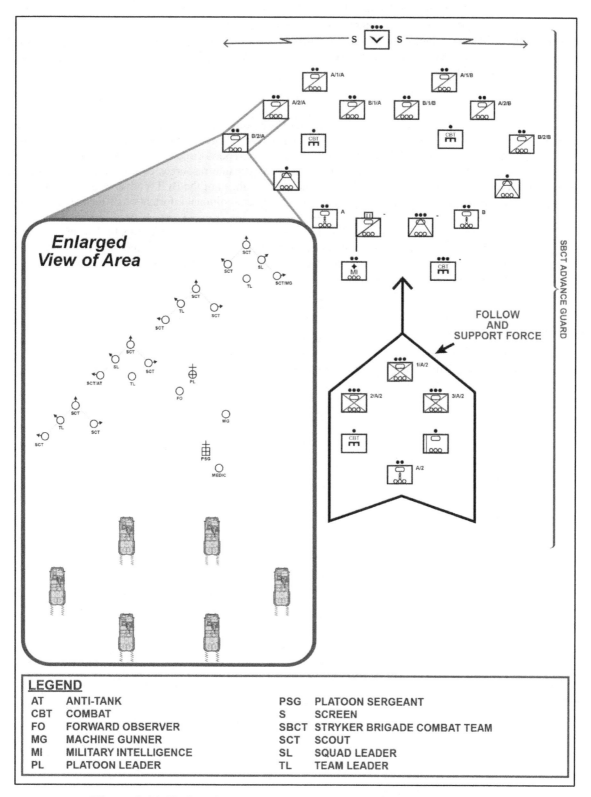

Figure 6-14. Notional organization of forces—SBCT advance guard

6-134.   The advance guard fixes the enemy to protect the deployment of the main body when the main body commits to action. The advance guard forces the enemy to withdraw, or destroys small enemy groups before they can disrupt the advance of the main body. When the advance guard encounters large enemy forces or heavily defended areas, it takes prompt and aggressive action to develop the situation and, within its capability, defeat the enemy. The commander of the advance guard force reports the location, strength, disposition, and composition of the enemy and tries to find the enemy's flanks, gaps, or other weaknesses in the enemy's position.

Covering Force

6-135.   A covering force's mission is to protect the main body, provide early warning, reaction time, and maneuver space before committing the main body. The covering force is task organized to accomplish tasks independent of the main body. The covering force commander reports directly to the establishing commander (division or corps). The BCT normally does not have the organic resources or capabilities to establish a covering force. A covering force, if established, moves well ahead of the BCT's advance guard and usually beyond the main body's fire support range. (See FM 3-90-1 for additional information on the covering force.)

> *Note.* In Army doctrine, a covering force is a self-contained force capable of operating independently of the main body, unlike a screening or guard force, to conduct the cover task. A covering force performs all the tasks of screening and guard forces. (See paragraph 5-85 for information specific to a division covering force.)

Flank and Rear Security

6-136.   The BCT establishes flank and rear security elements when their flanks or security area are unprotected. The BCT may use Cavalry organizations for flank security, or main body forces may provide flank (see figure 6-15) and rear security forces.

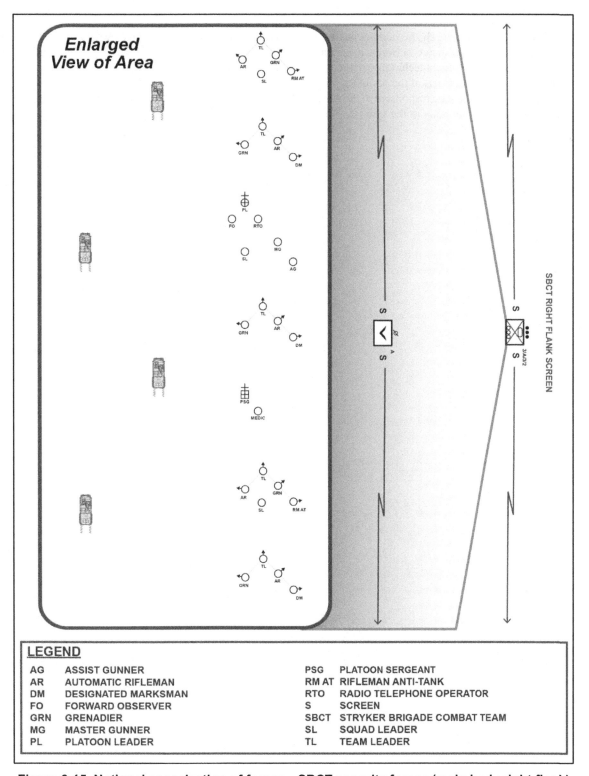

LEGEND

AG	ASSIST GUNNER	PSG	PLATOON SERGEANT
AR	AUTOMATIC RIFLEMAN	RM AT	RIFLEMAN ANTI-TANK
DM	DESIGNATED MARKSMAN	RTO	RADIO TELEPHONE OPERATOR
FO	FORWARD OBSERVER	S	SCREEN
GRN	GRENADIER	SBCT	STRYKER BRIGADE COMBAT TEAM
MG	MASTER GUNNER	SL	SQUAD LEADER
PL	PLATOON LEADER	TL	TEAM LEADER

Figure 6-15. Notional organization of forces—SBCT security forces (main body right flank)

Main Body

6-137. The bulk of the BCT's combat power is in the main body. The main body follows the advance guard and keeps enough distance between itself and the advance guard to maintain flexibility. The movement formations and movement techniques maneuver battalions/squadrons and subordinate companies/troops (see figure 6-16) use when part of the main body are always METT-TC dependent, keeping in mind the elements of the main body must be responsive to the actions of the advance guard. The BCT commander may designate a portion of the main body as the reserve.

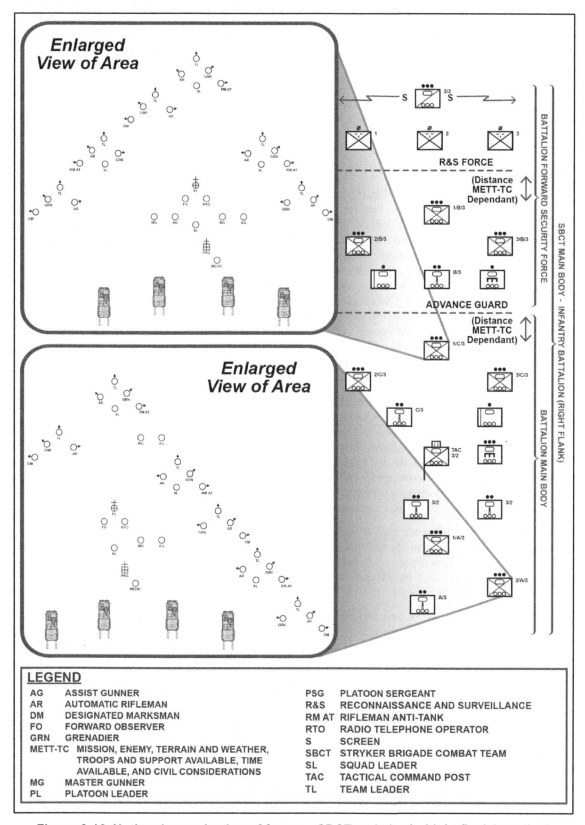

Figure 6-16. Notional organization of forces—SBCT main body (right flank battalion)

LEGEND

AG	ASSIST GUNNER	PSG	PLATOON SERGEANT
AR	AUTOMATIC RIFLEMAN	R&S	RECONNAISSANCE AND SURVEILLANCE
DM	DESIGNATED MARKSMAN	RM AT	RIFLEMAN ANTI-TANK
FO	FORWARD OBSERVER	RTO	RADIO TELEPHONE OPERATOR
GRN	GRENADIER	S	SCREEN
METT-TC	MISSION, ENEMY, TERRAIN AND WEATHER,	SBCT	STRYKER BRIGADE COMBAT TEAM
	TROOPS AND SUPPORT AVAILABLE, TIME	SL	SQUAD LEADER
	AVAILABLE, AND CIVIL CONSIDERATIONS	TAC	TACTICAL COMMAND POST
MG	MASTER GUNNER	TL	TEAM LEADER
PL	PLATOON LEADER		

6-138.   After the security force makes contact, the BCT commander receives information from the security force in contact. Based upon that information the commander directs a COA consistent with the higher commander's intent and within the main body's capability. Elements of the main body initiate direct and indirect fires to gain the initiative. The commander emplaces fires assets to respond immediately to calls for fire.

6-139.   A portion of the main body composes the BCT commander's sustaining base. The commander tailors the sustainment base to the mission. The commander decentralizes the execution of sustainment support, but that support must be continuously available to the main body. Sustainment support includes using preplanned logistics packages (LOGPACs).

PLANNING

6-140.   The BCT commander and staff plan for a movement to contact in the same manner as any offensive operations, however, time to plan may be constrained. Planning for a movement to contact begins, as with all operations, with a thorough understanding of the area of operations through a detailed IPB. The staff integrates the IPB, targeting, and RM (see chapter 4 for a detailed discussion) and information collection (see chapter 5 for a detailed discussion) throughout the military decision-making process (MDMP). IPB is the systematic process of analyzing the mission variables of enemy, terrain, weather, and civil considerations in an area of interest to determine their effect on operations. Targeting is the process by which the staff (specifically the BCT targeting work group) selects and prioritizes targets and matches the appropriate response to them considering operational requirements and capabilities. RM is the commander's and staff process to identify, assess, and control risks arising from operational factors and to make decisions that balance risk cost with mission benefits. The staff (specifically the BCT S-3, in coordination with the BCT S-2) integrates information collection into the concept of operations and manages the information collection effort through integrated staff processes and procedures.

6-141.   The BCT staff, in coordination with the commander, develops and executes Annex L (Information Collection) to the operation order. Annex L describes how information collection activities support the mission throughout the conduct of the operations described in the base order. Annex L synchronizes activities in time, space, and purpose to achieve objectives and accomplish the commander's intent for reconnaissance, surveillance, security operations, and intelligence operations (including military intelligence disciplines, see paragraph 5-90). The BCT commander and staff collaborate with the Cavalry squadron, military intelligence company, and security force assets to synchronize the information collection effort to allow the main body to focus on the conduct of the movement to contact.

6-142.   The BCT intelligence staff officer and S-2 section develop feasible threat COAs that address all aspects of the enemy's potential capabilities. The S-2 section, assisted by BCT engineer and air defense staff representatives, analyzes the terrain to include enemy air avenues of approach. The plan addresses actions the commander anticipates based upon available information and intelligence and probable times and locations of enemy engagements.

6-143.   BCT reconnaissance and security forces detect the enemy; then confirm or deny the enemy's presence making contact with the enemy using the smallest elements possible. A successful information collection effort integrates reconnaissance and security forces with HUMINT collection operations, signals intelligence collection, target acquisition assets, and aerial reconnaissance and surveillance (manned and unmanned) assets. The BCT commander may task organize reconnaissance and security forces with additional combat power allowing them to develop the situation on contact with the enemy. Additional combat power may include Abrams tanks, Bradley fighting vehicles, Stryker vehicles, weapons troop or Infantry weapons companies, or an Infantry rifle company. The unit's planned movement formation should contribute to the goal of making initial contact with the smallest force possible and provide for efficient movement of the force.

6-144.   The commander directs the establishment of decision points, branches, and sequels based upon the CCIRs to ensure flexibility in the plan. The commander controls the movement to contact by using control measures to provide the flexibility needed to respond to changes in the situation and to allow the commander to rapidly concentrate combat power at the decisive point.

6-145. The BCT commander may task a forward security force to conduct zone reconnaissance where the main body is to traverse. Based on the commander's decision points, the security force conducts reconnaissance or target handover with the main body to maintain contact with the enemy. This handover allows the BCT to manage transitions between phases of the operation, or follow-on tasks and allows the security force to conduct tasks that support the BCT scheme of maneuver.

6-146. The commander tailor's organic sustainment assets to the mission. Battalion and company trains may be combined and accompany the main body. METT-TC determines the type and amount of supplies transported in these trains. Locating the combat trains with the battalion permits rapid resupply of the maneuver units than if they were further to the rear. Commanders, however, may decide to assign combat units to combat trains for their security if they determine that the combat trains do not have sufficient combat power to counter the anticipated threat.

PREPARATION

6-147. The BCT staff, specifically the S-2, constantly refines the enemy situation during preparation based on information and integrated intelligence products. One of the primary concerns of the BCT S-3 and S-2 during preparation is to ensure the commander and staff have the latest information and that the COP is accurate, and the plan is still valid. The commander ensures, through confirmation briefs, backbriefs and rehearsals, that subordinates understand the commander's intent and their individual missions as new information becomes available. Simple, flexible plans that rely on standard operating procedures (SOPs) and battle drills, and plans that units rehearse against likely enemy COAs are essential to success.

6-148. The commander rehearses the operation from initiation to occupation of the final march objective or limit of advance. The commander prioritizes rehearsals of maneuver options, enemy COAs, and primary, secondary, and tertiary communications systems at all levels. Actions to consider during rehearsals include—

- Making enemy contact (advance guard).
- Making contact with an obstacle not identified and reported (advance guard).
- Making enemy contact (flank security force).
- Reporting requirements, engagement, and bypass criteria.
- Fire support.
- Maneuver.
- Unit transitions.
- Sustainment.

EXECUTION

6-149. The BCT maneuvers aggressively within its area of operations or along its axis of advance. Speed and security requirements must balance based on the effectiveness of the information collection effort, friendly mobility, effects of terrain, and enemy capabilities. The COP enables close tracking and control of the movement and location of units. The BCT (typically the tactical command post [TAC {graphic}]) continually monitors the location and movement of security forces. This monitoring of security forces ensures adequate security for the main body and ensures the security forces are within supporting range of main body maneuver forces and fire support assets. The BCT also controls the movement of sustainment assets, adjusting movement to meet support requirements, to avoid congestion of routes, and to ensure responsiveness.

Scheme of Maneuver

6-150. Movement to contact starts from a line of departure or a specified point(s) at the time specified in the operation order. A limit of advance or a forward boundary controls the depth of the movement to contact. PLs, contact points, and checkpoints control the rate of movement. Fire support is planned throughout the movement to contact to provide accurate and continuous fires. Actions on contact, (see paragraph 5-47), are planned for and rehearsed. Subordinate echelons must quickly react to contact, develop the situation, report, and gain a position of advantage over the enemy. Maneuvering unit commanders coordinate forward passage through friendly forces in contact as required.

6-151.   The primary focus of a movement to contact is the enemy force, which may be stationary or moving. Objectives can designate the movement of subordinate units and identify suspected enemy positions. Although an axis of advance can guide movement, there is the risk of enemy forces outside the axis being undetected and inadvertently bypassed. During a movement to contact, the intent of the commander is to maneuver quickly to defeat the enemy before the enemy can react. The commander avoids piecemeal commitment of the main body unless failure to do results in mission failure or prevented by restricted or severely restricted terrain. The BCT commander uses the advance guard to fix the enemy while the main body maneuvers to seek the assailable flank. The commander focuses on the enemy's flanks and rear before the enemy can counter these actions.

Maneuver Options

6-152.   The commander makes the decision to execute a maneuver option based on the progress of the advance guard's initial engagement. The movement to contact generally ends with the commitment of the main body. The tactical options available to the BCT after contact are addressed in the following paragraphs.

6-153.   Attack. The commander directs an attack when the BCT has greater combat power than the enemy does or when the commander assesses that the BCT can reach a decisive outcome. The commander can direct an ambush against a moving or infiltrating force that is not aware of the presence of the friendly force.

6-154.   Defend. The commander directs a defense when the BCT has insufficient combat power to attack. The commander also directs a defense when the enemy's superior strength forces the BCT to halt and prepare for a more deliberate operation.

6-155.   Bypass. The commander provides criteria detailing conditions for bypassing enemy forces. The unit in contact can bypass if authorized, but, if the bypassed force represents a threat, the unit must fix or contain it until released by the higher commander.

6-156.   Delay. A delaying force under pressure trades space for time by slowing the enemy's momentum and inflicting maximum damage on the enemy, without decisively engaging, in principle. Once the advance guard (fixing force) makes contact with the enemy, the enemy may attempt a frontal counterattack in response to the BCT's movement to contact. In this case, the fixing force defends itself or conducts a delay while the main body of the BCT maneuvers to attack.

6-157.   Withdraw. The commander directs a withdrawal when the BCT lacks the combat power to attack or defend, to improve a tactical situation, or to prevent a situation from worsening. Both direct- and indirect-fire assets from main body forces provide support to cover the withdrawal of the advance guard or lead elements of the main body. The commander also may employ obscuration to assist with breaking contact with the enemy.

Bypassed Forces

6-158.   Bypassed forces present a serious threat to forces that follow the maneuver elements, especially sustainment elements. Units conducting a movement to contact do not bypass enemy forces unless authorized by higher authority. Bypass criteria, if established, are measures established by higher headquarters that specify the conditions and size under which enemy units may be bypassed. The BCT distributes the location and strengths of enemy forces throughout the area of operations so following units can move around these threats. Bypassed enemy units are kept under observation unless otherwise directed by the commander. The destruction or containment of the bypassed enemy forces becomes the responsibility of the higher commander if the commander permits the lead elements to bypass.

Actions at Obstacles

6-159.   Once the unit detects an obstacle, the obstacle is immediately reported, and its location and description distributed. The element quickly seeks a bypass. If a bypass is available, the unit in contact with the obstacle marks the bypass; the unit reports the route of the bypass around the obstacle, also. The BCT breaches consistent with the breaching fundamentals of suppress, obscure, secure, reduce, and assault (described by the memory aid SOSRA) to create breach lanes and continue the movement to contact. Engineers support the breach effort by reducing the obstacle, improving the lanes, and guiding the main body through the obstacle. (See ATP 3-90.4 for additional information.) Civil affairs, military police, or tactical psychological operations (PSYOP) teams may redirect civilians away from the route of advance when the movement of displaced civilians causes reduced mobility.

Five Step Sequence

6-160. FM 3-90-1 discusses executing all four offensive operations in a five-step sequence, listed below. This sequence is for discussion purposes only and is not the only way of conducting offensive operations. Offensive operations tend to overlap each other during the conduct of offensive actions. Normally the first three of these steps are shaping operations or supporting efforts, while the maneuver step is the decisive operation or main effort. Follow through is normally a sequel or a branch to the plan based on the current situation.

Step 1, Gain and Maintain Enemy Contact

6-161. The advance guard focuses on identifying the enemy's composition, strength, and dispositions. The forces provide the commander with combat information. The commander can then maneuver units to positions of advantage to commit friendly forces under optimal conditions.

Step 2, Disrupt the Enemy

6-162. On contact, the advance guard maneuvers to disrupt or defeat the enemy to prevent enemy from conducting a spoiling attack or organizing a coherent defense. The advance guard commander gathers as much information as possible about the enemy's dispositions, composition, strengths, capabilities, and probable course(s) of action.

Step 3, Fix the Enemy

6-163. The advance guard prevents the enemy from maneuvering against the main body. If unable to defeat the enemy, the advance guard reports the enemy strength and disposition and establishes a base of fire for the subsequent attack by the main body.

Step 4, Maneuver

6-164. If the advance guard cannot defeat the enemy with a frontal or flank assault or an engagement of an enemy's assailable flank, the commander quickly maneuvers the main body to attack. This offensive maneuver seeks to achieve a decisive massing of effects at the decisive point, or at several decisive points if adequate combat power is available. The commander aims the decisive operation toward the decisive point, which can consist of the immediate and decisive destruction of the enemy force, its will to resist, seizure of a terrain objective, or the defeat of the enemy's plan. The commander attempts to defeat the enemy while still maintaining the momentum of the advance. The main body commander resumes the movement to contact after a successful attack. The intent is to deliver an assault before the enemy can deploy or reinforce their engaged forces.

Step 5, Follow Through

6-165. The unit transitions back to a movement to contact and continue to advance if the enemy is defeated. The movement to contact terminates when the unit reaches its final objective or limit of advance; otherwise, it must transition to another offensive or defensive operation. After committing the reserve, the commander develops a plan to reconstitute another reserve force once the original reserve force is committed, most often accomplished with a unit out of contact.

SEARCH AND ATTACK

6-166. *Search and attack* is a technique for conducting a movement to contact that shares many of the characteristics of an area security mission (FM 3-90-1). The BCT conducts a search and attack to destroy enemy forces, deny the enemy certain areas, protect the force, or collect information. Although the battalion is the echelon, that usually conducts a search and attack, the BCT assists its subordinate battalions by ensuring the availability of indirect fires and other support.

CORDON AND SEARCH

6-167. *Cordon and search* is a technique of conducting a movement to contact that involves isolating a target area and searching suspected locations within that target area to capture or destroy possible enemy forces and contraband (FM 3-90-1). The BCT normally assigns a cordon and search mission to a battalion. The BCT supports the cordon and search by conducting shaping operations and providing additional resources to the unit conducting the cordon and search. A cordon and search may support site exploitation (see ATP 3-90.15).

ATTACK

6-168. An *attack* is a type of offensive operation that destroys or defeats enemy forces, seizes and secures terrain, or both (ADP 3-90). Although an attack may be a deliberate operation or a hasty operation, both synchronize all available warfighting functions to defeat the enemy. The main difference between a hasty and a deliberate operation (see chapter 2) is preparation and planning time.

6-169. The key difference between a movement to contact and an attack is the amount of information known about the enemy. Information enables the commander to have more control, to better synchronize the operation, and to employ combat power more effectively than in a movement to contact. The commander has the advantage of being extremely deliberate and refined in task organization, assignment of tactical mission tasks, and the scheme of maneuver.

6-170. The BCT executes subordinate forms of the attack to achieve different results. These subordinate forms of the attack have special purposes and include the ambush, counterattack, demonstration, feint, raid, and spoiling attack (see paragraph 6-197). The commander's intent and the mission variables of METT-TC determine the specific attack form. The commander can conduct these forms of attack, except for a raid, as a hasty or a deliberate operation.

ORGANIZATION OF FORCES

6-171. The BCT commander determines the scheme of maneuver and task organizes the force to give each subordinate unit the combat power to accomplish its assigned missions. The commander normally organizes the attacking force into a security force, a main body, and a reserve. The commander completes any changes in task organization early in the process to allow subordinate units to conduct rehearsals with their attached and supporting elements.

Security Forces

6-172. The BCT executes most attacks while in contact with the enemy which reduces the requirement for a separate forward security force. The commander commits security forces during an attack only if the attack is likely to uncover one or more flanks or the rear of the attacking force as it advances. The commander designates a flank or rear security force and assigns it a guard or screen mission depending on METT-TC.

Main Body

6-173. The BCT commander allocates forces based on the assigned tasks, the terrain, and the size of the enemy force that each *avenue of approach*—a path used by an attacking force leading to its objective or to key terrain. Avenues of approach exist in all domains (ADP 3-90)—can support (probable force ratio). The BCT attacks to destroy enemy forces or to seize key terrain. The scheme of maneuver identifies the decisive operation or main effort. During the course of the attack, the unit(s) executing the decisive operation or main effort may change based upon conditions or plans.

6-174. Maintaining mobility in an attack is critical. The assistant brigade engineer (known as ABE) officer must plan and allocate mobility resources to the main body and security forces. The commander designates a breach, assault, and support force as the initial decisive operation if the commander anticipates or has identified the need to conduct a breach during the attack. The breaching fundamentals applied to ensure success when breaching against a defending enemy are SOSRA. These obstacle reduction fundamentals always apply, but they may vary based on METT-TC. The commander isolates and secures the breach area, breaches the enemy's defensive obstacles, seizes the point of penetration, and rapidly passes through follow-on forces to continue the attack. (See ATP 3-90.4 for additional information.)

6-175.   The commander arranges forces in-depth and designates a reserve. The commander controls the field artillery battalion, long-range fire support systems, and any breaching assets to retain flexibility until the point of breach is identified. The commander focuses all available resources to support achievement of the decisive operation.

6-176.   The commander designates subordinate units to conduct shaping operations for the execution of the decisive operation. The commander allocates only the combat power needed to accomplish the mission since overwhelming combat power cannot be executed everywhere. Shaping operations disrupt enemy defensive preparations through aggressive combat patrolling, feints, limited-objective attacks, harassing indirect fires, and air strikes. The commander uses shaping operations to isolate the enemy and destroy the enemy's ability to mutually support or reinforce enemy positions. (See figures 6-17 and 6-18 on page 6-48.)

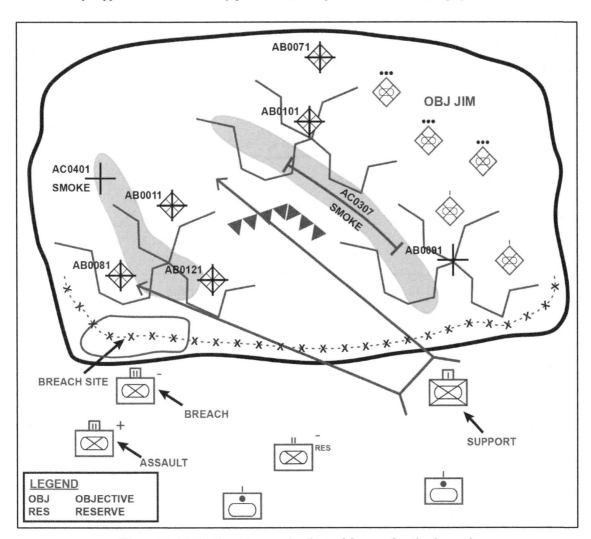

Figure 6-17. Notional organization of forces for the breach

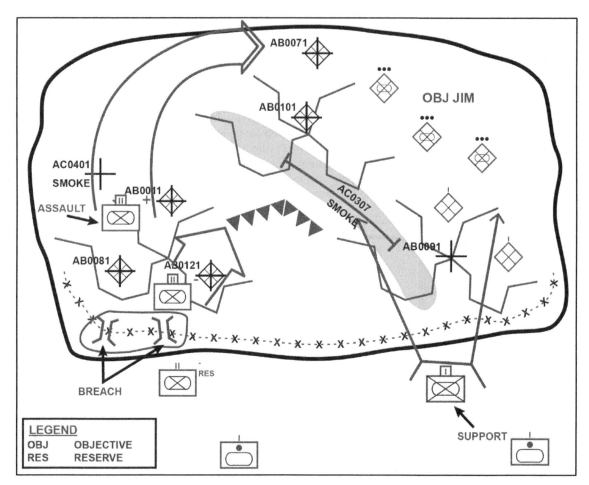

Figure 6-18. Notional organization of forces for the assault

Reserve

6-177.   A *reserve* is that portion of a body of troops that is withheld from action at the beginning of an engagement, to be available for a decisive movement (ADP 3-90). The reserve is not a committed force and is not used as a follow and support force or a follow and assume force. The commander uses the reserve to exploit success, to defeat enemy counterattacks, or to restore momentum to a stalled attack.

6-178.   Once committed, the reserve's actions normally become or reinforce the BCTs decisive operation. The commander makes every effort to reconstitute another reserve from units made available by the revised situation. Often a commander's most difficult and important decision concerns the time, place, and conditions for committing the reserve.

6-179.   In an attack, the commander prioritizes the positioning of the reserve to reinforce the success of the decisive operation, then to counter enemy counterattacks. The reserve must be able to move quickly to areas where it is needed in different contingencies. This is most likely to occur if the enemy has strong counterattack forces. Once committed, the reserve's actions normally become or reinforce the echelon's decisive operation or main effort, and the commander makes every effort to reconstitute another reserve from units made available by the revised situation.

Sustainment

6-180.   The BCT commander resources sustainment assets to support the attacking force. The BSB commander and BCT subordinate maneuver commanders organize sustainment assets to support the BCT's concept of support. The BSB commander controls the sustainment for the BCT with priority of support to the decisive operation or main effort. The BSB commander positions sustainment units well forward in an attack whenever possible to provide immediate support. As the BCT advances, sustainment units and capabilities echelon support forward to ensure uninterrupted support to maneuver units (see chapter 9).

PLANNING

6-181. The BCT commander and staff plan for an attack in the same manner as discussed in paragraph 6-125 for a movement to contact. As with the movement to contact, planning for an attack begins with a thorough understanding of the area of operations through a detailed IPB. The staff integrates the IPB, targeting, and RM (see chapter 4 for a detailed discussion) and information collection (see chapter 5 for a detailed discussion) throughout the MDMP. The BCT commander allocates resources as required to provide the maximum possible combat power to the decisive operation or main effort. Units conducting shaping operations or supporting operations should have sufficient combat power to conduct their mission.

6-182. Fire support planning (see chapter 4) directly supports the BCT's concept of operations to engage enemy forces, movement formations, and facilities in pursuit of tactical and operational objectives. The commander uses a blend of friendly information management, knowledge management, and information collection operations to take advantage of the range, precision, and lethality of available weapon systems and information superiority, thus achieving fire superiority. The commander focuses fire support effects to gain and maintain fire superiority at critical points during the attack and to maintain freedom of maneuver. Responsiveness and flexibility require that the BCT must have the ability to rapidly clear fires.

6-183. Army attack reconnaissance aviation units (see chapter 4) conduct shaping operation attacks to assist the BCT in finding, fixing, and destroying the enemy. Attack reconnaissance aviation units support ground forces in contact through Army aviation attacks. During a meeting engagement, attack reconnaissance aviation units provide information to help develop the situation. Assault helicopter battalions support ground force maneuver through air movement and air assault missions.

PREPARATION

6-184. The BCT uses the available time before the attack to conduct reconnaissance, precombat checks and inspections, and rehearsals. The BCT conceals attack preparations from the enemy. The commander and staff refine the plan based on continuously updated intelligence. Subordinates conduct parallel planning and start their preparation for the attack immediately after the BCT issues a fragmentary order. As more intelligence becomes available, the commander revises orders and distributes them; thereby giving subordinates more time to prepare for the attack.

EXECUTION

6-185. For discussion purposes, the execution of the attack is addressed using the following five-step sequence, gain and maintain enemy contact, disrupt the enemy, fix the enemy, maneuver, and follow through. This sequence is not the only method of executing an attack. These steps may overlap or be conducted simultaneously. Normally the first three of these steps are shaping operations or supporting efforts, while the maneuver step is the decisive operation or main effort. Follow through is normally a sequel or a branch to the plan based on the current situation.

Step 1, Gain and Maintain Enemy Contact

6-186. The commander positions maneuver forces and information collection assets to maintain observation of enemy reactions to maneuver on the objective. Information collection focuses on areas the enemy may use to reposition forces, commit reserves, and counterattack. For example, the commander may infiltrate or insert reconnaissance and security forces to observe the objective or routes that an enemy reserve may use. As the BCT attacks, reconnaissance and security forces report enemy reactions, repositioning, and battle damage assessment. The BCT may task the reconnaissance forces to target and engage enemy repositioning forces, reserves, counterattacking forces, and other HPTs with indirect fires. Early identification of enemy reactions is essential for the BCT to maintain momentum and the initiative during the attack. To regain contact with the enemy during an attack, the BCT commander may use the Cavalry squadron to regain contact and provide information on the enemy's current location, disposition, and movement.

Step 2, Disrupt the Enemy

6-187. Disrupting one or more parts of the enemy weakens their entire force and allows the BCT commander to attack the remaining weakened enemy force. The commander can disrupt the enemy's defenses using a variety of methods including:

- Gaining surprise.
- Avoiding enemy security forces.
- Using suppressive, interdiction, preparation, and counterair fires against enemy formations, strong points, and assembly areas.
- Destroying target acquisition systems.
- Taking advantage of limited visibility, concealment, and cover by masking the approach.
- Using augmented cyberspace operations and EW assets to degrade enemy command and control systems.
- Using military deception to conceal the exact time and location of the attack.
- Using precision fires (precision guide munitions, multiple launch rocket system/high mobility artillery rocket system, M982 Excalibur) against HPTs in-depth coordinated with long-range surveillance and precision observation teams.

Step 3, Fix the Enemy

6-188. The primary purpose in fixing the enemy is to prevent the enemy from maneuvering to reinforce the unit targeted for destruction. Fixing the enemy into a given position or COA limits the enemy's ability to respond to the attack effectively. Fixing the enemy usually is a shaping operation. To conserve combat power, the BCT commander carefully considers which enemy elements to fix and targets only the elements that can affect the point of attack.

6-189. The BCT commander fires on supporting and rear positions to isolate the objective. The commander uses the fires to suppress the enemy's suspected command and control centers, fire support systems, and reserve. The commander degrades the enemy's command and control systems through cyberspace electromagnetic activities (CEMA), also.

Step 4, Maneuver

6-190. The BCT commander maneuvers forces to gain positional advantage to seize, retain, and exploit the initiative and consolidate gains. The commander avoids the enemy's strength, employing tactics that defeat the enemy by attacking through a point of relative weakness, such as a flank or rear. The key to success is to strike hard and fast, overwhelm a portion of the enemy force, and quickly transition to the next objective or phase, thus maintaining the momentum of an attack without reducing pressure on the enemy.

6-191. The coordination between fire and movement is critical to massing combat power. As maneuver forces approach the enemy defense, the commander shifts fires and obscurants to suppress and obscure the enemy. Proper timing and adjustment of fires enable the maneuver force to close on the enemy's positions. The commander echelons fires to maintain effective suppression on the objective(s) up to the last possible moment while reducing any possibility of fratricide. The key to a successful attack is the suppression of the enemy force by indirect and direct fires that shift in the front of the assault force as it reaches its limit of advance. Maneuver forces and information collection assets provide battle damage assessment to the commander. The commander may need to adjust the speed of the approach to the objective based on reports from forward reconnaissance and surveillance assets.

6-192. The BCT employs fires to weaken the enemy's position. The BCT sets conditions for success before closure within direct fire range of the enemy. Initially, fires focus on the destruction of key enemy forces that affect the concept of operations, such as to destroy the enemy positions at the point of penetration during an attack.

6-193. Fires allow the commander to destroy enemy security forces and weaken or neutralize enemy reserves. Fires can emplace artillery delivered obstacles to block enemy reserve routes to the objective, support breaching operations, isolate the objective, and suppress enemy positions. The commander can employ obscuration and screening fires to deceive the enemy of the BCT's actual intentions. Obscuration fires (placed on or near enemy positions) decrease an enemy's capability to visually sight friendly forces. Screening fires (delivered in areas between friendly and an enemy force) degrade enemy detection, observation, and engagement capabilities to enable friendly maneuver and action. The commander employs fires to disrupt enemy counterattacks and neutralize bypassed enemy combat forces. The commander employs fires to conduct CEMA to degrade, neutralize, or destroy enemy combat capability. The BCT neutralizes the enemy's indirect fires through counterfire.

6-194. Fires assets are usually positioned forward so they can cover the objective and beyond without having to displace. The field artillery battalion positions its batteries as close as possible to the line of departure. The battalion heavy mortars position themselves close to assault units and are prepared to displace forward as required. Attached platoons from Infantry weapons companies or SBCT weapons troop may displace by sections and closely follow the maneuver companies. Close air support and Army aviation attacks identify and attack preplanned targets.

Step 5, Follow Through

6-195. After seizing an objective, the BCT commander has two alternatives: exploit success and continue the attack or terminate the operation. Normally, the BCT maintains contact and attempts to exploit its success. Indirect and direct fires may continue to suppress other enemy positions. Follow-on forces, which may or may not be part of the BCT, can conduct a forward passage of lines to continue the attack.

6-196. The most likely on-order mission is to continue the attack after seizing an objective. During consolidation, the commander continues the MDMP (or rapid decision-making and synchronization process, chapter 3) in preparation for any on order missions assigned by a higher headquarters.

SUBORDINATE FORMS OF THE ATTACK

6-197. The BCT can launch subordinate forms of the attack with various purposes to achieve different results. Special purpose attacks are ambush, counterattack, demonstration, feint, raid, and spoiling attack.

6-198. An *ambush* is an attack by fire or other destructive means from concealed positions on a moving or temporarily halted enemy (FM 3-90-1). The three forms of an ambush are point ambush, area ambush, and antiarmor ambush. An ambush is generally conducted at the small unit level and takes the form of an assault to close with and destroy the enemy, or it might be an attack by fire only, executed from concealed positions. An ambush does not require seizing or holding the ground. Ambushes are generally executed to reduce the enemy force's overall combat effectiveness through destruction, although other reasons could be to harass and capture the enemy or capture enemy equipment and supplies.

6-199. A *counterattack* is an attack by part or all of a defending force against an enemy attacking force, for such specific purposes as regaining ground lost or cutting off or destroying enemy advance units, and with the general objective of denying to the enemy the attainment of the enemy's purpose in attacking. In sustained defensive actions, it is undertaken to restore the battle position and is directed at limited objectives (FM 3-90-1). The commander plans counterattacks as part of the BCT's defensive plan, or the BCT might be the counterattack force for the higher headquarters. The BCT must provide the counterattack force with enough combat power and mobility to affect the enemy's offense.

6-200. A *demonstration* in military deception, is a show of force similar to a feint without actual contact with the adversary, in an area where a decision is not sought that is made to deceive an adversary (JP 3-13.4). The BCT commander uses demonstrations and feints in conjunction with other military deception activities. The commander generally attempts to deceive the enemy and induce the enemy commander to move reserves and shift fire support assets to locations where they cannot immediately affect the friendly decisive operation or take other actions not conducive to the enemy's best interests during the defense. The BCT commander must synchronize the conduct of these forms of attack with higher and lower echelon plans and operations to prevent inadvertently placing another unit at risk. Both forms are always shaping operations, but a feint will require more combat power and usually requires ground combat units for execution. (See FM 3-90-1 for additional information.)

6-201. A *feint* in military deception, is an offensive action involving contact with the adversary conducted for the purpose of deceiving the adversary as to the location and/or time of the actual main offensive action (JP 3-13.4). The principal difference between a feint and a demonstration is that in a feint the BCT commander assigns the force an objective limited in size, scope, or some other measure. The force conducting the feint makes direct fire contact with the enemy but avoids decisive engagement. The planning, preparing, and executing considerations for demonstrations and feints are the same as for the other forms of attack. The commander assigns the operation to a subordinate unit and approves plans to assess the effects generated by the feint, to support the operation. (See FM 3-90-1 for additional information.)

6-202. A *raid* is an operation to temporarily seize an area in order to secure information, confuse an adversary (Army uses the term enemy instead of adversary), capture personnel or equipment, or to destroy a capability culminating with a planned withdrawal (JP 3-0). The BCT plans raids and usually executes them at battalion level and below. The raiding force may operate within or outside of the BCT's supporting range, and it moves to its objective by infiltration. The raiding force quickly withdraws along a different route once the raid mission is completed.

6-203. A *spoiling attack* is a tactical maneuver employed to seriously impair a hostile attack while the enemy is in the process of forming or assembling for an attack (FM 3-90-1). The BCT commander conducts a spoiling attack during the defense to strike the enemy while in assembly areas or attack positions preparing for offensive mission or has temporarily stopped. The BCT commander employs organic fires, as well as other available units, to attack the enemy's assembly areas or other positions. (See FM 3-90-1 for additional information.)

EXPLOITATION

6-204. *Exploitation* is a type of offensive operation that usually follows a successful attack and is designed to disorganize the enemy in depth (ADP 3-90). Exploitation is the bold continuation of an attack designed to increase success and take advantage of weakened or collapsed enemy defenses. The purpose of exploitation can vary, but generally, an exploitation capitalizes on a temporary advantage, on preventing the enemy from establishing an organized defense, or preventing the enemy from conducting an orderly withdrawal. An exploitation should prevent reconstitution of enemy defenses, prevent enemy withdrawal, secure deep objectives, and destroy enemy command and control facilities, logistics, and forces.

6-205. The conditions for exploitation develop very quickly. The commander capitalizes on opportunities using information collected to seize, retain, and exploit the initiative. The commander designates priority intelligence requirements tied to decision points that seek out the following:

- A significant increase in enemy prisoners of war.
- An increase in abandoned enemy equipment and material.
- The overrunning of enemy artillery, command and control facilities, and logistics sites.
- A significant decrease in enemy resistance or in organized fires and maneuver.
- A mixture of support and combat vehicles in formations and columns.
- An increase in enemy movement rearward, including reserves and fire support units.

6-206. The commander plans the exploitation to maintain pressure on the enemy. To accomplish this, the BCT attacks over a broad front to prevent the enemy from establishing a defense, organizing an effective rear guard, withdrawing, or regaining balance. The BCT secures objectives, severs escape routes, and destroys all enemy forces. (See figure 6-19.) The commander may employ the reserve as an exploitation force.

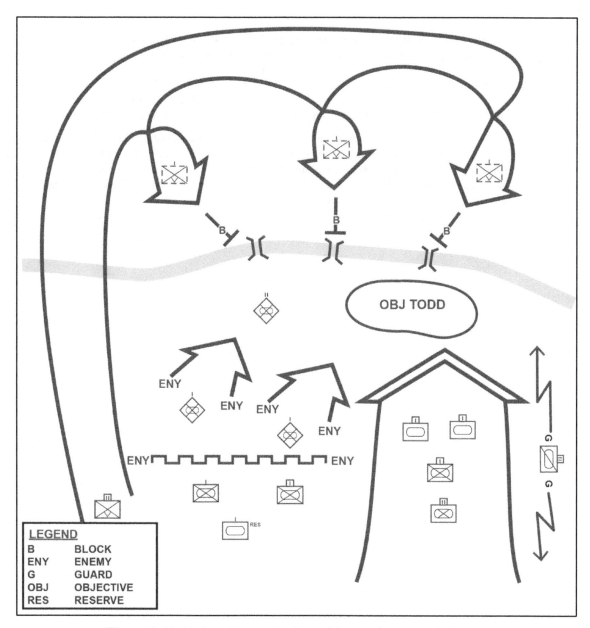

Figure 6-19. Notional organization of forces for an exploitation

6-207. Decentralized execution is characteristic of the exploitation; however, the commander maintains enough control to prevent overextension of the command. Minimum control measures are used. Tactical air reconnaissance and Army aircraft maintain contact with enemy movements and advise the commander of enemy activities. Interdiction, close air support, Army aviation attacks, and deep artillery fires can attack moving enemy reserves, withdrawing enemy columns, enemy constrictions at choke points, and enemy forces that threaten the flanks of the exploiting force. The commander must consider the security of ground supply columns and an aerial resupply may be necessary. Exploiting forces take advantage of captured supplies whenever possible.

6-208. Failure to exploit success aggressively gives the enemy time to reconstitute an effective defense or regain the initiative using a counterattack. BCT mounted elements may move rapidly to positions of advantage to block enemy forces. If available, Army aviation assets can move forces to blocking positions and UASs can maintain contact. (See FM 3-90-1 for additional information.)

PURSUIT

6-209. *Pursuit* is a type of offensive operation designed to catch or cut off a hostile force attempting to escape, with the aim of destroying it (ADP 3-90). The commander orders a pursuit when the enemy force can no longer maintain its position and tries to escape. Normally, the commander does not organize specifically for pursuit operations ahead of time, although the unit staff may plan for a pursuit mission as a branch or sequel to the current order. The plan must be flexible for subordinate elements of the BCT to react when the situation presents itself. Subordinate elements are made as self-sufficient as resources will permit.

6-210. Two options exist when conducting a pursuit. Both pursuit options involve assigning a subordinate the mission of maintaining direct pressure on the rearward moving enemy force. The first option is a frontal pursuit that employs only direct pressure. (See figure 6-20.) The second is a combination that uses one subordinate element to maintain direct pressure and one or more other subordinate elements to encircle the retrograding enemy. (See figure 6-21 on page 6-56.) The combination pursuit is more effective, generally. The subordinate applying direct-pressure or the subordinate conducting the encirclement can conduct the decisive operation in a combination pursuit.

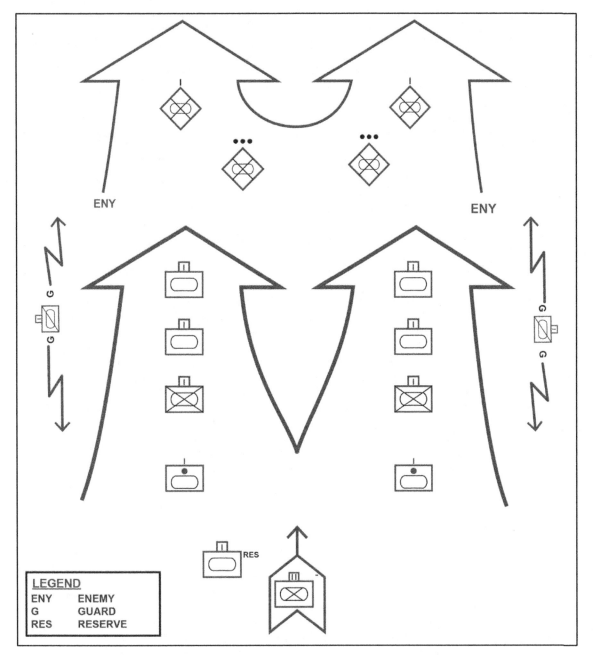

Figure 6-20. Notional organization of forces for a frontal pursuit

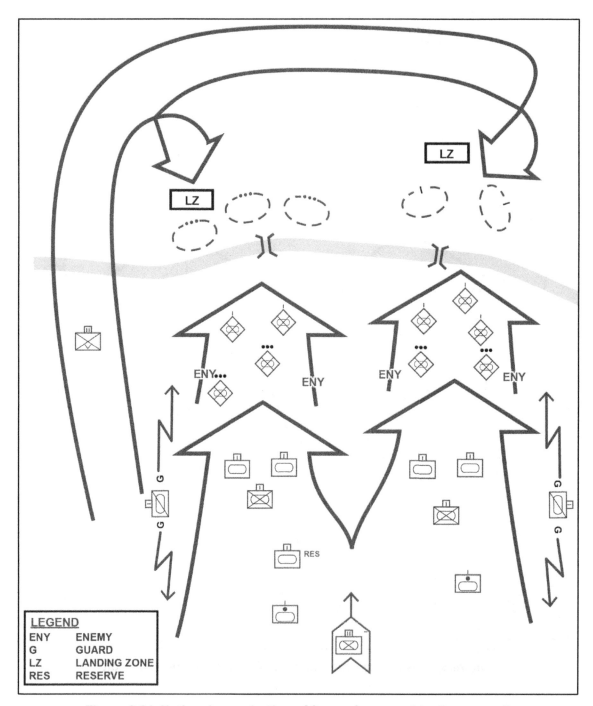

Figure 6-21. Notional organization of forces for a combination pursuit

6-211.   During the pursuit, the commander exerts unrelenting pressure to keep the enemy force from reorganizing and preparing its defenses. The BCT may be a part of a corps or division pursuit, either functioning as the direct pressure or encircling force. An aggressive pursuit leaves the enemy faced with the options of surrendering or facing complete destruction. Pursuits require swift maneuvers and attacks.

6-212.   The pursuit normally follows a successful exploitation. The primary function of a pursuit is to complete the destruction of the enemy force. Although the BCT may pursue a physical objective, the mission is the destruction of the enemy's main force. Pursuits include the rapid shifting of units, continuous day and night movements, hasty operations, containment of bypassed enemy forces, and large numbers of prisoners.

A pursuit includes a willingness to forego some synchronization to maintain contact and pressure on a fleeing enemy.

6-213.   A mobility advantage over the enemy is vital to the BCT's effectiveness in pursuit. A combination of Armored or Stryker forces, combined with Infantry conducting air assaults, can be extremely effective when cutting off the enemy forcing them to either surrender or be destroyed. The range, speed, and weapons load of attack reconnaissance aviation units makes them uniquely useful in an exploitation or pursuit to extend the ground commander's reach. Dismounted movement over difficult terrain allows Infantry units to seize blocking positions. (See FM 3-90-1 for additional information.)

SECTION VI – TRANSITIONS

6-214.   Decisive action involves more than simultaneous execution of all tasks. Decisive action requires the commander and staff to consider the BCT's capabilities and capacities relative to each assigned task. The commander considers the mission, determines which tactics to use, and balances the tasks of decisive action while preparing the commander's intent and concept of operations. The commander determines which tasks the force can accomplish simultaneously, if phasing is required, what additional resources the force may need, and how to transition from one task to another.

6-215.   Transitions between tasks of decisive action require careful assessment, prior planning, and unit preparation as the commander shifts the combination of offensive, defensive, and stability operations. Commanders first assess the situation to determine applicable tasks and the priority for each. When conditions change, commanders adjust the combination of tasks of decisive action in the concept of operations.

6-216.   A transition occurs when the commander makes an assessment that the unit must change its focus from one element of decisive action to another. A commander halts the offense when the offense results in complete victory and the end of hostilities reaches a culminating point, the BCT is approaching a culminating point due to operational reach (see chapter 9), or the commander receives a change in mission from a higher commander. This change in mission may be a result of the interrelationship of the other instruments of national power, such as a political decision.

6-217.   All offensive actions that do not achieve complete victory reach a culminating point when the balance of strength shifts from the attacking force to its opponent. Usually, offensive actions lose momentum when friendly forces encounter and cannot bypass heavily defended areas. Offensive actions also reach a culminating point when the resupply of fuel, ammunition, and other supplies fails to keep up with expenditures, Soldiers become physically exhausted, casualties and equipment losses mount, and repairs and replacements do not keep pace with losses. Offensive actions also stall when reserves are not available to continue the advance, the defender receives reinforcements, or the defender counterattacks with fresh troops. Several of these actions may combine to halt an offense. When the offensive action halts, the attacking unit can regain its momentum, but normally this only happens after difficult fighting or after an operational pause.

6-218.   The commander plans a pause to replenish combat power and phases the operation accordingly, if the commander cannot anticipate securing decisive objectives before subordinate forces reach their culminating points. Simultaneously, the commander attempts to prevent the enemy from knowing when friendly forces become overextended.

TRANSITION TO THE CONDUCT OF DEFENSIVE OPERATIONS

6-219.   Once offensive actions begin, the attacking commander tries to sense when subordinate units reach, or are about to reach, their respective culminating points. The commander must transition to a focus on the defense (see chapter 7) before subordinate units reach this point. The commander has more freedom to choose where and when to halt the attack, if the commander can sense that subordinate forces are approaching culmination. The commander can plan future activities to aid the defense, minimize vulnerability to attack, and facilitate renewal of the offense as the force transitions to branches or sequels of the ongoing operation. For example, some subordinate units may move into battle positions before the entire unit terminates its offensive actions to start preparing for ensuing defensive operations. The commander can echelon sustainment assets forward to establish a new echelon support area.

6-220. A lull in combat operations often accompanies a transition. The commander cannot forget about stability operations because the civilian populations of the unit's area of operations tend to come out of their hiding positions and request assistance from friendly forces during these lulls. The commander must consider how to minimize civilian interference with the force's combat operations while protecting civilians from future hostile actions according to the law of war. The commander must also consider the threat enemy agents or saboteurs pose when infiltrating or operating form within civilian populations. (See chapter 8.)

6-221. A commander anticipating the termination of unit offensive actions prepares orders that include the time or circumstances under which the current offense transitions to the defense, the missions and locations of subordinate units, and control measures. As the unit transitions from an offensive focus to a defensive focus, the commander maintains contact with the enemy, using a combination information collection assets to develop the information required to plan future actions. The commander also establishes a security area and local security measures.

TRANSITION TO THE CONDUCT OF STABILITY OPERATIONS

6-222. A transition to stability centric operations occurs for several reasons. A transition may occur from an operation dominated by large-scale combat operations to one dominated by the consolidation of gains. Transitions also occur with the delivery of essential services or retention of infrastructure needed for reconstruction. An unexpected change in conditions may require commanders to direct an abrupt transition between phases. In such cases, the overall composition of the force remains unchanged despite sudden changes in mission, task organization, and rules of engagement. Typically, task organization evolves to meet changing conditions; however, transition planning must account for changes in mission, also. Commanders continuously assess the situation, task organize, and cycle their forces to retain the initiative. Commanders strive to achieve changes in emphasis without incurring an operational pause.

6-223. Planning for operations focused on stability begins the moment the BCT receives the mission. Coordinated early planning between the military and the interagency for post-conflict operations is vitally important. When coordinated planning to transition responsibility from military to civilian entities does not occur, the result is always the development of military and civilian parallel efforts, which seek to either secure or develop the host nation. The end state of the offense is the eventual transfer of all security operations to host nation control. Transferring security operations does not allow the commander to abdicate the role of providing security for the host nation, facilities, or friendly units. The commander must work in concert with host-nation security forces to ensure a smooth transition to host nation control.

6-224. Building partner capacity is the outcome of comprehensive interorganizational activities, programs, and engagements. Building partner capacity enhances security, rule of law, essential services, governance, economic development, and other critical government functions. Army forces support host nation ownership when planning and implementing capacity building as part of a comprehensive approach.

6-225. All actors involved in decisive action integrate with the operation from the onset of planning. Together, they complete detailed analyses of the situation and operational environments, develop integrated COAs, and continuously assess the situation. Integrating civilian and military efforts into a whole of government approach has challenges. First, the efforts have differing capacities and differing perspectives. Second, the two efforts use different approaches and decision-making processes.

6-226. A comprehensive approach integrates the cooperative efforts of the departments and agencies of the U.S. Government, other unified action partners, and private sector entities to achieve unity of effort toward a shared goal. A comprehensive approach builds from the cooperative spirit of unity of effort. Successful operations use this approach, even for those operations involving actors participating at their own discretion or present in the operational area but not acting as a unified action partner member. Integration and collaboration among actors with different agendas and experience is challenging. A comprehensive approach achieves unity of effort to forge a shared understanding of a common goal. Mandates, experiences, structures, and bureaucratic cultures make it difficult to sustain a comprehensive approach. Commanders overcome and mitigate this challenge with extensive cooperation and coordination.

6-227. Five broad conditions provide the underpinnings for strategic, whole-of-government planning and serve as a focal point for integrating operational and tactical level tasks. The end state conditions are flexible and adaptive to support activities across the range of military operations but rely on concrete principles and fundamentals in application. (See chapter 8.) End state conditions are—

- A safe and secure environment.
- Established rule of law.
- Social well-being.
- Stable governance.
- A sustainable economy.

This page intentionally left blank.

Chapter 7
Defense

The brigade combat team (BCT) conducts defensive operations to defeat enemy attacks, gain time, control key terrain, protect critical infrastructure, secure the population, and economize forces. Most importantly, the BCT sets conditions to transition to the offense or operations focused on stability. Defensive operations alone are not decisive unless combined with offensive operations to surprise the enemy, attack enemy weaknesses, and pursue or exploit enemy vulnerabilities. This chapter addresses the characteristics of the defense, common defensive planning considerations, forms of the defense, defensive control measures, defensive operations, and planning considerations when transitioning to other tactical operations.

SECTION I – CHARACTERISTICS OF THE DEFENSE

7-1.   Successful defenses share the following characteristics: disruption, flexibility, maneuver, mass and concentration, operations in depth, preparation, and security. Defenses are aggressive. Defending commanders use all available means to disrupt enemy forces. Commanders disrupt attackers and isolate them from mutual support to defeat them in detail. Defenders seek to increase their freedom of maneuver while denying it to attackers. Defending commanders use every opportunity to transition to the offense, even if only temporarily. As attackers' losses increase, they falter, and the initiative shifts to the defenders. These situations are favorable for counterattacks. Counterattack opportunities rarely last long. Defenders strike swiftly when the attackers reach their decisive point. Surprise and speed enable counterattacking forces to seize the initiative and overwhelm the attackers.

7-2.   The Battle of Kasserine Pass, described below, is an example of neglecting the characteristics of the defense. Prior to the Battle of Kasserine Pass, II Corps failed to adequately resource and prepare defensive positions; ensure defensive positions could mass effects of direct and indirect fires; adequately include flexibility, depth, and maneuver in planning, and conduct continuous reconnaissance and security operations to provide early and accurate warning.

The Battle of Kasserine Pass

The Battle of Kasserine Pass, Tunisia in February 1943, served as a rude awakening for the American Army in World War II. Over the course of the month, German Field Marshal Erwin Rommel's veteran Armeegruppe Afrika delivered a series of defeats to the relatively inexperienced American II Corps under Major General Lloyd Fredendall. Kasserine Pass was a tremendous blow to American pride and a loss of confidence in the eyes of II Corps' British and French allies. However, lessons learned from the battle led to changes in leadership, tactics, and training, resulting in a competent force in the African theater, as well as more realistic and effective training in America. Kasserine Pass remains a bitterly poignant example of the disasters that befall a force that neglects the characteristics of the defense.

The Anglo-American advance into Tunisia transitioned to a defense in December 1942 due to poor weather and logistical challenges. Major General Fredendall's II Corps was tasked with reinforcing the French defenses around several mountain passes and a road junction in southern Tunisia. Fredendall, headquartered some 70 miles from the forward line of troops, personally directed the dispersion of his subordinate elements over a large area of operations. Omar Bradley later noted in A Soldier's Story that, "American Infantry had been lumped on isolated [hills] ...and mobile reserves were scattered in bits and pieces along the line." Simultaneously, engineer assets needed for improving of defensive positions were instead constructing a cavernous bunker for the corps command post.

Lack of cooperation with allied forces and inadequate reconnaissance and security operations allowed German and Italian elements of Armeegruppe Afrika to achieve tactical surprise at the onset and throughout the battle. During the following weeks, French and American units routinely found themselves surprised by enemy contact, fighting from non-mutually supporting positions, and unable to mass direct and indirect fires. German and Italian forces, enjoying local air superiority as allowed by weather, rarely experienced disruption at the hands of isolated and easily bypassed defensive positions. Retrograde operations frequently degenerated into routes with significant losses of manpower and equipment. Allied attempts to maneuver against Axis forces were poorly coordinated due to failures to incorporate flexibility into the array of forces as well as command failures at multiple echelons. Brigadier General Paul R. Robinette, commander of Combat Command A of the 1ˢᵗ Armored Division, would later record the Soldiers' observation that during this battle "never were so few commanded by so many from so far away."

The poor performance of the American Army left a bitter legacy for the American Soldier that would haunt Anglo American relations in the theater. At the human level, II Corps sustained approximately six thousand casualties during the February 1943 engagements as well as the loss of 183 tanks, 104 half-tracks, 208 artillery pieces, 500 other vehicles, and vast amounts of supplies. Conversely, Rommel's forces sustained approximately 1000 casualties and a tenth of the material losses. (See figure 7-1.)

Various

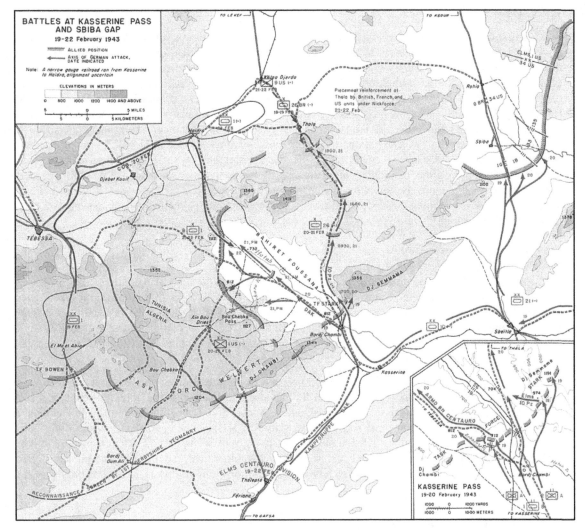

Figure 7-1. Kasserine Pass and Sbiba Gap map

DISTRUPTION

7-3. The BCT must disrupt the tempo and synchronization of the enemy's operation to counter the enemy's initiative, to prevent enemy from concentrating combat power against a part of the defense, and to force the enemy to go where the commander wants the enemy to go. The commander achieves disruption by defeating or misleading enemy reconnaissance forces, impeding maneuver, disrupting reserves, neutralizing fire support, and interrupting command and control. Defensive techniques vary with circumstances, but all defensive concepts of operation aim to spoil the attacker's synchronization. Strong security forces to defeat enemy reconnaissance, phony initial positions or dummy positions, and obstacles are some of the measures used to increase BCT security in the defense.

7-4. The commander uses counterattack, counterbattery, and countermortar fires; spoiling attacks; electromagnetic attacks (see chapter 4); obstacles; and retention of key or decisive terrain to prevent the enemy from concentrating overwhelming strength against portions of the defense. *Decisive terrain* is key terrain whose seizure and retention is mandatory for successful mission accomplishment (ADP 3-90). An analysis of friendly force networks will inform the development of critical information and provide a basis for establishing key terrain in cyberspace. Key terrain in the defense is those physical and logical entities in friendly force technical networks of such extraordinary importance that any disruption in their operation would have debilitating effects upon accomplishment of the mission.

FLEXIBILITY

7-5. The BCT commander uses detailed planning, sound preparation, operations in depth, retaining reserves, and command and control to maintain flexibility. Flexibility requires the commander to visualize the battlefield to detect the enemy's scheme of maneuver in time to direct fire and movement against it. The commander does not limit information collection efforts only to the forces in contact. The commander also concentrates on formations arrayed in-depth. The enemy may try to bypass areas where the defense is strong. Hence, the BCT commander ensures detection and the defeat of the enemy along all possible avenues of approach. The commander uses aviation reconnaissance and surveillance assets and cyberspace and electromagnetic warfare (EW) operations to support information collection. The BCT's plan allows the commander to shift the decisive operation or the main effort quickly, if the situation changes, while maintaining synchronization. In addition, alternate and subsequent positions provide the flexibility needed to execute the defense, effectively. Small reserves may position near critical terrain or likely avenues of attack to enable rapid deployment to those areas. Blocking positions can be established to deny the enemy a chance for a rapid breakthrough.

MANEUVER

7-6. Maneuver allows the commander to take full advantage of the area of operations and to mass and concentrate resources where required. The BCT arrays and allocates forces in relationship to likely enemy courses of action (COAs). The BCT uses allocations based on the results of the relative combat power analysis of the BCT and enemy forces' assigned tasks and the terrain. The commander accepts risk along less likely avenues of approach to ensure that adequate combat power is available for more likely avenues of approach.

7-7. Maneuver also encompasses defensive actions such as security and support area operations. In some cases, the commander must accept gaps within the defense, but must take measures to maintain security within these risk areas. The BCT integrates assigned or attached and echelon above brigade enablers, for example, surveillance assets, reconnaissance and security forces, space and cyberspace capabilities, patrols (dismounted and mounted), combat outpost, observation posts, sensor outposts, or listening posts. Additional enablers include engineer reconnaissance teams, or chemical, biological, radiological, and nuclear (CBRN) reconnaissance forces, observation outposts, forward observer or spotter outposts, or other economy of force effort for these areas.

MASS AND CONCENTRATION

7-8. The BCT masses its combat power to overwhelm the enemy and regain the initiative. The commander must be able to concentrate forces and mass the effects of fires at the decisive point and time. To accomplish this, the commander may economize forces in some areas, retain a reserve, shift priority of fires, and maneuver repeatedly to concentrate combat power. Commanders accept risks in some areas to concentrate for decisive action elsewhere. Obstacles, security forces, and fires assist in reducing these risks as forces economize.

7-9. Dependent on the operational framework, the commander designates a main effort to achieve concentration, and directs all other elements and assets to support and sustain this effort. The commander may reprioritize forces, designating a new main effort as the situation changes. The commander directs the task and purpose of supporting elements to create the conditions necessary for the main effort to accomplish its task and purpose. The commander narrows the width of subordinate areas of operations, focusing counterattack plans to support the main effort; assigns the main effort unit priority of obstacle preparation; gives the unit priority of indirect fire; and positions the reserve to influence the main effort's area.

7-10. Targets determined during the BCT's planning process and refined during preparation is described broadly as physical and logical entities in cyberspace consisting of one or more networked devices used by enemy and adversary actors. These targets may be established as named area of interests (NAIs) and target area of interests (TAIs) as appropriate. As part of cyberspace electromagnetic activities (CEMA), the division staff, in coordination with the corps staff, performs a key role in target network node analysis (see chapter 2) supporting the BCT's cyberspace and EW effort.

7-11. Concealment and deception must mask the concentrating forces since concentration increases the risk level of large losses from enemy fires. The strategy is to concentrate the effects of the forces, not to physically concentrate the forces themselves. Defending units use engagement areas (EAs) to concentrate combat power from mutually supporting positions. Reconnaissance, surveillance, and security operations, organic and nonorganic to the BCT, are vital to gaining the information and time needed to concentrate the forces and fires of the BCT.

OPERATIONS IN DEPTH

7-12. *Operations in depth* are the simultaneous application of combat power throughout an area of operations (ADP 3-90). Integration of all combat power throughout the area of operations, as well as the BCT's area of influence and area of interest, improves the chances for success while minimizing friendly casualties. Quick, violent, and simultaneous action throughout the depth of the BCT's area of operations can hurt, confuse, and even paralyze an enemy force when most exposed and vulnerable. Such actions weaken the enemy's morale and do not allow any early successes to build their confidence. Operations in depth prevent the enemy from gaining momentum in the attack. Synchronization of actions within the division's operational framework facilitates mission success.

7-13. Alternate and supplementary positions, combat outposts, observation posts, and mutually supporting strong points extend the depth of the defense. The commander plans fires throughout the defensive area up to the maximum range of available weapons. Fire support units and observers move and reposition to maintain contact with enemy forces and observe TAIs in-depth as the battle develops. The commander plans for the emplacement of obstacles around critical locations to disrupt the enemy's most dangerous and most likely COAs.

PREPARATION

7-14. The commander must be familiar with the enemy's abilities and limitations to prepare the defense properly. The enemy's abilities and limitations include their organization, offensive doctrine (tactics, techniques, and procedures), weapons systems, and equipment. Collection means (reconnaissance, surveillance, security operations, and intelligence operations) inform the commander and staff to enable understanding and multiply the effectiveness of the defense.

7-15. The commander analyzes the terrain in detail from all perspectives and then verifies on the ground to select EAs and positions that allow for the massing of fires and the concentration of forces on likely enemy avenues of approach. Emphasis is on preparing and concealing positions, routes, obstacles, logistical support, and command and control facilities and networks. The commander plans, coordinates, and prepares military deceptions and uses rehearsals to ensure staffs and subordinates understand the concept of operations and commander's intent.

7-16. During preparation, aerial (manned and unmanned) reconnaissance and surveillance collection efforts (internal and external to the BCT) complement ground efforts by increasing speed and depth with which reconnaissance can be conducted over an area. Ground reconnaissance and security forces employ, and supplies are pre-positioned. Counterattack plans to support the defense and to place the BCT on the offense are key to retaining the initiative. Counterattack routes must be reconnoitered, improved, secured, and rehearsed. Defensive preparations within the main battle area (MBA) continue in-depth even as close engagement begins.

SECURITY

7-17. The BCT commander establishes security areas forward of the MBA, on the flanks, and within the BCT's support area to protect the force while in the defense. Security operations forward of the MBA normally include screen, guard, and cover. The presence of a security force forward of the MBA does not relieve the MBA units from their own security responsibilities (area security and local security tasks). All units must maintain security, for example civil reconnaissance and security patrolling, within assigned areas and contribute to counterreconnaissance.

7-18. The BCT may defend to conserve combat power for use elsewhere at a later time. The commander secures the force through integrated security operations throughout the depth and breadth of its assigned area of operations. Long-range reconnaissance and surveillance assets task organized at the division and corps level conduct information collection to define and confirm the enemy at extended ranges and in time and manner. The commander plans for and employs information-related capabilities and CEMA to confuse the enemy as to the BCT's manner of defense and to aid in securing the force.

SECTION II – COMMON DEFENSIVE PLANNING CONSIDERATIONS

7-19. The commander in the defense exploits prepared, mutually supporting positions even though the initiative is yielded to the enemy. The commander uses knowledge of the terrain to slow the enemy's momentum. The defending force maintains its security and disrupts the enemy's attack at every opportunity. The defending commander uses long-range fires to reduce the force of the enemy's initial blow, hinder enemy offensive preparations and wrest the initiative from the enemy. The commander draws the enemy into EAs to surprise the enemy with concentrated and integrated fires from concealed and protected positions. The commander then counterattacks the enemy, repeatedly imposing blows from unexpected directions. The following discussion uses the warfighting functions (command and control, movement and maneuver, intelligence, fires, sustainment, and protection) and specific operational environments as the framework for planning considerations that apply to defensive operations.

COMMAND AND CONTROL

7-20. The BCT commander understands, visualizes, and describes the anticipated enemy actions and issues commander's guidance to the staff. Based upon the commander's guidance, the staff refines the higher headquarters' products to enable the BCT commander to visualize the operational environment. The BCT commander and staff refine the higher headquarters' intelligence preparation of the battlefield (IPB) products to focus on the details of the operation in the BCT's area of operations. The higher commander normally defines where and how the BCT defeats or destroys the enemy and the operational framework. The BCT commander defines (envisions) the BCT's executions of its portion of the higher (division and corps) echelon fight. As each decision is made and each action is taken in the presence of risk and uncertainty, commanders must anticipate and prevent or mitigate risk, including ethical risks, to support mission accomplishment.

DEFEAT THE ENEMY

7-21. The BCT commander and staff analyze how and where to defeat the enemy. The BCT commander may define a defeat mechanism that includes use of single or multiple counterattacks to achieve success. Subordinate commanders and staffs analyze their unit's role in the fight and determine how to achieve success. In an area defense, usually the BCT achieves success by massing the effects of obstacles and fires to defeat the enemy forward of a designated area, often in conjunction with a higher echelon's counterattack. In a delay operation, the BCT achieves success by combining maneuver, fire support, obstacles, and the avoidance of decisive engagement until conditions are right to gain time or shape the battlefield for a higher echelon's counterattack.

ORGANIZE THE DEFENSE

7-22. The BCT commander organizes in the defense to facilitate the execution of a defensive operation. The commander and staff use an operational framework, and associated vocabulary, to help conceptualize and describe the concept of operations in time, space, purpose, and resources (see figure 2-3 on page 2-24). An operational framework is a cognitive tool used to assist commanders and staffs in clearly visualizing and describing the application of combat power in time, space, purpose, and resources in the concept of operations (see chapter 2). An operational framework establishes an area of geographic and operational responsibility for the commander and provides a way to visualize how the commander will employ forces against the enemy. To understand this framework is to understand the relationship between the area of operations and operations in depth. Proper relationships allow for simultaneous operations and massing of effects against the enemy. (See FM 3-0 for additional information.)

7-23. As stated in chapter 2, the operational framework has four components. First, the BCT commander is assigned an area of operations for the conduct of operations. Second, the commander can designate deep, close, rear, and support areas to describe the physical arrangement of forces in time and space. Third, within this area, the commander conducts decisive, shaping, and sustaining operations to articulate the operation in terms of purpose. Finally, the commander designates the main and supporting efforts to designate the shifting prioritization of resources.

Note. The BCT does not conduct operationally significant consolidate gains activities unless tasked to do so, usually within a division consolidation area.

7-24. As an example, deep, close, and rear areas historically have been associated with terrain orientation, but this framework can apply to temporal and organizational orientations as well. The BCT can use the deep, close, and rear area component to engage simultaneously the enemy in three distinct areas—deep area, close area, and rear area. (See figure 7-2.)

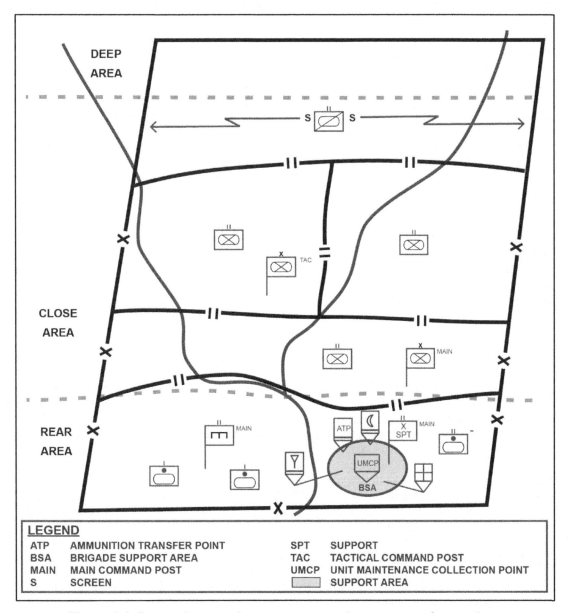

LEGEND
ATP	AMMUNITION TRANSFER POINT	SPT	SUPPORT
BSA	BRIGADE SUPPORT AREA	TAC	TACTICAL COMMAND POST
MAIN	MAIN COMMAND POST	UMCP	UNIT MAINTENANCE COLLECTION POINT
S	SCREEN		SUPPORT AREA

Figure 7-2. Deep, close, and rear areas—contiguous area of operations

7-25. Deep area is the portion of the commander's area of operations that is not assigned to subordinate units. Within this area, the BCT commander conducts deep operations against uncommitted enemy forces to set conditions for subordinate commanders conducting operations in the close area. In noncontiguous areas of operations, the deep area is the area between noncontiguous areas of operations or beyond contiguous areas of operations.

7-26. Close area is the portion of a commander's area of operations assigned to subordinate maneuver forces. Operations in the close area are operations within a subordinate commander's area of operations. In contiguous areas of operations, a close area assigned to a maneuver force extends from its subordinates' rear boundaries to its own forward boundary. In noncontiguous areas of operations, the close area is the area within the subordinate commanders' areas of operations.

7-27. Rear area is the portion of the commander's area of operations that is designated to facilitate the positioning, employment, and protection of assets required to sustain, enable, and control operations. A rear area in contiguous areas of operations is an area for any command that extends from its rear boundary forward to the rear boundary of the next lower level of command. In noncontiguous areas of operations, the rear area is that area defined within the higher commander's area of operations providing a location to base sustainment assets and provide sustainment to the force. A support area(s), generally located within the rear area, facilitates the positioning, employment, protection, and control of base sustainment assets required to sustain and enable combat operations (see chapter 9).

Note. Corps and division commanders may establish a consolidation area to exploit tactical success while enabling freedom of action for forces operating in the other areas. When designated, a consolidation area refers to an area of operations assigned to an organization, generally a BCT or task force, where forces have established a level of control and large-scale combat operations have ceased. Consolidation area activities require a balancing of area security and stability operations tasks. (See chapter 2, section IV for a detailed discussion.)

TASKS ASSIGNMENT

7-28. The BCT commander assigns tasks to subordinate units through the staff. The assignment of a task includes not only the task (what), but also the unit (who), place (where), time (when), and purpose (why). The commander and staff develop obstacle fire support plans concurrently with the defensive force array, again defining a task and purpose for each obstacle and target in keeping with the commander's stated fire support tasks and intended obstacle effects. The desired end state is a plan that defines how the commander intends to mass the effects of direct and indirect fires with obstacles and use of terrain to shape the battlefield and defeat or destroy the enemy.

CONTROL MEASURES

7-29. The BCT plans control measures to provide the flexibility needed to respond to changes in the situation and allow the BCT to concentrate combat power at the decisive point. Defensive control measures within the BCTs area of operations include designating the security area, the battle handover line, the MBA with its associated forward edge of the battle area, and the echelon support area. The BCT and subordinate units use battle positions (primary, alternate, supplemental, subsequent, and strong point), direct fire control, and fire support coordination measures to conduct defensive operations. The commander designates disengagement lines to trigger the displacement of subordinate forces when required. A *disengagement line* is a phase line located on identifiable terrain that, when crossed by the enemy, signals to defending elements that it is time to displace to their next position (ADP 3-90). (See paragraph 7-94 for a discussion of control measures associated with defensive operations.)

MOVEMENT AND MANEUVER

7-30. The BCT can conduct defensive operations with units out of range and in mutual support of each other. Defensive operations with out of range units require a judicious effort by the BCT commander and staff to determine the positioning and priority of support assets and capabilities. During the terrain analysis, the commander and staff must look closely for key and decisive terrain, EAs, choke points, intervisibility lines, and reverse slope opportunities to take full advantage of the BCT's capabilities to mass firepower to support defensive maneuvers.

7-31. The BCT commander must determine any potential gaps between units once maneuver units are assigned area of operations. The BCT should plan to cover any gaps with reconnaissance and security forces and surveillance assets. The BCT must plan local counterattacks to isolate and destroy any enemy that penetrates a gap in the area of operations. The commander should also plan to reposition units not in contact to mass the effects of combat power against an attacking enemy.

7-32. The BCT commander identifies EAs (see paragraph 7-130) to contain or destroy the enemy force with the massed effect of all available weapons and supporting systems with the assignment of area of operations. The commander determines the size and shape of the EA by the visibility of the weapons systems in their firing positions and the maximum range of those weapons. The commander designates EAs to cover each enemy avenue of approach into the area of operations. Elements, deliberately left behind or inserted through infiltration or helicopter, can report and call in fires on an approaching enemy.

7-33. The BCT combines fires, defensive positions, countermobility obstacles, and counterattacks to disrupt the enemy's attack and break the enemy's will. The BCT must disrupt the synchronization of the enemy's operation to counter the enemy's initiative, prevent the enemy's concentrating combat power against a part of the defense, and force the enemy where the commander wants the enemy to go. The commander causes disruption defeating or misleading the enemy's reconnaissance forces, impeding maneuver, disrupting reserve, neutralizing fire support, and interrupting command and control.

7-34. Defensive techniques vary with circumstances, but all defensive concepts of operation aim to spoil the attacker's synchronization. Strong security forces to defeat enemy reconnaissance, phony initial positions or dummy positions, and obstacles are some of the measures used to increase security in the defense. Repositioning forces, aggressive local protection measures, and employment of roadblocks and ambushes combine to disrupt the threat of an attack. Counterattack, counterbattery fires, obstacles, and retention of key or decisive terrain prevent the enemy from concentrating overwhelming strength against portions of the defense.

7-35. The information environment supports the commander's mission and desired end state using information-related capabilities, techniques, or activities. These capabilities include, but are not limited to, public affairs operations, psychological operations (PSYOP), combat camera, Soldier and leader engagement, civil affairs operations, civil and cultural considerations, operations security, military deception, and CEMA. CEMA (see chapter 4) at the BCT level include cyberspace operations, EW, and spectrum management operations. (See FM 3-13 and FM 3-12 for additional information.)

7-36. The BCT commander considers mutual support when task organizing forces, assigning areas of operations, and positioning units. Mutual support is that support which units render each other against an enemy, because of their assigned tasks, their position relative to each other and to the enemy, and their inherent capabilities. Mutual support has two aspects—supporting range and supporting distance.

7-37. Supporting range is the distance one unit may be geographically separated from a second unit yet remain within the maximum range of the second unit's weapons systems. Mutual support exists when positions and units are in supporting range by direct or indirect fires, thus preventing the enemy from attacking one position without subjecting themselves to fire from one or more adjacent positions. Supporting distance is the distance between two units that can be traveled in time for one to come to the aid of the other and prevent its defeat by an enemy or ensure it retains control of a civil situation. When friendly forces are static, supporting range equals supporting distance.

7-38. Mutual support increases the strength of all defensive positions, prevents defeat in detail, and helps prevent infiltration between positions. Tactical positions achieve the maximum degree of mutual support between them when they are located to observe or monitor the ground between them or conduct patrols to prevent any enemy infiltration. At night or during periods of limited visibility, the commander may position small tactical units closer together to retain the advantages of mutual support. Unit leaders must coordinate the nature and extent of their mutual support.

7-39. Capabilities of supported and supporting units affect supporting distance. Units may be within supporting distance, but if the supported unit cannot communicate with the supporting unit, the supporting unit may not be able to affect the operation's outcome. In such cases, the units are not within supporting distance, regardless of their proximity to each other. The following factors affect supporting distance: terrain and mobility, distance, enemy capabilities, friendly capabilities, and reaction time. (See ADP 3-0 for additional information.)

7-40. The need for flexibility through mobility requires the use of graphic control measures to assist command and control during counterattacks and repositioning of forces. Specified routes, phase lines (PLs), attack-and support by fire positions, battle positions, EAs, target reference points (TRPs), and other fire support coordination measures are required to synchronize maneuver effectively.

7-41. Army aviation conducts offensive operations to support the maneuver commander's defensive operation. Manned and unmanned aircraft can provide reconnaissance, surveillance, and security for ground forces. Aviation quick response force can respond to a counterattack during the maneuver commander's transition from offensive to defensive operations, allowing ground forces to focus on consolidation and reorganization. Additionally, once in an established defensive position, aviation assets can conduct information collection and delay advancing enemy forces. Aviation allows the maneuver commander to mass reserves by air to reinforce a defensive position. Additional aviation considerations include—

- Conduct reconnaissance to identify bypasses, adequate sites and routes, and provide overwatch for security force operations.
- Provide direct fires or call for fires to cover obstacles.
- Provide security and early warning for ground movement, assembly areas, and fixed base operations.
- Transport air defense teams, CBRN teams, and supplies.
- Conduct aerial surveys of known or suspected CBRN contaminated areas.
- Provide information collection for targeting.

7-42. The speed and mobility of aviation can help maximize concentration and flexibility. Attack reconnaissance helicopters routinely support security area operations and mass fires within the MBA. Synchronization and integration of aviation assets into the defensive ground maneuver plan is important to ensure engagement as a whole. If the BCT augments with aviation assets, it must involve the direct fire planning processes of the supporting aviation unit through its aviation liaison officer, the air defense airspace management (ADAM) element, and brigade aviation element (BAE) within the fire support cell.

7-43. Air assets provide direct fire, observation, and the rapid movement of supplies and personnel during the conduct of the defense. Attack reconnaissance helicopters and fixed-wing aircraft can employ guided and unguided munitions that provide Army aviation attack and close air support to ground forces in direct contact with enemy elements. Through reconnaissance, intelligence, surveillance, and target acquisition planning (see FM 3-09), these assets can conduct interdiction missions to destroy high-value targets (HVTs) and high-payoff targets (HPTs) before their employment to shape the operation. Attack reconnaissance helicopters can assist the BCT reserve in exploiting opportunities to attack an enemy weakness or to support restructuring of friendly lines in the event of enemy penetration. Rotary and fixed-wing aircraft can provide additional observation and control indirect fires directed at enemy formations prior to contact with the BCT defense and enhance situational awareness for the commander and staff. Utility and cargo rotary-wing aircraft can provide casualty evacuation (see ATP 4-25.13) and conduct emergency resupply operations depending on the enemy's air defense capabilities.

7-44. The ground commander is responsible for the priority, effects, and timing of fires and maneuver within their area of operations (see JP 3-09). Air-ground operations require detailed planning and synchronization timelines, aviation tasks and purposes, and airspace control. Analysis of enemy COAs and timelines allow the BCT staff to synchronize aircraft operational times to match expected enemy contact. Security forces forward of the BCT MBA assist in synchronizing aircraft employment at the decisive point.

7-45. Development of detailed task and purpose for the supporting aviation is essential as it enables the aviation commander and staff to employ the right platforms and munitions. Understanding the threat and the BCT commander's desired aviation effects drives the aviation units' task organization of air elements and selection of weapon systems. (See FM 3-04 for additional information.)

7-46. Effective division airspace control is contingent on the development of a unit airspace plan that includes subordinate BCT airspace coordinating measures that synchronize airspace users and activities supporting the BCT. Airspace control is essential for deconflicting manned and unmanned aircraft from indirect fires. Properly developed airspace coordinating measures enable the BCT to mass aerial and surface-based fires simultaneously while using unmanned assets to maintain surveillance. (See JP 3-52, FM 3-52, and ATP 3-52.1 for additional information on airspace control and ATP 3-91.1 for information on the joint air-ground integration center [JAGIC].)

7-47. During the conduct of defensive operations, situations requiring denial operations, defending encircled, and stay-behind operations have their own unique planning, preparation, and execution considerations. In the defense, denial operations conducted to deprive the enemy of some or all of the short-term benefits of capturing an area may be required. Denying the enemy, the use of space, personnel, supplies, or facilities may include destroying, removing, and contaminating those supplies and facilities or erecting obstacles. Subordinate units of the BCT when encircled can continue to defend, conduct a *breakout*—an operation conducted by an encircled force to regain freedom of movement or contact with friendly units (ADP 3-90)— from encirclement, exfiltrate toward other friendly forces, or attack deeper into enemy-controlled territory. In other defensive situations, subordinate units may be directed to conduct operations as a stay behind force. These actions may be planned or forced by the enemy.

7-48. A common additional action planned or forced by the enemy can include a relief in place. A *relief in place* is an operation in which, by direction of higher authority, all or part of a unit is replaced in an area by the incoming unit and the responsibilities of the replaced elements for the mission and the assigned zone of operations are transferred to the incoming unit (JP 3-07.3).

Note. The Army uses an area of operations instead of a zone of operations.

7-49. The BCT normally conducts a relief in place as part of a larger operation, primarily to maintain the combat effectiveness of committed forces. The higher headquarters directs when and where to conduct the relief and establishes the appropriate control measures. Normally, during the conduct of combat operations, the unit relieved is defending. However, a relief in place may set the stage for resuming offensive operations or serve to free the relieved unit for other tasks.

7-50. There are three types of relief in place operations:

- A *sequential relief in place* occurs when each element within the relieved unit is relieved in succession, from right to left or left to right, depending on how it is deployed (ADP 3-90).

- A *simultaneous relief in place* occurs when all elements are relieved at the same time (ADP 3-90).

- A *staggered relief in place* occurs when a commander relieves each element in a sequence determined by the tactical situation, not its geographical orientation (ADP 3-90).

INTELLIGENCE

7-51. IPB helps the BCT commander determine where to concentrate combat power, where to accept risk, and where to plan the potential decisive operation. The staff integrates intelligence from the higher echelon's collection efforts (see chapter 5, section I) and from units operating forward of the BCT's area of operations. Information collection includes collection from spot reports, tactical unmanned aircraft systems (known as TUASs), and other higher-level collection assets. Early warning of enemy air attack, airborne or helicopter assault or insertion, and dismounted infiltration are vitally important to provide adequate reaction time to counter these threats as far forward as possible. The essential areas of focus are terrain analysis, determination of enemy force size and likely COAs with associated decision points, and determination of enemy vulnerabilities.

7-52. Intelligence operations, conducted by the military intelligence company, collect information about the intent, activities, and capabilities of threats and relevant aspects of the operational environment to support commanders' decision-making (see chapter 5, section IV). The commander uses intelligence products to identify probable enemy objectives and approaches and develops NAIs and TAIs from probable objectives and approaches. The commander studies the enemy operation patterns and the enemy's vulnerability to counterattack, interdiction, EW, air attacks, and canalization by obstacles. The commander examines the enemy's ability to conduct air attacks, insert forces behind friendly units, and employs nuclear, biological, and chemical weapons and determines how soon follow-on or reaction enemy forces can influence the operation.

7-53. The commander and staff use available reconnaissance and engineer assets to study the terrain. By studying the terrain, the commander tries to determine the principal enemy and friendly heavy, light, and air avenues of approach. The commander assesses the most advantageous area for the enemy's main attack, as well as other military aspects of terrain to include observation and fields of fire, avenues of approach, key terrain, obstacles, and cover and concealment (OAKOC). The BCT commander and staff assess ground and air mobility corridors and avenues of approach to determine where the enemy can maneuver to reach likely objectives and to identify limitations on friendly maneuver and positioning. Identification of terrain, such as chokepoints that create potential enemy vulnerabilities and opportunities for friendly attack, is critical. (See ATP 2-01.3 and ATP 3-34.80 for additional information.)

7-54. The BCT S-2, with the BCT geospatial engineer uses the Geospatial Intelligence Workstation to provide terrain analysis. The Geospatial Intelligence Workstation can identify critical terrain and position weapons systems and intelligence assets (see ATP 3-34.80). Once subordinate units know the area of operations, BCT units conduct their own terrain analysis using physical reconnaissance and the line of sight analysis function in Joint Capabilities Release (known as JCR). Terrain analysis must achieve a fidelity that allows for effective positioning of direct fire weapons systems and observers. The analysis must identify intervisibility lines, fields of fire, dead spaces, and integrate the effects of weather.

7-55. The staff weather officer, or higher headquarters staff if a staff weather officer is not assigned, can assist the BCT staff by supplying predictive and descriptive weather information for specific time periods and locations within the BCT's area of operations. In addition, the weather program of record (for example, the Distributed Common Ground System-Army [DCGS-A]) can provide weather predictions and weather effects for a specific mission, desired area of operations, or particular weapons system.

7-56. The result of the terrain analysis is a modified, combined obstacle overlay and identification of defensible areas. The BCT staff should transmit results of the analysis by any available means to subordinate units. When the staff has analyzed the BCT's assigned area of operations, the staff should expand its analysis to adjacent area of operations and areas forward and to the rear of the BCT.

7-57. The staff determines enemy force sizes, likely COAs, and decision points through analysis. The staff determines the size of the enemy force that each avenue of approach and mobility corridor can support. The expected size of the enemy force drives the determination of friendly force allocation, fires, and obstacle efforts. The commander and staff use the enemy force's size to understand how the enemy intends to utilize its forces and the terrain. The enemy COAs developed must be feasible and must reflect the enemy's flexibility and true potential. All COAs, at a minimum, should analyze the following:

- Likely enemy objectives.
- Enemy composition, disposition, and strength.
- Schemes of maneuver including—
  - Routes.
  - Formations.
  - Locations and times the enemy may change formations.
  - Possible maneuver options available to the enemy.
  - Key decision points.
- Time and distance factors for the enemy's maneuver through the area of operations.
- Likely employment of all enemy combat multipliers including—
  - Artillery.
  - Air defense.
  - Obstacles.
  - CBRN strikes.
  - Dynamic obstacles.

- Attack aircraft.
- Likely use of all enemy reconnaissance assets and organizations including likely reconnaissance objectives, reconnaissance avenues of approach, and times to expect enemy reconnaissance, based on doctrinal rates of march.
- Likely use of all reconnaissance assets to locate observer locations and observation posts.
- Likely locations and identification of enemy HVTs—such as artillery formations, reserves, and command and control.
- Likely locations, compositions, strength, employment options, and time and distance factors for enemy reserves and follow-on forces.
- Locations of enemy decision points that determine selection of a specific COA.
- Likely breach sites, strike areas, and points of penetration.

7-58. The intelligence staff develops the enemy COA statement and sketch. The staff graphically depicts the enemy on a situation template based upon the results of the IPB. The intelligence staff officer (S-2) and staff use these items to develop the initial information collection plan. The intelligence staff should distribute all products by any available means to the entire staff and subordinate units to support parallel planning. (See ATP 2-01.3 for additional information.)

7-59. The intelligence staff observes the enemy's tactics, the terrain, the weather, and friendly and enemy capabilities to identify potential enemy vulnerabilities. To engage the enemy where the terrain puts the enemy at a disadvantage, the staff identifies restrictive terrain that may slow the enemy's attack, cause a separation of forces, create difficulties in command and control, or force the enemy to conduct defile drills; for example, narrow valleys, passes, or urban areas. The intelligence staff also identifies chokepoints or natural obstacles that may cause a loss of momentum, a potential fragmenting of forces, or a vulnerable concentration of forces (rivers and canals). The staff identifies terrain that canalizes enemy formations into areas that provide defending forces with good fields of fire, observation, and flanking fires. The intelligence staff also identifies areas dominated by key or defensible terrain that allows massing of fires.

7-60. The entire BCT staff must participate for IPB to develop successfully for the commander and subordinate units. Each staff member is responsible for analyzing the enemy based upon their warfighting function. Each staff member must be knowledgeable in friendly and enemy capabilities and terrain analysis. Each staff member must execute the process rapidly. The staff must ensure the results are detailed, legible, and disseminated quickly to support planning at all echelons.

7-61. The intelligence officer, supported by the entire BCT staff, provides the fire support officer and information operations officer information and intelligence for targeting and information capabilities. The intelligence officer supports targeting by providing accurate, current intelligence and information to the staff and ensures the information collection plan supports the finalized targeting plan. Intelligence support to targeting includes two tasks—providing intelligence support to target development and providing intelligence support to target detection. Intelligence support to information capabilities provides the commander with information and intelligence support for information tasks and targeting through nonlethal actions. It includes intelligence support to the planning, preparation, and execution of the information related activities, as well as assessing the effects of those activities. (See FM 2-0 and FM 3-13 for additional information.)

FIRES

7-62. Supporting the BCT commander's concept of operations during the defense involves attacking and engaging targets throughout the area of operations with massed or precision indirect fires, air and missile defense fires, defensive counterair, air support, and EW assets. As planning progresses, artillery counterbattery radar and counterfire radar employment is continually updated. Fire support planners must make maximum use of any preparation time available to plan and coordinate supporting fires. Planners must ensure fire support complements and supports all security forces and unit protection plans.

7-63. Fire support plays a key role in disrupting the attacker's tempo and synchronization during the defense. When required, massing overwhelming fires at critical places and times gains maximum efficiency and effectiveness in suppressing direct and indirect-fire systems and repelling an assault. Fire support planning and execution must address flexibility through operations in depth and support to defensive maneuver. Additional fire support considerations for supporting the commander's concept of operations include—

- Weight the main effort.
- Provide 360-degree air and missile defense coverage.
- Provide and disseminate early warning.
- Contribute targeting information.
- Engage critical enemy assets with fires before the attack.
- Plan counterfire against enemy indirect-fire systems attacking critical friendly elements.
- Use lethal and nonlethal means to apply constant pressure to the enemy's command and control structure.
- Provide fires to support defensive counterair operations to defeat enemy attacks.
- Plan the acquisition and attack of HPTs throughout the area of operations.
- Employ electromagnetic attack to degrade, neutralize, or destroy enemy combat capability.
- Concentrate fires to support decisive action.
- Provide fires to support counterattacks.
- Plan fires to support the barrier and obstacle plan.
- Plan for target acquisition and sensors to provide coverage of NAIs, TAIs, and critical assets.
- Plan for friendly force and allied force fratricide prevention measures.
- Plan for civilian noncombatant casualty prevention measures.
- Plan for un-intended collateral damage prevention measures.
- Request munitions authorities to ensure appropriate units on the ground have the operational and legal authority to employ munitions such as scatterable mine systems (air and ground volcanos).

7-64. The BCT may utilize unmanned aircraft systems (UASs), remote sensors, and reconnaissance and security forces to call for fire on the enemy throughout the area of operations. Quick, violent, and simultaneous action throughout the depth of the defender's area of operations can degrade, confuse, and paralyze an enemy force just as that enemy force is most exposed and vulnerable. (See FM 3-09 and FM 3-90-1 for additional information.)

SUSTAINMENT

7-65. Typically, sustaining operations in support of the defense requires more centralized control. Clear priorities of support, transportation, and maintenance are required. The BCT closely and continuously coordinates, controls, and monitors the movement of materiel and personnel within the operational environment based on the BCT's priorities and ensures their dissemination and enforcement. (See FM 4-0 for additional information.)

7-66. The routing function of movement control becomes an essential process for coordinating and directing movements on main supply routes or alternate supply routes and regulating movement on lines of communications to prevent conflict and congestion. Movement priorities must include throughput of echelons above brigade assets transporting additional engineer assets in preparation for the defense. Supply of class IV (construction and barrier materials) and class V (ammunition) normally have higher movement priorities during the defense. Planners may consider nighttime resupply operations to minimize enemy interference. (See ATP 4-16 for additional information.)

7-67. The BCT logistics staff officer (S-4) must ensure that the sustainment plan is coordinated fully with the rest of the staff. The S-4 coordinates with the brigade operations staff officer (S-3) to ensure that supply routes do not interfere with maneuver or obstacle plans but still support the full depth of the defense. Sustainment planners must consider prepositioning class IV, class V, and class III (bulk) far forward initially to support the security area during the counterreconnaissance fight, followed by the MBA so that the BCT can rapidly transition from defense to offense. Planning for sustainment operations throughout the security area is critical to sustaining reconnaissance and security operations to prevent enemy forces from determining friendly force disposition. Forces within the security area are configured prior to line of departure with a minimum of 72 hour logistics package (LOGPAC) of class I (subsistence), class III (petroleum, oils, and lubricants [POL]), and class V. Sustainment support to the security area must include planning for both ground and aerial medical evacuation of long duration observation posts. BCT sustainment planners also consider cross leveling classes of supply and sustainment assets upon transition from the offense to the defense.

7-68. Enemy actions and the maneuver of combat forces complicate forward area medical operations. Defensive operations must include health service support to medical personnel who have much less time to reach a patient, apply tactical combat casualty care (TCCC), and remove the patient from the battlefield. The enemy's initial attack and the BCT's counterattack produce the heaviest patient workload. These are also the most likely times for enemy use of artillery and CBRN weapons. The enemy attack can disrupt ground and air routes and delay evacuation of patients to and from treatment elements. The depth and dispersion of the defense create significant time distance problems for medical evacuation assets. (For additional information on the tactics, techniques, and procedures associated with health service support, see FM 4-02, ATP 4-02.2, and ATP 4-02.3.)

PROTECTION

7-69. The BCT must take measures to protect against all acts designed to impair its effectiveness and prevent the enemy from gaining an unexpected advantage. Because a force defends to conserve combat power for use elsewhere or later, the commander must secure the force. The BCT ensures security by employing security forces and surveillance assets throughout the depth and breadth of its assigned area of operations. The BCT may employ counterreconnaissance, combat outposts, a screen or guard force, and other security operations tasks to provide this security. Information related capabilities and CEMA aid in securing the force and confuse the enemy as to the manner of defense.

7-70. As discussed in chapter 6, personnel and physical assets have inherent *survivability*—a quality or capability of military forces which permits them to avoid or withstand hostile actions or environmental conditions while retaining the ability to fulfill their primary mission (ATP 3-37.34), which can be enhanced through various means and methods. One way to enhance survivability when existing terrain features offer insufficient *cover*—**protection from the effects of fires** and *concealment*—**protection from observation or surveillance** is to alter the physical environment to provide or improve cover and concealment. Similarly, natural or artificial materials may be used as camouflage to confuse, mislead, or evade the enemy. Together, these are called *survivability operations*—those protection activities that alter the physical environment by providing or improving cover, camouflage, and concealment (ATP 3-37.34).

7-71. All BCT units conduct survivability operations within the limits of their capabilities. Engineer and fire support assets have additional capabilities to support survivability operations. Engineer support to survivability operations is a major portion of the enhance protection line of engineer support (see FM 3-34). Fire support to survivability operations includes the employment of obscurants, which forces can use to enable survivability operations by concealing friendly positions and screening maneuvering forces from enemy observation and support to disengagement or movement of forces. (See ATP 3-11.50.)

7-72. CBRN support to survivability operations includes the ability to assess, protect and mitigate the effects of contamination. CBRN reconnaissance provides support by locating and marking contaminated areas and routes, allowing maneuver forces to avoid unnecessary exposure. Reconnaissance teams are focused on the collection of tactical and technical information to support survivability of friendly forces and facilities.

7-73. Although survivability encompasses capabilities of military forces both while on the move and when stationary, survivability operations focus more on stationary capabilities—constructing fighting and protective positions and hardening facilities. In the case of camouflage and concealment, however, survivability operations include both stationary and on-the-move capabilities. Conducting survivability operations is one of the tasks of the protection warfighting function, but forces can also use survivability operations to enable other warfighting functions. For example, military deception, part of the command and control warfighting function, can be enabled by the use of survivability operations intended to help mislead enemy decision makers. This may include the use of dummy or decoy positions or devices. (See FM 3-13 for additional information.)

7-74. Ground-based air defense artillery units execute most Army air and missile defense operations though air and missile defense support to the BCT may be limited. Subordinate units of the BCT should expect to use their organic weapons systems for self-defense against enemy air threats. When available air and missile defense protects the BCT from missile attack, air attack, and aerial surveillance by ballistic missiles, cruise missiles, conventional fixed- and rotary-wing aircraft, and UASs. Air and missile defense prevent the enemy from interdicting friendly forces, while freeing the commander to synchronize movement and firepower. (See ATP 3-01.8 for additional information.)

7-75. Indirect-fire protection systems protect the BCT from threats that are largely immune to air defense artillery systems. The indirect-fire protection intercept capability is designed to detect and destroy incoming rocket, artillery, and mortar fires. This capability assesses the threat to maintain friendly protection and destroys the incoming projectile at a safe distance from the intended target. The air and missile defense task consist of active and passive measures that protect the BCT from an air or missile attack. Passive measures include camouflage, cover, concealment, hardening, and operations security. Active measures are taken to destroy, neutralize, or reduce the effectiveness of hostile air and missile threats. The early warning of in-bound missile threats is provided in theater by the globally located, joint tactical ground stations. (See ADP 3-37 and ATP 3-09.42 for additional information.)

7-76. As stated in chapter 3, protection cell planners coordinate with the ADAM cell for air and missile defense for the protection of the critical asset list and defended asset list and for other air and missile defense protection as required. There is continuous coordination to refine the critical asset list and defended asset list throughout defensive and offensive operations, ensuring the protection of critical assets and forces from air and missile attack and surveillance. Air and missile defense assets integrate protective systems by using the six employment guidelines—mutual support, overlapping fires, balanced fires, weighted coverage, early engagement, and defense in-depth—and additional considerations necessary to mass and mix air and missile defense capabilities. (See ATP 3-01.50 for additional information.)

7-77. Military police planners, based upon the mission variables of mission, enemy, terrain and weather, troops and support available, time available, and civil considerations (METT-TC), identify requirements for military police support and augmentation. The BCT provost marshal and military police staff planners at division level coordinate military police activities and provide for the integration of military police focused considerations throughout the operations process. Military police operations require the use of military police specific technical skill sets to plan, manage, and execute the military police-specific disciplines. Liaisons may be needed in certain situations to ensure proper and complete staff planning. (See FM 3-39 for additional information.) During the defense, military police planners must—

- Understand the IPB, commander's critical information requirements (CCIRs), and priority intelligence requirements to facilitate the integration of police intelligence activities within all military police operations to support those requirements.
- Consider the type and size of the area of responsibility (AOR), line of communications security, and the threat and plan for detainee operations and dislocated civilians to determine how their presence may affect maneuver forces.
- Anticipate operational changes or transitions and prepare the military police effort toward that action.

7-78. In the defense, attached military police forces ensure movement of repositioning or counterattacking forces and provide and support the evacuation of captured or detained individuals. Military police missions provide freedom of movement for repositioning BCT forces and the reserve when it is committed. The mission variables of METT-TC determine priority of movement along main supply routes. Additional military police force activities include protection support to command posts (CPs), communications facilities, convoys, supply sites, support areas, and consolidation areas. Examples of expected missions include—

- Conduct detention operations.
- Establish a movement corridor.
- Conduct convoy escorts.
- Conduct response force operations.
- Conduct lines of communications security.
- Provide military working dog support (see ATP 3-39.34).
- Support to support area (see FM 3-81) and consolidation area (see chapter 2) operations.

7-79. *Chemical, biological, radiological, and nuclear environment* is an operational environment that includes chemical, biological, radiological, and nuclear threats and hazards and their potential resulting effects (JP 3-11). Within the BCT area of operations, CBRN environment conditions can be the result of deliberate enemy or terrorist actions or the result of an industrial accident. Possible CBRN threats include the intentional employment of, or intent to employ, weapons or improvised devices to produce CBRN hazards. CBRN hazards include those created from accidental or intentional releases of toxic industrial materials, biological pathogens, or radioactive matter. Toxic industrial material is a generic term for toxic or radioactive substances in solid, liquid, aerosolized, or gaseous form that may be used or stored for industrial, commercial, medical, military, or domestic purposes. Toxic industrial material may be chemical, biological, or radiological. (See FM 3-11 for more information on CBRN hazards.)

7-80. *Chemical, biological, radiological, and nuclear defense* are measures taken to minimize or negate the vulnerabilities to, and/or effects of, a chemical, biological, radiological, or nuclear hazard or incident (JP 3-11). The BCT commander integrates CBRN defense considerations into mission planning depending on the CBRN threat. (See FM 3-11 and ATP 3-11.37 for additional information.) Commanders at all echelons maintain the effectiveness of their force in CBRN environments by establishing CBRN defense plans that—

- Estimate enemy intent, capabilities, and effects for CBRN.
- Provide guidance to the force on necessary protective measures.
- Apply the IPB output to develop CBRN reconnaissance plans to answer priority intelligence requirements.
- Establish the employment criteria of CBRN enablers to counter CBRN threats.
- Establish a logistic support plan for long-term CBRN operations.
- Establish CBRN warning and reporting requirements.

7-81. Operationally, CBRN passive defense maintains the commander's ability to continue military operations in a CBRN environment while minimizing or eliminating the vulnerability of the force to the degrading effects of those CBRN threats and hazards. Tactical-level doctrine has traditionally segregated CBRN passive defense into the distinct principles of contamination avoidance, protection, and decontamination. While these principles remain valid, they are now recognized to be components of the more expansive concepts of hazard awareness and understanding and contamination mitigation. Since hazard awareness and understanding largely focuses strategic aspects of operations in a CBRN environment, tactical level doctrine is organized around the key activities (see figure 7-3 on page 7-18) of CBRN protection and contamination mitigation.

- *Chemical biological, radiological, and nuclear protection* consists of measures taken to keep chemical, biological, radiological, and nuclear threats and hazards from having an adverse effect on personnel, equipment, and facilities (ATP 3-11.32). CBRN protection encompasses the following activities: protect personnel, equipment, and facilities.

- *Contamination mitigation* is the planning and actions taken to prepare for, respond to, and recover from contamination associated with all chemical, biological, radiological, and nuclear threats and hazards to continue military operations (JP 3-11). The two subsets of contamination mitigation are contamination control and decontamination. (See ATP 3-11.32 for a detailed discussion of CBRN passive defense activities.)

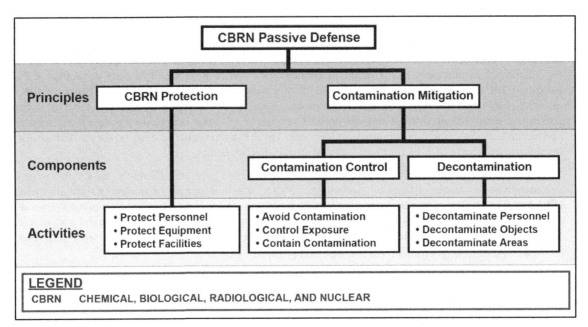

Figure 7-3. Chemical, biological, radiological, and nuclear passive defense architecture

7-82. When established, the CBRN working group led by the CBRN officer includes members from the protection-working group, subordinate commands, host-nation agencies, and other unified action partners. The CBRN working group—disseminates CBRN operations information, including trend analysis, defense best practices and mitigating measures, operations, the status of equipment and training issues, CBRN logistics, and contamination mitigation and remediation efforts and refines the CBRN threat, hazard, and vulnerability assessments. The working group helps to develop, train, and rehearse a CBRN defense plan to protect personnel and equipment from an attack or incident involving CBRN threats or hazards. CBRN threat and hazard assessments made by the working group help determine initial, individual protective equipment levels and the positioning of decontaminants. Force health personnel maintain the medical surveillance of personnel strength information for indications of force contamination, epidemic, or other anomalies apparent in force health trend data. (See FM 3-11 and ADP 3-37 for additional information.)

Note. (See chapter 6 for information on countering weapons of mass destruction [CWMD] and the explosive ordnance disposal company when supporting the BCT.)

7-83. Force health protection (see ATP 4-02.8), measures to promote, improve, conserve or restore the mental or physical well-being of Soldiers, enable a healthy and fit force, prevent injury and illness, and protect the force from health hazards. Defensive actions can result in prolonged occupation of static positions and corresponding exposure of personnel to diseases, weather and other health hazards and environmental affects that can quickly degrade readiness. The commander enforces environmental disciplines, such as hydration, sanitation, hygiene, protective clothing, and inspection of potable water supplies. Defensive actions also may entail sustained enemy bombardments or attacks resulting in dramatic effects on the mental and behavioral health of unit personnel. Soldiers can become combat ineffective from heavy indirect fire even if exposure is for short durations. Commanders deliberately emplace systems for combat stress identification and treatment to reduce the return to duty time of affected personnel. (See FM 4-02 and ATP 4-02.3 for additional information.)

7-84. When planning for base camp security and defense it is critical to remember that a properly designed perimeter security system should be an integrated, layered, defense in-depth that takes advantage of the security area. BCT commanders, supported by their staff, evaluate mission variables focusing on the threat to establish a viable perimeter defense plan. Planning for perimeter security and defense, like all protection measures integrates fires and obstacles, within the context of mission and operational variables and associated constraints, throughout the depth of the base camp area of operations to meet security and defense objectives. Commanders and staff with base camp security and defense responsibilities plan, coordinate, and synchronize actions using integrating processes to ensure full integration of their area security and base defense plans. (See ATP 3-37.10 for additional information.)

7-85. Refer to chapter 6 for a discussion of the following supporting tasks of the protection warfighting function:

- Implement physical security procedures.
- Apply antiterrorism measures.
- Provide explosive ordnance disposal support.
- Conduct personnel recovery.
- Conduct populace and resource control.
- Conduct risk management (RM).

SECTION III – FORMS OF THE DEFENSE

7-86. The three forms of the defense (defense of a linear obstacle, perimeter defense, and reverse-slope defense) have special purposes and require special planning and execution. The three forms of the defense provide distinct advantages for the BCT and its subordinate units and apply to the area defense and the operations of the fixing force during a mobile defense. (See FM 3-90-1 for additional information.)

DEFENSE OF A LINEAR OBSTACLE

7-87. The defense of a linear obstacle usually forces the enemy to deploy, concentrate forces, and conduct breaching operations. A defense of a linear obstacle generally favors the use of a forward defense (see paragraph 7-128). The defending unit constructs obstacles to stop the enemy forces and channel them into planned EAs. Maintaining the integrity of the linear obstacle is the key to this type of defense. When attacked, the defending force isolates the enemy, conducts counterattacks, and delivers fires onto the concentrated force to defeat attempts to breach the obstacle.

7-88. A defense of a linear obstacle often is used as part of an economy of force measure. In this situation, the defending force cannot allow the enemy to build up its forces on the friendly side of the obstacle because it may lack the required combat power to defeat the enemy forces. As forces to counterattack and destroy the enemy may not be available immediately; defending forces must be able to—

- Detect enemy penetrations early enough so that local counterattacks can defeat them.
- Defend after being isolated.
- Use reconnaissance elements, sniper teams, and other elements to detect enemy forces and call in fires.
- Bring the fight to the enemy side of the obstacle to destroy its forces and disrupt enemy preparations.
- Use fires to their maximum effect.
- Use its mobility to concentrate combat power.

Defense of a Linear Obstacle: Fredericksburg, VA 1862

By December 13, 1862, Confederate General Robert E. Lee established a strong defensive position behind the Rappahannock River in Northern Virginia against Union forces (see figure 7-4 on page 7-20). He only lightly defended the actual river line because the Union army artillery dominated both sides of the river. His main defensive position was directly west along a line of hills. The area between the river and the hills was generally open with scattered woods and streams or canals. The most concealed area was the town of Fredericksburg.

Early on December 13, the Union army crossed the river and formed for the attack. The plan was to conduct the main attack to the south with a supporting attack to the north. Despite repeated attacks, the Union forces were repulsed everywhere. Union casualties were approximately 10,000 while the confederate forces suffered approximately 5,000 casualties.

Vincent Esposito

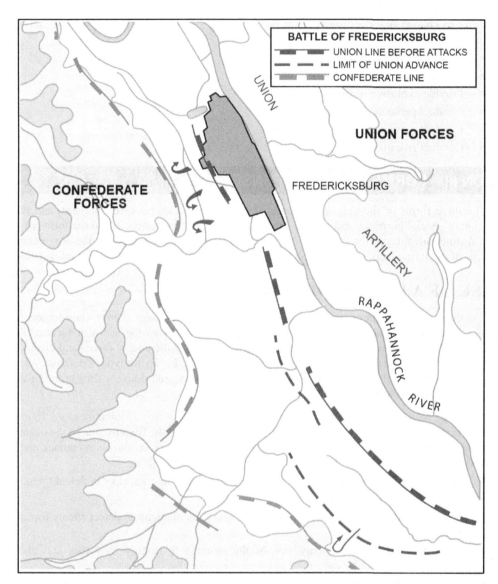

Figure 7-4. Historical example, defense of a linear obstacle, Fredericksburg, 1862

PERIMETER DEFENSE

7-89. The BCT and its subordinate units often use a perimeter defense when conducting airborne and air assault operations, as well as when conducting operations in noncontiguous areas of operations. The BCT presents no assailable flanks to the enemy and allows the defender to reinforce a threatened area rapidly. Some disadvantages of a perimeter defense include its isolation and the vulnerability of its concentrated units to enemy fires.

7-90. A commander establishes a perimeter defense when the unit must hold critical terrain, such as a strong point, or when it must defend itself in areas where the defense is not tied in with adjacent units. Depending on the situation, the commander maximizes the use of class IV barrier material, utilize engineer assets to create vehicle and crew served fighting positions, and manpower to emplace obstacles of many types (including wire, abates trees, antitank ditches). Additionally, spider mine obstacle groups can be emplaced and grouped under one or several system operators with appropriate triggers and rehearsals.

7-91. Commanders can organize a perimeter defense to accomplish a specific mission, such as protecting a base or to provide immediate self-protection, such as during resupply operations when all-around security is required. During a perimeter defense, leaders at all levels ensure that—

- Units physically tie into each other.
- Direct fire weapons use flanking fire to protect the perimeter.
- Field artillery and mortars are protected.
- Communications are secure and redundant systems in place.
- Obstacles are employed.
- Final protective fires are established.

7-92. After committing the reserve, the commander must reconstitute the reserve to meet other possible threats. This reconstitution force normally comes from an unengaged unit in another portion of the perimeter.

Perimeter Defense: Chip'yong Ni, Republic of Korea 1951

During the Chinese Fourth Phase offensive, the 23d Infantry Regimental Combat Team, reinforced, used a perimeter defense to defeat elements of three People's Republic of China armies, about 25,000 soldiers at Chip'yong Ni, Republic of Korea (see figure 7-5 on page 7-22). From 13 to 14 February 1951, the 23d established a perimeter defense around the town of Chip'yong-Ni and blunted a major People's Republic of China offensive. After a bitter fight, the Chinese forces withdrew at the cost of 51 United Nation's Soldiers and an estimated 2000 Peoples Republic of China soldiers killed. The battle was a major defeat for the Chinese forces and led to subsequent United Nation offensives that forced the Chinese back into the North.

Billy C. Mossman

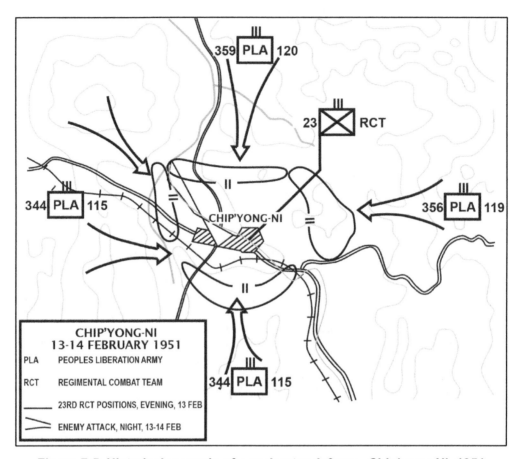

Figure 7-5. Historical example of a perimeter defense, Chip'yong Ni, 1951

REVERSE-SLOPE DEFENSE

7-93. The reverse-slope defense allows units to concentrate their direct fires into a relatively small area while being protected from the enemy's direct observation and supporting fires. The defender can destroy the enemy's isolated forward units through surprise and concentrated fires. The control of the forward slope is essential for success. Gaining control of the forward slope can be done by using dominating terrain behind the defenders or with the use of stay behind forces, such as reconnaissance and sniper teams, that can observe and call in fires on the attackers. Generally, a unit at battalion level and below conducts a reverse-slope defense even though the BCT may have areas within its area of operations that are conducive to the use of a reverse-slope defense.

Reverse-Slope Defense: Kakazu Ridge 1945

During the Okinawa campaign, Imperial Japanese forces conducted reverse-slope defenses along a series of ridges (see figure 7-6). This tactic was devastating and cost many American lives. The Japanese, dug in on the reverse slope, and able to maneuver through tunnels would immediately counterattack American forces that reached the crest of the defended ridge. The Japanese held their positions for many days against heavy American firepower and repeated American attacks. Some of the positions, such as Kakazu Ridge, were taken, lost, and retaken repeatedly until finally falling to American forces.

Various

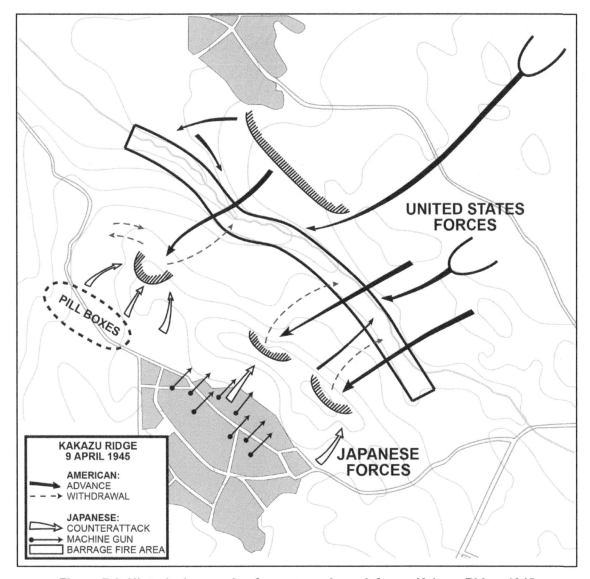

Figure 7-6. Historical example of a reverse slope defense, Kakazu Ridge, 1945

SECTION IV – DEFENSIVE CONTROL MEASURES

7-94. A control measure is a means of regulating forces or warfighting functions. Control measures provide control without requiring detailed explanations. Control measures can be permissive (which allows something to happen) or restrictive (which limits how something is done). Some control measures are graphic. A graphic control measure is a symbol used on maps and displays to regulate forces and warfighting functions. (See ADP 1-02 for illustrations of graphic control measures and rules for their use.)

COMMON DEFENSIVE CONTROL MEASURES

7-95. Control measures provide the ability to respond to changes in the situation. They allow the defending commander to concentrate combat power at the decisive point. Commanders use the minimum number to control their units and tailor their use to the higher commander's intent. Figure 7-7 on page 7-25 illustrates a BCT's use of the following control measures:

- Assembly area.

- Attack by fire position.
- Axis of advance.
- Battle handover line.
- Battle position (occupied and planed).
- Boundaries.
- Brigade support area.
- Contact point.
- EA.
- Coordinate fire line.
- Forward edge of the battle area.
- Lane.
- NAI.
- Observation post.
- Passage point.
- PL.
- Strong point battle position.
- TAI.
- TRP.
- Turning obstacle.

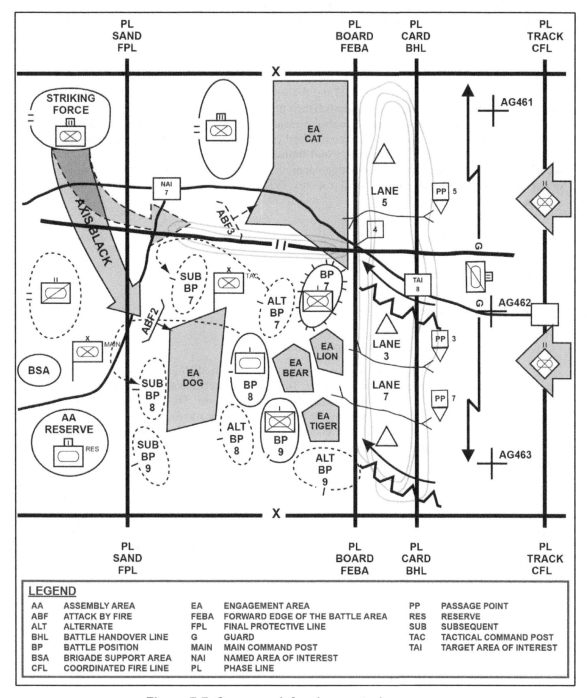

Figure 7-7. Common defensive control measures

EMPLOYING CONTROL MEASURES

7-96. Effectively employing control measures requires the BCT commander and staff to understand their purposes and ramifications, including the permissions or limitations imposed on subordinates' freedom of action and initiative. Each measure should have a specific purpose. Control measures include designating the security area, the MBA with its associated battle positions, the forward edge of the battle area, and the echelon support area. Commanders use graphic control measures to assign responsibilities, coordinate fires and maneuver, and control the use of airspace. Well-planned measures permit the proper distribution of fires and prevent multiple weapons from firing upon prominent targets while less prominent targets escape destruction. The BCT properly uses fire support coordination measures, direct fire control measures, maneuver control measures, and airspace coordinating measures to synchronize operations in the land and air domains.

AREA OF OPERATIONS

7-97. An area of operations is an operational area defined by the joint force commander for land and maritime forces that should be large enough to accomplish their missions and protect their forces. An area of operations is a basic tactical concept and the basic control measure for all types of operations. An area of operations gives the responsible unit freedom of maneuver and enables fire support planning within a specific area. Commanders employ control measures to avoid fratricide against forward deployed security forces in the security area and engineer assets constructing countermobility obstacles in EAs. All units assigned an area of operations have the following additional responsibilities within the boundaries of that area of operations:

- Terrain management.
- Information collection.
- Civil-military operations.
- Movement control.
- Clearance of fires.
- Security.
- Personnel recovery.
- Airspace management.
- Minimum-essential stability operations tasks.

7-98. A unit's area of operations should provide adequate depth based on its assigned tasks, the terrain, and the anticipated size of the attacking enemy force. To maintain security and a coherent defense, an area of operations generally requires continuous coordination with flank units. The BCT assigns control measures, such as PLs, coordinating points, EAs, obstacle belts, and battle positions, to coordinate subordinate unit defenses within the MBA. The BCT commander and staff use briefings, inspections, rehearsals, and supervision to ensure coordination among subordinate units, to eliminate any gaps, and to ensure a clear understanding of the defensive plan.

7-99. Subordinate unit area of operations may be contiguous or noncontiguous. A *contiguous area of operations* (see figure 7-2 on page 7-7) is where all a commander's subordinate forces' areas of operations share one or more common boundary (FM 3-90-1). A *noncontiguous area of operations* (see figure 7-8) is where one or more of the commander's subordinate force's areas of operation do not share a common boundary (FM 3-90-1). The higher headquarters is responsible for controlling the areas not assigned to subordinate forces within noncontiguous areas of operations. (See FM 3-90-1 for additional information.)

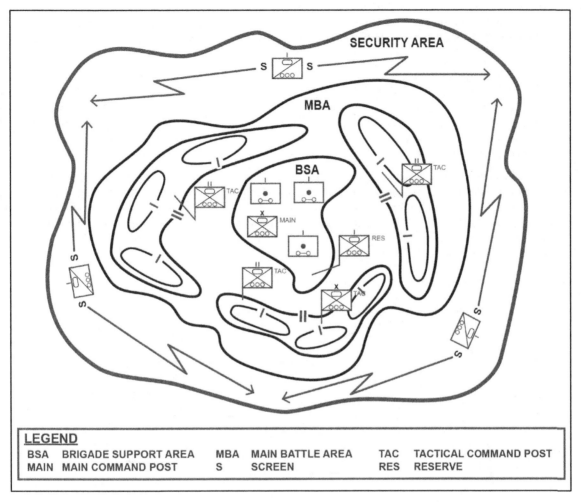

Figure 7-8. Area defense, noncontiguous area of operations

BATTLE POSITION

7-100.   A *battle position* is a defensive location oriented on a likely enemy avenue of approach (ADP 3-90). The BCT commander assigns a battle position to a battalion to control the battalion's fires, maneuver, and positioning. Usually, the commander assigns boundaries to provide space for the battalion security, support, and sustainment elements that operate outside a battle position. When the commander does not establish unit boundaries, the BCT is responsible for fires, security, terrain management, and maneuver between positions of adjacent battalions. The battle position prescribes a primary direction of fire by the orientation of the position. The commander defines when and under what conditions the battalion can displace from the battle position or maneuver outside it. The use of prepared or planned battle positions, with the associated tasks of prepare or reconnoiter, provides flexibility to rapidly concentrate forces and adds depth to the defense.

7-101.   There are five types of battle positions: primary, alternate, supplementary, subsequent, and strong point. The commander always designates the primary battle position. The commander designates and prepares alternate, supplementary, and subsequent positions as required.

7-102.   A *primary position* is the position that covers the enemy's most likely avenue of approach into the area of operations (ADP 3-90). Always designate this position. An *alternate position* is a defensive position that the commander assigns to a unit or weapon system for occupation when the primary position becomes untenable or unsuitable for carrying out the assigned task (ADP 3-90). The alternate position covers the same area as the primary position. A *supplementary position* is a defensive position located within a unit's assigned area of operations that provides the best sectors of fire and defensive terrain along an avenue of approach that is not the primary avenue where the enemy is expected to attack (ADP 3-90). Assigned when more than one avenue of approach into a unit's area of operations. A *subsequent position* is a position that a unit expects to move to during the course of battle (ADP 3-90). Subsequent positions can have primary, alternate, and supplementary positions associated with them.

7-103.   In accordance with the mission variables, units can conduct survivability moves between their primary, alternate, and supplementary positions. A *survivability move* is a move that involves rapidly displacing a unit, CP, or facility in response to direct and indirect fires, the approach of a threat or as a proactive measure based on intelligence, meteorological data, and risk assessment of enemy capabilities and intentions (ADP 3-90). A survivability move includes those based on the impending employment of weapons of mass destruction (WMD).

7-104.   A *strong point* is a heavily fortified battle position tied to a natural or reinforcing obstacle to create an anchor for the defense or to deny the enemy decisive or key terrain (ADP 3-90). A strong point implies retention of terrain to control key terrain and blocking, fixing, or canalizing enemy forces. Defending units require permission from the higher headquarters to withdraw from a strong point. All combat, maneuver enhancement, and sustainment assets within the strong point require fortified positions. In addition, extensive protective and tactical obstacles are required to provide an all-around defense. (See figure 7-9.)

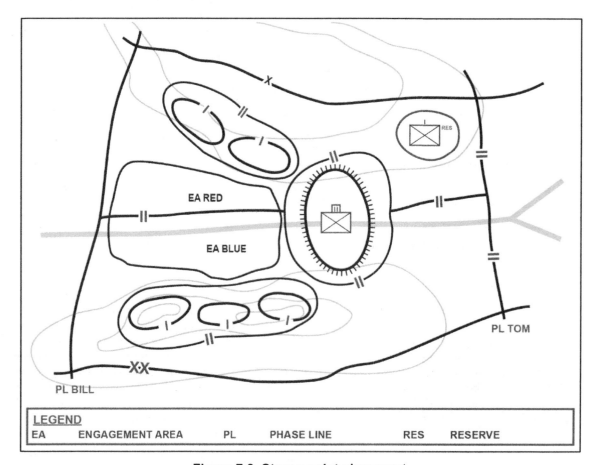

Figure 7-9. Strong point placement

7-105.   As a rule of thumb, a minimally effective strong point requires a one-day effort from an engineer unit the same size as the unit defending the strong point. Organic BCT engineers lack sufficient capacity to create a strong point within a reasonable amount of time; additional engineer assets from echelons above the BCT are required. Once the strong point is occupied, all units and equipment not essential to the defense are displaced from the strong point. This includes nonessential staff and elements from the forward support company (FSC). (See FM 3-90-1 for additional information.)

7-106. Planning considerations for a strong point, although not inclusive, may include—

- Establishment of outposts and observation posts.
- Development of integrated fires plans that include final protective fires.
- Priorities of work.
- Counterattack plans.
- Stockage of supplies.
- Integration and support of subordinate forces outside the strong point.
- Actions of adjacent units.

COMBAT OUTPOST

7-107. A *combat outpost* is a reinforced observation post capable of conducting limited combat operations (FM 3-90-2). Using combat outposts is a technique for employing security forces in restrictive terrain that precludes mounted security forces from covering the area. While the mission variables of METT-TC determine the size, location, and number of combat outposts established by a unit, a reinforced platoon typically occupies a combat outpost. Combat outposts normally are located far enough in front of the protected force to prevent enemy ground reconnaissance elements from directly observing the protected force. (See FM 3-90-2 and ATP 3-21.8 for additional information.)

SECTION V – DEFENSIVE OPERATIONS

7-108. A *defensive operation* is an operation to defeat an enemy attack, gain time, economize forces, and develop conditions favorable for offensive or stability operations (ADP 3-0). The three defensive operations are area defense, mobile defense, and retrograde. Planning and preparing an effective defense takes time. The commander uses security elements to provide early warning, reaction time, and maneuver space. Units establish a defense immediately upon occupation. Commanders refine the initial defense through planning and preparation and may require units to shift and adjust their positions after the plan is final. Defensive preparations and refinement are never complete. As required, defending units conduct consolidation and reorganization activities after each enemy engagement (see paragraph 6-30 for a discussion of these activities).

AREA DEFENSE

7-109. An *area defense* is a type of defensive operation that concentrates on denying enemy forces access to designated terrain for a specific time rather than destroying the enemy outright (ADP 3-90). The defender limits the enemy's freedom of maneuver and channels them into designated EAs. The focus of the area defense is to retain terrain where the bulk of the defending force positions itself in mutually supporting, prepared positions. Units maintain their positions and control the terrain between these positions. The decisive operation focuses on fires into EAs possibly supplemented by a counterattack. The commander can use the reserve to reinforce fires; add depth, block, or restore the position by counterattack; seize the initiative; or destroy enemy forces. The BCT conducts an area defense under the following conditions:

- When directed to defend or retain specified terrain.
- When forces available have less mobility than the enemy does.
- When the terrain affords natural lines of resistance.
- When the terrain limits the enemy to a few well-defined avenues of approach.
- When there is time to organize the position.
- When conditions require the preservation of forces.

ORGANIZATION OF FORCES

7-110. The BCT commander organizes an area defense around the static framework of the defensive positions seeking to destroy enemy forces by interlocking fire or local counterattacks. The commander has the option of defending forward or defending in-depth. The depth of the force positioning depends on the threat, task organization of the BCT, and nature of the terrain. When the commander defends forward within an area of operations, the force is organized so that most of the available combat power is committed early in the defensive effort. To accomplish this, the commander may deploy forces forward or plan counterattacks well forward in the MBA or even beyond the MBA. If the commander has the option of conducting a defense in-depth, security forces and forward MBA elements identify, define, and control the depth of the enemy's main effort while holding off secondary thrusts. Doing so allows the commander to conserve combat power, strengthen the reserve, and better resource the counterattack. In an area defense, the commander organizes the defending force to accomplish information collection, security, MBA, reserve, and sustainment missions. (See figure 7-10.)

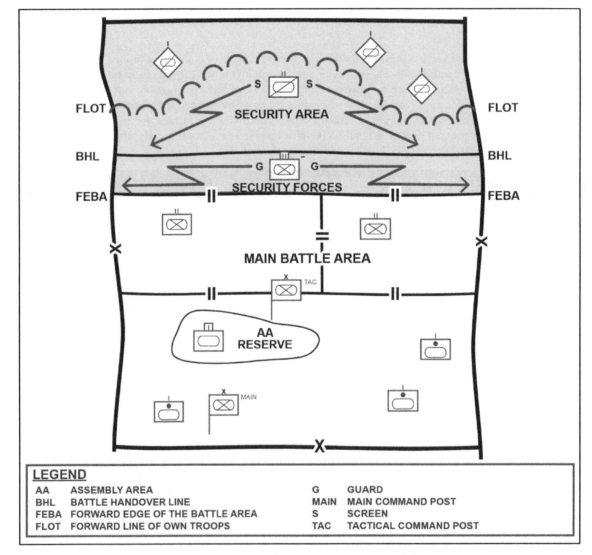

Figure 7-10. Area defense, organization of forces

Information Collection

7-111. The commander directs information collection assets to determine the locations, strengths, and probable intentions of the attacking enemy force. The commander places a high priority on early identification of the enemy's main effort. The commander ensures that the mission of reconnaissance forces and surveillance assets are coordinated with those of higher headquarters. In the defense, reconnaissance, surveillance, and security operations overlap the unit's planning and preparing phases.

7-112. BCT subordinate commanders and leaders performing reconnaissance, surveillance, and security missions understand that these missions often start before the commander fully develops the plan. Commanders and leaders have to be responsive to changes in orientation and mission. The commander ensures that the staff plans, prepares, and assesses the execution of the information collection portion of the overall plan. (See chapter 5 for a detailed discussion on reconnaissance operations.)

Security

7-113. The higher commander defines the depth of the BCT's security area. The BCT's security area extends from the forward edge of the battle area to the BCT's forward boundary. Depth in the security area gives the forces within the MBA more reaction time. Depth allows the security force more area to conduct security missions. A very shallow security area may require more forces and assets to provide the needed reaction time.

7-114. The BCT commander must clearly define the objective of the security area. The commander states the tasks of the security force in terms of time required or expected to maintain security, expected results, disengagement and withdrawal criteria, and follow-on tasks. The commander identifies specific avenues of approach and NAI on which the security force must focus. The BCT assists in the rearward passage of lines and movement through the BCT area of operations of any division and corps security force deployed beyond the BCT's forward boundary.

7-115. The BCT commander balances the need for a strong security force to shape the battle with the resulting diversion of combat power from the decisive operation in the MBA. The BCT frequently executes the forward security mission as a guard or screen. Typically, there are two options for organizing the security force. The BCT forward defending maneuver battalions establish their own security areas or the maneuver battalions provide security forces that operate with the Cavalry squadron under the BCT's direct control.

7-116. The BCT conducts counterreconnaissance and area security operations (see chapter 8), and implements local security measures, operations security, and information protection activities to deny the enemy information about friendly dispositions. BCT reconnaissance and security forces seek to confuse the enemy about the location of the BCT's main battle positions, to prevent enemy observation of preparations and positions, and to keep the enemy from delivering observed fire on the positions. The BCT conducts reconnaissance and security operations to gain and maintain contact with the enemy, develop the situation, answer CCIRs, retain freedom of maneuver, consolidate gains, secure the force, and protect the local population. (See chapter 5 for a detailed discussion on security force operations.)

Main Battle Area

7-117. The *main battle area* is the area where the commander intends to deploy the bulk of the unit's combat power and conduct decisive operations to defeat an attacking enemy (ADP 3-90). The BCT's MBA extends from the forward edge of the battle area to the unit's rear boundary. The commander selects the MBA based on the higher commander's concept of operations, IPB, results of initial information collection plan (reconnaissance, security operations, surveillance, and intelligence operations), and the commander's own assessment of the situation.

7-118. The BCT commander delegates responsibilities within the MBA by assigning areas of operations and establishing boundaries to and for subordinate battalions. The commander locates subordinate unit boundaries along identifiable terrain features and extends them beyond the forward line of own troops by establishing forward boundaries. Unit boundaries should not split avenues of approach or key terrain. The BCT is responsible for terrain management, security, clearance of fires, and coordination of maneuver among other doctrinal responsibilities within the entire area of operations if the commander does not assign area of operations to subordinate battalions. (See ADP 3-90 for additional information.)

Reserve

7-119. The reserve is not a committed force. The BCT commander can assign it a wide variety of tasks on its commitment, and it must be prepared to perform other missions. The reserve may be committed to restore the defense's integrity by blocking an enemy penetration, reinforcing fires into an EA, or conducting a counterattack against the flank or rear of an attacking enemy. The reserve gives the commander the flexibility to exploit success or to deal with a tactical setback.

7-120. The commander positions the reserve to respond quickly to unanticipated missions. The commander determines the reserve's size and position based on accurate knowledge about the enemy and whether the terrain can accommodate multiple enemy COAs. When the BCT has accurate knowledge about the enemy and the enemy's maneuver options are limited, the BCT can maintain a smaller reserve. If knowledge about the enemy is limited and the terrain allows the enemy multiple COAs, then the BCT needs a larger reserve positioned deeper into the area of operations. (See FM 3-90-1 for additional information.)

Sustainment

7-121. The sustainment mission in an area defense requires a balance among establishing forward supply stocks of ammunition, barrier material, and other supplies in sufficient amounts, and having the ability to move the supplies in conjunction with enemy advances. Proper forecasting of supply and support requirements is important to the success of the area defense. The location of sustainment units within the support area is METT-TC dependent. (See chapter 9.)

PLANNING AN AREA DEFENSE

7-122. An area defense requires detailed planning and extensive coordination. In the defense, synchronizing and integrating the BCTs combat and supporting capabilities enables a commander to apply overwhelming combat power against selected advancing enemy forces. A successful defense depends on knowing and understanding the enemy and its capabilities. The commander's situational understanding is critical in establishing the conditions that initiate the defensive action. As the situation develops, the commander reassesses the plan based on a revised situational understanding that results from an updated common operational picture (COP) as new intelligence and combat information becomes available. In planning an area defense, the commander may choose between two forms of defensive maneuver—a defense in-depth or a forward defense.

Understanding

7-123. The BCT commander considers the mission variables of METT-TC to determine how to concentrate efforts and economize forces in order to accomplish the mission. A detailed terrain analysis may be the most important process the BCT commander and staff complete. A successful defense relies on a complete understanding of terrain to determine likely enemy COAs and the best positioning of BCT assets to counter them.

7-124. The commander must understand the situation in-depth, develop the situation through action, and constantly reassess the situation to keep pace with the engagement. Defending forces must gain and maintain contact with the enemy to observe, assess, and interpret enemy reactions and the ensuing opportunities or threats to friendly forces, populations, or the mission. The commander must establish priority intelligence requirements to enable information collection through reconnaissance, surveillance, intelligence operations, and security operations to develop situational understanding.

7-125. Enemy forces counter friendly information collection efforts to prevent the BCT from gaining information. Enemy forces use other countermeasures such as dispersion, concealment, deception, and intermingling with the population to limit the BCT's ability to develop the situation out of contact. Reconnaissance and security forces (see chapter 5) fill in the gaps in commanders' understanding of the situation. Fighting for understanding and identifying opportunities to seize, retain, and exploit the initiative requires combined arms capabilities, access to joint capabilities, specialized training, and employing combinations of manned and unmanned air and ground systems.

Forms of Defensive Maneuver

7-126. The BCT commander may choose between two defensive maneuver forms when planning an area defense, a defense in-depth, or a forward defense. The commander usually selects the form of defensive maneuver, but the higher headquarters' commander may define the general defensive scheme for the BCT. These two deployment choices are not totally exclusory. Part of a defending commander's unit can conduct a forward defense, while the other part conducts a defense in-depth. The specific mission may also impose constraints such as time, security, and retention of certain areas, which are significant factors in determining how the BCT defends.

Defense In-Depth

7-127. A defense in-depth (see figure 7-11) reduces the risk of a quick penetration by the attacking enemy force. Even if initially successful, the enemy has to continue to attack through the depth of the defense to achieve a penetration. The defense in-depth provides more space and time to defeat the enemy attack. Dependent on the mission variables of METT-TC, it may require forces with at least the same mobility as the enemy to maneuver to alternate, supplementary, and subsequent positions. The mobility of the enemy force can determine the disengagement criteria of the defending forces as they seek to maintain depth. The BCT commander considers using a defense in-depth when—

- The mission allows the BCT to fight throughout the depth of the area of operations.
- The terrain does not favor a forward defense and there is better defensible terrain deeper in the area of operations.
- Sufficient depth is available in the area of operations.
- Cover and concealment forward in the area of operations is limited.
- CBRN weapons may be used.
- The terrain is restrictive and limits the enemy's maneuver and size of attack.

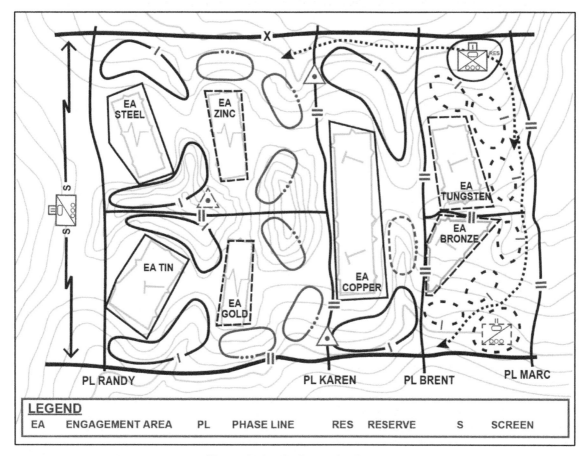

Figure 7-11. Defense in-depth

Forward Defense

7-128.   The intent of a forward defense (see figure 7-12) is to prevent enemy penetration of the defense. A forward defense is the least preferred form of the area defense due to its lack of depth. The BCT commander deploys the majority of combat power into defensive positions near the forward edge of the battle area. The commander fights to retain the forward position and may conduct spoiling attacks or counterattacks against enemy penetrations, or destroys enemy forces in forward EAs. Often, counterattacks are planned forward of the forward edge of the battle area to defeat the enemy. Commanders may use reconnaissance and security forces to find the enemy in vulnerable situations and exploit the opportunity to conduct a spoiling attack to weaken the enemy's main attacking force and to disrupt the enemy operation.

7-129.   The BCT commander uses a forward defense when a higher commander directs the commander to retain forward terrain for political, military, economic, and other reasons. Alternatively, a commander may choose to conduct a forward defense when the terrain in that part of the area of operations—including natural obstacles—favors the defending force because—

- Terrain forward in the area of operations favors the defense.
- Strong, existing natural or manmade obstacles, such as a river or a canal, are located forward in the area of operations.
- Assigned area of operations lacks depth due to the location of the protected area.
- Natural EAs occur near the forward edge of the battle area.
- Cover and concealment in the rear portion of the area of operations is limited.
- Directed by higher headquarters to retain or initially control forward terrain.

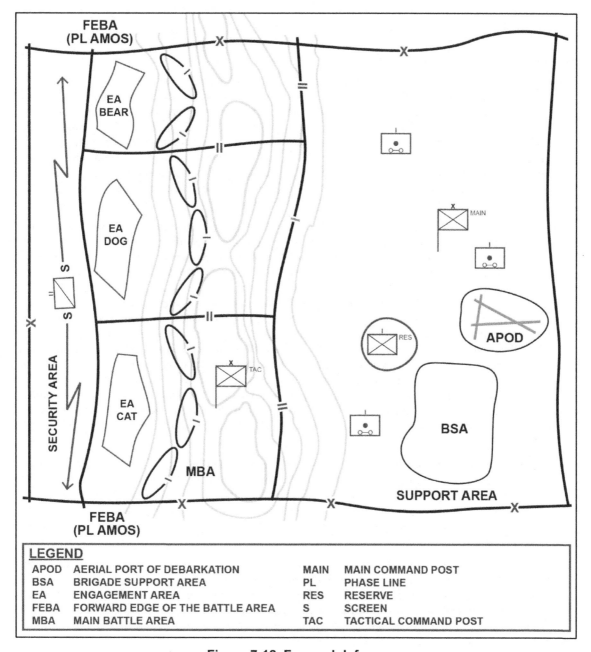

Figure 7-12. Forward defense

Engagement Area

7-130. An *engagement area* is an area where the commander intends to contain and destroy an enemy force with the massed effects of all available weapons and supporting systems (ADP 3-90). The success of any engagement depends on how effectively the BCT commander integrates the direct fire plan, the indirect-fire plan, the obstacle plan (see ATP 3-90.8), Army aviation fires, close air support, and the terrain within the EA to achieve the BCT's tactical purpose.

7-131. Effective use of terrain reduces the effects of enemy fires, increases the effects of friendly fires, and facilitates surprise. Terrain appreciation—the ability to predict its impact on operations—is an important skill for every leader. For tactical operations, commanders analyze terrain using the five military aspects of terrain, expressed as OAKOC. (See ATP 2-01.3 and ATP 3-34.80 for information on analyzing the military aspects of terrain.)

Engagement Area Development

7-132. The BCT commander and staff develop EAs, to include engagement criteria and priority, to cover each enemy avenue of approach. Within the BCT's MBA, the commander determines the size and shape of the EAs by the relatively unobstructed line of sight from the weapon systems firing positions and the maximum range of those weapon systems. Once the commander and staff select EAs, the commander arrays available forces and weapon systems in positions to concentrate overwhelming effects into these areas. The commander routinely subdivides EAs into smaller EAs for subordinates using one or more TRPs or by key terrain or prominent terrain feature. The commander assigns sector of fires to subordinates to ensure complete coverage of EAs and to prevent fratricide and friendly fire incidents. Never split the responsibility for an avenue of approach or key terrain.

7-133. Security area forces, to include field artillery fire support teams and observers, employ fires to support operations forward of the BCT's MBA using precision and other munitions to destroy enemy reconnaissance and security forces and identified HPTs, and to attrit enemy forces as they approach the BCT's MBA. The employment of fires within the security area also helps to deceive the enemy about the location of the BCT's MBA. The BCT fire support officer plans the delivery of fires at appropriate times and places throughout the area of operations to slow and canalize the enemy force as the enemy approaches. The employment of fires allows security area forces to engage the enemy without becoming decisively engaged. To prevent fratricide, allied casualties, civilian noncombatant casualties, and excessive unintended collateral damage, the commander designates fire support coordination measures (such as no-fire areas, restrictive fire areas, restrictive fire lines, fire support coordination lines, and restricted target lines) where security area forces are positioned. The commander establishes these measures in order to exercise restraint and balance the need for combat action and that of maintaining the legitimacy of the mission and to prevent unintended negative effects. The commander uses fires to support the withdrawal of security forces once shaping operations are completed within the security area and the defending unit is prepared to conduct MBA operations.

7-134. Engagement criteria are protocols that specify those circumstances for initiating engagement with an enemy force. Engagement criteria may be restrictive or permissive in nature. For example, the BCT commander may instruct a subordinate battalion commander not to engage an approaching enemy unit until the enemy commits to an avenue of approach. The commander establishes engagement criteria in the direct fire plan in conjunction with engagement priorities and other direct fire control measures to mass fires and control fire distribution.

7-135. *Engagement priority* specifies the order in which the unit engages enemy systems or functions (FM 3-90-1). The commander assigns engagement priorities based on the type or level of threat at different ranges to match organic weapon systems capabilities against enemy vulnerabilities. Engagement priorities are situationally dependent and used to distribute fires rapidly and effectively. Subordinate elements can have different engagement priorities but will normally engage the most dangerous targets first, followed by targets in-depth or specialized systems, such as engineer vehicles.

7-136. A *target reference point* is a predetermined point of reference, normally a permanent structure or terrain feature that can be used when describing a target location (JP 3-09.3). The BCT and subordinate units may designate TRPs to define unit or individual sectors of fire and observation, usually within the EA. TRPs, along with trigger lines, designate the center of an area where the commander plans to distribute or converge the fires of all weapons rapidly to further delineate sectors of fire within an EA. Once designated, target reference points may also constitute indirect-fire targets.

7-137. A *trigger line* is a phase line located on identifiable terrain that crosses the engagement area—used to initiate and mass fires into an engagement area at a predetermined range for all or like weapon systems (ATP 3-21.20). The BCT commander can designate one trigger line for all weapon systems or separate trigger lines for each weapon or type of weapon system. The commander specifies the engagement criteria for a specific situation. The criteria may be either time- or event-driven, such as a certain number or certain types of vehicles to cross the trigger line before initiating engagement. The commander can use a time-based fires delivery methodology or a geography-based fires delivery.

Note. The example below addresses the general steps to EA development for the area defense. In this example, a battalion task force within the BCT conducts an area defense (defense in-depth) against a motorized infantry and armor threat. The fictional scenario within this example, used for discussion purposes, is not the only way to develop an EA. For clarity, many graphic control measures, such as PLs, are not shown.

Engagement Area Development (Motorized Infantry/Armor Threat), Example

7-138. Although often identified as a method to defeat enemy armor, EAs are an effective method to defeat any enemy attack whether the attack is primarily an armor, Infantry, or a mixed armor and Infantry force. The key is the identification of the likely enemy avenues of approach and actions, and the placement of adequate friendly forces, obstacles, and fires to defeat the enemy. The following seven-step EA development process, used for discussion purposes, represents one way a BCT builds an EA. The BCT commander and staff (specifically the S-3) integrate these steps within the military decision-making process (MDMP) and the IPB. Steps (asterisks denote steps that occur simultaneously) include the following:

- Identify likely enemy avenues of approach.
- Identify most likely enemy COA.
- Determine where to kill the enemy.
- Position subordinate forces and weapons systems. *
- Plan and integrate obstacles. *
- Plan and integrate fires. *
- Rehearse the execution of operations within the EA.

Note. Within the scenario below, the BCT commander focuses the IPB effort on the characteristics of the operational environment that can influence enemy and friendly operations and how the operational environment influences friendly and enemy COAs. The BCT staff (specifically the S-2 and S-3) identified three likely enemy avenues of approach to and through the BCT's area of operations. Two enemy avenues of approach were identified within the Blue River Valley, Avenue of Approach 1 and Avenue of Approach 2 (area of operations assigned to Battalion 2). A third enemy Avenue of Approach (A and B) was identified north of the Blue River Valley, area of operations assigned to Battalion 1. Success, against these likely enemy avenues of approach, results in allowing the commander to quickly choose and exploit terrain, weather, and civil considerations to best support the mission. (See ATP 2-01.3 for a detailed discussion of the IPB.)

7-139. Step 1. Identify likely enemy avenues of approach. The brigade and battalion staffs identified significant characteristics of the operational environment to determine the effects of the terrain, weather, and civil considerations on enemy and friendly operations. The primary analytic tools used to aid in determining this effect, specific to terrain, are the modified combined obstacle overlay, and the terrain effects matrix. Figure 7-13 on page 7-38, identifies the three enemy avenues of approach and the terrain within the BCT's area of operations that impedes friendly and enemy movement (severely restricted and restricted areas) and the terrain where enemy and friendly forces can move unimpeded (unrestricted areas). Key terrain forward of the Green and Red Rivers is critical to the BCT's defense because occupying these position will allow the engagement of enemy forces forward of the river, preventing the establishment of an enemy force on the east bank and the use of crossing sites to support movement into less restrictive terrain west of the Green and Red Rivers.

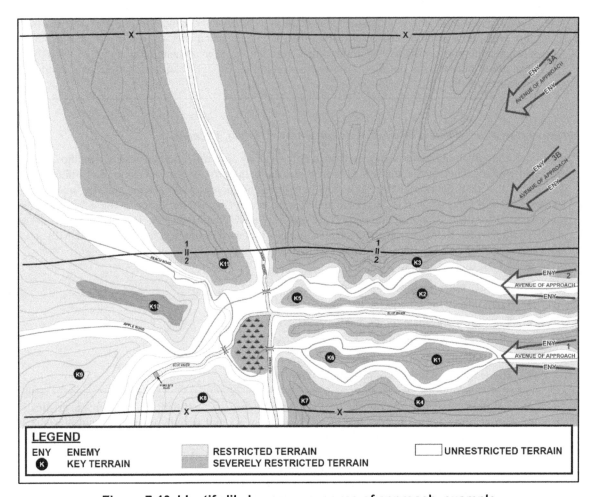

Figure 7-13. Identify likely enemy avenues of approach, example

7-140.   Enemy Avenue of Approach 1 and Avenue of Approach 2, within the scenario though restricted, support mounted movement though the terrain forward (east) of the Green and Red Rivers, thus hindering enemy movement to some degree. The terrain typically consists of moderate-to-steep slopes or moderate-to-densely spaced obstacles, such as trees or rocks. Enemy forces within this restricted area will have difficulty maintaining preferred speeds, moving in movement formations, and transitioning from one formation to another. Enemy movement will require zigzagging or frequent detours. A poorly developed road system will hinder the enemy's ability to sustain its attack along both enemy avenues of approach. The unrestricted terrain further west from Green and Red Rivers will allow the enemy to move unimpeded along enemy Avenue of Approach 1 and Avenue of Approach 2 once clear of the two rivers.

7-141.   The terrain along enemy Avenue of Approach 3 (A and B), identified to the north of the Blue River valley, is severely restricted. Steep slopes and large or densely spaced obstacles with little or no supporting roads characterize the terrain. Though suitable for dismounted movement and infiltration, the terrain within Avenue of Approach 3 (A and B) impedes motorized Infantry and armored movement. Swamps and the rugged terrain within this area are examples of restricted areas for dismounted Infantry forces. The road system utilized to sustain the enemy's attack is very limited along Avenue of Approach 3 (A and B). Due to the terrain, security patrols between battle positions will be key to impeding enemy infiltration efforts to division support and consolidation areas. (See ATP 2-01.3 for additional information on determining terrain characteristics and the terrain's effect on operations.)

Note. The BCT commander and staff, during step 4 of the IPB process (determine threat COAs), identify and develop possible enemy COAs that can affect the BCT's mission. (See chapter 4.) Enemy COA development requires identifying and understanding the significant characteristics related to enemy, terrain, weather, and civil considerations of the operational environment and how these characteristics affect friendly and enemy operations. Steps 1 and 2 of the IPB process are, respectively, define the operations environment and describe environmental effects on operations. The purpose of evaluating the enemy, step 3 of the IPB process (evaluate the threat), is to understand how an enemy can affect friendly operations. The commander, in order to plan for all possible contingencies, understands all COAs an enemy commander can use to accomplish the enemy objective(s). To aid in this understanding, the staff determines all valid enemy COAs and prioritizes them from most to least likely. The staff also determines which enemy COA is the most dangerous to friendly forces. To be valid, enemy COAs should be feasible, acceptable, suitable, distinguishable, and complete-the same criteria used to validate friendly COAs. (See ATP 2-01.3 for a detailed discussion.)

7-142. Step 2. Identify most likely enemy COA. The commander and staff (specifically the S-2 and S-3) determine the enemy's most likely COA, within the scenario, is to attack with two battalions (motorized Infantry battalion task forces) abreast, one along Avenue of Approach 1 and one along Avenue of Approach 2 (see figure 7-13 on page 7-38). The enemy's approach, compartmentalized forward of the Green and Red Rivers, restricts movement and prevents the attacking enemy force from fully exploiting its combat superiority. The terrain forward of the Green and Red Rivers allows for the massing of friendly fires with the enemy piecemeal commitment into friendly EAs. The terrain requires the enemy to zigzag and commit to frequent detours (due to the compartmentalization during movement), exposing portions of the enemy force for destruction without giving up the advantage of friendly forces fighting from protected positions.

7-143. The enemy's main effort, predicted to move along Avenue of Approach 2, requires crossing one river, the Green River. A secondary effort of the enemy, predicted to move along enemy Avenue of Approach 1, requires crossing both the Red and Blue Rivers. The least likely enemy avenue of approach, Avenue of Approach 3 (A and B) to the north, though the largest area of operations to defend requires the enemy to move through severely restricted terrain to the east and west of the Green River. The enemy, predicted to establish multiple infiltration lanes (company and platoon size elements) along this approach, infiltrates forces to the rear to disrupt friendly operations. Enemy follow-on forces, anticipated armor battalion task force, will attempt to exploit enemy successes along enemy Avenues of Approach 1 and 2 (see figure 7-13 on page 7-38).

Note. The desired end state of step 4 of the IPB process, determine threat COAs, is the development of graphic overlays (enemy situation templates) and narratives (enemy COA statements) for each possible enemy COA identified. Generally, there will not be enough time during the MDMP (see chapter 4) to develop enemy situation overlays for all COAs. A good technique is to develop alternate or secondary COAs, write a COA statement, and produce a list of HVTs to use during the mission analysis briefing and COA development during the MDMP. Once these tools and products are complete, the staff constructs overlays depicting the enemy's most likely and most dangerous COA to use during the friendly COA development and friendly COA analysis steps of the MDMP. (See ATP 2-01.3 and FM 6-0 for additional information.)

7-144. Step 3. Determine where to kill the enemy. Whether planning deliberately or rapidly when determining where to kill enemy, the BCT commander, subordinate commanders, and staffs maintain a shared understanding of the steps within the IPB process and the MDMP. Within the scenario that follows this paragraph, the BCT commander focuses this effort to determine the effects of the terrain, weather, and civil considerations on the enemy avenues of approach identified within and north of the Blue River Valley. During step 4 of the IPB, the BCT and battalion staffs (specifically the S-2 and S-3) identify and develop possible enemy COAs that can affect the BCT mission. Based on the results of this analysis, the commander concentrates efforts and economizes forces to kill the enemy east of the Green and Red Rivers along three enemy avenues of approach to best utilize the restricted and severely restricted areas forward in the BCT's area of operations.

EXAMPLE BCT AREA DEFENSE SCENARIO

This fictional scenario for steps 4, 5, and 6 of the EA development process, used for discussion purposes, is an example of how a BCT commander might build EAs within an area defense. In this scenario, an IBCT conducts an area defense against a motorized Infantry and armor threat. Army aviation attack and reconnaissance units, initially control by division and higher echelon, conducted attacks against enemy forces not in direct contact with the ground maneuver forces of the IBCT. As the enemy advanced, aviation attacks continued in close proximity or in direct support of IBCT and battalion security forces and main battle area forces. Artillery and mortar fire support plans were integrated into forward security area actions and the direct fire plans of maneuver companies in the main battle area. Engineer priorities of work were initially to countermobility, then to survivability. As the IBCT prepared for the defense, the brigade support battalion established the brigade support area (BSA) just forward of the division support area to support the BCT area defense.

The IBCT is task organized with two Infantry battalions, a combined arms battalion, a Cavalry squadron, a field artillery battalion, a brigade engineer battalion, and a brigade support battalion. Infantry Battalion 2 (main effort) is task organized with two Infantry rifle companies, a mechanized Infantry company team (two mechanized Infantry platoons and one tank platoon), and a weapons company. Company C, the third Infantry rifle company (mounted) from Infantry Battalion 2, was placed under IBCT control as the reserve for the BCT. Infantry Battalion 1 is task organized with its three Infantry rifle companies and weapons company. The combined arms battalion is task organized with two armor company teams, each with two tank platoons and one mechanized Infantry platoon. The commander weighted the main effort by attaching the mechanized Infantry company team from the combined arms battalion to Infantry Battalion 2, as stated above. No change in task organization for the IBCT Cavalry squadron, field artillery battalion, and brigade engineer battalion. The brigade support battalion is task organized with the logistical elements required to support the combined arms battalion. Subordinate unit task organization and scheme of maneuver are as follows:

- Infantry Battalion 2, main effort, conducts an area defense. The battalion defends in-depth with two Infantry rifle companies, with one assault platoon each attached, forward, and a mechanized Infantry company team (two mechanized Infantry platoons) and weapons company (two assault platoons) back. The tank platoon from the mechanized Infantry company team is the battalion reserve. (Illustrated within EA development example, see figure 7-14 on page 7-42.)

- Infantry Battalion 1, supporting effort, conducts an area defense to the north of Infantry Battalion 2. The battalion defends in-depth with two Infantry rifle companies forward, and one Infantry rifle company back. The weapons company (two assault platoons) is the battalion reserve. (Not illustrated.)

- The combined arms battalion, two armor company teams, is the counterattack force for the BCT. (Not illustrated.)

- Infantry rifle company C from Infantry Battalion 2 is the reserve for the BCT. Company C is mounted and has two attached assault platoons from Infantry Battalion 1. (Not illustrated.)

- The Cavalry squadron establishes the security area forward of the BCT main battle area. (Not illustrated.)

- The brigade engineer battalion priorities of work, countermobility, survivability, and then mobility. Priority of engineer effort initially to Infantry Battalion 2, then to the mobility of the BCT counterattack force.

- The field artillery battalion provides priority of fires initially to security area forces, then to Infantry Battalion 2, on order to the BCT counterattack force. (Not illustrated.)

- The brigade support battalion establishes the BSA just forward of the division support area. Priority of support initially to security area forces, then to Infantry Battalion 2, finally to the BCT counterattack force. (Not illustrated.)

7-145. **Step 4. Position subordinate forces and weapons systems.** Within the above scenario, the Infantry brigade combat team (IBCT) commander's concept for the area defense required the positioning of subordinate forces and weapon systems to accomplish their mission independently and in combination by means of fires, the employment of obstacles, and absorbing the strength of the attack within defensive battle positions. The commander assigned subordinate maneuver units an area of operations, based on the mission variables of METT-TC, to maximize decentralized execution empowering subordinate commanders to position battle positions within their assigned area of operations. At the same time, each subordinate commander addressed security requirements for the flanks of assigned area of operations by assigning responsibility to a subordinate element or organizing a security force or observation post(s) to accomplish that mission. The commander and subordinate commanders retained reserves to contain enemy penetrations between units and positions, to reinforce fires into an EA, or to help a portion of the security force or main body disengage from the enemy if required.

7-146. Step 5. Plan and integrate obstacles. During the conduct of the area defense, countermobility (see ATP 3-90.8) planning is the primary concern of the assistant brigade engineer (known as ABE) under the supervision of the brigade engineer battalion commander and battalion and squadron engineer staff noncommissioned officers, in coordination with the BCT and battalion and squadron S-3, S-2, and fire support officers. (External to the BCT, engineer planners [division and corps] coordinate with the BCT ABE and brigade engineer battalion). The plan addresses how security area and MBA forces reinforce the natural defensive characteristics of the terrain with the employment of obstacles to block, disrupt, fix, and turn attacking enemy forces into planned EAs. Countermobility planning also includes the positioning of protective obstacles to prevent the enemy from closing with defensive battle positons within subordinate unit area defenses.

7-147. Within the scenario, the IBCT commander's concept for the employment of obstacles within the area defense (Infantry Battalion 2) forces the enemy to enter established EAs positioned where the commander intends to kill the enemy. To succeed, the battalion through the employment of obstacles and the static positioning of company and platoon battle positions control, stop, or canalize attacking enemy forces to counteract the enemy's initiative. The commander, through dynamic actions of the battalion reserve, covers gaps between positions and takes advantage of available offensive opportunities such as a local attack or counterattack that do not risk the integrity of the defense (see figure 7-14 on page 7-42).

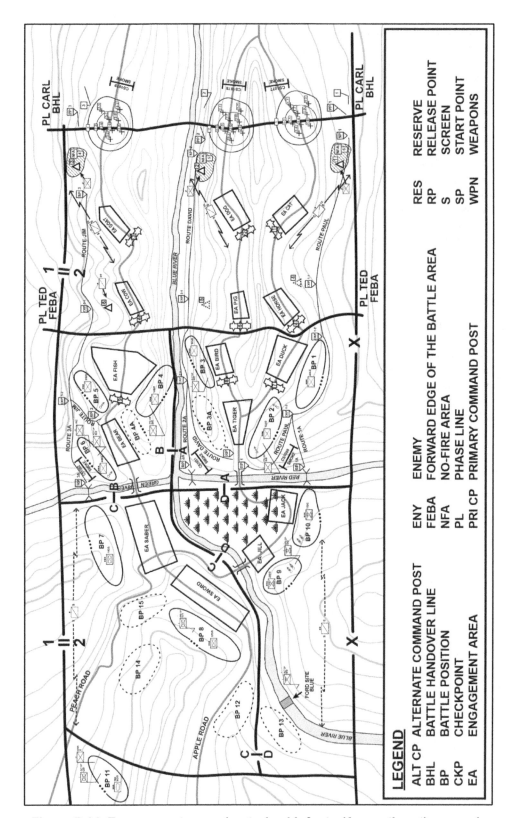

Figure 7-14. Engagement areas (motorized Infantry/Armor threat), example

7-148.   Step 6. Plan and integrate fires. During the conduct of the area defense, fire support planning is the primary concern of the BCT fire support officer (under the supervision of the BCT field artillery battalion commander) and battalion and squadron fire support officers, in coordination with the BCT and battalion and squadron S-3, S-2, and engineer staff officers. (External to the BCT, fire support planners [division and corps] coordinate with the BCT fire support officer and field artillery battalion.) Within the scenario, the IBCT and division higher area of interest and area of operations extend far enough beyond forward battalions' forward line of own troops that the BCT commander has the time and resources to identify approaching enemy forces, assess options, and recommend targets for attack to enable the mission. Fires conducted by joint fire assets or through the provision of mission orders to the division's attached combat aviation brigade and field artillery units at the BCT and division echelons further enable the ability to seize the initiative before the advancing enemy makes contact with forward defensive positions. The division JAGIC, in coordination with the BCT fire support cell, plans and coordinates joint fires, suppression of enemy air defenses, airspace coordination areas, ingress and egress routes, and other airspace requirements to deliver aerial and surface-delivered fires simultaneously into a given EA or target area.

7-149.   Before the enemy closes into direct fire EAs, in either the security area or the MBA, the BCT and subordinate units direct the initiation of fires. The commander and staff plan to provide the most effective fires resources and mitigate the risk of fratricide as the attacking enemy nears the designated EA while supporting air conducts army aviation and close air support attacks. During EA development, fire support coordination measures, such as TRPs, trigger lines, and final protective fires enable observed fires (see ATP 3-09.30) and the obstacle plan (see ATP 3-90.8) to force the enemy to use avenues of approach covered by friendly EAs. These shaping operations typically focus on enemy HPTs, such as command and control nodes, engineer, fire support, and air defense assets and follow-on forces for destruction or disruption.

7-150.   Step 7. Rehearse the execution of operations within the EA. The BCT and subordinate units coordinate and rehearse EA actions on the ground, gaining intimate familiarity with the terrain. The commander, the S-2, the S-3, engineer planner, and the fire support officer, at a minimum, rehearse the sequence of events with the subordinate commanders and separate element leaders for each EA.

7-151.   During rehearsals, the BCT commander confirms designated TRPs, trigger lines, final protective fires, EAs, and other direct- and indirect-fire control measures in each EA within the BCT's area of operations. Once in position, the commander may modify subordinate unit positions and preplanned control measures during rehearsals to improve defensive capabilities as required. The commander ensures the integration of fires by adjusting the planned positions of weapon systems to obtain maximum effectiveness against targets in the planned EA. The commander coordinates all fires, including those of supporting Army aviation and close air support, used to isolate the targeted enemy force in the planned EA while preventing the target's escape or reinforcement. The BCT and subordinate headquarters rehearse the confliction of fires to ensure maximum damage before the enemy can respond. The commander rehearses the actions of the reserve and counterattack forces to reinforce fires, add depth, or block, to restore a position by counterattack, or to reinforce the destruction of enemy forces within planned EAs.

7-152.   Commanders at each echelon rehearse their planned actions within EAs. Subordinate commanders and leaders reconnoiter and identify positions and identify movement or withdrawal routes and revise them as required. Subordinate units rehearse assigned weapon system primary sectors of fire and secondary sectors of fire to increase the capability of concentrating fire in certain areas in accordance with established criteria and priorities for engagement. Secondary sectors of fire, when there are no targets in the primary sector or when commanders need to cover the movement of another friendly element, correspond to another element's primary sector of fire to obtain mutual support. Secondary sectors of fire are rehearsed and confirmed depending on the availability of time before execution. Subordinate commanders may impose and rehearse additional fire support coordination measures as required and as time permits.

PREPARING AN AREA DEFENSE

7-153.   The BCT uses time available to build the defense and to refine counterattack plans. The commander and staff assess unit preparations while maintaining situational awareness of developments in the BCT's areas of interest. Collection activities (see chapter 5) begin soon after receipt of the mission and continue throughout preparation and execution. Security operations are conducted aggressively while units occupy and prepare assigned positions and rehearse defensive actions. During preparation, surveillance, reconnaissance, and intelligence operations help improve understanding of the enemy, terrain, and civil considerations.

7-154. Revising and refining the plan is a key activity of preparation. The commander's situational understanding may change over the course of operations, enemy actions may require revision of the plan, or unforeseen opportunities may arise. During preparation, assumptions made during planning may be proven true or false. Intelligence analysis may confirm or deny enemy actions or show changed conditions in the area of operations because of shaping operations.

Establish Security

7-155. The first priority in the defense is to establish security. During the defense, effective security requires the establishment of the security area (forward of the MBA), the employment of patrols and observation posts, the use of manned and UASs and sensors, and the use of the terrain (cover and concealment). Security operations, counterreconnaissance, survivability operations, military deception, information-related capabilities, and CEMAs (specifically EW) counter enemy intelligence, surveillance, and reconnaissance from determining friendly locations, strengths, and weaknesses.

7-156. Potential threats to the defense may include noncombatant access to communications, digital cameras, and similar devices. Security measures, such as shutting down telephone exchanges and cell telephone towers and preventing unauthorized personnel from moving in the BCT's area of operations may be required. The BCT should request guidance from higher headquarters before implementing any security measures that could affect the civilian population.

7-157. As part of the defense, higher headquarters may have created a military deception operation and associated information operations to protect the force, cause early committal of the enemy, or mislead the enemy as to the defender's true intentions, composition, and disposition of friendly forces. The BCT aids in the preparation and execution of the military deception plan to—

- Exploit enemy prebattle force allocation and sustainment decisions.
- Exploit the potential for favorable outcomes of protracted minor engagements and battles.
- Lure the enemy into friendly territory exposing the enemy's flanks and rear to attacks.
- Mask the level of the sustaining and operational forces committed to the defense.

7-158. A defense containing branches and sequels gives the commander preplanned opportunities to exploit the situation and around these branches and sequels that deception potentials exist. Specific deceptive actions the BCT commander can take to hasten exhaustion of the enemy offensive include but are not limited to—

- Masking the conditions under which the enemy will accept decisive engagement.
- Luring the enemy into a decisive engagement that facilitates the transition sequence.
- Employing camouflage, decoys, false radio traffic, movement of forces, and the digging of false positions and obstacles.

Occupation of Positions

7-159. The BCT commander and staff monitor and deconflict any positioning problems with BCT or higher headquarters' reconnaissance and security efforts as units move into their assigned areas of operation and occupy positions. The BCT may have to make minor adjustments to areas of operation, EAs, battle positions, and other defensive control measures based on unanticipated conditions the occupying units encountered as they begin preparing the defense.

7-160. The ABE monitors units assigned to close gaps or to execute directed obstacles such as demolition of bridges or dams to assure the units are ready to execute their mission. The ABE also ensures the units site and complete all obstacle emplacements within the BCT according to the obstacle plan.

Rehearsals

7-161. The BCT conducts defensive rehearsals as time permits. The commander uses any, or combinations of, the four types of rehearsals: backbrief, combined arms rehearsal, support rehearsal, and battle drill or standard operating procedures (SOPs) rehearsal. Each rehearsal type achieves a different result and has a specific place in the preparation timeline. The commander's imagination and available resources are the only limits restricting methods of conducting rehearsals. The BCT commander ensures the integration of attached enabling forces into the defensive scheme of maneuver through rehearsals. (See chapter 4, section II of this manual and FM 6-0 for additional information.)

EXECUTING AN AREA DEFENSE

7-162. In an area defense, the BCT concentrates combat power effects against attempted enemy breakthroughs and flanking movements from prepared and protected positions. The commander uses the reserve to cover gaps between defensive positions, to reinforce those positions as necessary, and to counterattack to seal penetrations or block enemy attempts at flanking movements. For discussion purposes, the following paragraphs divide execution of an area defense into a five-step sequence:

Step 1, Gain and Maintain Enemy Contact

7-163. Gaining and maintaining contact with the enemy is vital to the success of the defense. As the enemy's attack begins, the BCT's initial goals are to identify committed enemy units' positions and capabilities, determine the enemy's intent and direction of attack, and gain time to react. Initially, the commander accomplishes these goals in the security area. The sources of this type of information include reconnaissance and security forces, surveillance assets, intelligence operations, and supporting echelons above the BCT. The commander ensures the distribution of a COP throughout the BCT during the battle to form a shared basis for subordinate commanders' actions. The commander uses available information, in conjunction with judgment, to determine the point at which the enemy is committed to a COA.

Step 2, Disrupt the Enemy

7-164. The commander executes shaping operations to disrupt the enemy. After making contact with the enemy, the commander seeks to disrupt the enemy's plan and ability to control its forces. Ideally, the commander's shaping operations result in a disorganized enemy force conducting a movement to contact against a prepared defense. Once the process of disrupting the enemy begins, it continues throughout the defense. An enemy airborne assault or air assault in the BCT area of operations must be attacked immediately with available ground forces and fires before the enemy airhead can be organized and reinforced. The BCT uses indirect fires, close air support, Army aviation attacks, and other available fires and nonlethal effects during this phase of the battle to—

- Support the security force.
- Disrupt or limit the momentum of the enemy's attack.
- Destroy HPTs.
- Divert the enemy's attack.
- Deceive the enemy's knowledge of the BCT's MBA.
- Reduce the enemy's combat power.
- Separate enemy formations.

Step 3, Fix the Enemy

7-165. The commander has several options to fix an attacking enemy force. The commander can design shaping operations, such as securing the flanks and point of penetration, to fix the enemy and allow friendly forces to execute decisive maneuver elsewhere. Combat outposts and strong points can deny enemy movement to or through a given location and as to the exact location of the BCT's MBA. A properly executed deception operation can constrain the enemy to a given COA.

Step 4, Maneuver

7-166. The decisive operation occurs in the MBA. This is where the effects of shaping operations, coupled with sustaining operations, combine with the decisive operation of the MBA force defeat the enemy. The commander's goal is to prevent the enemy's further advance using a combination of fires from prepared positions, obstacles and reserve forces. To accomplish this, the commander masses effects by maneuvering forces to focus direct and indirect fires at a critical point to counter the enemy's attack.

7-167. In an area defense, the need for flexibility through movement and maneuver requires the use of graphic control measures to assist command and control during the repositioning forces and counterattacks. Specified routes, PLs, attack and support by fire positions, battle positions, EAs, TRPs, and other fire support coordination measures are required to synchronize movement and maneuver.

7-168. During the defense, the BCT commander must prepare to quickly take advantage of fleeting opportunities, seize the initiative, and assume the offense. Although the BCT commander plans for the counterattack, the plan may not correspond exactly with the existing situation when the commander launches the counterattack. As the situation develops, the commander reassesses the plan based on a revised situational understanding that results from an updated COP.

7-169. Ideally, the commander has a counterattack plan appropriate to the existing situation. When this is not the case, the commander must rapidly reorganize and refit selected units, move them to attack positions, and attack or the commander must conduct an attack using those units already in contact with the enemy, which is normally the least favorable COA.

Step 5, Follow Through

7-170. Three conditions may result from the initial enemy attack: friendly forces achieve their objectives, friendly forces do not achieve their objectives, or both forces are in a stalemate with neither side gaining a decisive advantage over the other. A successful area defense allows the commander to transition to an attack. An area defense resulting in the defender being overcome by the enemy attack and needing to transition to a retrograde operation must consider the current situation in adjacent defensive areas. Only the commander who ordered the defense can designate a new forward edge of the battle area or authorize a retrograde operation.

7-171. As the purpose of a defensive action is to retain terrain and create conditions for a counteroffensive that regains the initiative. A successful area defense causes the enemy to sustain unacceptable losses short of any decisive objectives. During follow through, time is critical. Unless the commander has a large, uncommitted reserve prepared to quickly exploit or reverse the situation, the commander must reset the defense as well as maintain contact with the enemy. Time is also critical to the enemy to reorganize, establish a security area, and fortify positions.

7-172. The BCT commander plans and conducts a counterattack to attack the enemy when and where the enemy is most vulnerable. There is a difference between local counterattacks designed to restore the defense and a decisive operation designed to wrest the initiative from the enemy force and then defeat it. To conduct a decisive counterattack, the defending force must bring the enemy attack to or past its culminating point before it results in an unacceptable level of degradation to the defending force. To do this, the defending force must disrupt the enemy's ability to mass, causing the enemy to disperse its combat power into small groups or attrit enemy forces to gain a favorable combat power ratio. The defending force must continue to disrupt the enemy's ability to introduce follow-on forces and attack the defender's sustainment system. (See figure 7-15.) As the objective of the counterattack is reached, the BCT consolidates and continues reorganization that is more extensive and begins preparation to resume the offense.

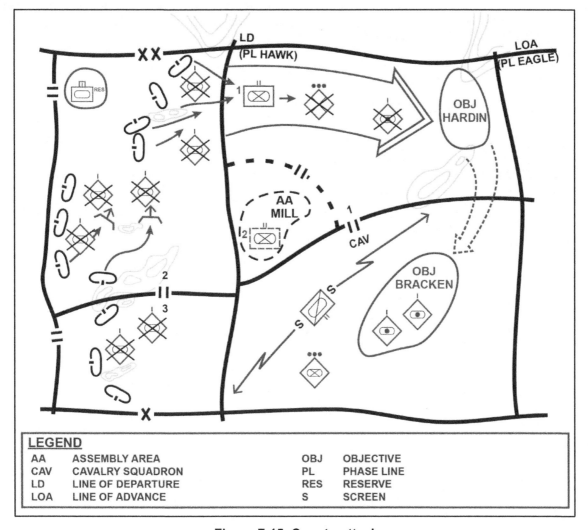

Figure 7-15. Counterattack

7-173.   In a successful defense, the enemy's attack is defeated, and the defensive plan must address missions following successful operations. The division's follow-on missions for the BCT governs this plan. The staff must begin planning for future offensive operations as they develop defensive plans. The commander and staff must develop maneuver plans, control measures, obstacle restrictions, and sustainment plans that enable the BCT to quickly transition to follow-on offensive missions or to pass follow-on forces.

MOBILE DEFENSE

7-174.   The *mobile defense* is a type of defensive operation that concentrates on the destruction or defeat of the enemy through a decisive attack by a striking force (ADP 3-90). The mobile defense focuses on defeating or destroying the enemy by allowing enemy forces to advance to a position that exposes them to a decisive counterattack by the *striking force*—a dedicated counterattack force in a mobile defense constituted with the bulk of available combat power (ADP 3-90). The commander uses the *fixing force*—a force designated to supplement the striking force by preventing the enemy from moving from a specific area for a specific time (ADP 3-90)—to help channel attacking enemy forces into EAs and to retain areas from which to launch the striking force. (See figure 7-16 on page 7-48.) A mobile defense requires an area of operations of considerable depth. The commander must be able to shape the battlefield, causing an enemy force to overextend its lines of communication, expose its flanks, and risk its combat power. Likewise, the commander must be able to move friendly forces around and behind the enemy force, cut them off, and destroy them. (See figure 7-17 on page 7-49.) Divisions and larger formations normally execute mobile defenses. However, BCTs and maneuver battalions may participate in a mobile defense as part of the fixing force or the striking force. (See FM 3-90-1 for additional information.)

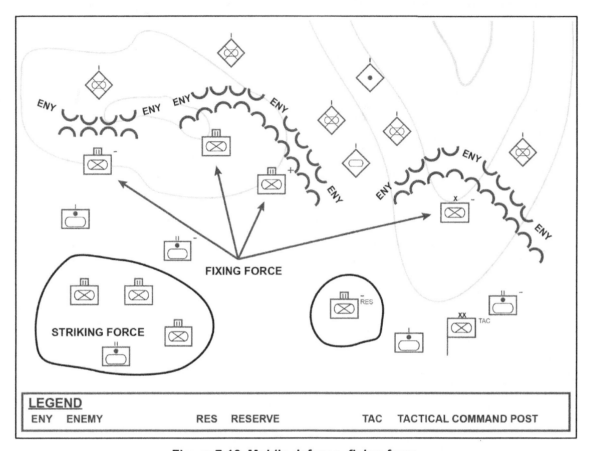

Figure 7-16. Mobile defense, fixing force

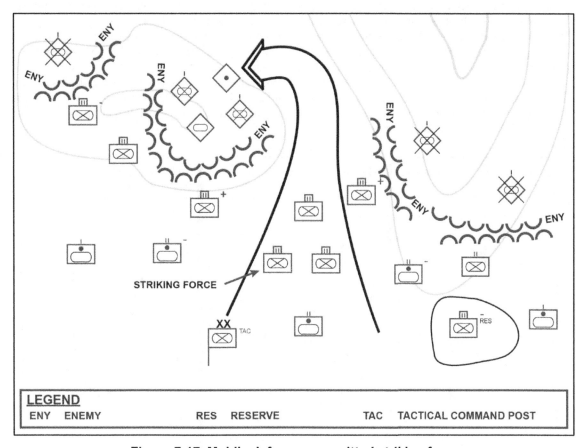

Figure 7-17. Mobile defense, committed striking force

RETROGRADE

7-175. A *retrograde* is a type of defensive operation that involves organized movement away from the enemy (ADP 3-90). The enemy may force the retrograde or a commander may execute it voluntarily. The three variations of the retrograde are: delay, withdraw, and retirement.

DELAY

7-176. A *delay* is when a force under pressure trades space for time by slowing down the enemy's momentum and inflicting maximum damage on enemy forces without becoming decisively engaged (ADP 3-90). Delays allow units to yield ground to gain time while retaining flexibility and freedom of action to inflict the maximum damage on the enemy. The methods are delay from successive positions and delay from alternate positions. The BCT conducts the delay by using one or a combination of the two methods. The method selected depends on the width of the front, the terrain, the forces available, the enemy, and the amount of time required of the delay. In either method, a mobility advantage over the enemy is required.

Delay from Successive Positions

7-177. A delay from successive positions involves fighting rearward from one position to the next, holding each as long as possible or for a specified time (figure 7-18 on page 7-50). In this type of delay, all maneuver battalions are committed on each of the BCT delay positions or across the area of operations on the same PL. The BCT commander uses a delay from successive positions when an area of operations is so wide that available forces cannot occupy more than a single line of positions. The disadvantages of this delay are lack of depth, less time to prepare successive positions, and the possibility of gaps between units.

7-178.   When ordered to move, the BCT disengages, then moves and occupies the next designated position. A part of the unit displaces directly to the rear when the order to begin the delay is received and occupies the next designated position. The rest of the unit maintains contact with the enemy between the first and second delay positions. As these elements pass through the second position, the forces on that position engage the enemy at the greatest effective range. When the BCT can no longer hold the position without becoming decisively engaged, it moves to the next successive position. When conducting a delay from successive positions, the BCT may retain a reserve if the division has none. The reserve will frequently be small and employed as a counterattacking force. It protects a threatened flank, secures vital rear areas, or provides overwatch fires to a withdrawing unit.

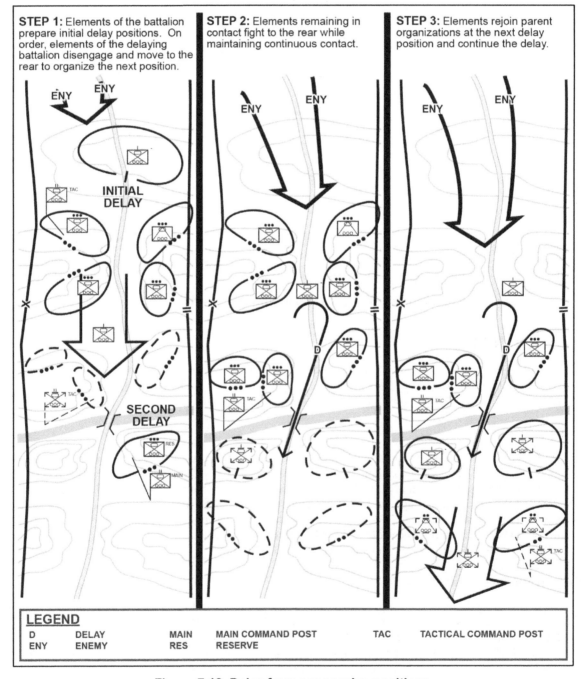

Figure 7-18. Delay from successive positions

Delay from Alternate Positions

7-179.   Delay from alternate positions can be used when a force has a narrow area of operations or has been reinforced to allow positioning in-depth (see figure 7-19). This is the preferred method of delay. One or more maneuver units employ this method to occupy the initial delay position and engage the enemy. Other maneuver units occupy a prepared second delay position. These elements alternate movement in the delay. While one element is fighting, the other occupies the next position in-depth and prepares to assume responsibility for the fight.

7-180.   Units occupying the initial delay position can delay between it and the second position. When the delaying units arrive at the second delay position, they move through or around the units that occupy the second delay position. The units on the second delay position assume responsibility for delaying the enemy; the delaying procedure is then repeated. Moving around the unit on the next delay is preferred because this simplifies passage of lines. The alternate method provides greater security to the delay force and more time to prepare and improve delay positions. Normally, when delaying from alternate positions, the BCT commander does not maintain a reserve. The forces not in contact with the enemy are available to function in the role of a reserve if needed.

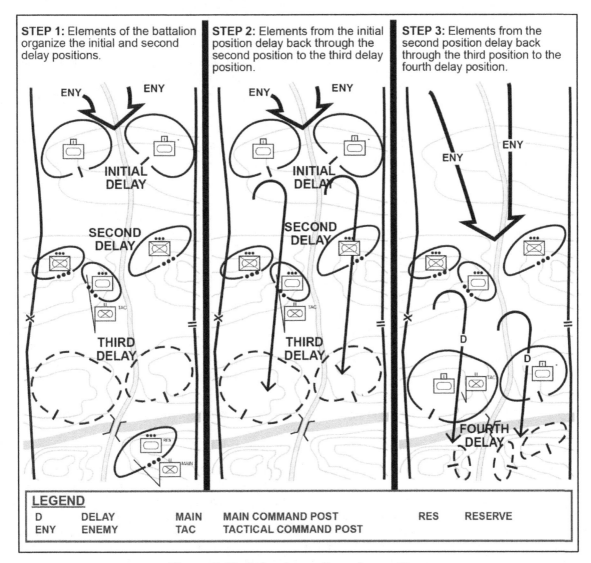

Figure 7-19. Delay from alternate positions

WITHDRAW

7-181.   *Withdraw* is to disengage from an enemy force and moves in a direction away from the enemy (ADP 3-90). Withdrawing units, whether all or part of a committed force, voluntarily disengage from the enemy to preserve the force or release it for a new mission. The purpose of a withdrawal is to remove a unit from combat, adjust defensive positions, or relocate forces. A withdrawal may free a unit for a new mission. A unit may execute a withdrawal at any time and during any type of operation.

7-182.   BCTs normally withdraw using a security force, a main body, and a reserve. There are two types of withdrawals, assisted and unassisted. In an assisted withdrawal, the next higher headquarters provides the security forces that facilitate the move away from the enemy. In an unassisted withdrawal, the BCT provides its own security force. Withdrawals are generally conducted under one of two conditions, under enemy pressure and not under enemy pressure. Regardless of the type or condition under which it is conducted, all withdrawals share the following planning considerations:

- Keep enemy pressure off the withdrawing force. Position security elements to delay the enemy. Emplace obstacles and cover by fire to slow its advance.
- Maintain security. Know the enemy's location and its possible COAs. Observe possible enemy avenues of approach.
- Gain a mobility advantage. Gain the advantage by increasing the mobility of the BCT, reducing the mobility of the enemy, or both.
- Reconnoiter and prepare routes. Each unit must know the routes or lanes of withdrawal. Establish priority of movement and traffic control if two or more units move on the same route.
- Withdraw nonessential elements early. Withdrawing nonessential elements early may include some command and control and sustainment elements.
- Move during limited visibility. Movement under limited visibility provides concealment for the moving units and reduces the effectiveness of enemy fires.
- Concentrate all available fires on the enemy. Alternate movement between elements so some of the force can always place direct or indirect fires on the enemy.

7-183.   Withdrawing under enemy pressure demands superior maneuver, firepower, and control. The BCT executes a withdrawal in the same manner as a delay, although its ultimate purpose is to break contact with the enemy rather than maintain it as in the delay. When conducting a withdrawal under enemy pressure, the commander organizes the BCT into a security force and a main body. Use control measures that facilitate the accomplishment of the commander's intent. These control measures may include areas of operations, battle positions, PLs, routes, passage points and lanes, contact points, checkpoints, and battle handover lines.

7-184.   Success depends on facilitating the disengagement of the main body by massing its own fires and the overwatch fires provided by the security element. The division commander may place adjacent units in overwatch or require them to conduct security operations or limited counterattacks to support the withdrawing BCT. To assist withdrawing elements, the security force must be strong enough to detect and engage the enemy on all avenues of approach. The BCT may form its own security force from forward maneuver battalion elements or the Cavalry squadron. The security force must:

- Stop, disrupt, disengage, or reduce the enemy's ability to pursue.
- Reduce, through smoke and suppressive fires, the enemy's capability to observe the movement of the main body.
- Rapidly concentrate additional combat power in critical areas.

7-185.   As the commander gives the order to withdraw, the BCT must engage the enemy with concentrated direct and indirect fire to enable the withdrawing force to disengage, conduct a rearward passage through the security force, assemble, and move to their next position. The security force assumes the fight from the forward elements. This includes delaying the enemy advance while the bulk of the BCT conducts movement to the rear. On order, or when the BCT meets other predetermined criteria, the security force disengages itself and moves to the rear as a rear guard. Depending on the BCT's next mission, the security force may be required to maintain contact with the enemy throughout the operation.

7-186.   When the BCT conducts a withdrawal not under enemy pressure, it must plan as though enemy pressure is expected, and then plan for a withdrawal without pressure. Withdrawal not under enemy pressure requires the formation of a detachment left in contact (DLIC). Reconnaissance, security, and deception are critical to conducting a withdrawal not under enemy pressure. The commander must know the activities and movements of any enemy in the area that could influence the operation. The commander ensures the security of the force against surprise and projects the impression of conducting operations other than a withdrawal. If the enemy becomes aware that the BCT is withdrawing, the enemy may choose to exploit the BCT's relative vulnerability by attacking or employing indirect fires against elements in movement. Therefore, deception and operations security measures are essential to the success of a withdrawal not under enemy pressure. These measures include—

- Maintaining the same level of communications.
- Continuing the use of patrols.
- Moving during limited visibility.
- Maintaining the same level of indirect fires.
- Avoiding compromise of the operation by radio.
- Maintaining noise and light discipline.
- Using the DLIC to simulate or continue normal activities.

7-187.   A *detachment left in contact* is an element left in contact as part of the previously designated (usually rear) security force while the main body conducts its withdrawal (FM 3-90-1). The DLIC is a force organized from within the BCT that maintains contact with the enemy while the majority of the BCT withdraws. The DLIC usually comprises one third of the available combat power. A BCT may direct that three maneuver companies, augmented with the necessary combat power and increased mobility and firepower, form the DLIC.

7-188.   Two techniques for organizing the DLIC are designating one maneuver battalion as the DLIC or forming a new organization under the BCT S-3. When one maneuver battalion forms the DLIC, it repositions its force through a series of company-sized relief in place operations with companies in the other maneuver battalion's areas of operations. The advantages of this technique are that command and control is facilitated by the organic nature of the units involved and the focus of the force as a whole is dedicated toward one mission maintaining contact and preparing to fight a delay, if necessary. The disadvantages are the time needed to reposition and the increased amount of movement in the area of operations that may signal a vulnerability to the enemy.

7-189.   Forming a new organization under a new controlling headquarters also has advantages and disadvantages. The advantages are that the units of the DLIC may have to do less repositioning and some may not have to move at all. This advantage helps to deceive the enemy as to the intentions of the BCT. The disadvantage is that the unit's ability to fight as a team decreases if the organization is ad hoc in nature. This organization must train together to avoid this disadvantage.

RETIREMENT

7-190.   A *retirement* is when a force out of contact moves away from the enemy (ADP 3-90). A retirement is conducted as a tactical movement to the rear. The brigade may move on one or more routes depending on the routes available. Security for the main body is similar to that for a movement to contact using advance, flank, and rear guards. As in all tactical movements, all round security must be maintained. In all retrograde operations, control of friendly maneuver elements is a prerequisite for success. A withdrawal may become a retirement once forces have disengaged from the enemy, and the main body forms march columns.

7-191.   The BCT conducts retrograde operations to improve a tactical situation or to prevent defeat. Retrograde operations accomplish the following:

- Resist, exhaust, and defeat enemy forces.
- Draw the enemy into an unfavorable situation.
- Avoid contact in undesirable conditions.
- Gain time.
- Disengage a force from battle for use elsewhere in other missions.

- Reposition forces, shorten lines of communication, or conform to movements of other friendly units.
- Secure terrain that is more favorable.

SECTION VI – TRANSITION

7-192. Transitions between tasks during decisive action whether anticipated or unanticipated require adaptability as the BCT commander copes with changes in the operational environment. During transition and operating with mission orders, subordinate leaders take disciplined initiative within the commander's intent, bounded by the Army Ethic. The commander considers the concurrent conduct of each task—offensive, defensive, and stability—in every phase and ongoing operation. Transition between tasks during decisive action require careful assessment, prior planning, and unit preparation as the commander shifts the combinations of offensive, defensive, and stability operations.

7-193. The BCT commander halts the defense only when the operation accomplishes the desired end state, reaches a culminating point or receives a change of mission from higher headquarters. Transitions mark a change of focus between phases or between the ongoing operation and execution of a branch or sequel. In the defense, the BCT and subordinate units often transition from one phase of the operation to another sequentially or simultaneously. In decisive action, it is common for subordinate units of the BCT to transition to the offense and operations focused on stability, while maintaining the defense with other subordinate units.

7-194. The commander deliberately plans for sequential operations, assisting the transition process and allowing the setting of the conditions necessary for a successful transition. Such planning addresses the need to control the tempo of operations, maintain contact with both enemy and friendly forces, and keep the enemy off balance. The BCT establishes the required organization of forces and control measures based on the mission variables of METT-TC.

7-195. Prior contingency planning decreases the time needed to adjust the tempo of combat operations when a unit transitions from the defense to the offense or operations focused on stability. It does this by allowing subordinate units to simultaneously plan and prepare for subsequent operations.

TRANSITION TO THE CONDUCT OF OFFENSIVE OPERATIONS

7-196. The BCT or higher commander may order an attack, a movement to contact, or participate in an exploitation and subsequent pursuit if conditions are suitable. The commander transitions to the offense as soon as possible to attack the enemy when it is most vulnerable. The commander does not want to give the enemy time to prepare.

7-197. A defending commander transitioning to the offense anticipates when and where the enemy force will reach its culminating point or when it will require an operational pause before it can continue. At those moments, the combat power ratios most likely favor the defending force. The actions which may indicate the enemy has reached its culminating point include transitioning to the defense, heavy losses, lack of sustainment to continue the mission, unexpected success of friendly operations, increased enemy prisoners of war, and a lack of coherence and reduced combat power in the enemy's attacks.

7-198. The BCT commander must be careful not to be the target of enemy information activities designed to encourage the commander to abandon the advantages of fighting from prepared defensive positions. The commander ensures the force has the assets necessary to complete its assigned offensive mission. The commander should not wait too long to transition from the defense to the offense as the enemy force approaches its culminating point. The BCT must disperse, extend in-depth, and weaken enemy forces. At that time, any enemy defensive preparations will be hasty and enemy forces will not be adequately disposed to defend. The BCT commander wants the enemy in this posture when the force transitions to the offense. The commander does not want to give the enemy force time to prepare the defense. Additionally, the psychological shock on enemy soldiers will be greater if they suddenly find themselves desperately defending on new and often unfavorable terms, while the commander's own Soldiers will enjoy a psychological boost by going on the offense.

7-199. A commander can use two basic methods when transitioning to the offense. The first, and generally preferred, method is to attack using forces not previously committed to the defense. This method has the advantage of using rested units at a high operational strength. A drawback to this method is the requirement to conduct a forward passage of lines. Additionally, enemy intelligence assets are likely to detect the arrival of significant reinforcements. Another consideration of using units not in contact occurs when they are operating in noncontiguous areas of operations. The commander rapidly masses overwhelming combat power in the decisive operation. This might require the commander to adopt economy of force measures in some areas of operations while temporarily abandoning others to generate sufficient combat power.

7-200. The other method is to conduct offensive actions using the currently defending forces. This method has the advantage of being more rapidly executed and thus more likely to catch the enemy by surprise. Speed of execution in this method results from not having to conduct an approach or tactical road march from reserve assembly areas or, in the case of reinforcements, move from other area of operations and reception, staging, organization, and integration locations. Speed also results from not having to conduct a forward passage of lines and perform the liaison necessary to establish a COP that includes knowledge of the enemy force's patterns of operation. The primary disadvantage of this method is that the attacking force generally lacks stamina and must be quickly replaced if friendly offensive actions are not to culminate quickly.

7-201. If units in contact participate in the attack, the commander must retain sufficient forces in contact to fix the enemy. The commander concentrates the attack by reinforcing select subordinate units so they can execute the attack and, if necessary, maintain the existing defense. The commander can also adjust the defensive boundaries of subordinate units so entire units can withdraw and concentrate for the attack.

7-202. The commander conducts any required reorganization and resupply concurrently with transition activities. This requires a transition in the sustainment effort, with a shift in emphasis from ensuring a capability to defend from a chosen location to an emphasis on ensuring the force's ability to advance and maneuver. For example, in the defense, the sustainment effort may have focused on the forward stockage of class IV (construction and barrier materials) and class V (ammunition) items and the rapid evacuation of combat damaged systems. In the offense, the sustainment effort may need to focus on providing POL and forward repair of maintenance and combat losses. Transition is often a time in which forces perform deferred equipment maintenance. Additional assets may also be available on a temporary basis for casualty evacuation and medical treatment because of a reduction in the tempo of operations.

TRANSITION TO THE CONDUCT OF STABILITY OPERATIONS

7-203. During the transition to operations focused on stability, the role of the BCT varies greatly depending upon the security environment, the authority and responsibility of the BCT, and the presence and capacities of other nonmilitary actors. When transitioning from the defense, these other actors will normally be less established before stability operations tasks begin. The BCT in this case will operate before other actors have a significant presence. Generally, the BCT will focus on meeting the immediate essential service and civil security needs of the civilian inhabitants of the area of operations in coordination with any existing host-nation government and nongovernmental organizations before addressing the other stability operations tasks. (See chapter 8.) Support requirements may change dramatically. During transition, the commander may adjust rules of engagement or their implementation. The commander must effectively convey these changes to the lowest level.

7-204. The BCT must remain versatile and retain flexibility when transitioning from the defense to operations focused on stability. The commander may plan on order transition to a stability-focused mission when certain conditions are met. These conditions may include a sharp reduction of the enemy's offensive capabilities or deterioration in civilian governance and security. These conditions may require the rapid occupation and security of civilian areas. The commander must make every attempt to begin transition operations as soon as subordinate units of the BCT arrive within an assigned area of operations.

7-205. BCT subordinate units and Soldiers must be aware that during the transition to operations focused on stability, there may be events that escalate to combat. The BCT must always retain the ability to conduct offense and defense during transition. Preserving the ability to transition allows the commander to maintain initiative while providing security. The commander should consider planning an on-order offensive and defensive contingency in case the transition to operations focused on stability deteriorates. Subordinate commanders and leaders must be well-rehearsed to recognize activities that would initiate these contingences.

Chapter 8

Stability

The requirement for military formations to conduct operations focused on stability is not new. Our involvement in military conflict from the Revolutionary War to Operation Enduring Freedom consists of only eleven conventional military operations. Conversely, that same history reveals hundreds of operations focused on stability with recent history proving no different. Since the fall of the Berlin Wall, the United States led or participated in over fifteen operations in places such as Haiti, Liberia, Somalia, the Balkans, Iraq, and Afghanistan. While the magnitude of violence may not match conventional operations, history often measures the duration of stability operations in decades. This fact combines with the disturbing spread of international terrorism, fragile states allowing safe haven to terrorist organizations and or possessing weapons of mass destruction (WMD), along with an endless array of humanitarian and natural disasters illustrates the increasing requirement for operations focused on stability.

Military formations conduct operations focused on stability to transition the security and governance of populations to legitimate civilian authorities. The brigade combat team (BCT) lacks the organic capability to stabilize an assigned area of operations independently. The BCT's central role in operations focused on stability is to establish and maintain unity of effort towards achieving the political objectives of the operation. To do this the BCT employs combined arms formations that execute offensive and defensive operations, and stability operations tasks to identify and mitigate critical sources of instability. Essentially, the BCT unifies governmental, nongovernmental, and elements of the private sector activities with military operations to seize, retain, and exploit the initiative and consolidate gains.

The first three sections of this chapter discuss the doctrinal foundation, stability environment, and the Army's six stability operations tasks for operations focus on the stability element of decisive action. Sections IV and V focus on the challenges confronting the BCT commander and staff and subordinate commanders and leaders in accomplishing stability-focused missions or tasks, specifically area security operations and security force assistance (SFA). The final section addresses transition to offensive or defensive operations if the focus of the operation changes from stability. This section concludes with a discussion of transitions during SFA.

SECTION I – FOUNDATION FOR OPERATIONS FOCUSED ON STABILITY

8-1. Stability ultimately aims to establish conditions the local populace regards as legitimate, acceptable, and predictable. Stabilization is a process in which personnel identify and mitigate underlying sources of instability to establish the conditions for long-term stability. Stability operations tasks focus on identifying and targeting the root causes of instability and building the capacity of local institutions. Army forces accomplish stability missions and perform tasks across the range of military operations and in coordination with other instruments of national power. Stability missions and tasks are part of broader efforts to establish and maintain the conditions for stability in an unstable area before or during hostilities, or to re-establish enduring peace and stability after open hostilities cease.

FUNDAMENTALS OF STABILIZATION

8-2. The BCT applies the fundamentals of stabilization to the offense, defense, and operations focused on stability to achieve political and military objectives. (See ADP 3-07 for additional information.) The following fundamentals of stabilization lay the foundation for long-term stability:

- Conflict transformation.
- Unity of effort.
- Build host-nation capacity and capabilities.
- Host-nation ownership and legitimacy.

CONFLICT TRANSFORMATION

8-3. Conflict transformation is the process of converting the actors and conditions that motivate violent conflict into the governmental process to address the causes of instability. Conflict transformation sets the host nation on a sustainable, positive trajectory in which transformational processes directly address the dynamics causing instability. The use of the BCT in a combat role serves as a temporary solution until the situation is stabilized and host-nation forces are able to provide security for the populace. In all cases, the combat role supports the host nation's ability to provide for its internal security and external defense.

UNITY OF EFFORT

8-4. Military operations typically demand unity of command. The challenge for military and civilian leaders is to forge unity of effort or unity of purpose among the diverse array of actors involved in an operation focused on stability. This is the essence of *unified action*—the synchronization, coordination, and/or integration of the activities of governmental and nongovernmental entities with military operations to achieve unity of effort (JP 1). *Unity of effort* is the coordination and cooperation toward common objectives, even if the participants are not necessarily part of the same command or organization, which is the product of successful unified action (JP 1). Unity of effort is fundamental to successfully incorporating all the instruments of national power in a collaborative approach when conducting stability operations tasks during military operations.

8-5. When countering insurgency an example of unity of effort could be a military commander and a civilian leader ensuring that governance and economic lines of effort are fully coordinated with military operations. Unity of effort among nationally, culturally, and organizationally distinct partners is difficult to maintain, given their different layers of command. Achieving unity of effort requires participants to overcome cultural barriers and set aside parochial agendas. It also requires that each organization understand the capabilities and limitations of the others.

BUILDING HOST-NATION CAPACITY AND CAPABILITIES

8-6. Building host-nation capacity and capabilities is the outcome of comprehensive inter-organizational activities, programs, and military-to-military engagements that enhance the ability of partners to establish security, governance, economic development, essential services, rule of law, and other critical government functions. The Army integrates capabilities of operating forces and the institutional force to support interorganizational capacity and capabilities-building efforts, primarily through security cooperation interactions. The institutional force advises and trains partner army activities to build institutional capacity for professional education, force generation, and force sustainment. Army integrates capabilities of operating forces and the institutional force to support interorganizational capacity and capabilities-building efforts, primarily through security cooperation interactions. BCTs apply a comprehensive approach to sustained engagement with foreign and domestic partners to co-develop mutually beneficial capacities and capabilities to address shared interests.

8-7. Unified action is an indispensable feature of building host-nation capacity and capabilities. In operations characterized by stability operations tasks, unified action to enhance the ability of partners for security, governance, economic development, essential services, rule of law, and other critical government functions exemplifies building host-nation capacity and capabilities. Building the capacity and capability, during SFA, of foreign security forces (FSF) and their supporting institutions is normally the primary focus of security force assistance brigades (known as SFABs). On occasion, the BCT as a whole or selected unit(s) of the BCT may support SFA activities. SFA will encompass various activities related to the organizing, equipping, training, advising, and assessing of FSF and their supporting institutions. SFA activities conducted by SFABs and BCTs build host nation capacity and capabilities to defend against internal, external, and transnational threats to stability. (See FM 3-22 for additional information.)

HOST-NATION OWNERSHIP AND LEGITIMACY

8-8. Ownership and legitimacy is a condition based upon the perception by specific audiences of the legality, morality, or rightness of a set of actions, and of the propriety of the authority of the individuals or organizations in taking them. Legitimacy enables host nation ownership by building trust and confidence among the people. The foundation of ownership and legitimacy affects every aspect of operations from every conceivable perspective. Ownership of the mission and legitimacy of the host-nation government enables successful operations characterized by stability operations tasks.

8-9. *Security sector reform* is a comprehensive set of programs and activities undertaken by way a host nation to improve the way it provides safety, security, and justice (JP 3-07).

> *Note.* Security sector reform can be an activity conducted during security cooperation (see paragraph 8-112).

8-10. The BCT's primary role in security sector reform is to support the reform, restructuring, or re-establishment of the armed forces and the defense sector across the range of military operations. The overall objective is to support in a way that promotes an effective and legitimate host-nation government and its ownership of the mission that is transparent, accountable, and responsive to civilian authority. (See ADP 3-07 for additional information.)

STABILIZATION FRAMEWORK

8-11. A stabilization framework based on conditions within the area of operations of initial response, transformation, and fostering sustainability, helps the BCT determine the required training and task organization of forces before initial deployment, and serves as a guide to actions in an operation focused on stability operations tasks. A BCT deployed into an area of operations where the local government is nonexistent may conduct a set of tasks while another BCT may conduct another set of tasks in an area of operations with a functioning local government. The phases described in the following paragraphs facilitate identifying lead responsibilities and determining priorities. (See FM 3-07 for additional information.)

INITIAL RESPONSE PHASE

8-12. Initial response actions generally reflect activity executed to stabilize a crisis state in the area of operations. The BCT typically performs initial response actions during, or directly after, a conflict or disaster in which the security situation prohibits the introduction of civilian personnel. Initial response actions aim to provide a secure environment that allows relief forces to attend to the immediate humanitarian needs of the local population. They reduce the level of violence and human suffering while creating conditions that enable other actors to participate safely in relief efforts.

TRANSFORMATION PHASE

8-13. Stabilization, reconstruction, and capacity building are transformation actions performed in a relatively secure environment. Transformation actions occur in either crisis or vulnerable states. These actions aim to build host nation capacity across multiple sectors. Transformation actions are essential to the continuing stability of the environment and foster sustainability within the BCT's area of operations.

FOSTERING SUSTAINABILITY PHASE

8-14. Fostering sustainability actions are those activities that encompass long-term efforts, which capitalize on capacity building and reconstruction. Successful accomplishment of these actions establishes conditions that enable sustainable development. Usually military forces perform fostering sustainability phase actions only when the security environment is stable enough to support efforts to implement the long-term programs that commit to the viability of the institutions and economy of the host nation. Often military forces conduct these long-term efforts to support broader, civilian led efforts.

COMPREHENSIVE APPROACH

8-15. A comprehensive approach to achieve unity of effort during stability operations tasks requires contributions from a variety of partners outside the United States and the U.S. Government. These partners include foreign military and police forces, nongovernmental organizations, international organizations, host-nation organizations, news media, and businesses. Many partners have no formal relationship with Army units but are, nevertheless, instrumental in achieving the desired outcomes. Army units must interact effectively with these partners to exchange information and strive for unified action. Army units demonstrate to the host nation and international community through action its character, competence, and commitment to adhere to and uphold the Army Ethic.

8-16. When developing an operational approach, commanders consider methods to employ a combination of defeat mechanisms and stability mechanisms. Defeat mechanisms relate to offensive and defensive operations (see chapters 6 and 7). Stability mechanisms relate to stability operations tasks, security, and consolidating gains in an area of operations. Planning operations related to stability mechanisms employed by the BCT requires a comprehensive approach, as well as an in-depth understanding of the stability environment (see paragraph 8-18). Planning must be nested within policy, internal defense and development (IDAD) strategy, the campaign plan, and any other higher-echelon plans. Continuous and open to change, planning includes identifying how to best assist the FSF and developing a sequence of actions to change the situation. Planning involves anticipating consequences of actions and developing ways to mitigate them to attain conditions that support establishing a lasting, stable peace.

Note. IDAD focuses on building viable institutions (political, economic, social, and military) that respond to the needs of society. Ideally, IDAD is a preemptive strategy. However, if an insurgency or other threat develops, it becomes an active strategy to combat that threat. To support the host nation effectively, U.S. forces, especially planners, consider the host-nation's IDAD strategy.

8-17. Considering the elements of operational art (the cognitive approach by commanders and staffs—supported by their skill, knowledge, experience, creativity, and judgement, see ADP 3-0) provides the BCT commander and staff with a combination of conventional forces while leveraging the unique capabilities of special operations forces, to assist in achieving operations focused on stability. The planning for and selection of the appropriate mix of military forces, civilian expeditionary workforce, or civilian personnel and contractors should be a deliberate decision based on thorough mission analysis and a pairing of available capabilities to requirements. Important factors to consider in these decisions include the nature of the host-nation force, the nature of the skills or competencies required by the host-nation force, and the nature of the situation and environment into which the BCT will deploy.

SECTION II – STABILITY ENVIRONMENT

8-18. Operations focused on stability, range across all military operations and offer perhaps the most diverse set of circumstances the BCT faces. The objective of operations focused on stability is to create conditions that the local populace regards as acceptable in terms of violence; the functioning of governmental, economic, and societal institutions; and adhere to local laws, rules, and norms of behavior. During unified land operations, the BCT provides the means for seizing and retaining initiative through partnership with associated enabling organizations that are better suited to bring stability to the operational environment. To successfully seize, retain, and exploit the initiative and consolidate gains in operations focused on stability:

the BCT must identify and mitigate sources of instability, understand and nest operations within political objectives, and achieve unity of effort across diverse organizations.

SOURCES OF INSTABILITY

8-19. The BCT conducts information collection to gain a detailed understanding of the sources of instability, and the capability and intentions of key actors within its area of operations. Sources of instability are actors, actions, or conditions that exceed the legitimate authority's capacity to exercise effective governance, maintain civil control, and ensure economic development. Enemy forces leverage sources of instability to create conflict, exacerbate existing conditions, or threaten to collapse failing or recovering states. Examples of sources of instability include but are not limited to—

- Ungoverned areas.
- Religious, ethnic, economic, political differences among the local population.
- Natural disasters.
- Resource scarcity.
- Individual disrupting legitimate governance.
- Degraded infrastructure.
- Economic strife.
- Immature, undeveloped or atrophied systems.
- Ineffective or corrupt host-nation security forces.

8-20. The BCT commander and staff must apply the same fundamental planning processes in the military decision-making process (MDMP) and the intelligence preparation of the battlefield (IPB) process to identify the tactical problem, and conduct information collection to fulfill priority intelligence requirements or identified information gaps. Critical thinking, innovative problem solving, and leveraging different tools to address these tactical problems assists the BCT commander and staff in identifying sources of instability. Thorough analysis, engaging with local leaders and populations, leveraging unified action partners, and research are standard methods used to identify sources of instability. The commander and staff consider operational variables (in coordination with division and corps staffs) and mission variables, with emphasis on civil considerations, to gain an understanding of the interests and motivations particular to different groups and individuals to enhance situational understanding.

8-21. Interactions of various actors affect the BCT's operational environment in terms of operational and mission variables. Some of these actors include the following:

- Unified action partners.
- Nongovernmental organizations.
- Private volunteer organizations.
- International and private security organizations.
- Media.
- Multinational corporations.
- Transnational criminal organizations.
- Insurgents.
- Violent extremist organizations.
- Tribes, clans, and ethnic groups indigenous to the area of operations.
- Regional influences such as other nation states.

8-22. The commander and staff consider alternative perspectives and approaches to the ones used in offense and defense. The BCT analyzes sources of instability from both the local, indigenous perspective and the U.S. military perspective to understand the differences between viewpoints. During the IPB, the BCT identifies key actors and their interests or agendas. Additionally, the BCT analyzes how these key actors influence the local civil capacity; this analysis drives the BCT's planning effort that addresses accomplishment of stability operations tasks. Staffs conduct preparation to understand unique aspects of operations focused on stability. For example, the brigade assistant engineer might conduct an assessment on the local electrical grid system of an assigned area of operations or the brigade surgeon or medical planner may conduct an assessment on host-nation medical facilities and their capacity before employment to an assigned area of operations. This staff specific assessment further enables the conduct of the six stability operations tasks (see section III) and makes the BCT's planning effort during operations focused on stability more informed and efficient.

8-23. Once the commander and staff possess an understanding of the operational environment, the BCT applies a mixture of stability mechanisms to set conditions to retain and exploit stabilizing factors. A stability mechanism is the primary method through which friendly forces affect civilians in order to attain conditions that support establishing a lasting, stable peace. Stability mechanisms relate to stability operations, security, and consolidating gains in an area of operations. The four stability mechanisms are compel, control, influence, and support. Combinations of stability mechanisms produce complementary and reinforcing outcomes that accomplish the mission more effectively and efficiently than single mechanisms do alone.

8-24. The BCT simultaneously uses stability mechanisms such as compel and control to assist with seizing initiative. Compel means to use, or threaten to use, lethal force to establish control and dominance, effect behavioral change, or enforce compliance with mandates, agreements, or civil authority. Control means to impose civil order. Offensive operations reveal and exploit enemy weaknesses by defeating, destroying, or neutralizing threat forces. These actions disrupt threat forces, prevent them from negatively influencing populations, and provide opportunities to continue exploiting weaknesses—but they are not decisive by themselves. The design of these actions should consider how and what they compel the population to do and whether or not the action will result in positive, neutral or negative support by the population in the long- and short-term. (See ADP 3-0 for additional information.)

UNDERSTANDING POLITICAL OBJECTIVES

8-25. Understanding political objectives frames the unique operations required to conduct stability operations. General political objectives are broad and conceptual in nature, but they give contextual guidance that informs the expanded purpose of the echelons above the BCT commander's intent. Political objectives may shift and change as the operational environment changes. That same guidance unifies or alienates partners that may fall outside of the military chain of command.

8-26. Given the inherently complex and uncertain nature of political objectives, the BCT commander and staff use the Army design methodology (see chapter 4) to help understand the root cause of instability and approaches to solve problems. The Army design methodology entails framing an operational environment, framing the problem, and developing an operational approach to solve the problem. The Army design methodology results in an improved understanding of an operational environment. Based on this improved understanding, the commander issues planning guidance, to include an operational approach, to guide more detailed planning using the MDMP (see chapter 4).

8-27. Incorporating political objectives into the planning process (see chapter 4) is a shared task amongst the BCT staff. Each staff officer understands the general and specific political objectives and the commander's intent two levels up of an assigned operation focused on stability and considers the implications and effects of the political objectives when presenting the commander with running estimates, courses of action (COAs) and other decision support staff products. The BCT staff must understand how to communicate general and specific political objectives into the themes and messages delivered in the operation order and fragmentary orders that Soldiers display through their actions. These political objectives must translate across the entirety of the operational environment yet be understood by subordinate units allowing them to affect the local population's perception.

ACHIEVE UNITY OF EFFORT ACROSS DIVERSE ORGANIZATIONS

8-28. BCT commanders and staffs must understand how to build relationships with many diverse organizations within an area of operations. These relationships allow the BCT to nest operations with both their higher headquarters and with the overall U.S. effort within the joint operational area. A whole-of-government approach, along with collaboration and cooperation with *unified action partners*— those military forces, governmental and nongovernmental organizations, and elements of the private sector with whom Army forces plan, coordinate, synchronize, and integrate during the conduct of operations

(ADP 3-0)—are key components of operations focused on stability. The BCT staff incorporates personnel from these organizations into the operations process as soon as possible.

8-29. By building relationships, the BCT reinforces the legitimacy of the BCT operation. Legitimacy is of great importance with stability operations. Military activities must sustain the legitimacy of the operation and of the local emerging or host government. Having a just cause, and establishing and sustaining trust affects several relationships: trust with the American people; trust within the unified force; trust with allies, governmental and nongovernmental organizations, and coalition partners; trust with the host nation government; and trust with the indigenous population. Restraint in the disciplined and ethical application of lethal force has a significant influence on those relationships of trust and legitimacy of the operation. There is a direct relationship between restraint, protection of noncombatant civilians and legitimacy in any military operation, but especially in stability operations.

MILITARY POLICE

8-30. Mission tailored military police support to the BCT integrates police intelligence operations (see ATP 3-39.20) throughout the offense, the defense, and operations focused on stability. Police intelligence operations address the reality that, in some operational environments, the threat is more criminal than conventional in nature. In those environments, it is not uncommon for members of armed groups, insurgents, and other belligerents to use or mimic established criminal networks, activities, and practices to move contraband, raise funds, or generally or specifically further their goals and objectives. Police intelligence can provide relevant, actionable police information or police intelligence to the BCT through integration into the operations process and fusion with other intelligence data. U.S. Army criminal investigations division and provost marshal staffs provide police intelligence analysis to the commander that identifies indicators of potential crimes and criminal threats against facilities or personnel.

SPECIAL OPERATIONS FORCES

8-31. BCT will typically coordinate with a Special Operations Task Force or a Joint Special Operations Task Force. Army special operations forces operating in a BCT's area of operations must coordinate their activities with the BCT regardless of command or supporting relationships. To best support integration efforts, and generate and sustain interdependence, the BCT and the special operations units should exchange a variety of liaison and coordination elements. This is especially important for the intelligence and fires warfighting functions. They range in size from individual liaisons to small coordination elements. Whatever their size or location, these elements coordinate, synchronize, and deconflict missions in the other unit's area of operation. By exchanging liaisons, conditions are created that foster interdependence. The exercise of interdependence facilitates shared understanding between the BCT and special operations forces and provides a conduit by which the two units can provide each other relevant, useful and timely information during the operations processes. Liaisons should attend and participate in all planning efforts, update briefs, and working groups. (See ADP 3-05 for additional information.)

8-32. Civil affairs operations are essential to the conduct of operations focused on stability. The full capability of the civil affairs force manifests itself in the conduct of stability operations tasks in every environment across the range of military operations. Civil affairs support to stability operations tasks include the execution of all five civil affairs core tasks, employment of civil affairs functional specialists, and continuous analysis of the civil component of the operational environment in terms of both operational and mission variables by civil affairs staff elements. Civil affairs activities are civil reconnaissance, civil engagement, civil-military operations center, civil information management and civil affairs operation staff support. Civil affairs forces also conduct military government operations that include transitional military authority and support to civil administration and provide support to civil affairs supported activities such as populace and resources control, foreign humanitarian assistance, and civil foreign assistance

8-33. Civil affairs support to operations focused on stability depends on the nature of the operation and the condition of the affected indigenous population and institutions. The civil affairs staff continually monitors the condition of the host nation throughout the operation, applies available resources to affect the civilian component, and recommends functional skills required to support this critical phase of the operation. Civil affairs support the BCT, U.S. Government agencies, and the host-nation civil administration in transitioning power back to the local government. During the transition from offense or defense to operations focused on stability, civil affairs units place greater emphasis on infrastructure, economic stability, and governance expertise. (See FM 3-57 for additional information.)

PROVINCIAL RECONSTRUCTION TEAM

8-34. A provincial reconstruction team, when established, can be part of a long-term strategy to transition the functions of security, governance, and economics to provincial governments. It is a potential combat multiplier for maneuver commanders performing governance and economics functions and providing expertise to programs designed to strengthen infrastructure and the institutions of local governments. The provincial reconstruction team leverages the principles of reconstruction and development to build host nation capacity while speeding the transition of security, justice, and economic development to the control of the host nation. Depending on the situation, a provincial reconstruction team is manned between 60 to 90 personnel. A provincial reconstruction team may have the following complement of personnel:

- Provincial reconstruction team leader.
- Deputy team leader.
- Multinational force liaison officer.
- Rule of law coordinator.
- Provincial action officer.
- Public diplomacy officer.
- Agricultural advisor.
- Engineer.
- Development officer.
- Governance team.
- Civil affairs team.
- Bilingual cultural advisor.

INTERGOVERNMENTAL AND NONGOVERNMENTAL ORGANIZATIONS

8-35. BCTs also must recognize the value of intergovernmental and nongovernmental organizations and build effective relationships with these actors. These organizations may have the most extensive amounts of resources to conduct stability operations tasks within the BCT's area of operations. Intergovernmental organizations and nongovernmental organizations are the primary sources of subject matter expertise in many essential services and governance topics. They also are the primary provider of humanitarian, infrastructure and essential services in immature operational environments. Intergovernmental and nongovernmental organizations potentially have experienced and detailed knowledge of the civil environment. Usually the intergovernmental and nongovernmental organizations will have a better understanding of the civil considerations than any other actors other than host-nation personnel will. This insight can assist the BCT in the continual process to understand and shape the environment.

8-36. Building relationships with intergovernmental and nongovernmental organizations is unique, as opposed to host-nation forces and interagency actors, who often have different mandates and alternative perspectives to operations, focused on stability. The BCT commander and staff utilize these differences to see the operational environment and tactical problems from different perspectives. Additionally, understanding where intergovernmental and nongovernmental organizations are in the area of operations and the nature of their activities helps develop a common operational picture (COP). This COP enables the BCT to anticipate changes to the operational environment, the effects of intergovernmental and nongovernmental organizations on stability operations tasks and BCT operations, and future friction points between the organization's interests and the BCT's interests.

8-37. Building relationships with intergovernmental and nongovernmental organizations might also be difficult because these organizations are reluctant to establish associations with U.S. forces. The BCT must be cognizant of this and establish these relationships on terms beneficial to all parties involved. Intergovernmental and nongovernmental organizations can bring valuable resources, information regarding the civil populace and the operational environment, and alternative perspectives to the BCT's operation. Examples of interagency personnel that can provide the BCT valuable information are members of a United

States Department of State Embassy or Consulate country team, chiefs of stations or bases, defense attaches, and subject matter expertise from other governmental departments (Departments of Agriculture, Justice, Treasury, and so forth).

8-38. Humanitarian organizations avoid any blurring of the distinction between neutral, independent, and impartial humanitarian action and development aid derived from political engagement, as the latter is potentially linked to security concerns or support to one side. The BCT often works through civilian representatives from United States Agency for International Development, the United Nations, or the host nation when coordinating with nongovernmental organizations. The BCT avoids publicly citing nongovernmental organizations as information sources, as that might jeopardize their neutrality and invite retaliation by adversaries. The BCT primarily uses civil-military operations centers operated by civil affairs units to coordinate with nongovernmental organizations. (See ATP 3-07.5 for additional information.) BCTs operating with nongovernmental organizations follow these guidelines:

- Military personnel wear uniforms when conducting relief activities.
- Military personnel make prior arrangements before visiting nongovernmental organizations.
- Military personnel do not refer to nongovernmental organizations as force multipliers or partners or other similar terms.
- U.S. forces respect a nongovernmental organization's decision not to serve as an implementing partner.

SEIZING THE INITIATIVE

8-39. The enduring theme of seizing the initiative is as applicable in operations focused on stability as in the offense and defense. What is significantly different is the context in which the operational framework occurs. Operations focused on stability have fewer specified applications of tactics and procedures. The BCT commander must study and use critical thinking and creativity to address the tactical problems in this complex environment. Operations focused on stability have broader temporal considerations; this operational framework occurs before, during and after conflict as well as simultaneous to offensive and defensive operations. The BCT commander must consider the effects of this temporal aspect and manage it appropriately.

8-40. Operations focused on stability are by nature conceptual. The BCT commander must understand how each action affects the other elements in the complex system of host-nation governmental institutions, civil society and local economies. Success in seizing the initiative from a stability-focused perspective is critical to preventing conflict, setting conditions for success during the offense and defense and securing hard won successes in a post conflict environment.

RETAINING THE INITIATIVE

8-41. Retaining initiative gained through the offense and operations focused on stability requires the BCT to anticipate and act on civil requirements while actively averting threat actions. Influencing the population towards the legitimacy of the civil authority is critical to retaining initiative. In many circumstances, security is the most influential element affecting the population beyond their basic needs of food, water, and shelter. A secure environment fosters a functioning economy, which provides employment and gradually transfers the population's dependence from military to civilian authorities and host-nation governments.

8-42. The BCT executes defensive operations and operations focused on stability to retain key terrain, guard populations, and protect critical capabilities that inhibit threat actions while fostering conditions to increase the impact of stabilizing efforts. Often the BCT assigns these tasks as an economy of force to conduct offensive operations and operations focused on stability operations tasks. These tasks take form in such as actions as partnership with other indigenous security forces but must be executed to prevent instability. The objective is that the population feels that the level of security promotes evolving and often sequential growth and stabilization. Host nation actors are often the best and most informed sources on the local environment.

EXPLOITING THE INITIATIVE

8-43. A secure operational environment enables unified action partners to capitalize on their unique capabilities—thus exploiting the initiative gained in earlier operations. Governmental, nongovernmental, and other actors must be unified in purpose for this to be effective. Effective civilian-military teaming starts with the development of shared goals, aims, and objectives and a unity of purpose, which leads to a relationship of shared trust and a unity of effort. The BCT commander's responsibilities include creating and fostering this dynamic and culture among organizations.

8-44. Unity of effort is more than working with other U.S. Governmental agencies. Political leaders, governmental agencies, security forces, and local businesses are examples of host-nation actors that a BCT works with during operations focused on stability. The BCT also leverages relationships with host-nation actors to develop their understanding of the information environment and to answer information requirements.

8-45. Actors that encompass unified action partners are not limited to host nation and interagency personnel. The BCT's area of operations may include allied and multi-national forces. Understanding capabilities, constraints and limitations, and command relationships amongst the allied and multi-national forces facilitate the mission preparation and execution of operations in a constantly changing operational environment. The BCT cannot conduct successful operations focused on stability without building relationships. Each actor brings expertise, perspective, resources and capabilities that are necessary to ensure stability operations tasks are accomplished and that they achieve the end state for the operation.

STABILIZING THE ENVIRONMENT

8-46. Operations focused on stability seek to stabilize the environment enough so that the host nation can begin to resolve the root causes of conflict and state failure. During consolidation of gains, these operations will focus on security and stability operations tasks to establish conditions that support the transition to legitimate authorities. Initially, this is accomplished by performing the minimum essential stability operations tasks of providing security, food, water, shelter, and medical treatment. Once conditions allow, these tasks are a legal responsibility of U.S. Army forces. However, the BCT commander may not need to have the BCT conduct all of these essential tasks. Other military units or appropriate civilian organizations may be available to adequately perform these tasks. As the operational environment and time allow, the effort will transition to the more deliberate of execution of the six stability operations tasks.

SECTION III – STABILITY OPERATIONS

8-47. A *stability operation* is an operation conducted outside the United States in coordination with other instruments of national power to establish or maintain secure environment and provide essential governmental services, emergency infrastructure reconstruction, and humanitarian relief (ADP 3-0). The BCT executes operations focused on stability operations tasks against destabilizing factors by establishing unity of effort among diverse organizations, and then task organizing and partnering with other elements to mitigate sources of instability. (See ADP 3-0, ADP 3-07, FM 3-07, ADRP 1-03, and ATP 3-07.5 for additional information.)

SIX STABILITY OPERATIONS TASKS

8-48. U.S. Army forces often seek to stabilize an area of operations by performing stability operations tasks. A single action taken by a BCT or partnered element can support multiple stability operations tasks because they are interrelated and interdependent. Each stability operations task carries unique considerations, but actions taken affect each differently. The BCT plans, prepares, executes, and assesses operations to determine impacts on the area of operations as positively, negatively or neutral considering long-term and short-term effects. In operations focused on stability, planning and assessing require significant analysis supported through information collection activities (see chapter 5) focused on identifiable indicators within and external to the BCT's area of operations. As the commander considers each stability operations task within the context of the stability principles, the BCT staff analyzes measures of performance (MOPs) and measures of effectiveness (MOEs) during assessment to plan for the next operation asking, "What is needed to accomplish the intended outcome?" The Army's six stability operations tasks are—

- Establish civil security.
- Support to civil control.
- Restore essential services.
- Support to governance.
- Support to economic and infrastructure development.
- Conduct security cooperation.

8-49. The combination of stability operations tasks conducted during operations depends on the situation. In some operations, the host nation can meet most or all of the population's requirements. In those cases, Army forces work with and through host-nation authorities. Commanders use civil affairs operations to mitigate how the military presence affects the population and vice versa. Conversely, Army forces operating in a failed state may need to support the well-being of the local population. That situation requires Army forces to work with civilian organizations to restore basic capabilities. Civil affairs operations prove essential in establishing the trust between Army forces and civilian organizations required for effective, working relationships.

8-50. Six Army stability operations tasks (see figure 8-1 on page 8-12) correspond directly to the five stability sectors, used by the U.S. Department of State, Office of the Coordinator for Reconstruction and Stabilization, and directly support the broader efforts within the stability sectors. Together these six stability operations tasks and the U.S. Department of State stability sectors provide a mechanism for interagency tactical integration, linking the execution of discreet tasks among the instruments of national power required to establish end state conditions that define success. Tasks performed in one sector inevitably create related effects in another sector; planned and performed appropriately, carefully sequenced activities complement and reinforce these effects. The subordinate tasks performed by the BCT under the six stability operations tasks directly support broader efforts within stability executed as part of unified action.

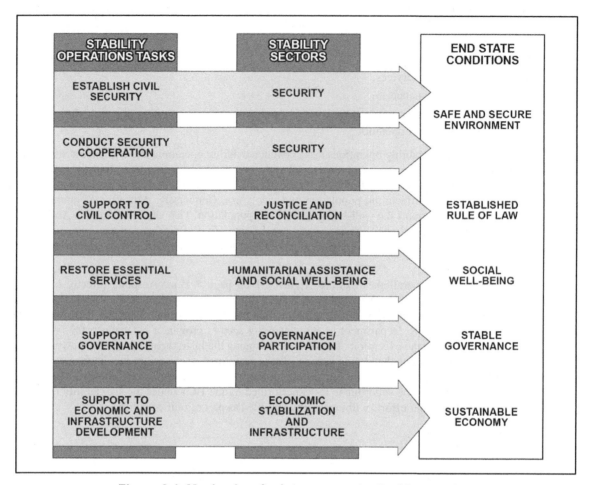

Figure 8-1. Mechanism for interagency tactical integration

8-51. Over time, to ensure safety and security are sustained, unified action partners perform numerous tasks across all stability sectors. As part of a joint team working with unified action partners, achieving a specific objective or setting certain conditions often requires the BCT to perform a number of related tasks among the six stability operations tasks. An example of this is the effort required to provide a safe, secure environment for the local populace. Rather than the outcome of a single task focused solely on the local populace, safety and security are broad effects. The BCT can help achieve safety and security by performing a number of related tasks to assist in ending hostilities, isolating belligerents and criminal elements, demobilizing armed groups, eliminating explosives and other hazards, and providing public order and safety.

ESTABLISH CIVIL SECURITY

8-52. Civil security provides for the safety of the host nation and its population, including protection from internal and external threats. The BCT coordinates operations to restore order, halt violence and to support, reinstate, or create civil authority by establishing a safe, secure, and stable environment for the local populace supporting the overall stability operation. (See ATP 3-07.5 for additional information.) Establishing civil security subtasks include the following:

- Enforce cessation of hostilities, peace agreements, and other arrangements.
- Determine disposition and composition of host nation armed and intelligence services.
- Conduct disarmament, demobilization, and reintegration.
- Conduct border control, boundary security, and freedom of movement.
- Support identification programs.

- Protect key personnel and facilities.
- Clear explosive and other hazards.

8-53. The BCT conducts operations that directly support subtasks: enforce cessation of hostilities, peace agreements, and other arrangements; conduct border control, boundary security, and freedom of movement; support identification; protect key personnel, and facilities. The BCT has limited capability to support, determine disposition and composition of national armed and intelligence services; conduct disarmament, demobilization, and reintegration; and to clear explosives and other hazards.

INITIAL RESPONSE

8-54. During the initial response phase, the BCT often executes subordinate tasks because the host nation lacks the capability. BCT subordinate units occupy areas of operation in accordance with geographical, political, socioeconomic, task, or supported actor boundary considerations. Information collection activities will develop further understanding of boundaries requiring the BCT to shift unit assets and resources to better align with unified action partners to mitigate sources of instability. BCTs may be required to identify and segregate combatants and noncombatants, search them, safeguard them, and move them out of the immediate area of operations. The BCT commander establishes priorities for protection of civil and military personnel, facilities, installations, and key terrain within the area of operations and initiates the stability principles of conflict transformation, unity of effort, and building partner capacity during the initial response.

8-55. The BCT conducts operations that safeguard the local population and prevent factions or actors contributing to sources of instability. Ultimately, these operations convince rival factions and actors to secure their interests through negotiation and peaceful political processes rather than violence, intimidation, coercion, or corruption. BCT units must remain neutral during this period. Supporting one of more factions or leveraging one faction against another may contribute to instability. Perception from the local population must be that the U.S. forces are neutral and have the best interests of the population and are providing security to the area allowing further development to occur.

8-56. Identifying actors and their intentions during this phase through information collection allows the BCT to seize the initiative. The BCT commander and subordinate leaders must reach, through engagements, binding agreements or understandings with unified action partners to determine the best way to divide labor and deconflict efforts so that partners are not working at cross-purposes. Actor agendas or intents are not known or understood in their entirety during this phase. Psychological operations (PSYOP) staff planners and intelligence staff sections develop indicators that commanders and leaders can clearly understand and identify to reveal actor agendas or intents. Information collection, with and without unified action partners, along with continuous assessment enables the commander's understanding of unity of effort and unity of purpose.

TRANSFORMATION

8-57. In the transformation phase, host-nation security forces and, potentially, intergovernmental organization peacekeepers begin to contribute. The BCT focuses more on SFA, particularly on the systems required to professionalize the host-nation security forces. The BCT continues in partnership with unified action partners according to the legitimate government binding agreements. Information collection will develop further understanding of boundaries requiring the BCT to shift unit assets and resources to better align with unified action partners to mitigate sources of instability. The BCT advises and assists the security force leadership empowering them to assume as much of the security effort as possible.

8-58. Host-nation security forces prominently work on security efforts so local populations do not perceive the BCT as an invading force, to resist. Host-nation security forces stay involved to promote the legitimacy of their government and progress with unified action partners. The BCT will assist host-nation security force information operations, sustainment in support of the host nation, protection and area security operations as needed.

8-59. As soon as possible, the BCT transfers host nation infrastructure security to host-nation organizations. Host-nation military units may temporarily be committed to securing public infrastructure, but eventually police forces or dedicated security organizations conduct this function. The BCT continues coordination between unified action partners to help mediate any disagreements among them.

8-60. The BCT ensures host-nation forces act in accordance with respect to human rights; failure to do so undermines popular support for the host-nation government and can quickly revert conditions back to those experienced during the initial response phase. Partnered security operations that place the host-nation forces in the lead or independent of the BCT that result in combating instability are the most credible to the population and build legitimacy of government through ownership.

8-61. Initially the BCT may simply be a support apparatus or, contrastingly, may make most security related decisions and perform most tasks. Nevertheless, host-nation actors support and increasingly take ownership in such matters. The BCT develops clear MOPs and MOEs leading to the security efforts shifting from the BCT as the lead to the host-nation forces in the lead. The BCT develops indicators of the host-nation unit's actions and conduct as well as the local population sentiment that their units can identify while conducting partnered actions.

FOSTERING SUSTAINABILITY

8-62. In the fostering sustainability phase, the BCT transitions to a steady state posture focused on advisory duties and security cooperation. The BCT commander implements additional peace measures depending upon further negotiations. During this phase, the BCT enables the host nation to sustain the peace.

SUPPORT TO CIVIL CONTROL

8-63. Civil control centers on rule of law by promoting efforts to rebuild host nation judiciary and corrections systems by providing training and support to law enforcement and judicial personnel. Civil control tasks focus on building temporary or interim capabilities to pave the way for the host nation or international organizations to implement permanent capabilities. (See ATP 3-07.5 for additional information.) Support to civil control subtasks include—

- Establish public order and safety.
- Establish an interim criminal justice system.
- Support law enforcement and police reform.
- Support judicial reform.
- Support a civil property dispute resolution process.
- Support criminal justice system reform.
- Support corrections reform.
- Support war crimes courts and tribunals.
- Support public outreach and community rebuilding programs.

8-64. Establishing security and rebuilding justice institutions can help to develop the necessary climate for reconciliation, public confidence, and subsequent economic growth. The BCT supports civil control tasks directly by conducting operations that support subtasks establishing public order and safety and supporting public outreach and community rebuilding efforts. The BCT supports the remaining civil control subtasks indirectly.

INITIAL RESPONSE

8-65. During initial response, the BCT conducts area security to protect the population, facilitate access to critical resources for endangered populations, and secure vital resources and infrastructure for the interim and future criminal justice institutions to strengthen the legitimacy of the operation and maintain the trust of the host nation. The BCT initiates the principles of conflict transformation and unity of effort to begin establishment of civil control.

8-66. The BCT conducts operations that safeguard the local population and prevent factions or actors from contributing to sources of instability, in this case actors or groups enacting their form of justice. The legal and justice system will be in disarray during the initial response from the interim and host-nation government. The BCT respects and implements laws established by the host nation in support of political objectives and directly addresses sources of instability. The BCT staff analyzes these laws to recommend COAs and develop rules of engagement or other mission parameters.

8-67. The BCT conducts information operations related to the development of judicial systems that are outside of the interim or established government and disrupt the organizations that control them. Ultimately, these operations convince rival factions and actors to secure their interests through negotiation and peaceful political processes rather than violence, intimidation, coercion, or corruption.

8-68. The BCT develops plans for coordinating the security, safety, and care for displaced communities in camps and settlements. This includes the movement of displaced people, the screening of returnees at checkpoints, the protection of relief convoys, and public safety in returnee communities that lack local law and order.

8-69. The BCT requests partnerships with U.S. Army and other unified action partners to assist in the initial establishment of civil control. For example, the BCT request military police forces with the technical skills to conduct investigations, collect and handle evidence, and undertake correction reform (see ATP 3-39.12). The BCT seeks additional judge advocates since the BCT lacks the legal manpower to assist the host nation in judicial reform, as well as support to war crime courts and tribunals.

TRANSFORMATION

8-70. During transformation, the BCT continues to conduct operations with unified action partners, however, shifts efforts to legitimacy and host-nation ownership, and building partner capacity. In the transformation phase of the stabilization framework, host-nation police forces and inter organizational entities take the lead with Army units focusing on SFA, particularly the professionalization of host-nation security forces.

8-71. Legitimate political authorities pass laws and orders that are binding to the local population during this phase. The BCT continues its operations and partnership but must know the laws and orders to properly mentor and guide its partner to support legitimacy. These laws and orders may also cause an adjustment to rules of engagement and the conduct of operations.

8-72. Building partner capacity to protect military and public infrastructure and facilitate emergency response is the primary goal during transformation phase for civil control. Security measures should be integrated into broader programs that foster good order and discipline, including personnel accountability, property accountability, and maintenance. The BCT continues its engagements with unified action partners to establish timelines and measurable standards as capacity develops for conditions improving or regressing.

FOSTERING SUSTAINABILITY

8-73. In fostering sustainability, the BCT transfers all public security responsibilities to host-nation forces while monitoring and reporting on progress as well as identifying modernization needs and the means to achieve them. Through engagements, the BCT ensures political authorities do not abuse their institutions and maintain civil control.

RESTORE ESSENTIAL SERVICES

8-74. Restoring essential services addresses the fundamental needs of the populace, beyond the provision of security. The BCT normally supports subtasks of restore essential services, for example, conducting tasks related to civilian dislocation and support to food relief and public health programs within its area of operations. (See ATP 3-07.5 for additional information.) Restore essential services subtasks include—

- Provide essential civil services.
- Perform tasks related to civilian dislocation.
- Support famine prevention and emergency food relief programs.
- Support nonfood relief programs.
- Support humanitarian demining.
- Support human rights initiatives.
- Support public health programs.
- Support education programs.

8-75. The BCT works to transfer responsibility to a transitional intergovernmental, nongovernmental organization, or host-nation authority as quickly as possible. However, maintains responsibility for security in the area of operations so that the transitional authority can best meet the needs of the population. The BCT prioritizes restoration of essential services based solely on need and mitigate unnecessary suffering. The BCT commander and staff consider location, security, and quantity of distribution of humanitarian aid, as well as who and how the aid is distributed. The BCT staff must determine the perception of the local population, mitigate corruption of a partnered element, and ensure these actions mitigate instability for the short and long term.

8-76. The BCT continuously coordinates with unified action partners assigned by the joint task force headquarters and may be partnered with the civil-military operations center, host-nation ministry of health and agriculture, and relief organizations such as U.S. Agency for International Development and the United Nations World Food Program to ensure the population has access to food and water. Partnership with special operations forces may also augment the BCT by providing additional geographic and cultural knowledge and sharing intelligence to address potential sources of instability.

INITIAL RESPONSE

8-77. The BCT's primarily focus on essential services is to alleviate unnecessary suffering among the population. This includes providing basic humanitarian needs such as food, water, and shelter, along with providing support for displaced civilians and preventing the spread of epidemic disease. Efforts to restore essential services in any operation contribute to the social well-being of the population. The BCT supports government efforts to establish or restore basic civil services, including food, water, shelter, and medical support with such actions as:

- Conducting area and zone reconnaissance to identify areas that require immediate assistance.
- Conducting area security to ensure those affected receive the goods, services, and protection.
- Sustaining operations to deliver class I; water, food, and emergency shelter (tents).
- Engineering operations to assess and repair infrastructure within capabilities.
- Providing medical treatment to civilians that are at immediate risk of losing life, limb, or eyesight.
- Identifying requirements and request enablers, augmentation or integration of theater assets for specified requirements (for example, repair or rebuild infrastructure, task organization, request for forces, project request, or operational needs statement processes).

8-78. A BCT may provide for immediate humanitarian needs of the population within its organic capabilities to restore essential services following the conduct of offensive or defensive operations or in response to disaster as part of a humanitarian relief effort. The BCT applies the principles of conflict transformation and unity of effort and sets the conditions to build partner capacity when restoring essential services in the initial response phase.

8-79. Beyond security considerations, the BCT relies on its information collection and sustainment capability to apply conflict transformation. This includes operations that identify the sources of instability and support the restoration of essential civil service as defined in terms of immediate humanitarian needs (such as providing food, water, shelter, and medical support) necessary to sustain the population until local civil services are restored. Contracted services are often a viable option to fulfill essential civil services, therefore, trained contracting officer representatives (CORs) and pay agents are critical enablers at the battalion and company level that can ensure that contracted support meets its intended purpose in the manner as agreed to in contracts or other binding agreements. (See chapter 9.)

8-80. Operations focused on stability require unique sustainment considerations. The brigade support battalion (BSB) must always plan to support the BCT, but also may be required to plan sustainment beyond the BCT when supporting stability operations tasks. The BCT logistics staff officer (S-4), during the development of the BCT's concept of support, must remember that the design of the BSB is only to support the BCT's assigned Soldiers and equipment. Prior to the execution of the sustainment plan, the BSB commander must ensure that sustainment estimates differentiate what is supporting internal BCT requirements and what is supporting external support mission requirements. Sustainment during operations focused on stability often involves supporting U.S. forces, multinational forces, and other contributing partners in a wide range of missions and tasks.

8-81. The BCT remains responsible for achieving unity of effort in all subtasks, but other agencies, including host nation, U.S. Department of State, or nongovernment organizations will most often assume responsibility for execution. Credibility of the legitimate authority or illegitimate actors is not the primary consideration during the initial response phase of restoring essential services. Credibility will become vital later in the initial response phase, as actors will seize the initiative to influence their political position when an interim authority is established and as the host nation begins the transformation phase. The BCT monitors and keeps a record of actions taken by actors to influence the population perception of legitimacy. This allows the BCT to credit and discredit legitimate and illegitimate actors through factual information.

8-82. BCTs also focus on maintaining civil security to allow other agency and host nation entities to meet these needs, thus ensuring the BCT builds capacity to transition this responsibility quickly. The activities associated with this stability operations task extend beyond simply restoring local civil services and addressing the effects of humanitarian crises. While military forces generally center efforts on immediate needs of the populace, other civilian agencies and organizations focus on broader humanitarian issues and social well-being.

8-83. The presence of dislocated civilians can threaten success in any operation. A number of factors may displace civilians, which will prompt the BCT to assist meeting the immediate needs of the displaced civilians until other better-equipped organizations establish control in the BCT area of operations. Dislocated civilians may indicate symptoms of broader issues such as conflict, insecurity, and disparities among the population. The BCT must address and assess the impact of displaced civilians immediately to establish conditions for stabilization and reconstruction of a traumatized population. Local and international aid organizations are most often best equipped to handle the needs of the local populace but require a secure environment in which to operate. Through close cooperation, military forces can enable the success of these organizations by providing critical assistance to the populace.

8-84. Understanding the location, disposition, and composition of displaced civilians must remain a high priority and consideration throughout operations focused on stability. A large number of all dislocated civilians are women and children. Most suffer from some form of posttraumatic stress disorder, and all require food, shelter, and medical care. External groups may target the displaced civilians as a continuation of earlier conflict, thus escalating the conflict and destabilizing the environment. Internal unrest within the displaced civilian population can create de-stabilizing actors as well. Thus, the BCT coordinates with unified action partners and augments with available assets from a higher headquarters, such as medical support, security, establishing secure facilities and possibly evacuation to maintain initiative over potential destabilizing factors.

8-85. The BCT plans for health threats and diseases prevalent in the region to provide support for deployed forces and affected civilians within the BCT's area of operations. Roles 1 and 2 medical assets support force health protection (see ATP 4-02.8) measures to mitigate the adverse effects of disease and nonbattle injuries and promote the health of deployed forces. The BCT conducts area and zone reconnaissance to gain information on public health hazards by collecting information on sewage, water, electricity, academics, trash, medical, safety, and other considerations. If necessary, the BCT assesses existing medical infrastructure including preventative health services and may temporarily operate or augment operations of existing medical facilities in extreme circumstances. (See FM 4-02 and ATP 4-02.3 for additional information.)

TRANSFORMATION

8-86. Operations conducted during the transformation phase establish the foundation for long-term development, resolving the root causes of conflict that lead to events such as famine, dislocated civilians, and human trafficking. The BCT primarily conducts operations that secure the environment to enable other agencies and host nation to meet the needs of the populace. The BCT commander and staff assess related activities and missions to achieve unity of effort within the operational environment and continually identify potential sources of instability. The commander and staff apply the principles of unity of effort, legitimacy and host nation ownership, and building partner capacity during transformation that enables a combined effort toward sustained social well-being for the population and achieving progress towards fostering sustainability.

8-87. As other organizations, nongovernmental organizations, U.S. Agency for International Development, and other interagency partners and the host nation assume responsibilities for restoring essential services, the BCT continues its partnership with those organizations. The BCT supports the efforts by ensuring that security exists, identifying needs that have been overlooked, facilitating the activities of these other partners, and continuing to provide critically needed humanitarian assistance.

8-88. To achieve legitimacy in the eyes of the population host-nation authorities must demonstrate the ability to restore essential services. The BCT includes host nation legitimate authorities in assessments and in establishing priorities. The BCT transitions its relationship from direct support and coordination to coaching, teaching, mentoring, and facilitating unified action partners so they can continue to make progress toward essential services.

8-89. The BCT scales activities to local capacity for sustainment. Proper scaling also creates opportunity for the local populace to generate small-scale enterprise to provide for services as much as possible. The BCT must not initiate large-scale projects until the necessary infrastructure is in place to support such efforts.

FOSTERING SUSTAINABILITY

8-90. Fostering sustainability tasks ensures the permanence of those efforts by institutionalizing positive change in society. Conditions for sustained social well-being depend on the ability of the legitimate authority to meet basic needs of the population, ensure right of return, address instances of civilian harm, promote transitional justice, and support peaceful coexistence.

SUPPORT TO GOVERNANCE

8-91. *Governance* is the state's ability to serve the citizens through the rules, processes, and behavior by which interests are articulated, resources are managed, and power is exercised in a society (JP 3-24). Support to governance subtasks include—

- Support transitional administrations.
- Support development of local governance.
- Support anticorruption initiatives.
- Support elections.

8-92. The BCT's support to governance varies over the range of military operations and area of operations. Support to transitional authorities and development of local governance can be supported by the BCT as a primary or subordinate role with the partnering entities. The BCT's support to anticorruption and elections does not change its role throughout the area of operations; instead, it varies depending on how it may be conducted when considering the mission variables of mission, enemy, terrain and weather, troops and support available, civil considerations (METT-TC). The establishment of civil security and civil control provide a foundation for transitioning authority to civilian agencies and eventually to the host nation. (See ATP 3-07.5 for additional information.)

INITIAL RESPONSE

8-93. The initial response phase sets the conditions on how governance can be supported. Information collection during this phase must be continuous and leveraged through Soldier and leader engagements to identify sources of instability, understanding of unified action partners motivations and agendas, and identification of local civic leaders. The BCT commander and subordinate commanders and leaders apply adroit diplomatic and communications skills to build constructive relationships during this phase. The BCT focuses on providing a secure environment allowing relief forces to focus on the immediate local population humanitarian needs.

8-94. The BCT sets the conditions for civic leaders and factions to address grievances and sources of instability peacefully and openly. The BCT can set conditions regarding:

- Identification of leaders of factions, legitimate authority, popular and minority support, or aligned with coalition forces political interests.
- Provide a safe and secure environment for these leaders to meet and communicate.

- Influence leaders to participate in political system and not through violence.
- Influence local population to participate in elections.
- Remain actively neutral but keeping the local populations interests in mind, allowing leaders to communicate, but not tolerating intimidation, violence, corruption, or sources of instability to escalate.

8-95. A thorough information collection plan provides the commander and subordinates units with an understanding of the area of operations allowing them to engage unified action partners from a position of advantage. The commander must reach clear communication, agreements, understandings, or accords to conduct operations for laying the foundation of governance during the initial response phase. Conducting operations after Soldier and leader engagements confirm or deny unified action partners motivations and agendas. Assessing the outcomes of unified action partners' operations and actions reveal their relationship to the BCT and provide direction on how to support governance further.

8-96. The initial response phase normally ends with the election of officials and the establishment of a safe and secure environment capable of mitigating sources of instability for a longer period time. Elections may require an increased presence by BCT and host-nation security forces, and additional information-related capabilities to encourage participation and responsible behavior. BCT and host-nation forces may conduct security and area security for polling sites, political rallies, media centers, international observers, and candidates.

TRANSFORMATION

8-97. The transformation phase begins after establishment of some form of a government becomes operational. This may include oversight by the transitional military authority but must include a relatively safe environment in which a tolerable level of instability can be mitigated. Transformation ends with the host nation capable of conducting good governance with minimal assistance, acceptance by the local population, and prepared for long-term development.

8-98. The BCT supports governance during transformation by advising, assisting, supporting, and monitoring other actors. During this phase, BCT commanders and subordinate commanders and leaders continue to build constructive relationships even through changes in unit or organizational leaderships in efforts to progress the stability principles of unity of effort, legitimate and host nation ownership, and building partner capacity.

8-99. The BCT commander, subordinate leaders, and unified action partners advise civil servants during administrative actions. They continue to conduct engagements within their area of operations and assess actors' agendas or intentions through operations ensuring unity of effort. BCT subordinate units interact with multiple host-nation actors helping them coordinate efforts more effectively.

8-100. Legitimate and host nation ownership must retain support of the local population. The BCT conducts operations to build the local population's perception that the local government and unified action partners are capable, willing, and progressing towards mitigating sources of instability with little assistance if not independently. The local population must be convinced that the BCT and host nation can sustain this achievement for the duration and not become unstable or corrupt.

8-101. The BCT commander builds partner capacity by leveraging unified action partners providing for government needs and demonstrating the host-nation's government legitimacy and capability. The BCT commander and staff identify gaps in capability to address sources of instability gained through information collection and engagements with the local government. The BCT shares this information between the local population and government leveraging unified action partners resources to build capacity mitigating instability where capability did not previously exist. See an example of building partner capacity on page 8-20.

A patrol identifies two villages that are in contention with one another because of a lack of water. One has an abundance of water due to building a levee and the other does not. The patrol shares this information to the company commander. The company commander engages the local government leader about the issue and the local leader does not have the capability to solve the problem. The company commander engages the battalion commander and staff who arranges an engagement with the provincial reconstruction team and unified action partners. An engineer from the provincial reconstruction team coordinates with U.S. Army Corps of Engineers and local regional development planners discover that they can provide water to the other village by digging wells that access underground water sources. The provincial reconstruction team provides a proposal for a contract to the company commander and local government leaders. The local government leader reaches a decision to commission a contract to build wells. While the wells are being built, the levee must be lowered to allow some water to flow downstream for certain periods of time. In return, both villages provide labor to the project equally. The company commander coordinates with the provincial reconstruction team, sends patrols with a qualified COR and ensures fulfillment of the contract terms. The BCT commander and staff, and subordinate units conduct information operations to show a partnership with the host nation.

8-102.   The BCT supports good governance by ensuring that local governments adhere to the rule of law and to the law established by the higher level of government. The incoming legitimate authority, with the population's support, establishes (or re-establishes) and supports the rule of law during the transformation phase. The BCT commander and subordinate leaders must be familiar with the current state of the rule of law to ensure their partnered elements support the rule of law and act accordingly. Operations conducted during this phase support the messages and themes that support the rule of law as well as reports of violations of the law by actors and how the government acts consistently with the rule of law in addressing violations of the law.

FOSTERING SUSTAINABILITY

8-103.   The fostering sustainability phase begins with the host nation capable of conducting good governance with minimal assistance, accepted by the local population, and prepared for long-term development. This phase ends with the complete withdrawal of BCTs interacting with local government on a routine basis. The BCT support to governance during fostering sustainability oversees transfer responsibility of governance to an enduring host nation authority.

SUPPORT TO ECONOMIC AND INFRASTRUCTURE DEVELOPMENT

8-104.   The BCT assists host-nation actors to begin the process of achieving sustainable economic development by establishing a safe and secure environment. Other U.S. Government agencies, intergovernmental organizations, and civilian relief agencies often have the best qualifications to lead efforts to restore and help develop host-nation economic capabilities. Ultimately, the goal is to establish conditions so that the host nation can generate its own revenues and not rely upon outside aid. The desired end state is for the host nation to achieve a robust, entrepreneurial, and sustainable economy. All economic development actions build upon and enhance host nation economic and management capacity. (See ATP 3-07.5 for additional information.) Support to economic and infrastructure development subtasks include—

- Support economic generation and enterprise creation.
- Support monetary institutions and programs.
- Support national treasury operations.
- Support public sector investment programs.
- Support private sector development.
- Protect natural resources and environment.
- Support agricultural development programs.
- Restore transportation infrastructure.

- Restore telecommunications infrastructure.
- Support general infrastructure reconstruction programs.

8-105. The BCT's support to economic and infrastructure development varies over the range of military operations and area of operations but the principles of conflict transformation, unity of effort, legitimacy and host nation ownership, and building partner capacity are enduring throughout all phases.

INITIAL RESPONSE

8-106. The BCT supports economic and infrastructure development at the local level during the initial response phase by establishing areas of operation, and task organizing and empowering its subordinate units by allowing them to coordinate with unified action partners. The BCT synchronizes efforts regionally and shifts assets and resources as main efforts change or as opportunity to exploit the initiative arise. The building blocks for broad national recovery and development are set at the local-level and the BCT's information collection effort should focus on identifying microeconomic information such as changes in cost of a commodity, number of unemployed males 15 to 45 years of age, changes in costs of services such as medical treatment, and so forth. Although the BCT maintains responsibility for security, secondary efforts will include facilitating the emergence of employment opportunities, infusing monetary resources into the local economy, stimulating market activity, fostering recovery through microeconomics, and supporting the restoration of physical infrastructure to help retain and exploit the initiative.

8-107. The BCT may have to take the lead in responding to immediate economic needs, including assessing the critical micro- and macro-economic conditions, during the initial response phase. These economic needs include ensuring host-nation civilians can bring agricultural products and other goods to safe and secure marketplaces, generating jobs that can be filled with qualified laborers, and others. Unity of effort is essential for the BCT to identify and engage all relevant actors from the host nation, U.S. civil agencies, and international organizations. These evolving partnerships and assessments will significantly enhance the economic development management transition tasks from the BCT to the U.S. Government civil agencies and host-nation actors. Hostile individuals and groups can take advantage of gaps if the actors do not engage, and exploit opportunities for profit, contributing to long-term instability.

8-108. Infrastructure reconnaissance is a multidiscipline variant of reconnaissance to collect detailed technical information on various categories of the public systems, services, and facilities of a country or region. The infrastructure reconnaissance develops the situational understanding of the local capability to support the infrastructure requirements of the local populace or military operations within a specific area. Infrastructure reconnaissance is accomplished in stages: the infrastructure assessment and the infrastructure survey. (See ATP 3-34.81 for additional information.)

8-109. Coordinating with the combat engineer units for an on-site visit, an engineer reconnaissance team can be expected to conduct the initial assessment with available expertise from the supported unit. The initial assessment provides information to confirm or deny planning assumptions, update running estimates/staff estimates, determine immediate needs, develop priorities, obtain resources, and refine a plan. As operations continue, general engineer and other supporting technical support elements provide teams that are qualified to perform an infrastructure survey. These infrastructure survey teams use the infrastructure assessments from the engineer reconnaissance teams to prioritize categories and identify those parts of the infrastructure to be reassessed in more detail. Technical capabilities required to perform a comprehensive reconnaissance include robust support from joint Service, multiagency, contractor, host nation, multinational, and reachback elements.

TRANSFORMATION

8-110. The goal of the transformation phase is to establish firmly the foundation for sustainable economic development and to begin to transition control of economic development to U.S. Government civil agencies, international civil agencies, and host nation economic officials and entrepreneurs. The collective emphasis is on establishing host nation institutions providing sustainable economic growth during this phase. Once a civilian administration assumes control, the primary economic development role of the BCT is to advise and assist local leaders.

FOSTERING SUSTAINABILITY

8-111. In fostering sustainability, the goal is to institutionalize a long-term sustainable economic development program and to transition control of the economy completely to host-nation officials, entrepreneurs, and civil society. This phase also includes steps that build on and reinforce the successes of the initial response and transformation phases. Steps taken during this phase support sustainable economic growth based on a healthy society supported by healthy communities and neighborhoods. The primary economic development role for the BCT is to continue to advise and assist host-nation civilian economic officials.

CONDUCT SECURITY COOPERATION

8-112. *Security cooperation* is all Department of Defense interactions with foreign security establishments to build security relationships that promote specific United States security interests, develop allied and partner nation military and security capabilities for self-defense and multinational operations, and provide United States forces with peacetime and contingency access to allied and partner nations (JP 3-20). Security cooperation provides the means to build partner capacity and achieve strategic objectives. These objectives include—

- Building defensive and security relationships that promotes specific U.S. security interests.
- Developing capabilities for self-defense and multinational operations.
- Providing U.S. forces with peacetime and contingency access to host nations to increase situational understanding of an operational environment.

8-113. Army forces support the objectives of the combatant commander's campaign plan in accordance with appropriate policy, legal frameworks, and authorities. The plan supports those objectives through security cooperation, specifically those involving SFA (see FM 3-22) and foreign internal defense (see ATP 3-05.2). *Security force assistance* is the Department of Defense activities that support the development of the capacity and capability of foreign security forces and their supporting institutions (JP 3-20). *Foreign internal defense* is participation by civilian and military agencies of a government in any of the action programs taken by another government or other designated organization to free and protect its society from subversion, lawlessness, insurgency, terrorism, and other threats to its security (JP 3-22).

8-114. SFA and foreign internal defense professionalize and develop security partner capacity to enable synchronized sustaining operations. Army security cooperation interactions enable other interorganizational efforts to build partner capacity. Army forces—including special operations forces—advise, assist, train, and equip partner units to develop unit and individual proficiency in security operations. The institutional Army advises and trains partner Army activities to build institutional capacity for professional education, force generation, and force sustainment. (See FM 3-22 for additional information on support to security cooperation.)

Notes. SFABs provide SFA to host-nation FSF. The SFAB provides organic forces to form the basis for the SFAB mission to support FSF. Within the SFAB, the company team is the foundation for the SFAB's mission and augmented with additional personnel and assets to accomplish the mission. (See ATP 3-96.1 for information on the SFAB.)

On occasion, the BCT as a whole or selected unit(s) of the BCT may support SFA activities, including potentially supporting multiple FSF organizations in or external to the BCT's area of operations. Additionally, these FSF organizations may each report through different host-nation government channels and even to different ministries. To synchronize efforts in this case, U.S. forces must achieve unity of effort. Similarly, each of the FSF organizational commanders should synchronize their efforts with the host-nation government representatives, as appropriate. (See ATP 3-21.20 for an example task organization, used for discussion purposes, for an Infantry battalion supporting multiple SFA activities.)

8-115.   The stability operations task of establishing security cooperation may include the BCT or selected subordinate unit(s), depending on the missions assigned, conducting SFA as a subset of security cooperation. SFA offers a means of support for security cooperation activities in support of building capacity of an FSF. As soon as the FSF can perform this task, the BCT or selected unit transitions this task within civil security to the host nation.

INITIAL RESPONSE

8-116.   As the initial response phase normally occurs during or immediately after a conflict where the operational environment prevents civilian personnel from operating effectively. The operational environment is typified as nonpermissive. The objective of this phase is to improve the security situation, reducing the threat to the populace and creating the conditions that allow civilian personnel to safely operate. SFA in the initial response phase is normally required when FSF lack the capability or capacity to provide the required level of security. This phase often requires SFA efforts to help generate and train or assist new and existing FSF. This phase may require a combination of the types of SFA and considerable support, sustainment, and medical resources. BCT activities during the initial response may have to be conducted with multinational combat operations to consolidate gains, to include providing a safe, secure environment for the local populace. SFA efforts during this phase focus on improving the FSF capability and capacity so all security forces— U.S., other, and FSF—provide a secure environment and reduce the threat. As security conditions improve, transition to the transformation phase begins.

TRANSFORMATION

8-117.   In the transformation phase, SFA activities seek to assist FSF to stabilize the operational environment in a crisis or vulnerable state. The operational environment in this phase is more permissive than the initial response phase; however, military forces will often be required to provide security to some actors. Activities in this phase normally include a broad range of post-conflict reconstruction, stabilization, and capacity-building efforts, which the embedded provincial reconstruction team is essential for long-term success. Objectives in this phase include continuing efforts to improve the security situation, reducing the threat to the populace, building host-nation capacity across the stability sectors, and facilitating the comprehensive approach to assist FSF.

8-118.   The transformation phase represents a broad range of SFA activities to support FSF. The initial response phase differs from the transformation phase in the FSF capability to provide for a safe and secure environment. More specifically, FSF may have a level of proficiency to no longer need a permanent United States and FSF relationship for tactical operations. However, they may still need full-time advisors and support, sustainment, and medical assistance. Embedded provincial reconstruction team members will continue to play a vital role in assisting governance and development efforts throughout this phase. SFA end state for this phase seeks to establish conditions so the host nation's security sector can provide a secure environment with its own security forces.

FOSTERING SUSTAINABILITY

8-119.   In this phase, the focus of SFA continues to shift toward assisting institutions required to sustain FSF. This phase encompasses long-term efforts to assist FSF. FSF conduct independent operations and can provide a safe, secure internal environment. While SFA activities may be initially required during this phase, activities reduce as FSF become more capable and viable. The determination for the BCT to receive a change of mission from SFA is based on the policy and conditions of the operational environment. Provincial reconstruction teams and other forces may remain to support a theater security cooperation plan.

SECTION IV – AREA SECURITY OPERATIONS

8-120.   The BCT engaged in area security operations is typically organized in a manner that emphasizes its mobility, lethality, and communications capability. Population-centric area security operations, for example to consolidate gains, are common across the range of military operations, but is almost a fixture during the conduct of stability-focused operations. Population-centric operational area security operations typically combine aspects of the area defense and offensive operations (for example, search and attack, cordon and search, raid, and ambush) to eliminate the efficacy of internal defense threats. During the conduct of area security operations, the BCT commander and staff must understand the relationship with host-nation authorities and the civilian population. A clear understanding of the commander's authority is essential in exercising that degree of control necessary to ensure security and safety to all military forces and the civilian population located within the BCT's area of operations.

OPERATIONAL OVERVIEW

8-121. Area security operations protect friendly forces, installations, routes, and actions within an area of operations. During the conduct of stability-focused operations, area security operations establish and maintain the conditions for stability in an unstable area before or during hostilities, or enduring peace and stability after open hostilities cease. Area security operations are often an effective method of providing civil security and civil control and supporting security cooperation during operations focused on stability. For example, an area security operation may have to be designed around numerous political constraints. This may include aligning unit areas of operation with the host nation's existing political boundaries. Security objectives, regardless of which element of decisive action (offense, defense, or stability) currently dominates, ensure freedom of action over a prolonged period in consonance with the BCT commander's concepts of operations and intent. (See ATP 3-91 for additional information.)

CIVIL CONSIDERATIONS

8-122. Civil considerations reflect the influence of manmade infrastructure, civilian institutions, and attitudes and activities of the civilian leaders, populations, and organizations within the operational environment on the conduct of military operations. Commanders and staffs analyze civil considerations within the characteristics of areas, structures, capabilities, organizations, people, and events (ASCOPE). (See ATP 2-01.3 for additional information on these characteristics.)

8-123. Since civilians are normally present in operations with a dominant stability component, the BCT normally restrains its use of force when conducting area security operations. However, the commander remains responsible for protecting the force and considers this responsibility when considering rules of engagement. Restrictions on conducting operations and using force must be clearly explained and understood by everyone. Subordinate commanders and leaders, and Soldiers must understand that their actions, no matter how minor, may have far-reaching positive or negative effects. Subordinate commanders and leaders, and Soldiers must realize that media (either hostile or neutral) and adversaries can quickly exploit their actions, especially the way they treat the civilian population.

AREA SECURITY

8-124. Area security, a security operation conducted to protect friendly forces, installations, routes, and actions within a specific area, takes advantage of the local security measures performed by all units, regardless of their location in the area of operations. Local security includes any local measure taken by units against enemy actions. Local security, dependent upon the situation, may involve avoiding enemy detection or deceiving the enemy about friendly positions and intentions. Local security may include finding any enemy forces in the immediate vicinity and knowing as much about their positions and intentions as possible. Local security prevents a unit from being surprised and is an important part of maintaining the initiative during area security.

8-125. The requirement for maintaining local security is an inherent part of any area security mission. Units use both passive and active measures to provide local security. Passive local security measures include using camouflage, movement control, noise and light discipline, operations security, and proper communications procedures. Measures also include employing available sensors, night-vision devices, and daylight sights to maintain surveillance over the area immediately around the unit. Active measures, dependent upon the situation, may include—

- Using observation posts, combat outposts, combat patrols, and reconnaissance and surveillance patrols.
- Establishing specific levels of alert based on the mission variables of METT-TC.

- Establishing stand-to times. (Unit standard operating procedures [SOPs] detail activities during the conduct of stand-to.)

LOCAL SECURITY

8-126. *Local security* is the low-level security activities conducted near a unit to prevent surprise by the enemy (ADP 3-90). Area security activities take advantage of the local security measures performed by all units (regardless of their location) in an area of operations, and all local security activities should be linked to the broader area security activities. Local security is closely associated with unit protection efforts (see ADP 3-37). Local security includes local measures that prevent or interdict enemy efforts. Local security is an enduring priority of work, is essential to maintaining initiative, and prevents units from being surprised. Local security involves avoiding detection and deceiving the enemy about friendly actions, positions, and intentions. Local security includes finding any enemy forces in the immediate vicinity and knowing as much about their positions and intentions as possible.

8-127. Local security can be part of the sustaining base or part of the area infrastructure. Local security protection ranges from echelon headquarters to reserve and sustainment forces using active and passive measures to provide local security. Active patrolling, unit SOPs, and continuous reconnaissance are active measures that help provide local security. Passive measures include using camouflage, movement control, noise and light discipline, proper communications procedures, ground sensors, night vision devices, and daylight sights.

ECONOMY-OF-FORCE MISSIONS

8-128. The BCT, charged with execution, conducts an area security operation as an economy-of-force mission. Area security missions are numerous, complex, and generally never ending. For this reason, the commander and staff synchronize and integrate security efforts, focusing on protected forces, installations, routes, and actions within the BCT's assigned area of operations. Protected forces within the BCT range from subordinate units and elements, echeloned command posts (CPs), and sustainment elements within the BCT's support area or consolidation area (when established). Protected installations can be part of the sustainment base, or they can constitute part of the area's civilian infrastructure within a consolidation area. Protected ground lines of communication include the route network to support the numbers, sizes, and weights of tactical and support area movement within the BCT's area of operations, for example a consolidation area. Actions range from securing key points (bridges and defiles) and terrain features (ridgelines and hills) to large civilian population centers and their adjacent areas.

OFFENSIVE AND DEFENSIVE ACTIVITIES

8-129. During the conduct of stability-focused tasks, area security missions are a mixture of offensive and defensive activities involving not only subordinate battalions, companies, and platoons, but also those host-nation security forces over which the BCT has a command relationship such as operational control (OPCON), or can otherwise influence. Offensive area security activities include subordinate tasks of movement to contact [search and attack (see ATP 3-21.20 and ATP 3-21.10) or cordon and search (see ATP 3-21.20 and ATP 3-21.10) missions] and combat patrols (see ATP 3-21.8), when required, designed to ambush detected enemy forces or to conduct raids within the BCT's area of operations. Defensive area security activities include the establishment of base perimeter security (see ATP 3-21.20, appendix I); combat outposts and observation posts (see ATP 3-21.20); moving and stationary screen and guard missions, and reconnaissance and counterreconnaissance missions (see ATP 3-20.96 and ATP 3-20.97).

8-130. During offensive or defensive-focused tasks, area security operations are often designed to ensure the continued conduct of sustainment operations to support decisive and shaping operations by generating and maintaining combat power. Area security operations may be the predominant method of protecting support areas that are necessary to facilitate the positioning, employment, and protection of resources required to sustain, enable, and control forces. (See chapter 9 for additional information.)

READINESS

8-131.   During area security operations, forces must retain readiness over longer periods without contact with the enemy. This occurs most often when the enemy commander knows that enemy forces or insurgents are seriously overmatched in available combat power. In this situation, the enemy commander normally tries to avoid engaging friendly forces unless it is on terms favorable to the enemy. Favorable terms include the use of mines and booby traps. Area security forces must not develop a false sense of security, even if the enemy appears to have ceased operations in the secured area. The commander must assume that the enemy is observing friendly operations and is seeking routines, weak points, and lax security for the opportunity to strike with minimum risk. This requires commanders at each echelon to influence subordinate small-unit leaders to maintain the vigilance and discipline of their Soldiers to preclude this opportunity from developing.

PLANNING CONSIDERATIONS

8-132.   During area security operations planning, the commander apportions combat power and dedicates assets to protection tasks and systems based on an analysis of the operational environment, the likelihood of threat action, and the relative value of friendly resources and populations. Based on an initial assessment of the operational environment, the commander task organizes subordinate units and elements and assigns security areas within the BCT's area of operations. Although all resources have value, the mission variables of METT-TC make some resources, assets, or locations more significant to successful mission accomplishment from enemy or adversary and friendly perspectives. Throughout the operations process the commander relies on the risk management (RM) process and other specific assessment methods to facilitate decision-making, issue guidance, and allocate resources (see chapter 4). Criticality, vulnerability, and recoverability are some of the most significant considerations in determining protection priorities that become the subject of the commander's guidance and the focus of area security operations.

COMMAND AND CONTROL

8-133.   During area security operations, the BCT commander devotes considerable time and energy to the problems of coordination and cooperation due to the joint, interagency, and multinational nature of stability-focused tasks. The BCT plans and conducts area security operations in concert with partner participants towards a unified effort, often as a supporting organization rather than the lead organization. The commander uses liaisons to enable unity of effort between partner elements and the coordination centers established by the division or higher commander.

Interagency and Multinational Organizations

8-134.   One factor that distinguishes the conduct of stability-focused tasks from the conduct of offensive- and defensive-focused tasks is the requirement for interagency coordination at battalion level and below. During area security operations with interagency partners, the commander has inherent responsibilities. These responsibilities include the requirement to clarify the mission; to determine the controlling legal and policy authorities; and to task, organize, direct, sustain, and care for the organizations and individuals for whom the BCT provides the interagency effort. The commander also ensures seamless termination of the mission under conditions that ensure the identified objectives are met and can be sustained after the operation.

8-135.   When operating inside or with multinational organizations, the BCT commander and subordinate commanders and leaders should expect to integrate foreign units down to the company level. SFA activities within an area security mission require carefully selected and properly trained and experienced personnel (as trainers or advisors) who are not only subject matter experts, but also have the sociocultural understanding, language skills, and seasoned maturity to more effectively relate to and train FSF. Additionally, commanders and subordinate leaders within the brigade support area (BSA) and battalion trains with the fact that they will routinely interact with multinational partners during other area security missions. SOPs at subordinate echelons will require modification to incorporate multinational small units that do not have compatible communications and information systems.

Desired End State

8-136. The BCT commander's definition of the desired end state is a required input to area security operations. While end state is normally described as a stable, safe, and secure environment during stability-focused operations, this description is not sufficient. Initial MOEs and MOPs quantifying that environment are determined during the planning process. (See chapter 3 for additional information.) Measures of effectiveness and performance are important in stability-focused area security operations since traditional combat measures, such as territory gained, enemy personnel killed or captured, and enemy combat vehicles destroyed or captured do not apply. The commander also ensures the desired end state reflects the prolonged time-period associated with many stability-focused area security operations.

8-137. Achieving the desired end state requires a knowledge of operational design (see chapter 4), the ability to achieve unity of effort, and a thorough depth of cultural awareness (see chapter 2) relating to the BCT's area of operations. Through economy of forces, the commander identifies a finite amount of available combat power to apply against the essential tasks associated with a given area security operation. Identifying essential tasks lays the foundation for the success of area security operations that represent the future stability of a state. Decisions about use of combat power are more than a factor of the size of the force deployed, its relative composition, and the anticipated nature and duration of the mission. Assuring the long-term stability depends on applying unity of effort to the tasks that are, in fact, essential.

Information Operations

8-138. The final success or failure of the BCT's area security operation rests with the perceptions of the inhabitants within and external to the BCT's area of operations and goes beyond defeating the enemy. Securing the trust and confidence of the civilian population is the chief aim of information operations, which integrates and synchronizes information-related capabilities to generate effects in the information environment necessary to influence enemy, adversary, neutral, and friendly audiences.

8-139. Information operations synchronization of information-related capabilities promotes the legitimacy of the mission and reduces bias, ignorance, and confusion by persuading, educating, coordinating, or influencing targeted audiences. Further, it promotes—through Soldier and leader engagement, civil affairs operations, and PSYOP, among other information-related capabilities—interaction at all echelons with these audiences so these target audiences understand the objectives and motives of BCT and that of higher headquarters, and the scope and duration of area security actions. Combined with broad efforts to build partner capacity, for example, SFA (see section V). Information operations are essential to achieving decisive results: a stable host-nation government and peaceful civilian population.

8-140. The BCT information operations officer or noncommissioned officer coordinates with the division (or higher) information operations officer to synchronize information-related capabilities into the BCT's information operations planning. Synchronization requires the BCT information operations officer (in coordination with the electromagnetic warfare officer [EWO]) to participate in targeting within the fire support cell as well as the various working groups and meetings chaired by the current and future operations and other integrating cells within the BCT. Participation allows for the development of a holistic understanding of the information environment within the problem sets facing the BCT staff. A staff-wide understanding helps synchronize the information-operations related planning and targeting and allows for shifts in priorities. This synchronization, in coordination with the information operations and civil-military operations managed at the BCT or division enables united action partners to be incorporated into planning.

Note. Within the BCT, the information operations officer is responsible for synchronizing and deconflicting information-related capabilities employed in support of BCT operations. The BCT information operations officer synchronizes capabilities within subordinate maneuver battalions and squadron, and subordinate companies and troops that communicate information to audiences and affect information content and flow of enemy or adversary decision-making while protecting friendly information flow. The information operations officer prepares appendix 15 and a portion of appendixes 12, 13, and 14 to Annex C (Operations) to the operation order. (See chapter 3 and FM 3-13 for additional information.)

8-141. Within the security environment, enemies, adversaries, and other organizations use propaganda and disinformation against the commander's efforts to influence various civilian populations within and external (area of interest) to the BCT's area of operations. The BCT's public affairs staff officer, in coordination with the division public affairs officer works closely with the intelligence staff officer to be proactive, rather than reactive, to such attacks. A coordinated information operations plan informs and counters the effects of propaganda and misinformation. The plan (generally developed at BCT level in coordination with the division information operations officer and public affairs officer) establishes mechanisms, such as a media center or editorial board, to educate and inform local and international media, which in turn, informs the public, with accurate and timely information. Additionally, civil affairs operations and PSYOP are integrated into counterpropaganda efforts at the BCT level through the division, or higher headquarters' information operations working group.

8-142. When needed, the BCT chaplain can play an important role in bridging gaps with religious leaders that set conditions for future successful key leader engagements and civil affairs operations. During planning, the chaplain advises the commander concerning matters of religion, culture, and religious key leaders in the area of operations and area of interest. The chaplain and religious affairs noncommissioned officer provide important, up-to-date perspectives concerning local, provincial, and national atmospherics not often included or clear in other sources. Their efforts should always be coordinated with the BCT information operations officer and BCT information operations working group (see FM 3-13), when established.

8-143. Without a detailed Soldier and leader engagement plan, different units and staff elements meet with and engage local leadership with different desired end states thereby undermining the ability of any or all forces to build capacity and work towards transition to host nation lead. Coordination between staff elements or units within the BCT, when working with the same host nation individual or office, enables unity of effort and the desired end state for the BCT's area security operation. The creation of a detailed engagement plan includes identifying differences between provinces or localities within the province and sets out the objectives to reach the desired end state. Host nation leaders in a city, district or province have face-to-face meetings with these leaders to advance the creation and building of host nation capacities.

8-144. Soldier and leader engagements have a significant impact on the human component during all operations that occur among the people. Human beings capture information and form perceptions based on inputs received through all the senses. Humans see actions and hear words. Humans compare gestures and expressions with the spoken word. Humans weigh the messages presented to them by the Soldiers and leaders of the BCT, and other sources with the conditions that surround them. When the local and national news media are unavailable or unreliable, people turn to alternative sources, such as the internet—where information flows freely at unimaginable speeds—or rumor and gossip. Perception equals truth to people lacking objective sources of information. Altering perceptions requires shaping information through engagements according to how people absorb and interpret information, molding the message for broad appeal and acceptance.

8-145. Operations security is as important during the conduct of stability-focused operations as it is during the conduct of offensive- and defensive-focused operations. Operations security contributes to the BCT's ability to achieve surprise during area security missions, thus enabling its chances for success. Within the BCT area of operations, human adversaries/enemies monitor the BCT's normal activities to detect variations in activity patterns that forecast future operations. They monitor the conversations of Soldiers both on duty and off duty to gain information and intelligence. Adversaries/enemies monitor commercial internet activity and phone calls from BCT operational and recreation facilities. They will look at trash created by BCT activities. The absence of operations security about BCT activities contributes to excessive friendly casualties and possible mission failure in area security operations just like it does in combat operations. The BCT's information superiority hinges in no small part on effective operations security; therefore, measures to protect essential element of friendly information (EEFI) cannot be an afterthought. (See FM 3-13 and ATP 3-13.1 for additional information.)

Note. The need to maintain transparency of the BCT's intentions during area security operations is a factor when balancing operations security with information release. Release authority for information—to include foreign disclosure rules—must be fully understood by commanders and staffs within the BCT. The public affairs and information operations officers (see FM 3-61 and FM 3-13, respectively) lead the coordination and synchronization processes within the BCT. Release authority for information rests with the commander at the appropriate level.

8-146. Multinational staffs result in additional security problems. Each nation has different access to U.S. information systems. Maintaining operations security with multinational staff members is difficult and sometimes the security rules restrict the ability of multinational partner staff officers to contribute. The chief of staff and foreign disclosure officer at division level develop workarounds when required. One such workaround is to provide the multinational staff officer a U.S. assistant to get on a U.S. secured information system to ensure the multinational staff officer has the information needed to contribute. The division assistant chief of staff, signal establishes and maintains two separate sets of different information systems when this occurs. (See ATP 3-91 and ATP 6-02.75 for additional information.)

MOVEMENT AND MANEUVER

8-147. During the conduct of stability-focused operations, the BCT plans its area security movement and maneuver simultaneously with offensive and defensive movement and maneuver, though with an extensive emphasis on security and engagement skills (negotiation, rapport building, cultural awareness, and critical language phrases). Movement and maneuver within the BCT area of operations is normally decentralized to the battalion and company. Through economy of force, the BCT commander determines the right mix of forces to quickly transition between operations as the situation requires. During area security operations, the commander plans for future movement within the BCT's area of operations and as required, in adjacent areas of operation. The BCT's lethal capabilities make the execution of area security operations possible even if the probability of combat is remote. When new requirements develop the BCT commander plans for the shifting of priorities when the need arises.

Fire and Movement

8-148. The application of fire and movement lends itself to several offensive and defensive operations (for example, search and attack, cordon and search, and area defense) within the civil security and civil control stability operations tasks. Across the range of military operations, the BCT and its subordinate units play a major role in ensuring the outcome of these stability operations tasks. The BCT and its subordinate units are useful in the conduct of other stability operations tasks (for example, security cooperation) because of their deterrence value and the flexibility and labor the BCT provides to the division or higher-level commander.

Mobility and Countermobility

8-149. Mobility (see chapter 6) and countermobility (see chapter 7) operations are key enablers to area security operations. In stability focused area security operations, mobility operations allow civilian traffic and commerce to continue or resume. Resuming normal civilian activities in the BCT's area of operations is an important objective within stability focused area security operations. Countermobility operations indirectly support stability focused area security operations in regard to offensive and defensive operations.

8-150. Mobility operations focus on keeping ground lines of communications open for both civilian and military activities and on reducing the threat of mines and other unexploded ordnance to the same. During area security operations, the commander and staff develop the countermobility plan concurrently with the fire support plan and defensive scheme of maneuver, guided by the commander's intent. When combat engineer support falls under the mobility and countermobility tasks, it can include—

- Constructing combat roads and trails.
- Breaching existing obstacles (including minefields).
- Marking minefields, including minefield fence maintenance.
- Clearing mines and debris from roads.

- Conducting route reconnaissance to support the main supply routes and civilian lines of communications.
- Creating obstacles between opposing factions to prevent easy movement between their positions.

8-151.   The BCT employs roadblocks not only to restrict traffic for security purposes, but also to control the movement of critical cargo. Cargo could be generators designed to restore electric power in a large area or items that support the population and resources within the BCT's area of operations.

Occupy an Area

8-152.   Planning for the occupation of an area or relief in place begins before the BCT deploys or when being relieved, redeploys. Planning includes not only BCT forces and their activities, but also other governmental agencies, multinational partners, host-nation agencies, and potential international organizations. The mission variables of METT-TC determine the occupation or relief in place that occurs. Sometimes occupation, much like occupying an initial area of operations, is appropriate. This can occur when the BCT's stability-focused area security operation occurs within limited intervention or peace operations. A relief in place may be appropriate during the conduct of an area defense (see chapter 7). However, a stability-focused area security transition by function may be more effective if the relief in place takes place with host-nation military forces and civil authorities within the range of military operations. Some of these functions include medical and engineer services, local security, communications, and sustainment. BCT plans do not remove a provided capability from the area of operations until the replacement capability is operating.

Surveillance Systems and Reconnaissance and Security Forces

8-153.   In restrictive (as well as unrestrictive) terrain, the commander relies on manned and unmanned surveillance systems and reconnaissance and security forces to collect information within the BCT's area of operations. Operation of ground and aerial surveillance systems in restricted terrain is often affected by interrupted line-of-sight, and extreme climate and weather variations. In restrictive terrain, reconnaissance and security forces within the BCT's area of operations focus on these areas to assist in collection when manned and unmanned surveillance capabilities are degraded. Using a combat outpost, a reinforced observation post capable of conducting limited combat operations, is a technique for employing reconnaissance and security forces in restrictive terrain that precludes mounted reconnaissance and security forces from covering the assigned area. While the mission variables of METT-TC determine the size, location, and number of combat outposts a unit establishes, a reinforced platoon typically occupies a combat outpost. (See chapter 7 and ATP 3-21.10 for additional information.) A combat outpost must have sufficient resources to accomplish its designated missions, such as conducting aggressive combat patrolling and reconnaissance patrolling. Combat outposts are established when observation posts (see chapter 5 for information on observation post activities) are threatened by insurgency or in danger of being attacked by enemy forces infiltrating into and through the BCT's assigned area of operations.

Note. During the conduct of defensive-focused operations, the commander uses a combat outpost to extend the depth of the security area, to keep friendly forward observation posts in place until they can observe the enemy's main body, or to secure friendly forward observation posts that will be encircled by enemy forces. Mounted and dismounted forces can employ combat outposts. (See chapter 7 for additional information.)

Army Aviation

8-154.   Army aviation attack and reconnaissance units with manned and unmanned systems—when deployed early with initial response forces—can be a significant deterrent on the indigenous combatants, particularly if factions or insurgences are not yet organized during the initial response phase. Attack and reconnaissance helicopters may be employed to act as a response force against enemy threats. Along with unmanned aircraft systems (UASs), attack and reconnaissance helicopters may conduct reconnaissance, surveillance, or security over wide areas and provide the BCT a means for visual route reconnaissance and early warning. Utility helicopters provide an excellent command and control capability to support stability focused area security operations and to transport patrols or security elements throughout the BCT's area of operations. Cargo helicopters provide the capable to move large numbers of military and civilian security force personnel and to conduct resupply when surface transportation is unavailable, or routes become impassable.

> *Note.* BCT plans include measures for the effective use of all resources, to include, exploiting airpower for transportation and resupply over extended distances and, where appropriate, tightly controlled close air support.

Reserve and Response Force Operations

8-155. Maintaining a reserve during any operation is difficult. Often, the commander finds that the BCT has more tasks than units to do, and stability focused area security operations are no exception. Nonetheless, contingencies or missions may arise that require establishing a reserve. Maintaining a reserve allows the establishing commander to plan for worst-case scenarios and to exploit opportunities, provide flexibility, and conserve the force during long-term operations.

8-156. The response force differs from a reserve in that it is not in support of a particular engagement. A response force is a dedicated force on a base with adequate tactical mobility and fire support designated to defeat Level I and Level II threats and shape Level III threats until a tactical combat force can defeat them or other available response forces. The response force answers to the establishing headquarters. (See ATP 3-91.) Considerations when establishing a response force include—

- Threats.
- Communication equipment and procedures.
- Alert procedures.
- Transportation.
- Training priorities.

8-157. To counter an indirect fire threat, the commander employs counterfire radars throughout and area of operations to locate hostile indirect fire systems. The use of quick reactionary forces, attack helicopter, or local friendly forces are ideal for response to counterfire radar acquisitions as clearance of fire procedures are often time-consuming and not necessarily reliable when determining locations for host-nation forces. Additionally, indiscriminate use of indirect fire on counterfire radar acquisitions can lead to unwanted collateral damage.

INTELLIGENCE

8-158. The conduct of stability focused tasks demands greater attention to civilian considerations—the political, social, economic, and cultural factors in an assigned area of operations—than does the conduct of conventional offensive and defensive focused tasks. Using the mission variable of civil considerations and its subordinate characteristics identified by the mnemonic ASCOPE, the BCT staff has a standardized baseline for analysis to generate understanding. This baseline is augmented by analyses conducted by organic and attached forces such as social-cultural analysis, target audience analyses, intelligence analyses, population, and area studies. During area security operations, the commander expands the IPB process beyond geographical and force capability considerations. (See ATP 2-01.3 for additional information on IPB for stability missions.)

Information Collection

8-159. Information collection, specifically plan requirements and assess collection, enables relevant, predictive, and tailored intelligence within an area of operations. (See ATP 2-01 for additional information on the specific functions for stability missions.) Intelligence cells and knowledge management elements within the BCT and division (or higher) headquarters and battalion and squadron headquarters develop procedures to share collected intelligence data and products, both vertically and horizontally, throughout the force. (See ATP 2-19.4 for additional information on intelligence techniques for stability missions.)

Understanding

8-160.   Area security operations require the integration of the division and BCT's information collection effort to develop a clear understanding of all potential threats and the populace. Success in the stability environment requires a cultural understanding to gauge the reaction of the civilian population within and external to the BCT's area of operations to a particular COA conducted, to avoid misunderstandings, and to improve the effectiveness of the execution of that COA by the BCT or division. Changes in the behavior of the populace may suggest needed change in tactics, techniques, or procedures or even strategy. Biographic information, leadership analysis, and methods of operation within the existing cultural matrix are keys to understanding the attitudes and ability of positional and reference civilian leaders to favorably or unfavorably influence the outcome of BCT area security operations.

Indicators of Change

8-161.   During area security operations, the commander and staff tie priority intelligence requirements to identifiable indicators of change within the operational environment, to include, civil inhabitants and their cultures, politics, crime, religion, economics, and related factors and any variances within affected groups of people. The commander often focuses on named areas of interest in an effort to answer critical information requirements to aid in tactical decision-making and to confirm or deny threat intentions regardless of which element of decisive action currently dominates. During area security operations, priority intelligence requirements related to identifying enemy and adversary activities are tracked, where appropriate.

Commander's Critical Information Requirements

8-162.   Due to the increased reliance on human intelligence (HUMINT), when conducting area security operations, the commander emphasizes the importance of commander's critical information requirements (CCIRs) to all personnel within the BCT. CCIRs are information requirements identified by the commander as being critical to facilitating timely decision-making, and answers to CCIRs can come from staff at all levels. All personnel must be given appropriate guidance to improve information-gathering capabilities throughout the BCT. Interpreters, speaking to local civilian personnel, security operations, and patrolling (combat and reconnaissance) are primary sources for assessing the economic and health needs, military capability, and political intent of those receiving assistance who or are otherwise a party to the area security operation. (See ADP 5-0 and ATP 3-55.4 for additional information.)

8-163.   Planners at the division and BCT ensure that any HUMINT assets assigned from outside the BCT are employed effectively, which is typically accomplished by integrating HUMINT collectors at the lowest level possible. The gaining unit accounts for HUMINT asset security and establishes tasking priorities and command relationships for temporary and long-term commitments. (See FM 2-22.3 for additional information.)

> *Note*. Medical personnel must know the Geneva Convention restrictions against medical personnel collecting information of intelligence value except that observed incidentally while accomplishing their humanitarian duties.

Employment and Control of Human Intelligence Collection Teams

8-164.   *Human intelligence* is the collection by a trained human intelligence collector of foreign information from people and multimedia to identify elements, intentions, composition, strength, dispositions, tactics, equipment, and capabilities (ADP 2-0). Commanders consider security when planning for the employment of HUMINT collection teams. (See FM 2-0.) Generally, three security conditions exist: permissive, uncertain, and hostile.

Note. The success of the HUMINT collection effort depends on a complex interrelationship between command and control elements, requirements, technical control, technical support, and collection assets. Each echelon of command has its supporting HUMINT elements to conduct sustained HUMINT operations under all operational environments using only its organic HUMINT assets. HUMINT units have specific support requirements to each echelon's commander though HUMINT units must be flexible, versatile, and prepared to conduct HUMINT collection and analysis operations in support of any echelon of command. A coherent command and control structure within these HUMINT organizations is necessary in order to ensure successful, disciplined, and legal HUMINT operations. This structure is part of a coherent architecture that includes organic HUMINT assets and HUMINT resources from national (for example, the Defense Intelligence Agency), theater, and non-Department of Defense HUMINT organizations. The corps, joint, division, and brigade and below intelligence staff officer is the primary advisor on HUMINT and counterintelligence within this structure, and is the focal point for all HUMINT and counterintelligence activities within a joint task force, an Army component task force or a BCT. The intelligence staff officer can be organic to the unit staff or can be attached or under OPCON to the staff from another organization such as the theater military intelligence brigade. (See FM 2-22.3 for additional information.)

Permissive Environment

8-165. In a permissive environment, HUMINT collection teams normally travel throughout the area of operations without escorts or a security element. HUMINT collectors may frequently make direct contact with overt sources, view the activity, or visit the area that is the subject of the information collection effort. They normally use debriefing and elicitation as their primary collection techniques to obtain firsthand information from local civilians and officials.

Uncertain Environment

8-166. In an uncertain environment, security considerations increase, but risk to the collector is weighed against the potential intelligence gain. An uncertain environment limits use of controlled sources and requires additional resources. HUMINT collection teams should still be used throughout the area of operations but normally are integrated into other ground reconnaissance or other missions. For example, a HUMINT collector may accompany a patrol visiting a village. Security for the team and their sources is a prime consideration. HUMINT collection teams are careful not to establish a fixed pattern of activity or arrange contacts in a manner that could compromise the source or the collector. Debriefing and elicitation are still the primary collection techniques. Teams are frequently deployed to conduct collection at checkpoints, dislocated civilian collection points, and detainee collection points. They may conduct interrogations of detainees within the limits of applicable laws and policies.

Note. The word "detainee" includes any person captured, detained, or otherwise under the control of Department of Defense personnel. This does not include Department of Defense personnel or Department of Defense contractor personnel or other persons being held primarily for law enforcement purposes except where the United States is the occupying power. As a matter of policy, all detainees will be treated as an enemy prisoner of war until the appropriate legal status is determined and granted by competent authority in accordance with the criteria enumerated in the Geneva Convention Relative to the Treatment of Prisoners of War (GPW). Detainees include enemy prisoners of war, retained persons, civilian internees, and detained persons. Detaining officials must recognize that detainees to include those who have not satisfied the applicable criteria in the GPW are still entitled to humane treatment. The inhumane treatment of detainees is prohibited and is not justified by the stress of combat or deep provocation (see JP 3-63).

Hostile Environment

8-167. In a hostile environment, three concerns for HUMINT collection are access to the sources of information, timeliness of reporting, and security for the HUMINT collectors. A hostile environment requires significant resource commitments to conduct controlled source operations. Prior to the entry of a force into a hostile area, HUMINT collectors may be used to debrief civilians, particularly dislocated civilians, and to interrogate other detainees who have been in the area. HUMINT collection teams are normally located with the friendly units to facilitate timely collection and reporting. HUMINT collectors accompany the BCT lead elements or ground reconnaissance and security forces during operations. They interrogate detainees and debrief dislocated civilians and friendly force patrols.

Security Missions

8-168. Due to the possibility of tying forces to fixed installations or sites, security missions may become defensive in nature. When this occurs the BCT commander carefully balances with the need for offensive action. Early warning of enemy activity through information collection is paramount in the conduct of area security missions to provide the commander with time to react to any threat or other type change identified within the stability environment. The BCT's IPB identifies the factors effecting security missions within the assigned area of operations. Factors, although not inclusive, include—

- The natural defensive characteristics of the terrain.
- The existing roads and waterways for military lines of communication and civilian commerce.
- The control of land and water areas and avenues of approach surrounding the area security.
- The airspace management.
- The proximity to critical sites such as airfields, power generation plants, and civic buildings.

FIRES

8-169. The conduct of fires in support of stability-focused tasks is essentially the same as for offensive- and defensive-focused tasks. However, constraint is vital in the conduct of fires during stability-focused tasks. Such constraint typically concerns the munitions employed and the targets engaged to obtain desired effects. Constraint increases the legitimacy of the organization that uses it while potentially damaging the legitimacy of an opponent.

Employment of Fires

8-170. Employment of fires provides continuous deterrents to hostile action and are a destructive force multiplier for the commander, regardless of which element of decisive action currently dominates. Within stability-focused tasks, the planning and delivering of fires precludes fires on protected targets, unwanted collateral damage, and the political ramifications of perceived excessive fire. In addition to lethal effects, the targeting functions of the BCT fire support cell includes nonlethal effects input to the information collection plan and the targeting working groups at the division and BCT headquarters (see chapter 4 for targeting functions within the BCT fire support cell).

8-171. During the employment of fires, the commander having the ability to employ a weapon does not mean it should be employed. In addition to collateral damage considerations, the employment of fires could have second and third order negative effects. Collateral damage could adversely affect efforts to gain or maintain legitimacy and impede the attainment of both short- and long-term goals. For example, excessive force can antagonize those friendly and neutral parties involved. The use of nonlethal capabilities should be considered to fill the gap between verbal warnings and deadly force to avoid unnecessarily raising the level of conflict. Key considerations for employment of fires in support of stability-focused tasks include—

- Stability-focused tasks conducted in noncontiguous areas of operation complicate the use of fire support coordination measures, the ability to mass and shift fires, and clearance of fires procedures.
- Key terrain may be based more on political, cultural or social considerations than physical features of the landscape; fires may be used more frequently to defend key sites than to seize them.

- Rules of engagement are often more restrictive than in combat operations; commander's guidance for fires requires careful consideration during development and wide dissemination to all levels.
- Precision-guided munitions or employment of nonlethal capabilities may be necessary to limit collateral damage.
- Fires that may be used to demonstrate capabilities, as a demonstration (see chapter 6), or during a denial operation (see chapter 7).

Note. Mortars at the BCT and below, due to their smaller bursting radius, reduce collateral damage. Mortars are generally more responsive to the small-unit operations common to area security missions. In addition to lethal fires, mortars may provide illumination to demonstrate deterrent capability, observe contested areas, or support area security missions (including patrolling [reconnaissance and combat]).

Application of Lethal and Nonlethal Capabilities

8-172.   Though highly effective for their intended purpose, lethal capabilities may not always be suitable. For example, during stability-focused tasks, the application of lethal fires is normally greatly restricted, making the use of nonlethal capabilities the dominant feasible option. The considerations for use of nonlethal capabilities in targeting should not pertain to only specific phases or missions but should be integrated throughout the area of operations. Escalation of force measures can be established in order to identify hostile intent and deter potential threats at checkpoints, entry control points and in convoys. Such measures remain distinct from other use of force guidance such as fire support coordination measures and are intended to protect the force, minimize the use of force against civilians while not interfering with self-defense if attacked by adversaries. One of the primary mechanisms for employing nonlethal capabilities and generating nonlethal effects is information operations. Participating in the targeting process, information operations synchronizes a range of nonlethal capabilities to produce nonlethal effects that advance the desired end state. Thus, information operations participate in the targeting process.

Fire Support Coordination Measures

8-173.   As during offensive- and defensive-focused tasks, fire support coordination measures are established for stability-focused tasks to facilitate the attack of high-payoff targets (HPTs) throughout the area of operations. Restrictive fire support coordination measures are those that provide safeguards for friendly forces and noncombatants, facilities, or terrain. For example, no-fire area, restrictive fire areas, restricted target lists, restrictive fire lines, and fire support coordination lines may be used not only to protect forces, but also to protect populations, critical infrastructure, and sites of religious or cultural significance. Regardless of which element of decisive action currently dominates, coordination measures are required to coordinate ongoing activities to create desired effects and avoid undesired effects.

Note. Fire support coordination, planning, and clearance demands special arrangements with joint and multinational forces and local authorities. These arrangements include communications and language requirements, liaison personnel, and procedures focused on interoperability. The North Atlantic Treaty Organization standardization agreements provide excellent examples of coordinated fire support arrangements. These arrangements provide participants with common terminology and procedures.

SUSTAINMENT

8-174.   The BCT commander's responsibilities during area security include support areas and extend to self-protection of BCT assets operating outside of the BCT echelon support areas. Forces engaged in area security operations protect the force, installation, route, area, or asset. Area security operations are often designed to ensure the continued conduct of sustainment operations to support decisive and shaping operations by generating and maintaining combat power. Area security operations may be the predominant method of protecting echelon support areas that are necessary to facilitate the positioning, employment, and protection of resources required to sustain, enable, and control forces. (See chapter 4 for additional information.)

PROTECTION

8-175.   BCT activities associated with executing security operations (see FM 3-90-2), physical security (see ATP 3-39.32), antiterrorism (see ATP 3-37.2), and operations security (see FM 3-13) tasks enhance the security of the command within an area of operations. In large part, the measures within these four tasks are the same or complementary. Stability-focused operations closely resemble BCT activities for these tasks during the conduct of offensive- and defensive-focused operations though the BCT generally works closer with civilian inhabitants. (See ATP 3-91 for additional information.)

Establish and Maintain Security

8-176.   The BCT conducts security operations to ensure freedom of movement and action and to deny the enemy the ability to disrupt operations. Commanders combine offensive, defensive, and stability operations tasks, and information collection means to protect friendly forces, populations, infrastructure, and activities critical to mission accomplishment. The BCT integrates with partner military, law enforcement, and civil capabilities to establish and maintain security. The ability to establish control is critical to consolidating gains in the wake of successful military operations.

8-177.   Security operations prevent surprise, reduce uncertainty, and provide early warning of enemy operations. Warning of enemy operations provides forces with time and maneuver space with which to react and develop the situation. Security operations prevent enemies from discovering the friendly plan and protect the force from unforeseen enemy actions. Security elements focus on preventing the enemy from gathering EEFI. Security is a dynamic effort that anticipates and thwarts enemy collection efforts. When successful, security operations allow the BCT to maintain the initiative within the stability environment.

8-178.   Protection is a continuous activity; it integrates all protection capabilities to safeguard bases, secure routes, and protect forces. Effective physical security and antiterrorism measures, like any stability measure, overlap and are employ in-depth. For example, planners determine how military police support enhances unit physical security and antiterrorism capabilities by performing area security operations inside and outside an echelon support area or consolidation area. Military police also conduct response force operations to defeat Level II threats against bases or critical assets and delay Level III threats in an echelon support area until a tactical combat force can respond. (See ATP 3-39.11 for information on military police special reaction teams.)

8-179.   The BCT commander pays attention to physical security and antiterrorism operations throughout the stability environment. This is especially true when subordinate units conduct noncontiguous operations, as the success of the BCT mission may depend on protecting support areas from enemy attacks. The commanders must address the early detection and immediate destruction of enemy forces attempting to attack support or consolidation areas. Enemy attacks against sustainment and other facilities can range in size from individual saboteurs to enemy airborne or air assault insertions targeted against key military and civilian facilities and capabilities. These enemy activities, especially at smaller unit levels, may even precede the onset of large-scale combat and be almost indistinguishable from terrorist acts. The BCT implements operations security and other information protection measures to deny the enemy force information about friendly dispositions.

Note. Within the stability environment, deploying battalions and higher echelons should have a trained Level II antiterrorism officer assigned. An assigned antiterrorism officer works to ensure that security considerations are integrated in base designs and unit operations. These individuals guide their units in conducting threat assessment, criticality assessments, and vulnerability analysis to determine each unit's vulnerability to terrorism. (See ATP 3-91 and ATP 3-37.2 for additional information.)

Assessments to Support Protection Prioritization

8-180. Initial protection planning by the BCT commander and staff requires various assessments to establish protection priorities. Assessments include threats, hazards, vulnerability, and criticality. These assessments are used to determine which assets can be protected given no constraints and which assets can be protected with available resources. There are seldom sufficient resources to simultaneously provide all assets the same level of protection. For this reason, the commander makes decisions on acceptable risks and provides guidance to the staff so that they can employ protection capabilities based on protection priorities.

8-181. Protection planning is a continuous process that includes an understanding of the threats and hazards that may impact operations throughout the BCT's area of operations. Protection capabilities are aligned to protect critical assets and mitigate effects from threats and hazards. The protection cell prioritizes the protection of critical assets that best supports the commander's end state. Protection prioritization lists are organized through the proper alignment of critical assets. The commander's priorities and intent and the impacts on mission planning determine critical assets. Critical assets can be people, property, equipment, activities, operations, information, facilities, or materials. For example, important communications facilities and utilities, analyzed through criticality assessments, provide information to prioritize resources while reducing the potential application of resources on lower-priority assets. Stationary weapons systems might be identified as critical to the execution of BCT operations and, therefore, receive additional protection. The lack of a replacement may cause a critical asset to become a top priority for protection.

Protection Template

8-182. The protection template lists and integrates all protection tasks in an appropriate way for use by subordinate units, and any base and base cluster operations envisioned to be established during the BCT's or subordinate battalion or squadron's area security operation. The protection cell when established within the BCT operations staff officer (S-3) section augments the staff with a small protection planning cell that maintains and publishes the template in coordination with the division protection cell. The template is used as a reference before or during employment. Battalion/squadron and base/base cluster situational modifications to this template, and their regular rehearsal of all parts of protection plans are inspected periodically by the BCT protection working group. During inspections, the protection working group identifies weak areas in subordinate protection plans, ensures that area of operations protection best practices are incorporated into the plans of the BCT, and provides protection-related observations, insights, and lessons learned to subordinate units, and any unit relieving the BCT or subordinate unit within its area of operations.

Note. When a protection cell officer or noncommissioned officer is not designated within or attached to a battalion/squadron, protection cell functions and tasks are the responsibility of the battalion/squadron operations officer or noncommissioned officer. Key protection tasks conducted within the BCT's area security operation include area security, chemical, biological, radiological, and nuclear (CBRN) operations, coordinating air and missile defense, personnel recovery, explosive ordnance disposal, and detainee operations. (See ATP 3-91 for additional information on integrating and synchronizing protection tasks.)

Protective Services

8-183. The commander may determine that it is necessary (or be required) to provide protective services from within the BCT to protect high-value host nation civil and military authorities or other selected individual(s). This requirement usually occurs when host-nation security forces have been so extensively penetrated by hostile elements that they cannot be trusted to provide protective services or when host-nation security forces lack the technical skills and capabilities to provide the desired degree of protection. The element(s) tasked to perform protective services for designated personnel receives as much training and specialized equipment as is possible before the mission. (See ATP 3-39.35 for additional information.)

Allocation of Combat Power

8-184. Protection of installations or areas of operation (including route and convoy security) by the BCT require significant allocation of combat power when a threat beyond organized crime exists. Conducting resupply from one base to another is treated as a tactical action and tracked in the BCT main CP current operations cell. When the BCT establishes a response force(s), care is taken so that the response force does not establish patterns when responding to incidents. Establishment of patterns-same route, same movement formation, configuration and order of vehicles, and same response force responding from the same base-allows an enemy to ambush the response force at a point of its choosing.

> *Note.* Dependent on the situation, host-nation security forces are involved as much as possible in the performance of the above protection tasks. Host-nation support is important in the variety of services and facilities that can support security and protection assets within the BCT's area of operations. Services provided by the host nation can relieve the BCT of the need to provide equivalent capabilities thereby reducing the BCT's sustainment and protection footprint. Key criteria in the decision-making process to utilize host-nation support is the trust in host-nation to provide the support and that host-nation forces have the technical skills and capabilities to provide the desired degree of protection.

Threat Levels

8-185. Threats within the BCT's area security operation are categorized by the three levels of defense required to counter them. Any or all threat levels may exist simultaneously in the BCT's area of operations. Emphasis on base defense and security measures may depend on the anticipated threat level. Within the BCT's area of operations all elements protect themselves from Level I threats. This includes medical elements although they have reduced defensive capabilities since they can only use their nonmedical personnel to provide their own local security. Locating medical elements and other support elements on bases with other units mitigate this factor.

8-186. The BCT commander positions response forces to respond to a Level II threat (enemy force or activities that can be defeated when augmented by a response force) in appreciation of time-distance factors so that no element is left outside supporting distance from a response force. The commander integrates fire support assets into the composition of the response because of the speed at which these assets can react over the extensive distances involved in area security operations. Where possible, host-nation security assets constitute part of the response to smooth the interactions of these forces with the civilian population.

> *Note.* A Level III threat is an enemy force or activities beyond the defensive capability of any local reserve or response force. The response to a Level III threat is a tactical combat force, generally established no lower than division level due to the inability to resource at lower echelons. (See chapter 9 for additional information on threat levels.)

Survivability

8-187. Precautions should be taken to protect positions, headquarters, support facilities, and accommodations including the construction of obstacles, protective bunkers, fighting positions, and shelters. BCT subordinate units practice alert procedures and develop drills to occupy positions. Engineer forces enable, when available, survivability needs. Units maintain proper camouflage and concealment based on the mission variables of METT-TC. Area security forces are vulnerable to personnel security risks from local employees and other personnel subject to bribes, threats, or compromise. The threat from local criminal elements is a constant threat and protection consideration. The most proactive measure for survivability is individual awareness by Soldiers in all circumstances. Soldiers look for things out of place and patterns preceding aggression. Commanders and subordinate leaders ensure Soldiers remain alert, do not establish routines, and maintain appearance and bearing. (See chapter 7 and ATP 3-37.34 for additional information.)

Notes. In stability-focused operations, the enemy sniper poses a significant threat to dismounted (or mounted) movement and marches. Counter-sniper drills should include rehearsed responses, reconnaissance and security operations, and the incorporation of cover and concealment. The BCT's rules of engagement provide instructions on how to react to sniper fire, including restrictions on weapons used depending on the circumstances. For example, rules of engagement may allow units to use weapon systems, such as a sniper rifle team, to eliminate a positively identified sniper even in a crowded urban setting because of the reduce possibility for collateral damage. (See ATP 3-21.20, appendix E and ATP 3-21.18 for additional information.)

An enemy improvised explosive device (IED) attack is another major threat to dismounted (or mounted) movement and marches. Prior to the conduct of any area security mission, commanders and subordinate leaders' brief personnel on the latest IED threat types, usage, and previous emplacements within an area of operations or along mounted and dismounted movement or march routes. All Soldiers maintain situational awareness by looking for IEDs and IED hiding places. Units vary routes and times, enter overpasses on one side of the road and exit out the other, train weapons on overpasses as the movement passes under, and avoid chokepoints to reduce risk. Units should expect an IED attack at any time during movements and expect an ambush immediately after an IED detonation. Early mornings and periods of reduced visibility are especially dangerous since the enemy has better opportunities to emplace IEDs without detection. (See ATP 3-21.18 and ATP 3-21.8 for additional information.)

Air and Missile Defense

8-188. Offensive and defensive air defense planning considerations continue to apply when the BCT conducts stability-focused operations. However, the air threat trends toward Group 1 and 2 UASs (see ATP 3-04.64) employed by enemy forces opposing the BCT's effort to provide a stable, safe, and secure environment. Air and missile defense sensors and command and control elements external to the BCT provide early warning against aerial attack, and populate the BCT's COP. Soldiers train in aircraft recognition and on rules of engagement due to multiple factions using the same or similar aircraft, to include international and private organizations employing their own or charter civilian aircraft. (See ATP 3-01.8 for additional information.)

Note. See ATP 3-01.15 for information on the tactics, techniques, and procedures for an integrated air defense system. See ATP 3-01.50 for information on the operations of the air defense and airspace management cell established within the BCT fire support cell. (See chapter 4 for additional information.)

8-189. Counterrocket, artillery, and mortar batteries may be located in or near the BCT's area of operations to support its area security mission. Battery sensors detect incoming rockets, artillery, and mortar shells and may be used to detect Group 1 and 2 UASs. The battery's fire control system predicts the flight path of incoming rockets and shells, prioritizes targets, and activates the supported area of operations' warning system according to established rules of engagement. Exposed elements within the area of operations then can take cover and provide cueing data that allows the battery's weapon system to defeat the target before the target can impact the area. The commander clearly defines command and support relationships between counter-rocket, artillery, and mortar elements and the BCT during planning. (See ATP 3-01.60 for additional information.)

8-190. The BCT commander and subordinate commanders and leaders ensure all passive and active air defense measures (see chapter 6) are well planned and implemented. Passive measures include use of concealed routes and assembly areas, movement on secure routes, marches at night, increased intervals between elements of the columns, and dispersion. Active measures include use of organic and attached weapons according to the operation order and unit SOP. Air guard duties assigned to specific Soldiers during dismounted (or mounted) movements and marches give each a specific search area. For movements and marches, seeing the enemy first gives the unit time to react. Leaders understand that scanning for long periods decreases the Soldier's ability to identify enemy aircraft. During extended or long movements and marches, Soldiers are assigned air guard duties in shifts. (See ATP 3-21.18 and ATP 3-21.8 for additional information.)

Force Health Protection

8-191. The nature of area security in support of stability-focused tasks requires the BCT surgeon to stress planning for the provision of preventive medicine, veterinary services, and combat and operational stress control over that inherent in supporting offensive- and defensive-focused tasks. The BCT's area security mission focused within the conduct of stability-focused tasks interacts with the civilian population of its area of operations to a far greater degree. Under these conditions, the probability of Soldiers exposure to zoonotic diseases, toxic industrial chemicals and other pollutants, and bad food and water increases. The prolonged tours of duty typically associated with these operations and the enemy's use of unconventional weapons, such as mines and suicide bombers, tends to increase psychiatric casualties. The BCT surgeon coordinates the employment of combat stress teams with the chaplain to best meet the needs of BCT Soldiers for stress control. (See ATP 3-91 and ATP 4-02.8 for additional information.)

Chemical, Biological, Radiological, and Nuclear Operations

8-192. CBRN operations are the employment of capabilities that assess, protect against, and mitigate the entire range of CBRN incidents to enable freedom of action. CBRN operations support operational and strategic objectives to counter WMD and operate safely in a CBRN environment. An effective CBRN defense by the BCT counters enemy threats and attacks and the presence of toxic industrial materials in its area of operations by minimizing vulnerabilities, protecting friendly forces, and maintaining an operational tempo that complicates enemy or terrorist targeting.

8-193. The BCT employs key CBRN passive defense activities organized within two overarching CBRN principles (protection and contamination mitigation, see figure 7-3 on page 7-18) to survive and sustain area security operations in a CBRN environment. The BCT commander and staff, in coordination with the division or higher headquarters, integrate these principles regardless of the mission type.

8-194. The commander considers the requirement for CBRN support if evidence exists that enemy forces or terrorists have employed CBRN agents or have the potential for doing so. A mix of different CBRN units—such as decontamination, hazard response, reconnaissance, and surveillance—are necessary to balance capabilities. The CBRN staff officer at the BCT and battalion/squadron participates in the intelligence process to advise the commander of commercial and toxic industrial materials in the local area. (See ATP 3-91 and FM 3-11 for additional information.)

Convoy Security

8-195. Convoy security is a specialized kind of area security operations conducted to protect convoys. Units conduct convoy security operations anytime there are insufficient friendly forces to secure routes continuously in an area of operations and there is a significant danger of enemy or adversary ground action directed against the convoy. The BCT may conduct convoy security operations in conjunction with route security operations within its area of operations. Planning includes designating units for convoy security; providing guidance on tactics, techniques, and procedures for units to provide for their own security during convoys; or establishing protection and security requirements for convoys carrying critical assets. Local or theater policy typically dictates when or which convoys receive security and protection. (See ATP 4-01.45 for additional information.)

PREPARATION

8-196. During preparation activities, the BCT continues to plan, train, organize, and equip for area security missions within its area of operations. The conduct of preparation activities in support of stability-focused tasks is essentially the same as for offensive- and defensive-focused tasks. (See ADP 5-0 for a complete discussion.) However, factors that distinguish stability-focused tasks are the increased requirement for interagency coordination at BCT level and below and the demands on the BCT and subordinate staffs to perform tasks or functions outside their traditional scope of duties. The commander's realignment of organizations and functions during area security operations reflect carefully weighing and accepting risk (for example, economy of force) to reflect the demands of the BCT's area security mission.

COMMAND AND CONTROL

8-197.   Stability-focused tasks within area security operations are more complex because they involve to a greater extent unified action partners, sister services, and host-nation forces. BCT preparatory activities stress clarity and transparency about the command relationship between the BCT and the other military service components or agencies that operate in assigned or projected areas of operations. Though difficult, the BCT commander strives to achieve unity of effort with unified action partners, spending a great deal of effort during preparations to clarify the roles and functions of the various, often competing agencies.

Inherent Responsibilities

8-198.   The BCT commander has inherent responsibilities—including the requirements to clarify the mission; to determine the controlling legal and policy authorities; and to organize, direct, sustain, and care for the organizations and individuals for whom they provide the effort in interagency and multinational operations. The commander serves as the unit's chief engager, responsible for informing and influencing audiences inside and outside the organization. For example, the commander often integrates host-nation security forces and interagency activities with subordinate battalion, companies and platoons and down to the individual Soldier level for support units. With this in mind, obtaining the necessary numbers of scalable communications packages and linguist to support the BCT's planned operations and training are important preparatory activities.

Continue to Coordinate and Conduct Liaison

8-199.   Coordinating and conducting liaison ensures that subordinate commanders and leaders internal and external to the BCT understand their unit's role in upcoming operations, and that they are prepared to perform that role. In addition to military forces, many civilian organizations may operate in the same area of operations. Their presence can both affect and be affected by BCT operations. Continuous coordination and liaison between the command and unified action partners helps to build unity of effort, especially with civilian organizations because of the variety of external organizations and the inherent coordination challenges.

8-200.   Available resources and the need for direct contact between sending and receiving headquarters determine when to establish liaison. Establishing and maintaining liaison enables direct communications between the sending and receiving units or headquarters beginning with planning and continue through preparing and executing, or it may start (although not preferred) as late as execution. The BCT commander and staff coordinate with higher, lower, adjacent, supporting, and supported units and civilian organizations. Coordination includes but is not limited to the following:

- Sending and receiving liaison teams.
- Establishing communication links that ensure continuous contact during execution.
- Exchanging SOPs.
- Synchronizing security operations with reconnaissance plans to prevent breaks in coverage.
- Facilitating civil-military coordination among those involved.

Continue to Build Partnerships and Teams

8-201.   As part of the BCT's coordination efforts, the commander may establish or utilize (from higher echelon) special negotiation elements that move wherever they are needed to build partnerships or teams and diffuse or negotiate confrontations within the BCT area of operations. Echeloned elements partner with linguist support and personnel with the authority to negotiate on behalf of the appropriate level chain of command. As the BCT and these elements conduct preparatory activities, subordinate units of the BCT rehearse activities supporting these operations and when required ensures that these elements have access to required transportation and communications assets.

Initiate the Information Network

8-202. During preparation, the information network is tailored and engineered to meet the specific needs of each operation and partnered participant. This includes not only communications, but also how the commander expects information to move between and be available for subordinate commanders and leaders and their units within an area of operations. During preparation, the staff and subordinate units prepare and rehearse the information network supporting the plan. Network considerations include the following:

- Management of available bandwidth.
- Availability and location of data and information.
- Positioning and structure of network assets.
- Tracking status of key network systems.
- Arraying sensors, weapons, and the information network to support the concept of operations.

Note. Defining the ground rules for sharing unclassified information between the BCT, other military forces and foreign governments, nongovernmental organizations and international agencies according to higher commander policy is an important function of the division and BCT knowledge management and foreign disclosure officers. The division assistant chief of staff, signal and BCT signal staff officer (S-6) staff sections are responsible for disseminating and implementing those ground rules to subordinate units of the BCT.

MOVEMENT AND MANEUVER

8-203. Success in area security operations hinge on protecting the BCT forces within the area of operations and their ability to act in support stability-focused tasks. The positioning and repositioning of forces address the early detection and defeat of enemy forces attempting to operate within the BCT's area of operations. Enemy attacks within the BCT's area security range from individual saboteurs and terrorist acts to enemy insurgent operations.

Assign and Define Responsibility

8-204. During preparation activities, the commander assigns and defines responsibilities for the security of units within the BCT's area of operations or respective base or base cluster. Subordinate areas of operation or base and base cluster commanders are responsible for the local security of their respective area or base and base cluster. Individual area of operations and base commanders' designate protection standards and defensive readiness conditions (in coordination with the BCT's security plan) for tenant units and units transiting through their area or base. Commanders coordinate with the BCT main CP to mitigate the effects of security operations on the primary functions of units located within the area of operations.

Degree of Risk

8-205. The degree of risk the BCT commander accepts within an area security operation, regarding the enemy threat, invariably passes to the subordinate unit commander assigned the area security mission. For example, the subordinate unit commander moves security forces to decrease the threat's impact on logistics and medical units to support the BCT's continued operations at the anticipated level. When available and to not divert any BCT assets from their primary area security missions, military police (see ATP 3-39.30) or other available security force (possibly host nation) screen or guard friendly CP facilities and critical sites from enemy observation or attack. Subordinate unit security plans, to protect CPs, critical sites, base, base clusters, and security corridors, are rehearsed and inspected by the commander. These plans address support unit, site, and base and convoy defense against Level I threats. Plans also address response force operations directed against Level II and Level III threats (see chapter 9 and ATP 3-91 for additional information on threat levels).

Terrain Management

8-206. *Terrain management* is the process of allocating terrain by establishing areas of operation, designating assembly areas, and specifying locations for units and activities to deconflict activities that might interfere with each other (ADP 3-90). The commander designates assembly areas and specifies locations for units and activities to deconflict movements and repositioning of units, and other activities that might interfere with each other. Subordinate commanders assigned an area security mission manage terrain within their boundaries and identify and locate key terrain in the area. The BCT operations officer, with support from others in the staff, deconflict operations, control movements, and deter fratricide as units move to execute planned area security missions. The commander and staff also track and monitor unified action partners and their activities in the BCT's area of operations.

Terrain Preparation

8-207. Terrain preparation starts with the situational understanding of the terrain through proper terrain analysis. Terrain preparation involves shaping the terrain to gain an advantage, such as improving cover, concealment and observation, fields of fire, new obstacle effects through reinforcing obstacles, or mobility operations for initial positioning of forces. Terrain preparation can make the difference between the area security operation's success and failure. Commanders must understand the terrain and the infrastructure of their area of operations as early as possible to identify potential for improvement and establish priorities of work, and to begin preparing the area.

INTELLIGENCE

8-208. As the BCT prepares, the commander takes every opportunity to improve their situational understanding before and during operations with an aggressive and continuous information collection plan (through reconnaissance, surveillance, intelligence operations, and security operations). The commander executes information collection early in planning and continues it through preparation and execution.

> *Note.* Intelligence operations are tasks undertaken by military intelligence units and Soldiers to obtain information to satisfy validated requirements (see chapter 5).

Information Collection

8-209. Through information collection, the commander and staff continuously plan, task, and employ collection forces and assets to collect timely and accurate information. Collection helps to satisfy the CCIRs, in addition to other information requirements. Collection efforts within the BCT worked through the BCT intelligence cell (specifically the intelligence staff officer) to the division intelligence cell. Intelligence cells, in coordination with the BCT provost marshal, work to develop a readily searchable database—including biometric data if possible—of potential insurgents, terrorists, and criminals within the BCT's area of operations. This information is use by patrols to identify individuals, according to applicable guidance, when encountered during civil reconnaissance patrols (see ATP 3-21.8) and other operations. (See chapter 4 for additional information.)

Analysis and Dissemination of Information and Intelligence

8-210. Intelligence analysis is the process by which collected information is evaluated and integrated with existing information to facilitate intelligence production. The commander and staff refine security requirements and plans (including counterterrorism and counterinsurgency) as answers to various requests for information become available. Timely, relevant, accurate, predictive, and tailored intelligence analysis; reporting; and products enable the commander to determine the best locations to place area security measures and to conduct area security missions in support of stability-focused tasks. Rehearsal of area security measures and missions enable subordinate units to understand how these measures and missions fit into the BCT's area security operation, and that of the host nation when applicable. (See ATP 2-19.4 for additional information.)

PROTECTION

8-211. As preparation activities continue, the commander's situational understanding may change over the course of the area security operation, enemy actions may require revision of the security plan, or unforeseen opportunities may arise. Protection assessments made during planning may be proven true or false. Intelligence analysis from reconnaissance, surveillance, and security operations may confirm or deny enemy actions or show changed security conditions in the area of operations because of shaping operations. The status of friendly forces may change as the situation changes. In any of these cases, the commander identifies the changed conditions and assesses how the changes might affect upcoming area security missions. Significant new information requires commanders to make one of three assessments listed below regarding the area security plans:

- The new information validates the plan with no further changes.
- The new information requires adjustments to the plan.
- The new information invalidates the plan, requiring the commander to reframe and develop a new plan.

8-212. Protecting information during preparation activities is a key factor in protecting BCT subordinate units and the overall BCT area security operation. The secure and uninterrupted flow of data and information allows the BCT to multiply its combat power and synchronize division and other unified action partner capabilities and activities. The need to be candid and responsive to requests for information balance the need to protect operational information, such as troop movements, security plans, and vulnerabilities identified during preparation (inspections and rehearsals). Working closely with all partners develops the EEFI to preclude inadvertent public disclosure of critical or sensitive information. Information protection includes cybersecurity, computer network defense, and electromagnetic protection (EP). All three are interrelated.

SUSTAINMENT

8-213. Resupplying, maintaining, and the issuing of supplies or equipment occur during temporary and long-term area security commitments. Repositioning of sustainment assets also occur. During preparation, sustainment planners take action to optimize means (force structure and resources) for supporting the commander's area security plan. These actions include, but are not limited to, identifying and preparing bases, host-nation infrastructure and capabilities, contract support requirements, and lines of communications. They also include forecasting and building operational stocks as well as identifying endemic health and environmental factors. Integrating environmental considerations will sustain vital resources and help reduce the logistics footprint. Planners focus on identifying the resources currently available and ensuring access to them. During preparation, sustainment planning continues to support operational planning (branch and sequel development) and the targeting (lethal and nonlethal) process.

8-214. Dependent on the mission variables of METT-TC, sustainment elements may support the BCT from within its area of operations or from echelon support areas located outside the area of operations. The threat within the assigned area of operations is generally the major consideration in determining the size and composition of forces (support and operational) arrayed during an area security operation. Support elements (and any other force) within the BCT's area of operations must be able to defend themselves against a Level I threat, a small enemy force that can be defeated by those units normally operating in the echelon support area or by the perimeter defenses established by friendly bases and base clusters. The BCT commander uses a response force to response to a Level II threat (see chapter 4). Host-nation security forces, when feasible, may be an effective means of reinforcing the security of sustainment elements supporting from within and external to the BCT's area of operations because of their knowledge of the area, its language, and customs. (See chapter 9 for additional information.)

Notes. Base and base cluster defense is the cornerstone of successful area security and support area efforts. The commander achieves the application of effective area security for base and base clusters and their tenant and transient units by developing a comprehensive plan linked to site selection, layout, and facility design. (ATP 3-21.20, appendix I outlines the organization of forces, control measures, and considerations about planning, preparing, and executing base and base cluster operations.)

The commander and staff assess the need for providing protection to contractors operating within the BCT's area of operations and designate forces to provide security to them when appropriate. The mission of, threat to, and location of each contractor determines the degree of protection needed. Protecting contractors involves not only active protection to provide escort or perimeter security, but also training and equipping of contractor personnel in self-protection (protective equipment and weapons). Under certain conditions, contract security forces may be another means of reinforcing the security of sustainment elements supporting from within and external to the BCT's area of operations, and base and cluster defenses.

EXECUTION

8-215. Though close combat dominance remains the principal means to influence enemy actions, the conditions and standards of performance are modified by the mission variables of METT-TC and the more restrictive rules of engagement required during the conduct of stability-focused tasks. The general scope of BCT missions supporting stability-focused tasks includes security operations, patrols and patrolling (reconnaissance and combat), intelligence operations (for example, HUMINT assets from the military intelligence company of the BCT, or higher headquarters), surveillance (ground forces and aerial assets), convoy security, and Soldier and leader engagements. Additionally, missions often require the establishment of static security posts, base and base clusters, searches, roadblocks, checkpoints, observation posts, and combat outposts supports the conduct of stability-focused tasks. The condition set surrounding each mission differs and requires detailed analysis and planning.

APPORTIONMENT OF COMBAT POWER AND DEDICATED ASSETS

8-216. The BCT commander, during area security operations, apportions combat power and dedicates assets to protection tasks based on an analysis of the operational environment, the likelihood of enemy action, and the relative value of friendly resources and populations. Although all resources have value, the mission variables of METT-TC make some resources, assets, or locations more significant from enemy or adversary and friendly perspectives. The commander relies on RM (see chapter 4) and other assessment methods to facilitate decision-making, issue guidance, and allocate resources. Criticality, vulnerability, and recoverability are some of the most significant considerations in determining protection priorities that become the subject of the commander's guidance and the focus of BCT's area security efforts.

8-217. Generally, the BCT conducts area security operations to establish stability after open hostilities cease. With the complex and dynamic nature of this area security operation, it is important to remember that area security operations and activities change from day to day, based upon the mission variables of METT-TC. ATP 3-21.20 provides two detailed scenarios, used for discussion purposes, representing ways an Infantry brigade combat team (IBCT) may employ forces during the conduct of an area security operation. ATP 3-21.10 discusses these same scenarios for the employment of forces (subordinate units of an Infantry battalion) during the conduct of area security.

CRITERIA TO JUDGE PROGRESS

8-218. During execution, *evaluating* is using criteria to judge progress toward desired conditions and determining why the current degree of progress exists (ADP 5-0). Evaluation is at the heart of the assessment process (see chapter 4) where most of the analysis occurs. Evaluation helps the BCT commander determine what is working and what is not working, and gain insights into how to better accomplish the mission.

8-219. Criteria in the form of MOE and MOP aid in determining progress toward attaining end state conditions, achieving objectives, and performing tasks. MOEs help determine if a task is achieving its intended results. MOPs help determine if a task is completed properly. MOEs and MOPs are simply criteria— they do not represent the assessment itself. MOEs and MOPs require relevant information as indicators for evaluation.

8-220. An MOE is a criterion used to assess changes in system behavior, capability, or operational environment that is tied to measuring the attainment of an end state, achievement of an objective, or creation of an effect. MOEs help measure changes in conditions, both positive and negative. MOEs help to answer the question "Are we doing the right things?" MOEs are commonly found and tracked in formal assessment plans. Examples of MOEs for the objective to "Provide a safe and secure environment" may include—

- Decrease in insurgent activity.
- Increase in population trust of host-nation security forces.

8-221. An MOP is a criterion used to assess friendly actions that is tied to measuring task accomplishment. MOPs help answer questions such as "Was the action taken?" or "Were the tasks completed to standard?" A MOP confirms or denies that a task has been properly performed. MOPs are commonly found and tracked at all levels in execution matrixes. MOPs are also commonly used to evaluate training during SFA missions conducted by the BCT. MOPs help to answer the question "Are we doing things rights?"

8-222. At the most basic level, every Soldier assigned a task maintains a formal or informal checklist to track task completion. The status of those tasks and subtasks are MOPs. Similarly, operations consist of a series of collective tasks sequenced in time, space, purpose and resources to accomplish missions. Current operations integration cells use MOPs in execution matrixes and running estimates to track completed tasks. The uses of MOPs are a primary element of battle tracking. MOPs focus on the friendly force. Evaluating task accomplishment using MOPs is relatively straightforward and often results in a yes or no answer. Examples of MOPs include—

- Route X cleared.
- Generators delivered, are operational and are secured at villages A, B, and C.

SECTION V – SECURITY FORCE ASSISTANCE

8-223. SFA contributes to unified action by the U.S. Government to support the development of the capacity and capability of FSF and their supporting institutions, whether of a partner nation or an intergovernmental organization (regional security organization). The development of capacity and capability is integral to successful stability missions and extends to all organizations and personnel under partner nation control that have a mission of securing its population and protecting its sovereignty from internal and external threats. FSF are considered to be duly constituted foreign military, paramilitary, police, and constabulary forces such as border police, coast guard, and customs organizations, as well as prison guards and correctional personnel, and their supporting institutions.

OPERATIONAL OVERVIEW

8-224. SFA activities are conducted to organize, train, equip, rebuild (or build), and advise FSF from the ministerial/department level down through the tactical units. The Department of Defense maintains capabilities for SFA through conventional forces, special operations forces, the civilian expeditionary workforce, and when necessary contractor personnel in both joint operational area and a non-joint operational area environment. SFA activities require carefully selected and properly trained and experienced personnel (as trainers or advisors) who are not only subject matter experts, but also have the sociocultural understanding, language skills, and seasoned maturity to more effectively relate to and train FSF. Ideally, SFA activities help build the FSF capacity to train their own forces independent of sustained U.S. Government efforts.

> *Note*. With, through, and by. Describes the process of interaction with FSF that initially involves training and assisting (interacting "with" the forces). The next step in the process is advising, which may include advising in combat situations (acting "through" the forces). The final phase is achieved when FSF operate independently (act "by" themselves) (see Department of Defense Instruction 5000.68 for additional information).

SUPPORT TO SERVICE AND JOINT OPERATIONS AND MISSIONS

8-225. SFA activities are conducted across the range of military operations and across the competition continuum (from peace through war), supporting Service and joint operations and missions (figure 8-2).

Significant security cooperation and military engagements are routinely conducted worldwide for peacetime theater and global shaping through the geographic combatant commanders' theater campaign plans. Some of those security cooperation activities are likely to include SFA activity efforts in the lower range of the competition continuum. Timely and effective execution of relevant SFA activities as part of security cooperation for shaping in the theater campaign may contribute to stabilization and perhaps a measure of deterrence to prevent the requirement for U.S. forces having to conduct a contingency operation. Joint forces must have the ability to conduct SFA activities throughout all phases of an operation or campaign to effectively partner with FSF supporting U.S. or U.S.-led multinational requirements. (See FM 3-22 for additional information.)

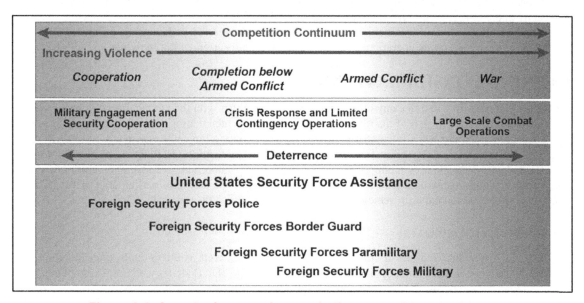

Figure 8-2. Security force assistance in the competition continuum

---

Note. For the purpose of the following discussion, SFA is addressed within the initial response, transformation, and fostering sustainability phases of the stabilization framework.

PHASING FOR SECURITY FORCE ASSISTANCE

8-226. Phasing for SFA, initial response, transformation, and fostering sustainability, mirrors the stabilization framework described in section I of the chapter, and are based on the operational environment. SFA can start in any phase or may even move to a previous phase due to changes in the conditions of the operational environment. Differences within and between phases may not change on the surface, but relationships with FSF can change drastically. For example, the latter stage of the transformation phase can differ greatly from the initial stages of the transformation phase. Span of control and the area of operations for SFA can expand within a phase and as operations continue within the stabilization framework. As the three phases are based on the operational environment, they provide a baseline for augmentation. Potential augmentation may require military police, legal, public affairs, civil affairs, PSYOP, engineering, sociocultural experts, sustainment, and SFA team personnel.

Note. A provincial reconstruction team embedded at BCT level is a key element during the conduct of SFA. The BCT leads the effort to establish civil security, support to civil control (when approved by Congress), and to develop and enable FSF. The embedded provincial reconstruction team has the lead for support to economic and infrastructure development, restore essential services, and support to governance. Together the BCT and an embedded provincial reconstruction team are able to effectively support the FSF and execute all six stability operations tasks.

TRANSITIONS

8-227.   Transitions during SFA are dependent upon the conditions within the operational environment. Transitions are initially identified during planning using a comprehensive approach (see paragraph 8-15). Transitions can occur simultaneously or sequentially in different levels of conflict and in separate echelons, to include having potential at the tactical level, transitions for different units within the BCT's area of operations. Major transitions can include the BCT in the beginning of an initial response phase being the supported unit with the FSF transitioning to the supported unit later on in the phase. At this point in the transformation phase, the area in which the BCT conducts SFA will expand. This expansion can occur multiple times during the transformation phase, which is based on conditions, especially the capability and capacity of the FSF. The commander, to facilitate flexibility, visualizes and incorporates branches and sequels into the overall plan to enable transitions. Unless planned, prepared for, and executed efficiently, transitions can reduce the tempo of the operation, slow its momentum, and surrender the initiative.

PLANNING CONSIDERATIONS

8-228.   Planning for SFA, like any other operation, begins either with the anticipation of a new mission or the receipt of mission as part of the MDMP. The Army design methodology is particularly useful as an aid to conceptual planning when integrated with the detailed planning typically associated with the MDMP to develop the capacity and capability of the FSF and its supporting institutions. Planning helps the commander create and communicate a common vision between the staff, subordinate commanders, and unified action partners. Planning results in a plan and order that synchronize the action of participating partners in time, space, purpose and resources to achieve objectives and accomplish the mission.

COMPREHENSIVE APPROACH

8-229.   SFA planning requires a comprehensive approach, as well as an in-depth understanding of the operational environment. Planning must be nested within policy, IDAD strategy, the campaign plan, and any other higher-echelon plan. Continuous and open to change, planning for SFA includes identifying how to best assist the FSF and developing a sequence of actions to change the situation. (See paragraphs 8-15 to 8-17 for a detailed discussion.)

UNDERSTANDING

8-230.   Understanding the operational environment is fundamental to all operations, and essential to SFA activities. An in-depth understanding of the operational environment includes the size, organization, capabilities, disposition, roles, functions, and mission of host-nation forces, opposing threats, regional players, transnational actors, joint operational area, or non-joint operational area of responsibility (AOR), especially the sociocultural factors of the indigenous and other relevant populations. Identifying all actors influencing the environment and their motivations will help planners and practitioners define the goals and methods for developing host-nation security forces and their institutions.

8-231.   The plan, which includes the commander's intent, provides understanding to United States and FSF on the actions to take. (SFA planning may involve the development of nonmilitary security forces and their supporting institutions.) Plans and orders provide decision points and branches that anticipate options that enable the force to adapt as the operation unfolds. This is especially important for SFA, as these operations tend to be prolonged efforts. Units conducting SFA often rotate before achieving all objectives. As a result, planning should establish objectives and milestones that can be achieved during the BCT's mission. These objectives and milestones must support higher echelon plans, including the campaign plan and IDAD strategy.

> *Note.* SFA planning may involve the development of nonmilitary security forces and their supporting institutions.

LEGITIMACY

8-232.   Legitimacy of the forces providing SFA may be tenuous during some phases of a complex operation, but it is an essential consideration for achieving long-term objectives. SFA should aim to ensure that all FSF operate within the bounds of domestic and international laws, respect human rights, and that they support wide-ranging efforts to enforce and promote the rule of law, thus supporting legitimacy and transparency. Legitimacy fosters transparency and confidence among host-nation government, FSF, host-nation population, and U.S. Government agencies. Another aspect of legitimacy is supporting host-nation ownership in the SFA effort, because it facilitates a sense of sustainability for building a capacity or security reform through acceptance by the host-nation population.

8-233.   Throughout planning, the BCT commander and staff consider how each SFA activity affect popular perceptions and focus on the activities that enable the legitimacy of the host-nation government and FSF, not just make them technically competent. Having a just cause and establishing and sustaining trust affect the relationship with the indigenous population. Commander and staff must ensure an appropriate information management plan is developed for SFA in coordination with interagency partners and the division or other higher headquarters. SFA advisors/trainers must work with the FSF to give a positive context and narrative to the FSF professionalization efforts and capacity to secure the population. Coordination of the information themes and messages among the United States, FSF, and the host-nation government, and the presentation or availability of information to the indigenous population can limit or mitigate the propaganda efforts of insurgents or hostile forces. This may serve to mitigate the potential for destabilizing influences of hostile forces or criminal elements to propagandize SFA efforts and damage the host-nation government's credibility and legitimacy.

SECURITY FORCE FUNCTIONS

8-234.   Security forces perform three generic functions: executive, generating, and operating. The executive function includes strategic and operational direction that provides oversight, policy, and resources for the FSF generating and operating functions. The generating function develops and sustains the capabilities of the operating forces. In the United States, the generating function is primarily performed by the services. For the United States, this function is performed by its military schools, training centers, and arsenals. FSF institutional forces refer to the capability and capacity of the FSF to organize, train, equip, and build operating force units. FSF operating forces form operating capabilities through the use of concepts similar to warfighting functions to achieve FSF security objectives.

Note. Employing operational forces to fill SFA capabilities associated with developing the FSF generating function (FSF tasks such as "develop FSF doctrine" or "stand up a staff officer's college"), and possibly in the FSF executive function (ministries) would likely be beyond the inherent capability of the operating force and would likely require special training or augmentation by subject matter experts drawn from U.S. generating organizations.

8-235.   U.S. operating forces are typically better suited to develop FSF operating force capabilities than they are to develop FSF institutional forces of generating capabilities. Typically, the BCT is tasked to train or advise FSF operating forces. The operating function employs military capabilities through application of the warfighting functions of command and control, movement and maneuver, intelligence, fires, sustainment, and protection during actual operations. Operating, as it applies to police security forces, may include training an actual operation with the integration of patrolling, forensics, apprehension, intelligence, investigations, incarceration, communications, and sustainment. Operating forces are responsible for collective training and performing missions assigned to the unit.

SECURITY FORCE ASSISTANCE TASKS

8-236.   SFA activities normally use the following general developmental tasks of organize, train, equip, rebuild and build, advise and assist, and assess (known as OTERA-A). These functional tasks, serving as SFA capability areas, are used to develop the capabilities required by the FSF. OTERA-A tasks are tools to develop, change, or improve the capability and capacity the FSF. Through a baseline assessment of the FSF, and considering U.S. interests and objectives, the BCT commander and staff planners determine which OTERA-A tasks are required to build the proper capability and capacity levels within the various units of the FSF. Assessments of the FSF against a desired set of capabilities will assist in developing an OTERA-A based plan to improve the FSF. (See FM 3-22 for additional information.) The following are basic descriptions of the OTERA-A tasks:

- Organize. All activities taken to create, improve, and integrate doctrinal principles, organizational structures, capability constructs, and personnel management. This may include doctrine development, unit or organization design, command and staff processes, and recruiting and manning functions.
- Train. All activities taken to create, improve, and integrate training, leader development, and education at the individual, leader, collective, and staff levels. This may include task analysis, the development and execution of programs of instruction, implementation of training events, and leader development activities.
- Equip. All activities to design, improve, and integrate materiel and equipment, procurement, fielding, accountability, and maintenance through life cycle management. This may also include fielding of new equipment, operational readiness processes, repair, and recapitalization.
- Rebuild or Build. All activities to create, improve, and integrate facilities. This may include physical infrastructures such as bases and stations, lines of communication, ranges and training complexes, and administrative structures.
- Advise/Assist. All activities to provide subject matter expertise, guidance, advice, and counsel to FSF while carrying out the missions assigned to the unit or organization. Advising may occur under combat or administrative conditions, at tactical through strategic levels, and in support of individuals or groups.
- Assess. All activities for determining progress toward accomplishing a task, creating an effect, or achieving an objective using MOEs and MOPs to evaluate FSF capability. Once an objective is achieved, the focus shifts to sustaining it.

DECISIONS TO REDUCE OR OFFSET RISK

8-237.   RM (see chapter 4) is the Army's process for helping organizations and individuals make informed decisions to reduce or offset risk. RM measures identified in SFA planning add to the plan's flexibility during execution. A flexible plan can mitigate risk by partially compensating for a lack of information. SFA planning requires a thorough, comprehensive approach to analyzing and agreeing upon risk reduction measures. Each SFA activity is distinct based on context and changes over time. There is a risk of focusing SFA efforts in one area or type of relationship at the expense of others based on short-term goals. To mitigate this risk, SFA activities should be regarded as the providing means and ways to achieve meaningful mid- to long-term objectives with partners as well as the end states.

8-238.   Reducing or alleviating risk does not only rely on the SFA force and supporting agencies but also on the FSF elements in question. Conditions determine when to use an element of the FSF. The BCT commander and staff use assessments to determine objectives and requirements for reducing or offsetting risk. Risk applies to how well the FSF, the BCT, and other host-nation and partner organizations agencies can tolerate changes in the operational environment, as well as the challenges and conditions inherent to the operation.

8-239.   The BCT commander and FSF commander assess the risk associated with the employment forces and mitigate that risk as much as possible. For example, advisors from the BCT play a significant role in security cooperation mission such as the SFA. They live, work, and sometimes are required to fight with their partner FSF. The relationship between advisors and FSF is vital. Advisors are not merely liaison officers. Though they do not command FSF units, they are a necessary element to understanding the human dimension, specifically managing relationships and mitigating risk between the SFA forces and FSF, across the range of military operations.

SUSTAINING ACTIVITIES

8-240. Sustaining SFA activities consists of two major components: the ability of the United States and other partners to sustain the SFA activities successfully and the ability of host-nation security forces to sustain their capabilities independently over the long term. The first component may be predicated on the host-nation maintaining legitimacy while the second component should be considered holistically when working with the host nation to build their security forces. It is important to consider the sociocultural factors, infrastructure, and education levels of prospective FSF when fielding weapons systems and maintaining organizations. Though this is not the BCT, or division commander's decision, a strong recommendation through the SFA chain should be made in regard to this consideration.

INTELLIGENCE

8-241. Intelligence provides an assessment of host nation and potential adversaries' capabilities, capacities, and shortfalls. It involves understanding sociocultural factors, information and intelligence sharing, and intelligence training. Information-sharing between the BCT and FSF must be an early consideration for planners. A continuous intelligence effort will gauge the reaction of the local populace and determine the effects on the infrastructure of SFA efforts as well as evaluate strengths, weaknesses, and disposition of opposition groups in the area. Ultimately, intelligence supports the SFA and FSF leaders' decision-making processes, and supports the protection of friendly forces and assets.

> *Note.* Train personnel two deep in every staff section or advisory subunit on foreign disclosure and derivative classification procedures before deployment. Interaction with host nation and FSF, even North Atlantic Treaty Organization or other coalition allies requires foreign disclosure officer approval. Authorized release or disclosure of classified information will require personnel trained in derivative classification and foreign disclosure. This will become a huge bottleneck if not trained for and decentralized.

PROTECTION

8-242. Protection is incumbent upon the commander to fully understand the threat environment within the BCT's area of operations. By having access to fused intelligence from local, regional and national resources, the commander can accurately assess threats and employ measures to safeguard SFA personnel and facilities. Protection planning considerations should address additional support requirements for the response force, emergency procedures, personnel recovery, or the requirement to integrate SFA personnel into the host-nation protection plan.

8-243. Nontraditional threats, such as the insider threat, can undermine SFA activities as well as the cohesion of U.S. forces and FSF. Tactically, the breakdown of trust, communication, and cooperation between host nation and U.S. forces can affect military capability. Adversaries may view attacks against U.S. forces as a particularly effective tactic, especially when using co-opted host-nation forces to conduct these attacks. While these types of insider or "green on blue" attacks have been context-specific to a particular theater, the commander should ensure that protection plans consider the potential for these types of attacks and plan appropriate countermeasures.

> *Note.* More stringent protection controls and measures that are overtly heavy handed must be well balanced yet culturally sensitive enough to not send the wrong message to the very people and organizations the United States is trying to assist.

LOGISTICS

8-244. Logistics planners at the BCT level must understand the division's concept of support and sustainment estimates that outline the responsibilities and requirements for maintaining logistics support for deployed forces within the division's area of operations. Logistics support might include support of SFA augmentees and FSF within the BCT's area of operations to conduct operational missions (supporting host-nation civilians or military forces with medical, construction, power generation, maintenance and supply, or transportation capabilities).

PREPARATION

8-245. Preparation for SFA creates conditions that improve the BCT's opportunities for success. The degree to which the BCT is tasked within SFA operations depends on preparation in terms of cultural knowledge, language, functional skills, and the ability to apply these skills within the operational environment. Preparation includes, but is not limited to, initiate security and information collection, continue to coordinate and conduct liaison, refine the plan, complete task organization, conduct pre-mission training, conduct rehearsals and inspections; continue to build partnerships and teams, and initial movement. Preparation facilitates and sustains plans-to-operations transitions, including those to branches and sequels, which are of vital importance for the often-dynamic operational environment for SFA.

PRE-MISSION PREPARATIONS

8-246. After receiving a mission, the BCT continues detailed preparation activities, prepares for and rehearses classes given in country, and conducts extensive briefings on the area of operations. Key staff and subordinate unit actions particular to SFA are address in the following paragraphs.

Current Operations

8-247. The BCT S-3 ensures predeployment training for Soldiers, to include preparation for training FSF and rehearsals for movement. The S-3 reviews the program of instruction for training FSF, to include getting approval from the commander, and higher headquarters if necessary. The S-3, in coordination with BCT civil affairs operations staff officer (S-9), ensures the operation plan minimizes how operations affect the civilian population and addresses ways to mitigate the civilian impact on military operations. The civil-military operations plan is coordinated with the indigenous population and institutions, unified action partners, other civil entities, and interagency as necessary. This coordination might include civil affairs battalions or brigades, provincial reconstruction teams, or United States Agency for International Development project officers in the area of operations.

Note. The primary staff officers of the current operations cell may be called upon to be the primary advisors to the host-nation forces staff sections and cells.

8-248. The BCT intelligence staff officer (S-2) supervises the dissemination of intelligence and other operationally pertinent information within the unit and, as applicable, to higher, lower, or adjacent units or agencies. The S-2 monitors the implementation of the intelligence collection plans to include updating the commander's priority intelligence requirements, conducting area assessment, and coordinating for additional intelligence support. The S-2 establishes liaison with FSF intelligence and security agencies (within the guidelines provided by applicable higher authority). The S-2 assesses the intelligence threat and resulting security requirements, including coordination with the S-3 on specific security and operations security measures.

8-249. The BCT personnel staff officer (S-1) supervises the battalion personnel staff section, in coordination with brigade and higher echelon manpower and personnel staff sections. The S-1 screens personnel files to review the records of identified Soldiers that might have specific skill sets useful to the BCT or higher echelon during the conduct of SFA operations. Skill sets include individuals with professional certification or work experience in those nonmilitary fields that might have utility during operations focused on the conduct of stability operations tasks.

8-250. The BCT S-4 supervises, as required, the logistics support of SFA augmentees and FSF within the BCT area of operations to conduct operational missions (supporting host-nation civilians or military forces with medical, construction, power generation, maintenance, supply, or transportation capabilities).

8-251. The BCT S-6, in coordination with the division assistant chief of staff, signal, ensures depth in communication and synchronization between organizations both horizontally and vertically within the BCT's proposed area of operations. In coordination with the BCT S-9, the S-6, establishes communications as early as possible upon arrival with the civil-military operations center (normally established at BCT level), civil liaison teams, civil information management architecture, and supporting networks to facilitate communication and coordination with the nonmilitary agencies.

Predeployment Training

8-252. During predeployment training, Soldiers receive training, materials, and briefings on the operational area. This training can cover the history, culture, religion, language, tribal affiliations, local politics, and cultural sensitivities as well as any significant nongovernmental organizations operating in the operational area. Advisors focus their pre-mission training on the specific requirements of developing an FSF. The training emphasizes the host-nation culture and language and provides cultural tips for developing a good rapport with foreign personnel.

8-253. Based on the BCT commander's, or higher commander's training guidance, subordinate unit commanders assign missions and approve the draft mission-essential task list that supports SFA. The staff plans, conducts, and evaluates training to support this guidance and the approved mission-essential task list for SFA missions. Subordinate commanders prioritize tasks that need training. Since there is never enough time to train in every area, commanders focus on tasks essential for mission accomplishment.

8-254. Once commanders select tasks for training, the staff builds the training schedule and plans on these tasks. The staff provides the training requirements to the commander. After approving the list of tasks to be trained, the commander includes the tasks in the unit training schedule. The staff then coordinates the support and resource requirements with the S-3 and S-4. Finally, the commander ensures standards are enforced during training.

Evaluation

8-255. Evaluations can be either internal or external. Internal evaluations occur at all levels, and they must be inherent in all training. External evaluations are usually more formal and conducted by a headquarters one or two levels above the unit being evaluated. This subject must be carefully planned and discussed with FSF leaders to account for cultural sensitivities and current capabilities. A critical weakness in training is the failure to evaluate each task every time it is executed. Every training exercise provides potential for evaluation feedback. Every evaluation is also a training session. Leaders and trainers must continually evaluate to optimize training. Evaluation must occur as the training takes place. Emphasis is on direct, on-the-spot evaluations. However, leaders allow Soldiers to complete the task first. Leaders plan after-action reviews at frequent, logical intervals during exercises. This technique allows the correction of shortcomings while they are still fresh in everyone's mind. The after-action review eliminates reinforcing bad habits.

Specified Training

8-256. Augmentation elements require area orientation, refresher combat training, field-training exercises, and the like. Unit training objectives are for developing capabilities to conduct internal and external defense activities for tactical operations, intelligence operations, PSYOP, populace and resources control operations, and civil affairs and advisory assistance operations in the host nation language. Units identified for SFA begin intensified training immediately upon deployment notification.

8-257. After deployment to the host nation and before commitment to operations, the unit may receive in-country training at host-nation training centers or at designated training locations. This training helps personnel become psychologically and physically acclimated to the host nation. This training also allows commanders and staffs some time to coordinate and plan within their own command and with civilian and military joint and multinational organizations. After commitment, training continues and is stressed between operations, using needed improvements identified in operations as the basis for training.

8-258. Insider attacks are a threat in any area of operations. The BCT commander ensures that military forces, civilian expeditionary workforce, or civilian personnel and contractors are trained to identify behavioral indicators of possible insider threats and the means to apply prevention tools to mitigate this threat. Cultural awareness yields situational awareness and leads to increased force protection for SFA personnel.

Eliminating or minimizing the insider threat, especially by proper preparation and training of forces, is critical to mission success.

> *Note.* To reduce the potential for insider attacks, FSF should be further vetted to identify individuals whose motivations toward the host nation and U.S. Government are in question.

BUILD PARTNERSHIPS AND TEAMS

8-259.   The BCT, augmented for SFA, will have subordinate units whose sole focus is working with the FSF. Advisor teams may be formed from BCT or battalion organic resources, external augmentation, or a combination. These teams optimally are embedded with the counterpart unit(s), or they may reside on a U.S. camp and commute to the FSF they support. The method depends on policy, direction from higher headquarters, the conditions of the operational environment, and capacity of the FSF camps to accommodate the U.S. forces.

8-260.   When the BCT has an SFA mission it can potentially support multiple FSF organizations in its area of operations. Additionally, these FSF organizations may each report through different host-nation government channels and even to different ministries. To synchronize efforts in this case, the BCT must achieve unity of effort within the different partner organizations. Similarly, each of the FSF organizational commanders should synchronize their efforts with the host-nation government representatives, as appropriate.

8-261.   Figure 8-3 depicts an example battalion task organization, used for discussion purposes, for a BCT supporting multiple SFA activities. Within the task organization, one company team acts as a response force with adequate tactical mobility and designated fire support to defeat Level I and Level II threats. The response force shapes Level III threats until a tactical combat force or other capable response force can defeat it. The response force is task organized with an intelligence support team consisting of two intelligence analysts (based on the situation) to support operations. Additionally, the task organization depicts how a company team may form the foundation for host-nation military and border support. Finally, the task organization depicts how a company team may provide police support. Support is in the form of two platoons supporting police assistance teams and a third platoon in a combined (multinational) security station providing support to a police assistance team.

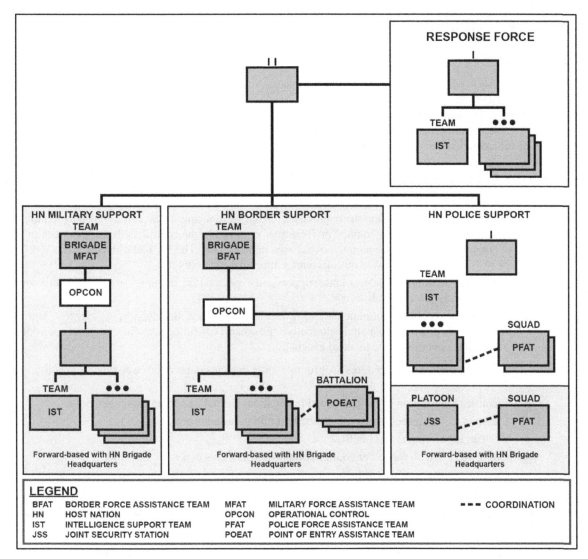

Figure 8-3. Support to security force assistance, example

Note. The designation of force assistance teams used in figure 8-3 is not to be prescriptive of how a battalion might support a particular SFA activity. Designations are intended to be used as guides illustrating one way the battalion may task organize to support multiple SFA activities within an area of operations.

8-262. Subordinate units of the BCT conducting SFA are best located inside the base of the FSF to be trained. Collocation facilitates integration with the FSF and allows the two forces to form mutual understanding and trust. Collocation and the close cooperation often facilitate and improves the population's perception of the legitimacy of United States and FSF, which can be an essential condition of the overall mission's end state.

8-263. When protection conditions require, a U.S. area may be established in the FSF base, although this is not optimal. Key considerations for collocation may include the threat, FSF acceptance, physical space inside the FSF base, sustainment capabilities, medical facilities, and availability of response forces.

8-264. When U.S. forces are operating out of smaller outposts in an urban environment, the local populace sees the integration and presence of the United States and FSF working together. This integration not only enhances overall operational effectiveness and trust, living and working together builds legitimacy of the two forces as well as FSF; it reinforces trust between the FSF and the people they are tasked to protect.

DEPLOYMENT

8-265. The BCT will often conduct SFA in operational environments in which it is a guest of the host nation or partner organization. When not already in country, the BCT moves into the operational area per its deployment order and within its established standing operating procedures. When located within the operational area, subordinate units conduct troop movement to their assigned area of operations. (See FM 3-22 for a detailed discussion of SFA deployment, redeployment, and post-deployment activities.)

IN-COUNTRY PREPARATIONS

8-266. Upon arrival, the BCT commander and S-3 brief the division or higher headquarters on the planned execution of the mission and reconfirm the required command relationship. Local conditions may require the BCT to confirm or establish its in-country and external command and control systems and sustainment functions relationships from outside its operational area upon arrival. The BCT establishes direct working relationships with the next higher in- or out-of-country supporting element to—

- Determine the limits of the available support and expected reaction time between the initiation of the support request and fulfillment.
- Confirm or establish communications procedures between the supporting element and the individual SFA unit(s), to include alternative and emergency procedures for command and control, all support operations, and medical evacuation.

8-267. The BCT establishes procedures to promote interagency cooperation and synchronization. The BCT—

- Identifies the location of the concerned host nation, United States, or other agencies.
- Contacts the concerned agency to establish initial coordination.
- Exchanges information or intelligence.
- Confirms or establishes other coordination protocols as necessary.
- Incorporates the newly established or changed procedures into the plans for mission execution.

8-268. The BCT immediately establishes operations security procedures to support its mission execution and identifies rally points incorporated into its protection (generally an area security mission[s]), evasion, and personnel recovery plans.

8-269. After receiving a detailed briefing and further guidance from the advance party (when established before the main body's arrival), BCT personnel continue to develop effective rapport with the FSF commander, staff, and other counterparts. (See chapter 4 for information early-entry CP.) They also assess their working, storage, and living areas for security and verify the location of the training site(s), communications center, dispensary area, and FSF troop area(s). With the FSF commander and staff, the BCT commander and staff (led by the S-3)—

- Establishes a rapport.
- Conducts introductions in a businesslike, congenial manner.
- Briefs on the BCT's mission, its capabilities, and the restrictions and limits imposed on the organization by its higher headquarters.
- Requests counterpart linkup(s) be made under the mutual supervision of the FSF commander and the BCT commander, or appropriate BCT subordinate commander.
- Ensures all current unit plans are tentative and that assistance is needed to finalize them.
- Deduces or solicits the actual estimate of unit capabilities and perceived advisory assistance and material requirements.
- Recommends the most desirable COAs while emphasizing how they satisfy present conditions, achieve the desired training, and meet advisory assistance goals.
- Prepares and briefs the plans for training and advisory assistance.

- Informs the division or higher in-country command of any significant changes in the BCT's plan to assist the FSF.

8-270. Through the BCT S-2, the commander's priority intelligence requirements are based on the latest information available and requirements for additional priority intelligence requirements that arise from modified estimates and plans. The S-2 also—

- Analyzes the FSF's status to finalize unit plans for advisory assistance. These plans can include task organization of unit with counterparts, staff functions for planning SFA, and advisory assistance for executing SFA.
- Explains analysis to counterparts and encourages them to help with—and participate in—analyzing, preparing, and briefing the analysis to the FSF commander.
- Helps (in coordination with the provost marshal) the FSF inspect the available facilities to identify deficiencies. If the unit finds deficiencies, the S-2 prepares estimates of COAs for the FSF commander to correct them.
- Supervises the preparation of the facilities with their counterparts and informs BCT and FSF commanders on the status of the facilities.

8-271. The BCT (led by the S-3 and provost marshal) ensures its security is based on the present or anticipated threat. Some recommended actions the BCT may take include—

- Hardening positions based on available means and requirements to maintain low visibility.
- Maintaining communications with all subordinate unit personnel deployed outside the immediate area controlled by the main body.
- Establishing plans for immediate defensive actions across the BCT in the event of an attack or a loss of rapport with hostile reaction.
- Discussing visible security measures with foreign counterparts to ensure understanding and to maintain effective rapport. BCT personnel do not divulge sensitive information for the sake of possible rapport benefits.
- Encouraging the FSF, through counterparts, to adopt additional security measures identified when analyzing the FSF's status and inspecting its facilities.
- Coordinating defensive measures with the FSF to develop a mutual defense plan. Encouraging the FSF to conduct mutual full-force rehearsals of defensive plans, if unsuccessful, the BCT still conducts its internal rehearsals of the plans.

EXECUTION

8-272. In execution, the BCT commander, staff, and subordinate commanders focus efforts on translating decisions, made during planning and preparation, into actions supporting the SFA mission. Once the BCT arrives in-country, it begins the employment of forces to support the development of FSF capabilities and capacities. Employment of the BCT occurs generally with the establishment of advising, assisting, and training teams and key individuals. These teams and key individuals' partner with foreign counterparts during FSF planning (preparing the FSF for the mission[s] itself) to increase the capability and capacity of FSF planning processes, as well as to increase the probability of success.

Note. SFA activities normally use the general developmental tasks (known as FSF development tasks) of OTERA-A to develop the functional capabilities required by the FSF. (See paragraph 8-236 for information on organize, equip, and rebuild and build developmental tasks.)

FOREIGN SECURITY FORCES DEVELOPMENT TASKS—ADVISE, ASSIST, AND TRAIN

8-273. The BCT conducting SFA missions normally task organizes into smaller rotational teams, and identifies key individuals, for execution. These teams and key individuals focus on advising, assisting, or training a specific partner individual, unit, or activity. These teams and key individuals include, but are not limited to, advising, or advisory teams and individuals within the BCT. Specialized teams and individuals may also be required for partner sustainment, engineer, or police units for example.

Advise and Assist Foreign Security Forces

8-274. The BCT, generally with additional augmentation teams, advises and assists FSF to improve their capability and capacity. Advising establishes a personal and a professional relationship where trust and confidence define how well the advisor will be able to influence the FSF. Assisting is providing the required supporting or sustaining capabilities so FSF can meet objectives and the end state. The level of advice and assistance is based on conditions and continues until FSF can establish required systems or until conditions no longer require it. Advising and assisting teams from within the BCT do not permit the FSF to fail critically at a point that would undermine the overall effort.

Security Force Assistance —Advise and Assist Activities

8-275. The following advise and assist activities, used for discussion purposes, present advise and assist actions conducted by the BCT. Key BCT advisors include the—

BCT Commander

8-276. Before the mission, the commander advises and assists the FSF commander. The FSF commander then issues planning guidance for planning the execution of the mission and clarifies commander's intent. The commander advises and assists the FSF commander throughout the operations process for the tactical operation(s) or training. By accompanying the FSF commander when the mission is received from higher headquarters, the commander assists any subsequent missions. The commander monitors how FSF subordinate units understand the commander's intent and all specified and implied tasks.

8-277. During the execution of the mission, the commander helps the FSF commander provide command and control during operations. While monitoring the tactical situation, the commander assesses activities and makes recommendations for changes to the chosen COAs to exploit the situation. After monitoring the flow of information, the commander assesses and makes recommended improvements to the use of intelligence collection assets and the processes used by subordinates to report required information.

BCT Executive Officer

8-278. The executive officer (XO) performs the organizational analysis of the FSF coordinating staff sections to ensure efficiency during the planning process according to the FSF commander's initial planning guidance. With the foreign counterpart, the XO advises and assists the counterpart in directing foreign staff sections as they develop estimates, plans, and orders. The XO monitors and assesses the liaison and coordination with FSF higher headquarters, recommending changes to improve efficiencies.

BCT Staff

8-279. Before the mission, members of the staff advise and assist foreign counterparts in preparing staff estimates and assess COAs for essential tasks. The staff helps write tentative plans or orders based on the FSF commander planning guidance and FSF standing operating procedures. Plans, depending on the situation, may include primary, alternate, contingency, and emergency (known as PACE) plans.

8-280. During execution, the staff helps foreign counterparts coordinate and assess the execution of FSF tasks. The staff assists in the dissemination of FSF plans or orders to senior and adjacent staff sections and supporting elements as required. The staff helps notify higher, lower, or adjacent staff sections of modified estimates and plans. The staff—led by the S-3 and S-2 and the S-3 and S-2 counterparts—helps update the CCIRs with the latest information and future requirements.

Personnel Staff Officer

8-281. The S-1 provides advice, assists, and assesses and makes recommendations to the foreign counterpart for all matters concerning human resources support. This includes monitoring the maintenance of foreign unit strength, pay, accountability of casualties, and unit morale. The S-1 must emphasize to subordinates the need to assist counterparts in paying troops and accounting for funds. Close observation of disbursement and unobtrusively polling FSF troops about their pay is a vital, but an unfamiliar, skill set amongst U.S. forces.

Note. U.S. forces' automated pay systems are nothing like the cash-only transactions in FSF. Graft, corruption (ghost soldiers/policemen), and extortion are rife in these circumstances.

Intelligence Staff Officer

8-282. The S-2 advises and assists the monitoring of FSF operations security to protect classified and sensitive material and operations and recommends improvements. By helping the foreign counterpart update the situation map, the S-2 helps to keep both commands up to date on the current situation. The S-2 assesses and recommends improvements to the standing operating procedures of the main CP (when established, the tactical command post [TAC {graphic}]) communications framework so the intelligence section receives situation reports. The S-2 helps the counterpart monitor the collection, evaluation, interpretation, and the dissemination of information. The S-2 assists in the examination of captured insurgent documents and material. The S-2 helps gather and disseminate intelligence reports from available sources to ensure the exploitation of all unit operations assets. The S-2 makes assessments to help the counterpart to brief and debrief patrols operating as a part of reconnaissance, surveillance, and security activities. The S-2 works with the advisor S-3 to develop an information collection plan with the FSF partner.

Note. Train personnel two deep in every staff section or advisory subunit on foreign disclosure before deployment. Interaction with host nation and FSF, even the North Atlantic Treaty Organization or other coalition allies requires foreign disclosure officer approval. This will become a huge bottleneck if not trained for and decentralized.

Operations Staff Officer

8-283. The S-3 helps the foreign counterpart to prepare tactical plans or orders using estimates, predictions, assessments, and information. The S-3 monitors command and communications nets, assists in preparing all plans and orders, and helps to supervise the training and preparation for operations. The S-3 monitors the planning process and makes recommendations for consistency with FSF partner objectives and goals.

Logistics Staff Officer

8-284. The S-4 advises and assists the foreign counterpart and makes assessments and recommendations to maintaining equipment readiness; monitoring the support provided to the foreign unit, its subunits, and attachments; and in recommending improvements. The S-4 helps to supervise the use of transportation assets.

Signal Staff Officer

8-285. The S-6 advises and assists the foreign counterpart for all matters concerning Department of Defense Information Network-Army (DODIN-A) operations, network transport, network sustainment, information services, and spectrum management operations within the BCT's SFA and FSF area of operations. The S-6 monitors communications security material throughout planning, preparation, and execution of SFA and FSF activities. The S-6 ensures SFA personnel are trained in the protection of sensitive communications equipment and cryptographic materials during the execution of FSF operations. The S-6 identifies SFA and FSF communications requirements, obtains communications resources for austere locations, and ensures redundant and backup systems are available and tested.

8-286. The S-6, in coordination with the division signal officer, or higher headquarters signal officer, continuously assesses and assists interorganizational information management coordination, normally required among participating interagency partners and the affected partner nation organizations. The S-6 uses assessments as part of the SFA and FSF communications synchronization plan. The S-6 uses foreign disclosure procedures and a tailored and responsive information-sharing process as part of the SFA and FSF assessment plan for dissemination with interagency partners and multinational audience.

Financial Management Staff Officer

8-287.   The brigade financial management staff officer (S-8) is the principal SFA financial management advisor to the BCT commander. The S-8, in coordination with legal representatives, advices the commander regarding laws and financial management regulations governing obligations, expenditures, and limitations on the use of funds within BCT's SFA area of operations. The S-8:

- Coordinates financial management policies and practices with U.S. counterparts in the contracting command.
- Identifies, certifies, and manages funds available for immediate SFA operations expenses.
- Integrates all financial management requirements into SFA planning.
- Analyses total cost to develop SFA funding requirements and submit requirements to the division or higher headquarters.
- Monitors and reports status of funding and spending plans.
- Coordinates contracting and financial management disbursing support for field ordering officers and pay agents.
- Monitors execution of the SFA contract expenditures.

Civil Affairs Operations Staff Officer

8-288.   The S-9, when assigned, is the principal and coordinating staff officer for synchronizing civil affairs operations and integrating civil-military operations within the BCT's SFA area of operations. The S-9 conducts the initial assessment during mission analysis that determines civil affairs force augmentation requirements. The S-9 provides direction and staff oversight of the supporting civil affairs units during planning, preparation, and execution, and throughout assessment. The S-9 ensures each COA effectively integrates civil considerations. The S-9's analysis considers the impact of SFA activities on public order and safety and enhances the relationship between military forces and civilian authorities and personnel in the BCT's area of operations. When the BCT is not assigned an S-9, the commander may assign these responsibilities to another staff member. The S-9 has SFA activities staff planning and oversight responsibility for—

- Advising the commander and staff and their counterparts on the effect on the civilian populations.
- Minimizing civilian interference, to include dislocated civilian operations, curfews, and movement restrictions.
- Deconflicting civilian and military activities with due regard for the safety and rights of dislocated civilians.
- Advising the commander on the long- and short-term effects (economic, environmental, and health) of SFA activities on civilian populations.
- Coordinating, synchronizing, and integrating civil-military plans, programs, and policies with that of the division or higher headquarters.
- Advising on the prioritization and monitoring of expenditures of allocated activities and other funds dedicated to civil affairs operation objectives.
- Coordinating and integrating area assessments and area studies in support of SFA activities.
- Integrating civil information from supporting civil affairs units into the COP.

Civil Affairs Team

8-289.   A civil affairs team, when assigned to a BCT deployed to an SFA mission, conducts civil affairs operations in support of the division or higher echelon headquarters civil-military operations plan. The civil affairs team is the basic civil affairs tactical support element provided to a supported commander. The civil affairs team executes civil affairs operations and is capable of conducting civil reconnaissance and civil engagement along with assessments of the civil component of the operational environment. The success of the overarching civil affairs operations plan is predicated on the actions of the civil affairs team at the lowest tactical levels. The civil affairs team, due to its limited capabilities, relies on its ability to leverage other civil affairs assets and capabilities through reachback to the civil affairs company civil-military operations center to shape operations. The civil-military operations center is a standing capability formed by civil affairs units and is tailored to the specific tasks associated with the mission and normally augmented by other enablers such as engineer, medical, and transportation resources available to the supported commander. The civil affairs team attached to the BCT will interface with the BCT S-9 (when assigned or S-3), civil-military operations center and civil affairs company at the BCT or division level to ensure all civil-military operations are nested with the higher commander's civil-military operations plan.

Civil-Military Teams

8-290. Upon deployment and when assigned, civil-military teams advise the SFA and FSF commanders and staffs on civil-military considerations and coordinate efforts of any civil affairs units supporting the FSF operation. Civil-military teams mentor counterpart teams and the supported foreign element staff on civil-military operations and the importance of respecting human rights. Civil-military teams may introduce counterparts to relevant nongovernmental organizations, U.S. Agency for International Development project officers, and provincial reconstruction team staff.

Note. The judge advocate (judge advocate general corps) mentors (provide legal mentorship) and coordinates the legal and moral obligations of BCT commanders to civilian populations under their control or when supporting FSF during SFA. (See AR 27-1 and FM 1-04.) The public affairs officer provides advice and counsel to the commander and the staff on how affected external and internal publics will accept and understand the BCT's operations. The public affairs officer understands and coordinates the flow of information to Soldiers, other U.S. units, and the public. (See FM 3-61.) BCT personal, coordinating, and special staff officers' mentor, and advise and assist foreign counterparts throughout planning, preparation, execution, and assessment of SFA and FSF activities.

Battalion and Company-Level and Below Advisors

8-291. Battalion and company-level and below advisors assist foreign counterparts to analyze the FSF mission and commander's intent from higher headquarters. Advisors assist FSF commanders and subordinate leaders restate the mission, conduct an initial risk assessment, identify a tentative decisive point, and define their own intent. Advisors assist their foreign counterparts to analyze the mission and operational variables. From these variables, advisors help their foreign counterparts to develop a COA that meets the higher headquarters concept of operations and commander's intent. Advisors assist in the conduct of operations and the flow of information to the FSF higher commander. (See ATP 3-21.20 and ATP 3-21.10 for additional information.)

Train Foreign Security Forces

8-292. Trainers (or advisors) within the BCT consistently provide and instill leadership at all levels of the FSF organization. Depending on the circumstances, the BCT or subordinate unit may execute an SFA training mission(s) unilaterally, or as part of a multinational force. In any case, leadership is especially important in the inherently dynamic and complex environment associated with SFA. SFA activities require the personal interaction of trainers (or advisors) and FSF trainees, and other military and civilians' organizations/agencies. A high premium is placed on effective leadership from junior, to the most senior noncommissioned and commissioned officers. This leadership must fully comprehend the operational environment and be prepared, fully involved, and supportive for FSF training to succeed. An effective FSF requires leadership from both the provider and the recipient sides throughout training to help build the FSF capacity to train their own forces.

Security Force Assistance—Training Activities

8-293. Trainers, within the BCT, work with the FSF to give a positive context and narrative to the FSF professionalization efforts and capacity to secure the population. Coordination of the information themes and messages among the BCT, FSF, and the host-nation government, and the presentation or availability of information to the indigenous population can limit or mitigate the propaganda efforts of insurgents or hostile forces. This may serve to mitigate the potential for destabilizing influences of hostile forces or criminal elements to propagandize the BCT's training effort and damage the FSF credibility and legitimacy. (See

FM 7-0 for additional information.) The following training activities, used for discussion purposes, present training actions to support FSF training.

Training Assessment

8-294. Prior to training the FSF, the BCT commander and subordinate commanders and leaders begin with a training assessment, in coordination with the FSF commander and subordinates, of the training plans designed before the BCT's employment. This assessment is important to evaluate the FSF and to exercise the working relationship between subordinate units of the BCT and the FSF. The training assessment covers all aspects of leadership, training, sustainment, and professionalization. To support an assessment, the commander analyzes the following specific foreign unit considerations:

- The unit's mission and mission-essential task list and capability to execute them.
- Staff capabilities.
- Personnel and equipment authorization.
- Physical condition.
- Any past or present foreign influence on training and combat operations.
- Operational deficiencies identified during recent operations or exercises with U.S. personnel.
- Sustainment capabilities, to include training programs.
- Internal training programs and personnel.
- Training facilities.

8-295. The commander assesses the level of professionalism of FSF, both units and individuals. Adhering to established rules of engagement, ethics that meet the established laws and regulations of the commanding authority, laws for land warfare, and human rights are key areas that require assessment. The FSF support of civilian leaders and political goals also fall within this assessment.

8-296. Subordinate commanders and leaders within the BCT, working with FSF leaders evaluate current members of the FSF for past military skills and positions. Often military reorganizations arbitrarily shift personnel to fill vacancies outside their knowledge and experience.

Analysis of the Prepared Training Plan

8-297. After completing the training assessment, the BCT commander and subordinate commanders and leaders analyze the prepared training plans and determine if changes are necessary. Training plans stress the deficiencies identified in the training assessment. The training plan identifies those in the host nation able to help train FSF to strengthen the legitimacy of the process. Using a comprehensive approach within the BCT's area of operations can provide support and expertise that enhance the training and operations process, and the FSF eventual self-sustainment. As the FSF gains sufficient capacity and capabilities to perform independently, trainers/advisors transition from a leading role to a mentoring role.

Program of Instruction

8-298. In coordination with the FSF staff and subordinate units, the staff and subordinate units of the BCT develop programs of instruction. These programs incorporate all training objectives that satisfy the training requirements identified during assessment. Training programs support these requirements. The FSF commander approves these programs of instruction before execution by subordinate units within the BCT. When executing programs of instruction, trainers/advisors adhere to training schedules consistent with changes in the mission variables. Trainers/advisors ensure through their counterparts and the FSF commander that all personnel receive training. Foreign counterpart trainers rehearse all classes approved on the programs of instruction.

Presentation of Instruction

8-299. Presenting the training material properly, trainers follow lesson outlines approved in the programs of instruction. All training clearly states the task, conditions, and standards desired during each lesson, ensuring the FSF understand them. Trainers/advisors state all warning and safety instructions (through interpreters when required) to the FSF. The training to reinforce the concepts includes demonstrations of the execution of each task, stressing the execution as a step-by-step process. Trainers monitor FSF progress during instruction and practical exercises, correcting mistakes as they are made.

Training Methodology (Crawl-Walk-Run)

8-300.   An effective method of training used is the crawl-walk-run training methodology to assist trainers in teaching individual tasks, battle drills, and collective tasks, and when conducting field exercises. This methodology is employed to develop well-trained leaders and units. Crawl-walk-run methodology is based on the three following characteristics in lane training:

- Crawl (explain and demonstrate). The trainer describes the task step-by-step, indicating what each individual does.
- Walk (practice). The trainer directs the unit to execute the task at a slow, step-by-step pace.
- Run (perform). The trainer requires the unit to perform the task at full speed, as if in an operation, under realistic conditions.

8-301.   During all phases, the training must include the mission of the unit in the context of the higher unit's mission to assist with the practical application of the training. Identifying the higher commander's mission and intent, as well as the tasks and purposes of other units in the area, adds context to the training. This method is expanded to include the role of other actors.

8-302.   Trainers continue individual training to improve and sustain individual task proficiency while units train on collective tasks. Collective training requires interaction among individuals or organizations to perform tasks, actions, and activities that contribute to achieving mission-essential task proficiency. Collective training includes performing collective, individual, and leader tasks associated with each training objective, action, or activity. (See FM 7-0 for additional information.)

Collective Training

8-303.   Collective training starts at squad level. Squad battle drills provide key building blocks to support FSF operations. Trainers within the BCT link battle drills and collective tasks through a logical, tactical scenario in situational training exercises. Although this exercise is mission-oriented, it results in more than mission proficiency. Battle drills and collective tasks support situational training exercises, while these exercises support operations. Trainers/advisors must understand the operational environment when training FSF; training incorporates how internal and external threats and civilians affect the environment.

8-304.   Flexibility in using Army doctrine in training enhances efforts to make training realistic. Trainers/advisors modify Army doctrine to fit the FSF level of expertise, command and control systems, the tactical situation, security measures, and sustainment base. Often the structure and capabilities of FSF differ from that required by Army doctrine. When FSF counter an insurgency, these exercises emphasize interplay among psychological and tactical, populace and resources control, intelligence, and civil affairs operations. (See FM 7-0 for additional information.)

Individual Training

8-305.   Individual training within the FSF by the BCT emphasizes physical and mental conditioning, tactical training, basic rifle marksmanship, first aid, combatives, and the operational environment. Individual training includes general tactics and techniques of security operations and the motivation, operations, and objectives of internal and external threats. Tough and realistic training conditions FSF troops mentally and physically to withstand the strain of continuous operations. BCT subordinate leaders' cross-train the FSF on weapons, communications systems, individual equipment, and skills particular to their unit. Personnel losses must never cause weapons, communications equipment, or essential skills to be lost due to a lack of fully trained replacement personnel.

Small-Unit Leader Training

8-306.   SFA activities frequently entail rapidly changing circumstances; thus, FSF small-unit leaders must be able to plan and execute operations with little guidance. Trainers/advisors stress small-unit leadership training concurrently with individual training. Tools the trainer uses to train leaders are manuals, previously established training, tactical exercises without troops, and unit missions. Small-unit leader training develops aggressiveness, tactical proficiency, and initiative. Small-unit leader training should include combined arms technical training procedures for forward observer and tactical air control party personnel. Leadership training includes land navigation in difficult terrain and under conditions of limited visibility. Mission readiness and the health and welfare of subordinates are continuous parts of training.

FOREIGN SECURITY FORCES DEVELOPMENT TASKS —ASSESS

8-307.   The functional tasks of OTERA-A (see paragraph 8-236) serve as SFA capability areas used by the BCT to develop, change, or improve the capability and capacity of the FSF. By conducting an assessment of the FSF, the BCT can determine which area or areas within the OTERA-A construct to use to improve the FSF to the desired capability and capacity. In essence, the BCT conducts an assessment of the FSF against desired capabilities and then develops an OTERA-A plan to help the FSF build capability and capacity.

Assess Foreign Security Forces

8-308.   During SFA assessments to evaluate the status of FSF capabilities and capacity, assessments by the BCT, establish a measurement at a particular time and can be compared to other assessments to observe differences and progress attributable to SFA activities. Activity assessment by the BCT involves deliberately comparing forecasted outcomes with actual events to determine the overall effectiveness of the BCT's employment. More specifically, assessment helps the BCT commander and staff determines progress toward attaining the desired end state, achieving objectives, and performing tasks.

Security Force Assistance —Assessment Activities

8-309.   The assessment developmental task, not limited to planning, preparing, or executing, by the BCT is ongoing throughout the operations process. Assessment involves continuously monitoring and evaluating the operational environment to determine what changes might affect the conduct of training and operations. The following paragraphs used for discussion purposes, present assessment actions to assess training and operations.

Foreign Security Forces Training and Evaluation

8-310.   In training, the after-action review provides the critical link between training and evaluation. The review is a professional discussion that includes the training participants and focuses directly on the training goals. An after-action review occurs after all collective FSF training. Effective after-action reviews review training goals with the responsible FSF commander or subordinate leader. During the review, SFA trainers/advisors ask leading questions, surface important tactical lessons, explore alternative COAs, assist the retention teaching points, and keep the after-action review positive.

Comprehensive Review

8-311.   The BCT commander encourages the FSF commander to conduct a comprehensive review of collective training events with the entire unit, or at a minimum, with key subordinate leaders. If possible, the review occurs during the field portion of the training when the unit assembles at logical stopping points. During the review, the commander and subordinate trainer/advisors avoid criticizing or embarrassing the FSF commander or subordinates. After-action reviews provide feedback to increase and reinforce learning, providing a database for key points. During reviews within subordinate echelons, evaluators draw information from FSF subordinate leaders to form possible alternative COAs for future activities.

Note. It is important to conduct comprehensive after-action reviews and reports, focusing on the specifics of the SFA activities, to gather information as soon as possible after execution.

Short-, Mid-, and Long-term Success

8-312.   During SFA activities, including FSF operations, success is defined within the context of three periods: short-, mid-, and long-term. In the short-term period, FSF make steady progress in fighting threats, meeting political milestones, building democratic institutions, and standing up security forces. In the midterm period, FSF lead fighting threats and provide security, have a functioning government, and work towards achieving economic potential. In the long-term period, FSF are peaceful, united, stable, and secure; integrated into the international community; and a full partner in international security concerns.

Monitor the Current Situation

8-313. The BCT commander and subordinate advisors help foreign counterparts monitor the current situation for unanticipated successes, failures, or enemy actions. As the commander assesses the progress of FSF operations, the commander looks for opportunities, threats, and acceptable progress. The commander considers, as part of the MDMP, the second- and third-order effects of the FSF operation. The commander and subordinate advisors develop a cultural awareness and use this awareness so that operations and relationships achieve the desired end state.

Operational Success

8-314. Throughout the operation, the BCT commander and staff assists the FSF commander and staff in addressing changes to the operation and the feeding the assessments of the progress or regression back into the planning process. The closer SFA and FSF commanders work with trainer/advisor teams and the more they interact with local political and cultural leaders, the better the overall chances of mission success. Keys to operational success within the SFA and FSF area of operations, although not all inclusive, include the following:

- Establish MOEs to provide benchmarks against which the commander assesses progress toward accomplishing the mission.
- Establish MOPs to determine whether a task or action was performed to standard.
- Establish close and continuing relationships with all advisor teams, other actors operating in its area of operations, and foreign area officers with local or regional expertise.
- Establish close and continuing relationships with all foreign units (military, police, and others) operating in the area of operations.
- Establish close and continuing relationships with all political entities and actors within the area of operations.
- Establish redundant communications within the area of operations, especially when the BCT shares its area of operations with other entities that have cultural differences and lack of or degraded communications.

SECTION VI – TRANSITIONS

8-315. The BCT commander and staff must always keep in mind the situation may escalate to combat operations at any time. The BCT may be ordered to transition to offensive or defensive operations if the focus of the operation changes from stability. The commander task organizes units to expeditiously transition to combat operations while maintaining a balance between conducting stability operations tasks and maintaining a combat posture. The BCT commander must consider transitions to outside authorities, including host nation, international government organizations, other allied coalition forces or another U.S. Government agency when a transition to offense or defense occurs. This section concludes with a discussion of transitions during SFA activities.

TRANSITION TO THE CONDUCT OF DEFENSIVE OPERATIONS

8-316. The primary focus on stability operations tasks in an operation may transition to a focus on defensive operations for three basic reasons. The situation within the BCT's area of operations has deteriorated so much that a primarily defensive orientation is required. An outside superior force threatens the BCT's area of operations, or higher orders the BCT to conduct a defense in a new area of operations.

8-317. The BCT commander's initial defensive scheme may be an area defense executed through smaller individual perimeter defenses. A mobile defense requires more time, deliberate planning, and organization of forces to accomplish. The commander performs the following actions in preparation for the transition to defense:

- Concentrates and orients forces on the enemy.
- Redirects BCT assets from current stability operations tasks to area security operations.
- Establishes a main battle area (MBA).
- Evacuates or secures critical facilities, organizations, and equipment with limited forces.
- Reconfigures sustainment operations to align with defensive operations.
- Informs partners of the change in operations and the plan to conduct stability operations tasks with limited resources.
- Conducts a battle handover when required with successor within the time constraints of the new mission.
- Ensures the mind-set of subordinate leaders and Soldiers has transitioned to the defense.

8-318.   Transitioning from supporting stability operations tasks to a retrograde normally occurs if civil strife escalates and the sources of instability are more overwhelming than the BCT and unified action partners can mitigate. The primary objective is for the BCT to preserve its forces, and gain time allowing conditions to change so that the BCT can continue its follow-on mission. The presence of the enemy, analyzed with time available, dictate what form of retrograde the BCT conducts. (See chapter 7 for additional information.)

TRANSITION TO THE CONDUCT OF OFFENSIVE OPERATIONS

8-319.   The BCT commander or higher command may order an offensive action such as an attack or movement to contact. The BCT commander and subordinate commanders must quickly orient their forces for the offense. (See chapter 6 for additional information.) This may include—

- Releasing Cavalry forces from current stability operations tasks to conduct reconnaissance and security operations tasks to seize the initiative.
- Concentrating forces in preparation for offensive actions.
- Securing critical facilities, organizations, and equipment with limited forces.
- Reconfiguring sustainment operations to align with the offense.
- Informing partners of the change in operations.
- Conducting a battle handover when required with successor within the time constraints of the new mission.
- Ensuring the mind-set of subordinate leaders and Soldiers has transitioned to the offense.

TRANSITIONS DURING SECURITY FORCE ASSISTANCE

8-320.   Transitions during SFA are dependent upon the conditions within the operational environment. Transitions are initially identified during planning using a comprehensive approach. Transitions can occur simultaneously or sequentially in different levels or war and in separate echelons, to include having potentially at the tactical level, transitions for different units within the BCT's area of operations. For example, a major transition can include a battalion or company in the beginning of an initial response phase being the supported unit with the FSF transitioning to the supported unit later on in the phase. At this point in the transformation phase, the area in which the battalion or company conducts SFA will expand. This expansion can occur multiple times during the transformation phase, which is based on conditions, especially the capability and capacity of FSF. The BCT commander, to facilitate flexibility, visualizes and incorporates branches and sequels into the overall plan to enable transitions. Unless planned, prepared for, and executed efficiently, transitions can reduce the tempo of the operation, slow its momentum, and surrender the initiative.

Chapter 9

Sustainment

Sustainment operations provide support and services to ensure freedom of action, extend operational reach, and prolong endurance. Brigade combat team (BCT) sustainment organizations synchronize and execute sustainment operations in support of the BCT under all conditions to allow the BCT to seize, retain, and exploit the initiative. BCT subordinate units and sustainment staffs anticipate future needs to retain freedom of movement and action at the end of extended and contested lines of operation. The brigade support battalion (BSB) commander is the BCT's senior logistician. The BSB commander is responsible for sustainment synchronization and execution across the BCT's area of operations. This chapter describes sustainment operations in support of the BCT, specifically the functions, command and staff roles and responsibilities, and unit relationships throughout high tempo and decentralized operations.

SECTION I – FUNDAMENTALS OF SUSTAINMENT

9-1. *Sustainment* is the provision of logistics, financial management, personnel services, and health service support necessary to maintain operations until successful mission completion (ADP 4-0). Sustainment within the BCT is a brigade wide responsibility; commanders at all levels and the various staffs, as trusted Army professional and stewards of the Army Profession, have a role to ensure sustainment support is ethically, effectively, and efficiently planned, understood, and executed. Sustainment must be coordinated and synchronized to facilitate the operational pace and support the commander's priorities before, during, and after operations. Staffs and planners must fully integrate sustainment throughout the operations process. (See ADP 4-0 and FM 4-0 for additional information.)

SUSTAINMENT WARFIGHTING FUNCTION

9-2. The sustainment warfighting function is the related tasks and systems that provide support and services to ensure freedom of action, extend operational reach, and to prolong endurance. The endurance of Army forces is primarily a function of their sustainment. Sustainment determines the depth and duration of Army operations. It is essential to retaining and exploiting the initiative. Sustainment provides the support necessary to maintain operations until mission accomplishment. The sustainment warfighting function consists of four major elements: logistics, financial management, personnel services, and health service support. Paragraphs 9-3 through 9-6 discuss the functional elements found in each of the sustainment elements applicable to the BCT.

LOGISTICS

9-3. *Logistics* is planning and executing the movement and support of forces. It includes those aspects of military operations that deal with: design and development, acquisition, storage, movement, distribution, maintenance, evacuation, and disposition of materiel, acquisition or construction, maintenance, operation, and disposition of facilities, and acquisition or furnishing of services (ADP 4-0). The elements of logistics make up the distinct function of logistics. The elements of logistics (see FM 4-0) within the BCT includes—

- Maintenance. (See ATP 4-33.)
- Transportation. (See FM 4-01.)
- Supply. (See FM 4-40.)

- Field services. (See FM 4-40.)
- Distribution. (See 4-0.1.)
- Operational contract support. (See ATP 4-10.)
- General engineering support. (See ATP 3-34.40.)

FINANCIAL MANAGEMENT

9-4. *Financial management* is defined as the sustainment of the United States Army and its unified action partners through the execution of Fund the Force, Banking and Disbursing, Accounting Support and Cost Management, Pay Support and Management Internal Controls (FM 1-06). The BCT commander leverages fiscal policy and economic power to enable decisive action across the range of military operations. Financial management encompasses finance operations and resource management to ensure that proper financial resources are available to accomplish the mission in accordance with the commander's priorities. Properly sized, modular financial management structures in the context of financial management operations planned and executed in consideration of operational variables and mission variables conducts these capabilities. Financial management capabilities reside within the BCT to sustain and support operations until successful mission accomplishment. Financial management operations extend the BCT's operational reach and prolong operational endurance, allowing the commander to accept risk and create opportunities for decisive results. (See FM 1-06 for additional information.)

PERSONNEL SERVICES

9-5. *Personnel services* are sustainment functions that man and fund the force, maintain Soldier and Family readiness, promote the moral and ethical values of the nation, and enable the fighting qualities of the Army (ADP 4-0). It includes essential personnel services such as evaluations, leaves and passes, awards and decorations, rest and recuperation, postal, personnel accountability, casualty operations, and personnel management. Personnel services within the BCT include the following:

- Human resources support. (See FM 1-0.)
- Legal support. (See FM 1-04.)
- Religious support. (See FM 1-05.)

HEALTH SERVICE SUPPORT

9-6. *Health service support* encompasses all support and services performed, provided, and arranged by the Army Medical Department to promote, improve, conserve, or restore the mental and physical well-being of personnel in the Army. Additionally, as directed, provide support in other Services, agencies, and organizations. This includes casualty care (encompassing a number of Army Medical Department functions—organic and area medical support, hospitalization, the treatment aspects of dental care and behavioral/neuropsychiatric treatment, clinical laboratory services, and treatment of chemical, biological, radiological, and nuclear patients), medical evacuation, and medical logistics (FM 4-02). Health services support elements provide health service support within maneuver units of the BCT and the brigade support medical company (known as BSMC) of the BSB. (See FM 4-02 for additional information.) Health service support within the BCT, and support and services to the BCT include the following:

- Casualty care— (See ATP 4-02.3 and ATP 4-02.5.)
  - Organic medical support.
  - Area medical support.
  - Hospitalization.
  - Dental treatment.
  - Behavioral health.
  - Clinical laboratory services.
  - Treatment of chemical, biological, radiological, and nuclear (CBRN) patients. (See ATP 4-02.7.)
- Medical evacuation. (See ATP 4-02.2.)

- Medical logistics. (See ATP 4-02.1.)

PRINCIPLES OF SUSTAINMENT

9-7. The principles of sustainment are essential to maintaining combat power, enabling strategic and operational reach, and providing Army forces with endurance. While these principles are independent, they are also interrelated. The BCT commander and staff use the eight guiding principles of sustainment (integration, anticipation, responsiveness, simplicity, economy, survivability, continuity, and improvisation) to shape the sustainment support to ensure freedom of action and prolonged endurance throughout the BCT. The principles of sustainment and the principles of logistics are the same. (See ADP 4-0 and FM 4-0 for additional information.)

PRINCIPLES OF FINANCIAL MANAGEMENT

9-8. Similar to sustainment, there are six financial management principles: stewardship, synchronization, anticipation, improvisation, simplicity, and consistency. These principles are critical to maintaining combat power, operational reach throughout the levels of war, and the endurance of the BCT. Although independent of one another, these principles must be integrated in the planning and execution of financial management operations at echelons above and below the BCT. This integration facilitates the optimal allocation of financial resources to accomplish the BCT's mission. (See FM 1-06 for additional information.)

PRINCIPLES OF PERSONNEL SERVICES

9-9. The principles of personnel services guide the functions for maintaining Soldier and Family support, establishing morale and welfare, funding the force, and providing personal legal services to personnel. They are in addition to the principles of sustainment and complement logistics by planning for and coordinating efforts that provide and sustain personnel. The following principles are unique to personnel services—synchronization, timeliness, accuracy, and consistency—and contribute to current and future BCT operations. These principles ensure personnel services effectively align with military actions in time, space, purpose and resources as well as ensuring decision makers within the BCT have access to relevant personnel services information and analysis. The stewardship of limited resources and the accuracy of information have an impact on the BCT commander and staff along with other decision makers within and above the BCT. Consistency ensures uniform and compatible guidance and personnel services to forces across all levels of operations. (See ADP 4-0 and FM 1-0 for additional information.)

PRINCIPLES OF THE ARMY HEALTH SYSTEM

9-10. The six principles of the Army Health System (AHS) are the foundation—enduring fundamentals—upon which the delivery of health care in a field environment is founded. Conformity, proximity, flexibility, mobility, continuity, and control are the principles that guide medical planning in developing health service support (see ATP 4-02.3) and force health protection (see ATP 4-02.8) missions, which are effective, efficient, flexible, and executable. These missions support the BCT commander's scheme of maneuver while retaining a focus on the delivery of health care. The AHS principles apply across all medical functions. They are synchronized through medical command and control and close coordination of all deployed medical assets through operational and medical channels.

> *Note.* The AHS includes both health service support and force health protection. The health service support mission is part of the sustainment warfighting function. The force health protection mission falls under the protection warfighting function. (See FM 4-02 for additional information on the principles of the AHS.)

SECTION II – SUSTAINING THE BRIGADE COMBAT TEAM

9-11. Sustainment based on an integrated process, (people, systems, materiel, health services, and other support) links sustainment to operations. Sustaining the BCT in austere environments, often at the ends of extended lines of communications, requires a logistics network capable of projecting and providing the support and services necessary to ensure freedom of action, extend operational reach, and prolong endurance. Success will require deployment and distribution systems capable of delivering and sustaining the BCT from strategic bases to points of employment within and throughout the operational area at the precise place and time of need.

SUSTAINMENT STAFF

9-12. The BCT commander and staff integrate forces, the operational plan, and existing and available logistics and services to ensure that the BCT can win across the range of military operations. The sustainment staff plans, directs, controls and coordinates sustainment in support of those operations. The following proponents make up the sustainment staff within the BCT headquarters.

EXECUTIVE OFFICER

9-13. The BCT executive officer (XO) provides oversight of operations and sustainment planning for the BCT commander. The XO directs, coordinates, supervises and synchronizes the work of the staff to ensure the staff is integrated and aligned with the BCT commander's priorities. (See chapter 4 for additional information.) The XO's primary sustainment duties and responsibilities in relation to sustainment operations include—

- Ensuring the concept of support is synchronized with the scheme of maneuver in-depth.
- Providing oversight over the maintenance status of the BCT.
- Setting priorities for the BCT staff sustainment cell (personnel staff officer [S-1], logistics staff officer [S-4], signal staff officer [S-6], specifically, all matters concerning sustainment network operations, financial management staff officer [S-8], surgeon, and chaplain).
- Monitoring contract operations for the BCT.

LOGISTICS STAFF OFFICER

9-14. The BCT S-4 is the coordinating staff officer for logistical planning and operations. The S-4 provides staff oversight to BCT units in the areas of supply, maintenance, transportation, and field services. The S-4 is the BCT staff integrator between the BCT commander and the BSB commander who executes sustainment operations for the BCT. (See chapter 4 for additional information.) Primary duties and responsibilities include but are not limited to—

- Developing the logistics plan, in coordination of the BSB commander and support operations office, to support BCT operations and determining support requirements necessary to sustain BCT operations.
- Coordinating support requirements with the division assistant chief of staff, logistics and BCT support operations section on current and future support requirements and capabilities.
- Conducting sustainment preparation of the operational environment.
- Managing the logistics status (known as LOGSTAT) report for the BCT.
- Monitoring and analyzing equipment readiness status of all BCT units.
- Requesting transportation to support special transportation requirements such as casualty evacuation and troop movement.
- Determining BCT requirements for all classes of supply, food preparation, water purification, mortuary affairs, aerial delivery, laundry, shower, and clothing/light textile repair. (See FM 4-0.)
- Recommending sustainment priorities and controlled supply rates to the commander.
- Monitoring and enforcing the BCT command supply discipline program throughout all phases of the operation.
- Managing organizational and theater provided equipment assigned to the BCT.
- Planning for inter-theater movement and the deployment of BCT personnel and equipment.

PERSONNEL STAFF OFFICER

9-15. The BCT S-1 is the principle staff advisor to the BCT commander for all matters concerning human resources support. The function of the BCT S-1 section is to plan, provide, and coordinate the delivery of human resources support, services, or information to all assigned and attached personnel within the BCT and subordinate units. The BCT S-1 may coordinate the staff efforts of the BCT equal opportunity, inspector general, and morale support activities. (See chapter 4 for additional information.) The S-1's primary duties and responsibilities include but are not limited to—

- Maintaining unit strength and personnel accountability statuses.
- Preparing personnel estimates and annexes.
- Planning casualty replacement operations.
- Assisting the support operation officer plan detainee, and dislocated civilian movement.
- Planning the BCT postal operation plan.
- Conducting essential personnel services for the BCT.

FINANCIAL MANAGEMENT OFFICER

9-16. As the principal financial management (resource management and finance operations) advisor to the commander, the financial management officer directs, prioritizes, and supervises the operations and functions of the BCT S-8 staff section. In coordination with the assistant chief of staff financial management, the S-8 establishes and implements command finance operations policy. The S-8 works with the servicing legal representative for advice regarding laws and financial management regulations governing obligations, expenditures, and limitations on the use of public funds. The S-8 coordinates financial management policies and practices with the U.S. Army financial management command to ensure guidance is according to Department of the Army mandates. (See chapter 4 for additional information.) Primary duties and responsibilities include but are not limited to—

- Identifying, certifying, and managing funds available for immediate expenses.
- Integrating all financial management requirements into operational planning.
- Utilizing staff, commanders, training calendar, fiscal triad, and analysis of total cost to develop funding requirements and submit requirements to higher headquarters.
- Receiving, developing, and disseminating financial management guidance at the BCT echelon.
- Monitoring and reporting status of funding.
- Submitting and monitoring the status of requirements packets and spending plans to the appropriate board.
- Coordinating contracting and financial management disbursing support for field ordering officers and pay agents.
- Managing the Government Purchase Card Program.
- Serving as the coordinator for the Managers' Internal Control Program.
- Monitoring execution of the BCTs contract expenditures.

SIGNAL STAFF OFFICER (MATTERS CONCERNING SUSTAINMENT NETWORK OPERATIONS)

9-17. The BCT S-6 is the principle staff officer for all matters concerning sustainment network operations, jointly consisting of Department of Defense information network (DODIN) operations, applicable portions of the defensive cyberspace operations, and sustainment. The S-6 is critical to ensure planning includes all considerations for maintaining communications throughout an operation. Primary network sustainment duties and responsibilities include but are not limited to—

- Modernization.
- Resource availability.
- Technical data and intellectual property.
- Supply chain-hardware and software.
- Network force structure implications.
- Interoperability (for unified action partners).

ASSISTANT BRIGADE ENGINEER

9-18. The assistant brigade engineer (known as ABE) oversees any contract construction activity planning, preparation, and execution in support of the S-4 contracting support plan. The ABE, in coordination with the brigade engineer battalion (BEB), and any additional supporting engineer units, assist in providing technical engineer oversight of contract construction activities within the BCT, unless assigned to other engineer assets outside of the BCT, for example—a forward engineer support team—main or the U.S. Army Corps of Engineers. The ABE, in coordination with the BEB commander, develops and oversees the details and scope of work, and submission of the BCT's contract construction requirements. (See ATP 3-34.22 for additional information.)

SURGEON

9-19. The BCT surgeon serves as the personal staff officer responsible for health service support and is the advisor to the commander on the physical and mental health of the BCT. The surgeon manages health service support activities and coordinates implementation through the BCT operations staff officer (S-3). The surgeon provides health service support and force health protection mission planning to support BCT operations. (See chapter 4 for additional information.) Primary duties and responsibilities include but are not limited to—

- Planning casualty care and area support medical treatment.
- Planning medical evacuation (ground and air).
- Planning dental care (operational dental care and emergency dental care).
- Coordinating medical logistics (class VIII, medical supplies, blood management, and field level and sustainment support medical maintenance, see ATP 4-02.1).
- Planning for brigade behavioral health/neuropsychiatric treatment.
- Advising the commander on treating patients contaminated with CBRN hazards and those potentially exposed.
- Planning and coordinating force health protection activities (preventive medicine, medical surveillance, occupational and environmental health, and field sanitation).
- Planning and coordinating for combat and operational stress control.
- Planning and coordinating veterinary services, dental services, and laboratory services.
- Advising on medical humanitarian assistance.
- Advising the command on the brigade health status, and the occupied or friendly territory's health situation within the command's assigned area of operations.
- Identifying potential medical hazards associated with the geographical locations and climatic conditions with the BCT's area of operations.

CHAPLAIN

9-20. The BCT chaplain and religious affairs noncommissioned officer provides religious support to the command group and brigade staff and exercises technical supervision over religious support by subordinate unit ministry teams. Chaplains personally deliver religious support. They have dual roles: religious leader and religious staff advisor. The chaplain as a religious leader executes the religious support mission to ensure the free exercise of religion for Soldiers, Families, and authorized civilians. As a personal staff officer, the chaplain advises the commander and staff on religion, morals, morale, and ethical issues, both within the command and throughout the area of operations. (See chapter 4 for additional information.) Primary duties and responsibilities include but are not limited to—

- Developing plans, policies, and programs for religious support.
- Coordinating and synchronizing area and denominational religious support coverage.
- Coordinating and synchronizing all tactical, logistical, and administrative actions for religious support operations.

BRIGADE SUPPORT BATTALION

9-21. As the BCT commander's primary sustainment organization, the BSB provides logistics and AHS to ensure freedom of action, extend operational reach, and prolong endurance to achieve success across the range of military operations. The BSB provides the BCT commander with increased flexibility to organize support for the BCT and to weight the sustainment effort by leveraging all BSB capabilities. The BSB in each of the different types of BCTs (Infantry brigade combat team [IBCT], Stryker brigade combat team [SBCT], and Armored brigade combat team [ABCT]) are similar in design with differences based on the type of BCT supported. Through the BSB's six forward support companies (FSCs), distribution company, field maintenance company, and BSMC, the BSB supports each maneuver battalion and squadron, the BEB, and the field artillery battalion within the BCT. Figure 9-1 depicts a typical BSB's task organization in support of a BCT.

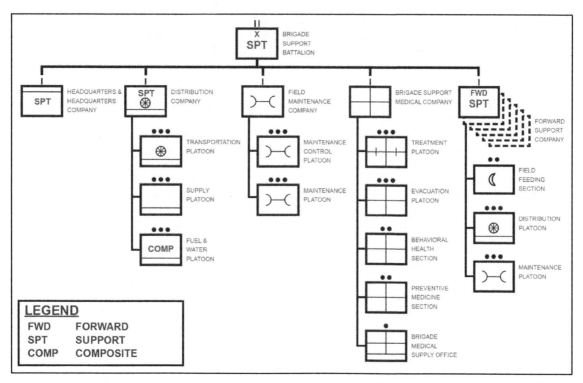

Figure 9-1. Brigade support battalion

9-22. The BSB supports the BCT's execution of all assigned operations. The BSB commander and staff plan, prepare, execute, and continuously assesses (in conjunction with the BCT commander and staff), sustainment operations in support of the BCT. The BSB provides supply class I (subsistence), class II (clothing, individual equipment, tentage, tool sets, and administrative and housekeeping supplies and equipment), class III (petroleum, oils, and lubricants [POL]), class IV (construction and barrier materiel), class V (ammunition), class VIII (medical), and class IX (repair parts); distribution support, food service support; and Roles 1 and 2 AHS support, and field maintenance and recovery. The BSB coordinates with division for sustainment requirements beyond its capability. (See ATP 4-90 for additional information.)

BRIGADE SUPPORT BATTALION COMMANDER

9-23. The BSB commander is the BCT's senior logistician. The BSB commander is responsible for sustainment synchronization and execution across the BCT's area of operations. The BSB commander, supported by the staff, uses the operations process to drive the conceptual and detailed planning necessary to understand, visualize, and describe the operational environment; make and articulate decisions; and direct, lead, and assess sustainment operations. The BSB commander executes the BCT's concept of support and advises the BCT commander on all aspects of sustainment support to the BCT. The BSB commander coaches both the BSB and BCT staff on the importance of synchronized logistics and health service support.

9-24. As the senior logistics commander charged with responsibility to sustain the BCT, the BSB commander must retain the ability to surge, mass, and reallocate logistics capabilities according to the BCT commander's intent and concept of the operation. The BSB commander makes recommendations to the BCT commander on the task organization for support to each maneuver battalion and squadron, the BEB, and the field artillery battalion.

SUPPORT OPERATIONS OFFICER

9-25. The support operations officer is assigned to the BSB and is not part of the BCT staff. However, the support operations officer serves as the principal staff officer responsible for synchronizing BSB sustainment operations for all units assigned or attached to the BCT. The support operations officer is responsible for applying sustainment capabilities against BCT requirements. The support operations officer conducts short and midrange planning (hours, days) and oversees the BSB's execution of the sustainment plan developed with the BCT S-4. The support operations officer also serves as the interface between supported units and the division sustainment brigade (known as DSB) and is responsible for coordinating support requirements with the DSB support operations section.

9-26. The support operations officer plans and coordinates orders published by the BSB S-3 for execution by all subordinate BSB units including the FSCs depending on the command relationship during the performance of current operations. These orders can include a synchronization matrix outlining the plan for execution. This enables the BCT S-4 and all subordinate BSB units to know the brigade support plan. The BSB support operations officer uses the LOGSTAT to update the logistics synchronization matrix. The updated LOGSTAT and logistics synchronization matrix complement paragraph 4 and Annex F of the operation order, or fragmentary order. (See ATP 4-90 for additional information.) The support operation officer's responsibilities include but are not limited to—

- Developing the BSB concept of support and the distribution or logistics package (LOGPAC) plan.
- Coordinating external support requirements with the BCT S-4, division assistant chief of staff, logistics, and supporting DSB.
- Planning, preparation, and oversight of logistics and AHS support tasks during BSB operations within the BCT's area of operations.
- Maintaining a common operational picture (COP) for logistics within each formation and throughout the BCT to ensure timely delivery of required support.
- Coordinates support for all units assigned or attached to the BCT.
- Advisor to the BCT commander for aerial delivery support.
- Plans and coordinates orders published by the BCT S-3 for execution by all subordinate BSB units, including the FSC, during current operations.
- Performs sustainment preparation of the operational environment and advises the commander on the relationship of support requirements.
- Plans and monitors support operations and makes necessary adjustments to ensure the BSB meets support requirements.
- Provides the status of commodities and materiel as required, updating LOGSTAT report.
- Providing centralized and integrated planning for all support operations within the BCT (structure varies by type of unit and generally includes transportation, maintenance, ammunition, AHS's support and distribution operations).
- Managing the BCT's maintenance readiness.

SUSTAINMENT AUTOMATION SUPPORT MANAGEMENT OFFICE

9-27. The sustainment automation support management office plans, prepares, executes, and sustains the tactical sustainment information systems network to meet the challenges in all environments. As the network administrator, the sustainment automation support management office will manage network configuration and supervise access operations related to supported units. Sustainment automation support management office coordinates with the S-6 to integrate into the BCT communications and electromagnetic warfare (EW) plan to ensure security and use of its vital functions.

FORWARD SUPPORT COMPANIES

9-28. The BSB has six organic FSCs that provide direct support to each of the BCT maneuver battalions and squadron, the field artillery battalion, and the BEB. The FSCs are the link from the BSB to the supported battalions and squadron and are the organizations that provide the BCT the greatest flexibility for providing logistics support. Each FSC is organized to support a specific combined arms, Infantry, Stryker, engineer, and field artillery battalion or Cavalry squadron. FSCs provide field feeding, bulk fuel distribution, general supply, ammunition, and field-level maintenance support to its supported unit. FSCs are structured similarly BCT types with the most significant differences in the maintenance capabilities.

9-29. The battalion or squadron S-4 is responsible to generate the support requirements for the battalion or squadron and creates their sustainment concept of support. The FSC commander can assist or provide feasibility guidance to the battalion or squadron S-4 for their concept of support as required and on a limited basis. The FSC commander is not a battalion staff officer. The FSC commander is responsible for executing logistics support in accordance with the BSB and supported commander's guidance and the BCT concept of support. Integrating the logistics plan early into the supported battalion's or squadron S-3's operational plan will help to mitigate logistic shortfalls and support the commander to seize, retain, and exploit gains.

9-30. FSCs receive technical logistic directions from the BSB commander. This allows the BSB commander and the BSB support operations officer to task organize the FSCs and cross level assets amongst FSCs when it is necessary to weight logistics support to the BCT. The task organization of the FSCs is a collaborative, coordinated effort that involves analysis by the staff and consensus amongst all commanders within the BCT. The BSB provides administrative support, limited logistic support, and technical oversight to the FSCs.

9-31. The BCT commander may attach or place an FSC (organic to the BSB) under operational control (OPCON) of their respective supported battalion or squadron based on the mission variables of mission, enemy, terrain and weather, troops and support available, time available, civil considerations (METT-TC). Upon the advice of the BSB commander, the BCT commander decides to establish these types of command relationships. The FSC attachment or OPCON to its supported battalion or squadron is generally limited in duration and may be for a specific mission or phase of an operation. Regardless of what command and support relationship the BCT commander determines for the FSCs, the BSB commander and staff retain the channels for technical supervision, advice, and support for logistics functions within the BCT (see ADP 3-0).

9-32. FSCs normally operate in close proximity to their supported battalion or squadron. The location of the FSC commander and the distance separating the FSC and the battalion is METT-TC dependent, with command and control, logistics asset protection, and required resupply turnaround times being key considerations.

9-33. The supported battalion may divide the FSC on the battlefield with some elements collocated with the supported unit and some elements located in the brigade support area (BSA). For example, it may be desirable to locate the FSC field maintenance teams with the supported unit and the remainder of the FSC in the BSA. The FSC commander in collaboration with the BSB commander and supported unit commander determines the task organization for the mission. FSC employment considerations include—

- Location, time, and distance of the FSC in relation to the supported battalion or squadron.
- Decision to separate elements of the FSC by platoon or other sub elements into multiple locations.
- Benefits of locating FSC elements in the BSA.
- Benefits of collocating battalion staff sections with the FSC.
- Benefits of collocating battalion medical elements with the FSC.
- Security of the FSC locations and during movement.
- Establishment and location of a maintenance collection point.

9-34. The FSCs have a headquarters section, a distribution platoon, and a maintenance platoon. The headquarters' food service section provides class I support, food service, and food preparation for the company and its supported battalion. The food service section prepares, serves, and distributes the full range of operational rations.

9-35. The distribution platoon of the FSC oversees LOGPAC operations and manages the distribution of supplies coming from or passing through the FSC in support of its battalion or squadron. The distribution platoon conducts replenishment operations and provides general supplies, fuel, and ammunition to its supported unit. The FSC distribution platoon consists of a platoon headquarters and four squads that can be task organized to distribute classes II, III, IV, V, and VII. FSCs' maintenance platoons vary based upon the equipment and major weapon systems of the supported unit.

9-36. The maintenance platoon of the FSC performs field-level maintenance, maintenance management functions, dispatching, and scheduled maintenance operations for their supported battalion and squadron and FSC vehicles and equipment. The platoon consists of the platoon headquarters section, maintenance control section, maintenance section, service and recovery section, and the field maintenance teams. The FSC maintenance platoon establishes the maintenance collection point and provides vehicle and equipment evacuation and maintenance support to the field maintenance teams. The maintenance collection point is normally located near or collocated with the combat trains for security and should be on or near a main axis or supply route. Field maintenance teams evacuate vehicles and equipment that require evacuation for repair and return, have an extended repair time, or when the vehicle or equipment exceeds its maintenance capabilities and augmentation is necessary.

9-37. Mechanics for combat systems (M1, M2, M109, and Strykers) are only found in the FSCs, and only by exception in the BSB field maintenance company. The FSC evacuates nonmission capable equipment from the forward line of troops, company trains, and combat trains to the FSC in the field trains or BSA, however, the BSB will need to task organize mechanics from the FSCs to be able to affect repairs.

DISTRIBUTION COMPANY

9-38. The BSB's distribution company is the primary supply and transportation hub of the BCT. It provides the supply and transportation components of logistics support to the BCT. The distribution company consists of a transportation platoon, a supply platoon, and a water and petroleum platoon and manages the distribution of supplies to the BCT. The company manages the distribution of supplies to the BCT and provides distribution capability for classes I, II, III (bulk and packaged), IV, V, VII, IX, and water.

9-39. The distribution company provides supply support through the FSCs and normally operates within the designated BSA. The BEB's FSC provides supply support to the BCT headquarters.

9-40. The transportation platoon of the distribution company provides transportation support to the BCT and distribution of supplies to the various FSCs. Of particular note, the transportation platoon cannot provide troop transport for the IBCT. When troop transport is required that is not within the capability of the transportation platoon, the support operations officer and BCT S-4 coordinates with the division assistant chief of staff, logistics and DSB for support.

9-41. The supply platoon of the distribution company provides classes I, II, III packaged, IV, V, VII, and IX support to the BCT through a multiclass supply support activity (SSA) and an ammunition transfer and holding point. The multiclass SSA receives, stores, and issues supply classes I, II, III packaged, IV, VII, and IX. The SSA is capable of handling retrograde of serviceable and unserviceable materiel.

9-42. The ammunition transfer and holding point section supports the BCT with class V and operates the BCT ammunition transfer and holding point. The ammunition transfer and holding point receives, temporarily stores, and issues class V. The ammunition transfer and holding point transfers munitions to BSB transportation assets and, if the situation dictates, holds ammunition for supported units and provides this ammunition to the supported units FSCs.

9-43. The water and petroleum platoon provides water and petroleum distribution for the BCT. The platoon does not provide a water purification or petroleum storage capability. If the BCT requires water purification or petroleum storage, the BSB must coordinate for this support. The lack of water purification or petroleum storage is particularly important in the planning phase of operations, and the BSB must plan for water and petroleum support from their supporting DSB. As the operational plan develops, the BSB must continually update their supporting division sustainment support battalion (known as DSSB) and DSB to ensure seamless water and petroleum support and continued momentum.

FIELD MAINTENANCE COMPANY

9-44. The field maintenance company provides field-level maintenance support to the BCT. Field-level maintenance is on or near system maintenance, often using line replaceable unit and component replacement, in the owning unit, using tools and test equipment found in the unit. Field-level maintenance is not limited to remove and replace actions, but also allows for repair of components or end items on or near system. Field-level maintenance includes adjustment, alignment, service, applying approved field-level modification work orders as directed, fault/failure diagnoses, battle damage assessment, repair, and recovery.

9-45. The field maintenance company provides lift capabilities, recovery of organic equipment, additional recovery support to supported units, and support of maintenance evacuation. Field-level maintenance is always repair and return to the user and includes maintenance actions performed by operators. The company provides limited maintenance support to the FSCs for low-density commodities such as communications, electronics, and armament equipment. The field maintenance company normally operates within the designated BSA.

BRIGADE SUPPORT MEDICAL COMPANY

9-46. The BSMC provides Role 1 (unit level medical care) and Role 2 (basic primary care) AHS support to all BCT units operating within the BCT area of operations as well as on an area basis to units outside the BCT. (See ATP 4-90.) The BSMC normally operates within the designated BSA.

9-47. The BSMC evacuates, receives, triages, treats, and determines the disposition of patients based upon their medical condition. This includes tactical combat casualty care (TCCC), primary, and emergency treatment, including basic primary care. The BSMC provides an increased medical capability with the addition of x-ray, laboratory, combat operational stress control, and dental services and has limited inpatient bed space (20 cots) for holding patients up to 72 hours. The BSMC may be augmented with a forward surgical team (known as FST) or forward resuscitative and surgical team capability based upon mission requirements.

Notes. TCCC is prehospital care provided in a tactical setting. TCCC (first responder capability) occurs during a combat mission and is the military counterpart to prehospital emergency medical treatment. TCCC is divided into three stages: care under fire, tactical field care, and tactical evacuation. (See FM 4-02 for additional information.)

The mission of the FST is to provide a rapidly deployable urgent initial surgical service forward in a BCT or at echelons above brigade. The FST is a 20-Soldier team, which provides far forward surgical intervention to render nontransportable patients sufficiently stable to allow for medical evacuation to a Role 3 hospital. Surgery performed by the FST is resuscitative surgery—urgent initial surgery required to render a patient transportable for further evacuation to a medical treatment facility staffed and equipped to provide for the patient's care. Patients remain with the FST until they recover from anesthesia and once stabilized, they are evacuated as soon as possible. The FST is not a self-sustaining unit and must be deployed with or attached to a medical company or hospital for support.

The mission of the forward resuscitative and surgical team is to provide a rapidly deployable damage control resuscitation process and damage control surgery forward in a BCT or at echelons above brigade. The forward resuscitative and surgical team is a 20-Soldier team, which provides far forward resuscitative surgical intervention to render nontransportable patients sufficiently stable to allow for medical evacuation to a Role 3 hospital. The forward resuscitative and surgical team provides the capability to perform resuscitative surgery (often referred to as damage control surgery) within the area of operations. Damage control resuscitation is a medical process that prevents or mitigates hypothermia, acidosis, and coagulopathy through combined treatment paradigms. Damage control surgery is rapid initial control of hemorrhage and contamination with surgical packing and temporary closure, followed by resuscitation in the intensive care unit and subsequent surgical exploration and definitive repair once normal physiology has been restored. (See FM 4-02 for additional information.)

SPECIAL CONSIDERATIONS FOR AREA SUPPORT

9-48. The BSB provides area support to units, not organic to the BCT, when tasked by the BCT commander. *Area support* is a task assigned to a sustainment unit directing it to support units in or passing through a specified location (ATP 4-90). The BSB is not organically equipped or intended to provide area support to non-BCT units for long-term operations, but mission will often require the BSB to provide support to units operating in the BCT area of operations. The BSB provides area support on an exception basis when they have the capability and capacity to do so. Units in the BCT's area of operations vary widely in type and size, such as aviation assets or special forces units. These increased support requirements put a greater burden on the BCT and BSB sustainment staffs and assets. Requirements to support various, and sometimes-unique elements, create complex problem sets for BSB commander or support operations officer. When tasked to provide support to non-BCT units and support requirements exceed capabilities, the S-4 and BSB must coordinate with division, and the DSB. (See ATP 4-90 and FM 4-0.)

> *Note.* While BSBs will often have a need to support non-BCT units, it is important to remember, a BSB's capacity is purposely limited in order to maintain its mobility. Any significant increase in support requirements could have negative effects on the BSB's primary mission to the BCT.

9-49. Army special operations forces are an example of units that may operate or transit through the BCT's area of operations but not in direct support of the BCT. Special operations units have organic support capabilities but are reliant upon regional or combatant command theater of operations infrastructure. These units may rely on the BSB to provide area support to special operations forces operating in the BCT area of operations. The BSB support operations officer, in conjunction with the BCT S-4, will coordinate support as required. (See ATP 3-05.40 for additional information.)

9-50. The BCT will often operate with unified action partners. When the BCT receives capabilities attached from the unified action partner, the BSB support operations officer must understand the task organization and the command relationship, often detailed in an Acquisition Cross-Servicing Agreement. The support operations officer coordinates with supporting organizations on what organic support they are bringing with them. The support operations officer employs those capabilities so that they integrate with BSB capabilities. In the event the unified action partners arrive with no support, the BSB support operations officer coordinates with the DSB for additional capabilities. (See FM 4-0, JP 3-08, and JP 3-16 for additional information.)

OPERATION PROCESS

9-51. BCT planners and staff fully integrate sustainment planning throughout the operations process with the sustainment concept of support synchronized within the BCT's concept of operations. Planning is continuous and concurrent with ongoing support preparation, execution, and assessment. The BSB must conduct parallel planning with the BCT staff in order to provide a supportability analysis for each course of action (COA) to ensure all COAs are feasible. The BSB commander develops mutual trust and cohesion by clearly communicating the BCT commander's intent through mission-type orders and encouraging acceptable risk-identified by the commander, while providing innovative solutions to logistic, financial, personnel, and health services support to the BCT. Key sustainment planners at all levels, ethically, effectively, and efficiently manage the resources entrusted to them. They actively participate in the military decision-making process (MDMP) to include war-gaming. Through a running estimate, sustainment planners continually assess the current situation to determine if the current operation is proceeding according to the commander's intent and if planned future operations are supportable. (See chapter 4 for information on the operations process.)

PLANNING

9-52. Sustainment planning supports operational planning (including branch and sequel development) and the targeting process. Sustainment planning is a collaborative function primarily performed by key members of the BCT and battalion staffs (XO, S-4, S-1, surgeon, and chaplain) and BSB staff (support operations officer and S-3). Sustainment planners and operators must understand the mission statement, the commander's intent, and the concept of operations to develop a viable and effective concept of support.

Sustainment preparation of the operational environment is the analysis to determine infrastructure, physical environment, and resources in the operational environment that will optimize or adversely impact friendly forces means for supporting and sustaining the commander's operations plan (ADP 4-0).

9-53. The BCT S-4 is the lead planner for sustainment within the BCT staff. The BCT S-1, the surgeon, and chaplain assist the S-4 in developing the BCT concept of support. Representatives from these and other sections form a sustainment planning cell at the BCT main command post (CP) ensure sustainment plans are integrated fully into all operations planning. Sustainment standard operating procedures (SOPs) within the BCT should be the basis for sustainment operations, with planning conducted to determine specific requirements and to prepare for contingencies. The BCT S-4 is responsible for producing the sustainment paragraph and annexes of the operation order. The BSB support operations officer may assist the S-4 in writing Annex F of an operation order.

Concept of Support

9-54. The BCT S-4 is responsible for developing the BCT sustainment concept of support. The BCT sustainment concept of support describes how sustainment support will be executed during the operation. Once approved by the BCT commander, the BCT S-4 briefs the concept of support to all commanders and staffs to ensure a shared understanding across the BCT. The BSB commander executes the BCT sustainment concept of support. The BSB commander (through the support operation officer) is responsible for the BSB's concept of support, which will ultimately tell subordinate BSB units (to include FSCs) how they are going to execute the BCT concept of support.

9-55. The sustainment concept of support is a written and graphical representation of how Army logisticians intend to provide sustainment and integrate support with the maneuver force's concept of operations for an operation or mission. The sustainment concept of support is the BSB's concept of operations for an operation or mission. It identifies the logistics requirements for an operation, the priority of support by phase of the operation (established by the BCT commander), and the forecasted receipt of resupply from the DSB.

9-56. The sustainment concept of support establishes priorities of support (by phase or before, during, and after) for the operation and gives the BSB commander the authority to weight support organizations and task organize accordingly. The commander sets these priorities for each level in the commander's intent statement and in the concept of operations. Priorities include such items as personnel replacements; maintenance and evacuation by unit and by system (air and surface systems are given separate priorities); fuel and ammunition; road network use by unit and commodity; and any resource subject to competing demands or constraints. To establish the concept of support, sustainment planners must know—

- Subordinate units' missions.
- Times missions are to occur.
- Desired end states.
- Schemes of movement and maneuver.
- Timing of critical events.
- BCT sustainment requirements.
- Unit capabilities.

Synchronization of Battle Rhythm and Sustainment Operations

9-57. Commanders and subordinate leaders fully integrate sustainment operations with the BCT battle rhythm through integrated planning and oversight of ongoing operations. Sustainment and operational planning, and the targeting process occur simultaneously rather than sequentially. Incremental adjustments to either the maneuver or the sustainment plan during its execution must be visible to all BCT elements. The sustainment synchronization matrix and LOGSTAT report initiate and maintain synchronization between operations and sustainment functions. (See ATP 4-90 for additional information.)

Fusion of Sustainment and Maneuver Situational Understanding

9-58. Effective sustainment operations by the BSB depend on a high level of situational understanding. Situational understanding enables the BSB commander and staff to maintain visibility of current and projected requirements; to synchronize movement and materiel management; and to maintain integrated visibility of transportation and supplies. The Joint Capabilities Release (known as JCR), Joint Capabilities Release-Logistics (known as JCR-Log) Joint Battle Command-Platform, Joint Battle Command-Platform Logistics, Global Combat Support System-Army (GCSS-Army), Command Post of the Future (known as CPOF or its replacement Command Post Computing Environment [known as CPCE]), Medical Communications for Combat Casualty Care (known as MC4) are some of the fielded systems the BSB uses to ensure effective situational understanding and logistics support. These systems enable sustainment commanders and staffs to exercise command and control, anticipate support requirements, and maximize battlefield distribution.

Reports

9-59. The LOGSTAT is an internal status report that identifies logistics requirements, provides visibility on critical shortages, allows commanders and staff to project mission capability, and informs the COP. Accurate reporting of the logistics and AHS support status is essential for keeping units combat ready. Brigade SOPs establish report formats, reporting times, redundancy requirements, and radio voice brevity codes to keep logistic nets manageable.

9-60. LOGSTAT reporting begins at the lowest level. The company first sergeant or XO compiles reports from subordinate elements and completes the unit's LOGSTAT report. Once completed, units forward reports to its higher headquarters and its supporting logistics headquarters, to include the FSC and the BSB. Normally LOGSTATs flow through S-4 channels. The BSB and its subordinate units report on-hand supply and supply point on-hand quantities.

> *Note.* (See FM 4-0 for an example LOGSTAT format that BCT units may adapt based on type of unit, on-hand equipment, type or phase of an operation, mission requirements, and commanders' requirements.) The format is an example spreadsheet for the report. It is not a prescribed format. Commanders can modify the format and tailor the report to their unit or mission.

9-61. The frequency of a LOGSTAT varies and is dependent on the operational tempo of the BCT or subordinate units. Typically, units complete a LOGSTAT report twice daily, but during periods of increased intensity, the commander may require status updates more frequently. As long as automation is available, LOGSTAT relayed via near-real time automation provides the commander with the most up-to-date information, ultimately improving the supporting unit's ability to anticipate requirements.

9-62. Units can complete the LOGSTAT reporting through any means of communication to include written reports, radio, email, JCR, or CPOF/CPCE. AHS status is typically reported through the MC4 system. The JCR system helps lower level commanders automate the sustainment data-gathering process. The system does this through logistics situation reports, personnel situation reports, logistics call for support, logistics task order messaging, situational understanding, and task management. This functionality affects the synchronization of all logistics support in the area of operations between the supported and the supporter.

9-63. Sustainment leaders utilize the GCSS-Army to track supplies, spare parts, and the operational readiness of organizational equipment. GCSS-Army is the tactical logistics and financial system of the U.S. Army. Within the BCT, supply rooms, motor pools, and the SSA platoon all use GCSS-Army to order supplies and repair parts, track maintenance status, and manage SSA operations.

9-64. The sustainment staff must proactively identify and solve sustainment issues. This includes—
- Using CPOF/CPCE, JCR, JCR-Log, GCSS-Army, and other Army command and control systems to maintain sustainment situational understanding.
- Working closely with higher headquarters staff to resolve sustainment problems.
- Recommending sustainment priorities that conform to mission requirements.
- Recommending sustainment-related commander's critical information requirements (CCIRs).
- Ensuring the staff keeps the commander aware of critical sustainment issues.
- Coordinating as required with key automated system operators and managers to assure focus and continuity of support.

9-65. The S-6 and the information systems technician work together to ensure that the CPOF/CPCE, JCR, JCR-Log, GCSS-Army, and other Army command and control sustainment information systems have interconnectivity. The BCT S-4, S-1, surgeon, sustainment automation support management office, and BSB support operations officer monitor the functionality of these systems and implement alternate means of reporting during degraded communications or as required. The MC4 system supports information management requirements for the BCT surgeon's section and the BCT medical units. The BCT uses sustainment information systems to support mission planning, coordinate orders and subordinate tasks, and to monitor and ensure mission execution.

PREPARATION

9-66. Preparation for sustainment consists of activities performed by units to improve their ability to execute an operation. Preparation includes but is not limited to plan refinement, rehearsals, information collection, coordination, inspections, and movements. Sustainment preparation of the operational environment identifies friendly resources (host-nation support, contractible, or accessible assets) or environmental factors (endemic diseases or climate) that affect sustainment. Factors to consider, although not inclusive, include geography information and the availability of supplies and services, facilities, transportation, maintenance, and general skills (such as translators or laborers).

9-67. Sustainment preparation of the operational environment assists planning staffs to refine the sustainment estimate and concept of support. Sustainment planners forecast and build operational stocks as well as identify endemic health and environmental factors. Integrating environmental considerations will sustain vital resources and help reduce the logistics footprint. Sustainment planners take action to optimize means (force structure and resources) for supporting the commander's plan. These actions include, resupplying, maintaining, and issuing supplies or equipment along with any repositioning of sustainment assets. Additional considerations may include identifying and preparing bases, host-nation infrastructure and capabilities, operational contract support requirements, and lines of communications.

9-68. Sustainment rehearsals help synchronize the sustainment warfighting function with the BCT's overall operation. These rehearsals typically involve coordination and procedure drills for transportation support, resupply, maintenance and vehicle recovery, and medical and casualty evacuation. Throughout preparation, sustainment units and staffs rehearse battle drills and SOPs. Leaders place priority on those drills or actions they anticipate occurring during the operation. For example, a transportation platoon may rehearse a battle drill on reacting to an ambush while waiting to begin movement. Sustainment rehearsals and combined arms rehearsals complement preparations for the operation. Units may conduct rehearsals separately and then combine them into full dress rehearsals. Although support rehearsals differ slightly by warfighting function, they achieve the same result.

9-69. The sustainment rehearsal validates the logistics synchronization matrix and BSB's concept of operations. The rehearsal focuses on the supported and supporting unit with respect to sustainment operations across time and space as well as the method of support for specific actions during the operation. The sustainment rehearsal typically occurs after the combined arms rehearsal. The BSB commander hosts the rehearsal for the BCT commander and XO. The support operations officer facilitates the rehearsal to ensure rehearsal of critical sustainment events. BCT attendees include the BCT XO, brigade S-1, surgeon, chaplain, intelligence staff officer (S-2) representatives, S-3 representatives, S-4 representatives, and S-6 representatives. Subordinate battalion representatives include the BSB commander, BSB command sergeant major, support operations officer, the BSMC, and each maneuver battalion XO, S-1, S-4, and medical platoon leader, as well as the FSC commanders, distribution company commander, and support maintenance company commander. The primary document used at the sustainment rehearsal is the logistics synchronization matrix. (See chapter 4 and FM 6-0 for additional information.)

EXECUTION

9-70. Sustainment plays a key role in enabling decisive action. The BCT commander plans and organizes sustainment operations to executive a rapid tempo of highly mobile and widely dispersed operations in every environment across the range of military operations. Sustainment determines the depth and duration of the BCT operation and is essential to retaining and exploiting the initiative to provide the support necessary to maintain operations until mission accomplishment. Failure to sustainment operations could cause a pause or culmination of an operation resulting in the loss of the initiative. Sustainment planners and operation planners work closely to synchronize all of the warfighting function, in particular sustainment, to allow commanders the maximum freedom of action.

Support to Offensive Operations

9-71. Support to offensive operations is by nature a high-intensity operation that requires anticipatory support as far forward as possible. The BCT commander and staff ensure adequate support as they plan and synchronize the operation. Plans should include flexible sustainment capabilities to follow exploiting forces and continue support. Considerations during execution include—

- Establish protection for sustainment units from bypassed enemy forces in a fluid, noncontiguous area of operations.
- Recover damaged vehicles from the main or alternate supply route.
- Preposition essential supplies far forward to minimize lines of communication interruptions.
- Plan increased consumption of petroleum, oils, lubricants, and ammunition.
- Anticipate longer lines of communications as the offensive moves forward.
- Anticipate poor trafficability for sustainment vehicles across fought over terrain.
- Consider preconfigured LOGPACs of essential items.
- Anticipate increased vehicular maintenance especially over rough terrain.
- Maximize field maintenance teams forward.
- Request distribution at forward locations, to include throughput.
- Increase use of meals-ready-to-eat or first strike rations.
- Use captured enemy supplies, equipment, support vehicles, and petroleum, oils, and lubricants (test for contamination before use).
- Suspend most field service functions except airdrop and mortuary affairs.
- Prepare for casualty evacuation (see ATP 4-25.13) and mortuary affairs (see ATP 4-46) requirements.
- Select potential and projected supply routes, logistics release points (known as LRPs), drop zones, landing zones and pickup zones, and support areas based on map reconnaissance.
- Plan and coordinate support for detainee operations.
- Plan replacement operations based on known or projected losses.
- Ensure that sustainment preparations do not compromise tactical plans such as excess stockpiles of vehicles and supplies as well as operations security.

Support to Defensive Operations

9-72. The BCT commander positions sustainment assets to support the forces in the defense. Sustainment requirements in the defense depend on the type of defense. Increased quantities of ammunition and decreased quantities of fuel characterize most area defenses. Barrier and fortification materiel to support the defense often has to move forward, placing increased demands on the transportation system. The following sustainment considerations will apply during operations:

- Pre-position ammunition, POL, and barrier materiel well forward.
- Make plans to destroy stocks if necessary.
- Resupply during limited visibility to reduce the chance of enemy interference.
- Plan to reconstitute lost sustainment capability.
- Use field maintenance teams from the maintenance collection point to reduce the need to recover equipment to the BSA.
- Consider and plan for the additional transportation requirements for movement of pre-position barrier materiel, mines, and ammunition.
- Consider and plan for sustainment requirements of additional engineer units assigned for preparation of the defense.

- Plan for pre-positioning and controlling ammunition on occupied and prepared defensive positions.

Support to Operations Focused on Stability

9-73. Sustainment while conducting operations focused on stability often involves supporting U.S. and multinational forces in a wide range of missions for an extended period. Tailoring supplies, personnel, and equipment to the specific needs of the task is essential for the BCT commander to accomplish the mission.

9-74. The BCT may utilize to a greater extent sustainment support from host nations, contractors, and local entities. This can reduce dependence on the logistics system, improve response time and free airlift and sealift for other priority needs. Support may include limited classes of supplies and services (field feeding, maintenance and repair, sanitation, laundry, and transportation).

9-75. The logistics civil augmentation program (LOGCAP) (see ATP 4-10.1) provides the ability to contract logistics support requirements in a theater of operations. (See AR 700-137 for additional information.) The BCT commander should expect contractors to be involved in operations focused on stability after the initial response phase. The terms and conditions of the contract establish relationships between the military and the contractor. The commander and staff planners must assess the need to provide security to a contractor and designate forces when appropriate. The mission of, threat to, and location of the contractor and designate forces determines the degree of protection needed.

DISTRIBUTION AND RESUPPLY OPERATIONS

9-76. The BSB support operations officer is the principal staff officer responsible for synchronizing BSB distribution or resupply operations for all units assigned or attached to the BCT. Distribution encompasses the movement of personnel, materiel, and equipment in support of decisive action. Resupply operations cover all classes of supply, water, mail, and any other items usually requested. The BSB support operations section is responsible for applying the BSB capabilities against the BCT's requirements. The BCT S-4 identifies requirements through daily LOGSTAT reports, running estimates, and mission analysis. Whenever possible, units conduct resupply on a regular basis, ideally during hours of limited visibility.

METHODS OF DISTRIBUTION

9-77. Distribution is the operational process of synchronizing all elements of the logistic system to deliver the "right things" to the "right place" at the "right time" to support the commander. The elements of logistics include maintenance, transportation, supply, field services, distribution, operational contract support, and general engineering. Distribution is the primary means that enables the other elements of logistics to provide operational reach, freedom of action, and prolonged endurance. The BSB executes distribution operations based on supply requirements communicated by their supported units. The BCT and subordinate maneuver forces communicate their requirements through LOGSTAT reports and other means, from battalion and squadron S-4s and BCT S-4 through the BSB support operations officer, to the BSB.

9-78. Methods of distribution integrate and synchronize materiel management and transportation. Logistics planners base the method of distribution decisions on the supported units' priorities and commodity priorities specified by the BSB and BCT commanders and described in the operation order and BCT sustainment concept of support. Sustainment units use the best distribution method dependent on the mission, the urgency of requirement, the threat, the supported unit's priority of support, time/distance, and other factors of mission and operational variables. The two methods of distribution are unit distribution (throughput is considered a subset of unit distribution) and supply point distribution.

Unit Distribution

9-79. Unit distribution is the routine distribution method the BSB uses to support the BCT. *Unit distribution* is a method of distributing supplies by which the receiving unit is issued supplies in its own area, with transportation furnished by the issuing agency (FM 4-40). In unit distribution, logisticians organize supplies in configured loads and deliver supplies to one or more central locations. Supply personnel can create unit load configurations to resupply specific battalion-, company-, or platoon-sized elements depending on the level of distribution needed and mission variables. Unit distribution maximizes the use of the BCT lift capacity of its transportation assets and minimizes the delivery and turnaround time.

9-80. In each method of distribution, there are multiple techniques for the distribution of supplies, personnel, and equipment. Logistics units use several techniques for unit distribution such as LRPs and aerial delivery. It is important to note that many of these techniques such as LRPs and aerial delivery can use a combination of unit distribution and supply point distribution and, in some cases, each technique can combine the two distribution methods in the same resupply mission.

Supply Point Distribution

9-81. *Supply point distribution* is a method of distributing supplies to the receiving unit at a supply point. The receiving unit then moves the supplies to its own area using its own transportation (FM 4-40). Supply point distribution requires unit representatives to move to a supply point to pick up their supplies. Units most commonly execute supply point distribution by means of an LRP.

Note. Within a maneuver company or troop, the first sergeant may replenish subordinate company elements using various resupply techniques depending on the situation. Subordinate elements may move from their positions to a designated site to feed, resupply, or turn-in damaged equipment. This is often referred to as a service station technique. This technique is normally used in assembly areas and when contact is not likely. This technique takes the least amount of time for the unit and sustainment operators. Conversely, the first sergeant may use unit or support personnel and vehicles to go to each subordinate element to replenish them. Soldiers can remain in position when using this technique. This technique is the lengthiest resupply method and may compromise friendly positions. This is often referred to as the tailgate technique or the in-position resupply.

Throughput Distribution

9-82. Throughput distribution (considered a subset of unit distribution) is a method of distribution which bypasses one or more intermediate supply echelons in the supply system to avoid multiple handling. The BSB or a DSSB may conduct throughput distribution in the BCT's area of operations when needed. An example of throughput distribution is when the BSB's distribution company bypasses the FSC to distribute supplies from the BSA directly to maneuver units. Additionally, a DSSB may distribute supplies from an echelon above brigade SSA to an FSC, bypassing the BSB. Mission variables are the major considerations for logisticians and operation planners when deciding whether to utilize throughput distribution.

METHODS OF RESUPPLY

9-83. Resupply operations require continuous and close coordination between the supporting and supported units. The two methods of resupply are planned resupply and emergency resupply. Planned resupply is the preferred method of resupply. The sustainment concept of support, synchronization matrix, LOGSTAT reports, and running estimates establish the requirement, timing, and frequency for planned resupply. Emergency resupply is the least preferred method of supply. While instances of emergency resupply may be required, especially when combat losses or a change in the enemy situation occurs, requests for emergency resupply often indicates a breakdown in coordination and collaboration between sustainment and maneuver forces.

Planned Resupply

9-84. Whenever possible, planned resupply by LOGPAC is conducted on a regular basis and is the preferred method for the distribution of supplies. Planned (routine) resupply, conducted based on intelligence provided by the BSB S-2, through LOGPAC covers all classes of supply, mail, and any other items usually requested. The LOGPAC, a grouping of multiple classes of supply and supply vehicles under the control of a single ground convoy commander (see ATP 4-01.45) or through aerial delivery under certain situations (see ATP 4-48), is an efficient method to accomplish routine resupply operations. The key feature is a centrally organized resupply operation carrying all items needed to sustain the force for a specific period, usually

24 hours or until the next scheduled LOGPAC. The BCT S-4, in coordination with the support operations officer, tailors a LOGPAC (commonly referred to as a push package) as much as possible to provide subordinate units with sufficient quantities of each supply item in anticipation of their requirements.

Emergency Resupply

9-85. Accurate reporting through LOGSTAT reports is critical to reduce the number of required emergency resupply operations. Poor logistics reporting from units places a burden on the sustainment system by needlessly putting personnel and equipment at risk through additional resupply operations and degrades the efficient distribution of supplies across the BCT. Emergency resupply can lead to excess materiel and needless LOGPAC operations. Emergency resupply that extends beyond BSB capabilities requires immediate intervention of the next higher command capable of executing the mission.

9-86. When a unit has an urgent need for resupply that cannot wait for a planned LOGPAC an emergency resupply may involve classes III, V, and VIII, and, on occasion, class I. In this situation, a maneuver battalion or squadron might use its FSC supply and transportation platoon located in the combat trains to conduct the resupply. An emergency resupply can be conducted using either supply point or unit distribution. The fastest appropriate means is normally used, although, procedures may have to be adjusted when in contact with the enemy.

TECHNIQUES OF RESUPPLY

9-87. In each method of resupply, there are multiple techniques. Logisticians and supported units can use several techniques for resupply during planned and emergency resupply operations. Units can utilize different techniques to conduct supply point and unit distribution operations. In many cases, units conduct both supply point and unit distribution operations during the same resupply technique.

Logistics Package

9-88. The LOGPAC, a grouping of multiple classes of supply and supply vehicles under the control of a single convoy commander, is a simple and efficient way to accomplish routine, planned resupply. The LOGPAC resupply convoy utilizes the combat and field trains to echelon sustainment across the battlefield. Before a LOGPAC, the BSB's distribution company configures loads for resupply to maneuver battalions in the BCT. Typically, a platoon leader from the BSB's distribution company leads a LOGPAC from the BSA. However, the distribution company from the BSB or the FSC supporting a maneuver battalion or squadron can conduct the LOGPAC from the BSA depending on mission variables. The BSB or a DSSB may conduct throughput distribution in the BCT's area of operations when needed. Scheduled LOGPACs typically contain a standardized allocation of supplies based on consumption rates of the supported force reported through LOGSTATS, the sustainment concept of support, and synchronization matrix. The BSB can dispatch an emergency (sometimes referred to as urgent or immediate) LOGPAC as needed.

9-89. Once received by the FSC, the platoon leader from the FSC's distribution platoon leads the battalion LOGPAC. The FSC often breaks the resupply into company-configured loads in the field or combat trains, and the maneuver battalion can reconfigure loads further at an LRP if necessary and mission variables of METT-TC allow. Maneuver company or troop representatives can accompany the LOGPAC. The maneuver company or troop XO or first sergeant meets the LOGPAC at the LRP and escorts the convoy to the maneuver company or troop's trains or positions.

9-90. When receiving resupply, FSCs must ensure they have resupplied the maneuver companies to allow space to receive as many classes of supply as possible. The FSC must especially synchronize classes III and V before receiving resupply. The length of time the unit must sustain itself in combat without resupply determines its combat load. The commander dictates minimum load requirements; however, the commander or the unit SOPs specifies most items. Specific combat loads vary by mission.

Contingency Resupply

9-91. Contingency resupply is the on-call delivery of prepackaged supplies during the execution phase of an operation. This type of on call delivery of a prepackaged resupply is generally used to support an operation of limited duration, such as an air assault or other limited engagement of short duration. Contingency resupply operations are identified during the MDMP, normally during war gaming as each COA is analyzed. Contingency resupply differs from a routine (planned) LOGPAC or emergency resupply, in that, before execution, triggers for delivery are developed to tie contingency resupply operations to the ground tactical plan. During the planning and preparation phases of the operations process units develop menus for prepackaged classes of supply to ensure their availability for expedited delivery as needed. A contingency resupply package can be as simple as a container or bag filled with a small amount of supplies or a unit basic load prepackaged for delivery when needed. Delivery means (see paragraph 9-100) vary between rotary-wing, fixed-wing, and ground delivery assets.

Logistics Release Point

9-92. Maneuver units most commonly execute supply point distribution by means of an LRP. The LRP may be any place on the ground where distribution unit vehicles take supplies met by the supported unit that then takes the supplies forward to their unit for subsequent distribution. Units can utilize both supply point and unit distribution when supplying a force at an LRP. Subsequent distribution below company and troop level generally involves using a service station or tailgate resupply technique or some combination of both.

9-93. Logisticians and maneuver units use an LRP to maximize efficient use of distribution assets and reduce how much time and distance the supported unit requires to travel in order to receive supplies. The LRP is often located between the maneuver battalion or squadron's combat trains and the company or troop trains. An LRP is normally established and secured for only a limited duration of time. Resupply at an LRP is a planned, coordinated, and synchronized operation.

9-94. The FSC commander and battalion or squadron S-4, in coordination with the S-3, plan the location, timing, and establishment of LRPs for the maneuver battalion and squadron. Planners must consider mission variables of METT-TC and security considerations when determining the LRP's location.

9-95. Finally, the maneuver force and sustainment planners must consider the timing of LRP operations. An FSC must deliver supplies to multiple companies during LOGPAC operations. The FSC could possibly deliver to multiple LRPs depending on the situation and mission variables of METT-TC. There may only be a small window of time before elements of the LOGPAC must meet to return to the combat trains or BSA. The maneuver company XO or first sergeant and FSC distribution platoon leader must consider timing of LRP operations, resupply of vehicles (particularly with classes III and V), and the download of supplies.

Pre-positioned Supplies

9-96. The pre-positioning of supplies is a planned resupply technique that reduces the reliance on traditional convoy operations and other resupply operations. Pre-positioned supplies build a stockage level on the battlefield of often-high demand, consumable supplies such as construction and barrier materials and water, and under certain security considerations-ammunition. BCT and sustainment units must carefully plan, prepare, and execute the pre-positioning of supplies. Commander's and subordinate leaders must know the exact locations of pre-positioned supply sites, which they verify during reconnaissance and rehearsals. The commander takes measures to ensure their survivability. These measures may include digging in pre-positioned supplies as well as selecting covered and concealed positions. The commander must also have a plan to remove or destroy pre-positioned supplies if required.

9-97. Based on the BCT's concept of operations and sustainment concept of support, commander's and logisticians consider using pre-positioned supplies along a planned axis of advance or within an area defense. Based on the BCT's scheme of maneuver, pre-positioned supplies can enable units during the conduct of retrograde operations that have extended lines of communication beyond a local haul resupply.

Cache

9-98. A cache is a pre-positioned and concealed supply point. Caches are different from standard pre-positioned supplies because the supported or supporting units conceal the supplies from the enemy whereas units may not conceal other pre-positioned supplies. Caches are an excellent tool for reducing the Soldier's load and can be set up for a specific mission or as a contingency measure. Cache sites have the same characteristics as an objective rally point (known as ORP) or patrol base, with the supplies concealed above or below ground. An above ground cache is easier to get to but is more likely for the enemy, civilians, or animals to discover. A security risk always exists when returning to a cache. A cache site is observed for signs of enemy presence and secured before unit's use it due to the potential of booby traps and enemy observation.

Aerial Delivery

9-99. Aerial delivery is a vital link in the distribution system and provides the capability of supplying the force even when enemy or other elements have disrupted the ground lines of communications or terrain is too hostile, thus adding flexibility to the distribution system. When applied together with surface distribution operations, aerial delivery enables maneuver forces to engage in a battle rhythm that is not as restricted by geography, supply routes, tactical situations, or operational pauses for logistic support. In order for effective aerial delivery, friendly forces must control the airspace in the area of operations and must neutralize enemy ground-based air defenses (see FM 3-99).

9-100. Aerial delivery includes airland, airdrop, and sling-load operations and can support units in various operational environments where terrain limits access. The BCT can use aerial delivery for both planned and emergency resupply of sustainment. Aerial delivery acts as a combat multiplier because it is an effective means of by-passing enemy activity and reduces the need for route clearance of ground lines of communications. (See ATP 4-48 for additional information.)

9-101. BCT units must be prepared to receive airland, airdrop, and sling-load resupplies. (See ATP 3-21.20 for a detailed discussion of aerial delivery means.) The receiving commander must consider the enemy's ability to locate friendly units by observing the aircraft. The receiving unit should establish the drop zone and landing zone away from the main unit and in an area that they can defend for a short time unless the resupply is conducted in an area under friendly control and away from direct enemy observation. The delivered supplies are immediately transported away from the drop zone and landing zone. Units must know how to select pickup zones and landing zones, how to receive aerial delivery of supplies and equipment, and have the ability to return any reusable rigging material to the owning or supporting unit. (See FM 3-21.38 for additional information.)

Refuel on the Move

9-102. Refuel on the move (known as ROM) can be tailored to many tactical situations but the primary purpose is to extend reach and tempo for the offensive operation. Any level unit, to meet mission requirements, can conduct ROM operations. Typically, an FSC will conduct ROM operations to support maneuver units between engagements or to increase time on target while maneuver units peel back and flow through the ROM and return to the current engagement. A ROM can be as simple as utilizing heavy expanded mobile tactical trucks or modular fuel systems, or as complex as needed utilizing any equipment available to support the largest of movements.

9-103. When vehicles enter a ROM site for refueling, fuel trucks issue a predetermined amount of fuel (usually timed) and the vehicles move out to return to their convoy or formation. The rapid employment of the ROM distinguishes it from routine convoy refueling operations. Planners do not intend a ROM to completely refuel a combat vehicle. Instead, they intend a ROM to rapidly resupply a set portion of fuel to extend the operational reach of ground maneuver forces.

9-104. Supported unit S-3 and S-4 staffs coordinate with the BCT S-4 and BSB support operations officer to set the time and place to conduct the ROM operations according to unit battle rhythm and establish how much fuel or time for fueling the BSB or FSC will give each vehicle. The concept can be extended based on the size and scope of the operation, for example, the DSSB can be the force conducting the ROM for the whole division, while the entirety of the BCT's fuel assets push through remaining topped off. In the BCT concept of operations, ideally the distribution company conducts the ROM, while the FSCs pass through maintaining full mobile storage capacity. (ATP 4-43 contains information about ROM operations. ATP 4-90 depicts an example of a ROM layout.)

Forward Arming and Refueling Point

9-105. A forward arming and refueling point (FARP) is a temporary facility that is organized, equipped, and deployed as far forward, or widely dispersed, as tactically feasible to provide fuel and ammunition necessary for the sustainment of aviation units in combat. Establishing a FARP allows commanders to extend the range of aircraft or significantly increase time on station by eliminating the need for aircraft to return to the aviation unit's central base of operations to refuel and rearm. FARPs may be task organized to provide maintenance support as well as air traffic control services, if required.

9-106. A FARP is an example of supply point distribution. Commanders employ FARPs in support of aviation operations, generally by the distribution company of an aviation support battalion, when the distance covered, or endurance requirements exceed normal capabilities of the aircraft. They may also use FARPs during rapid advances, when field trains cannot keep pace. (See ATP 4-43 for additional information.)

Modular System Exchange Operation

9-107. Modular system exchange operation is the resupply technique to distribute and exchange a full flatrack, multi-temperature refrigerated container system, modular fuel system, and modular water tank rack by the supporting unit and retrograding an empty flatrack, multi-temperature refrigerated container system, modular fuel system, and modular water tank rack from the supported unit. Logisticians can apply this method of exchange to any modular system for commodities. Modular system exchange increases distribution throughput capability, extends operational reach, and prolongs the endurance of maneuver forces. The use of flatrack distribution and exchange forward in the BCT area of operations increases the supported maneuver commander's tactical flexibility and decreases the sustainment transportation asset's time on station when resupplying. A DSSB can also conduct modular system exchange operations with a BSB or FSC. (See ATP 4-90 for additional information.)

OPERATIONAL CONTRACT SUPPORT

9-108. Operational contract support is the process of planning for and obtaining supplies, services, and construction from commercial sources in support of combatant commander directed operations. While varying in scope and scale, operational contract support is a critical force multiplier in unified land operations, especially long-term stability operations. (See ATP 4-10 for additional information.)

9-109. Contracting and purchasing will likely be a method of sustainment that helps to round out the BCT's concept of support. BCTs must have trained and ready contracting officer representatives (CORs), field ordering officers, and pay agents. These designated personnel must be carefully selected, as they will make up the acquisition team within the BCT. They must work closely together as these personnel are part of a larger acquisition team that includes the contract and financial management experts, external to the BCT, who will provide the guidance and direction to each COR, field ordering officer, and pay agent to meet unit needs. (See ATP 4-10 for additional information.)

9-110. The COR (sometimes referred to as a contracting officer's technical representative) is an individual appointed in writing by a contracting officer. Responsibilities include monitoring contract performance and performing other duties as specified by their appointment letter. The requiring unit or designated support unit normally nominates a COR. (See ATP 4-10.)

9-111. A field ordering officer is an individual who is trained to make micro purchases within established thresholds (normally with local vendors) and places orders for goods or services. A pay agent is an individual who is trained to account for government funds and make payments in relatively small amounts to local vendors. While performing as field ordering officers or pay agents, individuals work for and must respond to guidance from their appointing contracting and finance officials. One individual cannot serve as both field ordering officer and pay agent. Property book officers cannot serve as field ordering officers or pay agents. Field ordering officers and pay agents must be careful when dealing with local nationals because field ordering officers and paying agents have a ready source of cash, local nationals may overestimate the influence of field ordering officers and pay agent teams. (See ATP 1-06.1 for additional information.) Considerations for field ordering officers and pay agents include—

- Security (personal and cash).

- Unauthorized purchases.
- Type of purchase.
- Number of items purchased.
- Single item or extended dollar amount.
- Split purchases to get around limits.
- Poor record keeping.
- Accepting gifts of any kind and not reporting gifts.

9-112.   Though they involve a number of risks, contractors play an increasing role in providing sustainment during unified land operations. The BCT may use contractors to bridge sustainment gaps between required capabilities and the actual force sustainment structure available within an area of operations. The BCT legal section provides or coordinates any necessary legal reviews and is available to provide contract and fiscal law advice to the BCT.

9-113.   Contractors may be employed throughout the area of operations and in all conditions subject to the mission variables of METT-TC. Protecting contractors within the area of operations is the BCT commander's responsibility. (See ATP 4-10 for additional information.)

MAINTENANCE

9-114.   The primary purpose of maintenance is to ensure equipment readiness. Ideally, all equipment is fully mission capable, able for units to employ the equipment immediately, and operate fully of its intended purpose. The second purpose of maintenance is to generate combat power by repairing damaged equipment as quickly and as close to the point of failure as possible. Repairs should return the damaged equipment to fully-mission capable status or to a state, which allows mission accomplishment.

MAINTENANCE DURING COMBAT OPERATIONS

9-115.   Once units enter combat operations, maintenance is critical to maintain combat power and momentum. Replacement systems may not be immediately available. This is especially true during the early stages of an operation. Units must keep existing systems fully mission capable for the duration of the operation or until the system is clearly damaged beyond field-level maintenance repair capability.

9-116.   Maintenance and recovery planning is integrated into all aspects of the MDMP to ensure synchronization and unity of effort. Planning includes identifying requirements, reviewing available assets, preparing a maintenance estimate, comparing requirements to capabilities, and adjusting maintenance priorities to meet the mission requirement. Maintenance planning is included in the overall sustainment concept of support.

9-117.   Maintenance planners must understand the overall mission and concept of operations for maneuver forces in order to prioritize and weight maintenance support to the main effort. Maintenance planners must be able to recommend to the BSB commander, BCT XO, and BCT commander how to task organize for optimal maintenance capability. They must be able to recommend the cross leveling of system maintainers to ensure adequate maintenance capability is available to support the main effort. The BCT S-4 and support operations officer work together to determine how many key systems identified are mission ready and then work with the FSCs to prioritize their work. It is imperative that maintenance planners understand that there is no repair capability outside of the BCT for the main battle tank, Infantry fighting vehicles, or Stryker systems. The maintainers for the main battle tank, Infantry fighting vehicle or Stryker systems reside in the FSC in current force structure.

LEVELS OF MAINTENANCE

9-118.   The Army utilizes a tiered maintenance system. Two-level maintenance is a maintenance system comprised of field and sustainment-level maintenance. Two-level maintenance utilizes equipment design, diagnostic, and prognostic equipment and tools. It also employs mechanic and technician training as well as information systems in component repair or replacement taking full advantage of increased reliability. Two-level maintenance provides increased flexibility and depth of capability. In supporting the modular force, the goal of our maintenance system is to reduce repair times by repairing or replacing components, modules, and assemblies as far forward as possible.

Field-Level Maintenance

9-119. Field-level maintenance is on or near system maintenance focusing on the repair and return to the user. It includes maintenance actions performed by operators, crews, and ordnance maintainers. Units perform field-level maintenance as far forward as possible utilizing line replaceable units or modules and component replacement or repair. The owning or support unit most often performs field-level maintenance by using tools and test equipment found in the unit. Field-level maintenance is not limited to simply removing and replacing parts.

9-120. Field-level maintenance allows for repair of components or end items if the maintainers possess the requisite skills, proper tools, proper repair parts, references, and adequate time. Field maintenance includes adjustment, alignment, service, applying approved field-level modification work orders, fault and failure diagnoses, battle damage assessment and repair, and recovery. Field-level maintenance is always repair and return to the user and includes preventative maintenance checks and services.

9-121. The maneuver force organization's operators and crews have the responsibility to perform maintenance on their assigned equipment. Operators and crews receive formal training from their proponent typically through advanced individual training and new equipment training on a specific piece of equipment or weapon system. Operators and crew tasks consist of inspecting, servicing, lubricating, adjusting, and replacing minor components or assemblies using basic issue items and onboard spares. After operators have exhausted their maintenance capabilities, they rely on ordnance maintainers in field maintenance organizations or teams to conduct field-level maintenance on the item of equipment.

Sustainment-Level Maintenance

9-122. Sustainment-level maintenance is off-system component repair or end item repair, which returns the equipment back to the national supply system. National-level maintenance providers perform sustainment-level maintenance. Only in rare exceptions will sustainment-level maintenance personnel return an item back to the owning unit. One example is during reset. National-level maintenance providers include the U.S. Army Materiel Command and installation logistics readiness centers maintenance activities. Sustainment-level maintenance returns items to a national standard, providing a consistent and measurable level of reliability. Sustainment-level maintenance supports both operational forces and the Army supply system.

9-123. The Army conducts below depot sustainment-level maintenance on a component, accessory, assembly, subassembly, plug-in unit, or other portion after maintainers remove it from the system. The remove and replace authority for this level of maintenance is noted in the relevant maintenance allocation chart for the equipment. Sustainment-level maintainers return items to the supply system after they perform the maintenance. Below depot sustainment-level, maintenance can also apply to end item repair.

9-124. Depot level maintenance repairs end items or a component, accessory, assembly, subassembly, plug-in unit; either on the system or after maintainers have removed the inoperable or damaged item. Either depot personnel or contractor personnel, when authorized by the U.S. Army Materiel Command, perform depot sustainment-level maintenance. Depot level maintainers return items to the supply system after they perform the maintenance at this level.

RECOVERY OPERATIONS AND PLANNING

9-125. Recovery is the process of repairing, retrieving/freeing immobile, inoperative materiel from the point where it was disabled or abandoned. Maintenance planners should echelon dedicated recovery assets throughout the BSA, field, combat, and company trains for optimum support of the BCT.

9-126. Commanders must emphasize the use of self and like vehicle recovery methods to the greatest extent possible. These practices will minimize the use of dedicated recovery assets for routine recovery missions. Recovery managers and supervisors must ensure maneuver forces and logistics units use recovery vehicles only when necessary. The FSC commander, maintenance warrant officer, and supported battalion or squadron S-4 coordinate recovery operations supporting the commander's priorities by balancing the overall repair effort, available resources, and the tactical situation.

9-127. The FSC has recovery assets located in the recovery section and field maintenance teams within the field maintenance platoon. The FSC commander along with the maintenance warrant officer, or maintenance noncommissioned officer in-charge, and the battalion or squadron S-4 track and manage recovery operations. The field maintenance company is responsible for recovering the BSB's organic equipment and providing limited backup support with wreckers or tracked recovery vehicles when requirements exceed a supported unit's capability. They provide area support for recovery on a limited basis to units without a recovery capability.

9-128. Maintenance planners must establish recovery priorities when recovery assets are limited. These depend on the commander's need for an item and the tactical situation. The type of maintenance or repair required affects the priority when the FSC or field maintenance company must recover two or more like items.

9-129. The battalion or squadron S-4, the unit's maintenance warrant officer, and FSC commander are responsible for developing the maneuver unit's repair and recovery plan. They develop a plan of action for repair and recovery of the disabled equipment based on the subordinate units in the unit's request for assistance. The maintenance plan includes battle damage assessment, priority for support, tactical situation, forecasted workload, and availability of maintenance and recovery personnel.

MEDICAL SUPPORT

9-130. BCTs have organic medical resources within unit headquarters (BCT, battalion, and squadron surgeon's section), battalion and squadron unit (medical platoon), and the BSB (medical company). The medical command (deployment support) (known as MEDCOM [DS]) or the medical brigade (support) (known as MEDBDE [SPT]) serves as the medical force provider and is responsible for developing medical force packages for augmentation to the BCT, as required. Within each BCT (IBCT, SBCT, and ABCT), slight differences exist between the medical capabilities and resources. (See ATP 4-02.3 for these differences based upon the type of parent unit.)

9-131. Role 1 (also referred to as unit-level medical care) is the first medical care a Soldier receives. Nonmedical personnel performing first aid procedures assist the combat medic. An individual (self-aid and buddy aid) administers first aid and combat lifesavers administer enhanced first aid. If needed, the Soldier is evacuated to the Role 1 medical treatment facility (battalion aid station) at the battalion or squadron, or the Role 2 medical treatment facility (BSMC) in the BSB of the BCT. (See ATP 4-02.3 for additional information.)

COMBAT LIFESAVERS

9-132. The combat lifesaver is a nonmedical Soldier trained to provide enhanced first aid and lifesaving procedures beyond the level of self-aid or buddy-aid. As usually the first person on the scene of a medical emergency, the combat lifesaver provides enhanced first aid to wounded and injured personnel. The squad leader is responsible for ensuring that an injured Soldier receives immediate first aid and is responsible for informing the commander of the casualty.

COMBAT MEDIC

9-133. The combat medic is the first individual in the medical chain that makes medical decisions based on medical specialty-specific training. The platoon combat medic goes to the casualty and initiates TCCC or the casualty may be brought to the combat medic at the casualty collection point. The medic makes an assessment; administers initial medical care; initiates the DD Form 1380 (*Tactical Combat Casualty Care [TCCC] Card*), or other requisite forms; requests evacuation; or returns the Soldier to duty.

BATTALION AND SQUADRON AID STATION

9-134. The mission of the medical platoon is to provide Role 1 AHS support to the maneuver battalion or squadron and field artillery battalion. A medical treatment platoon is organic to each and is the unit level

Role 1 medical treatment facility, usually referred to as the battalion or squadron aid station. The medical platoon is dependent upon the maneuver elements to which it is assigned for all logistic support, except class VIII (medical) supplies. For information on class VIII coordination, synchronization, and execution of medical logistics support see paragraph 9-145.

9-135. Medical platoons within the various BCTs configure with a headquarters section, medical treatment squad, ambulance squad (ground), and combat medic section. Differences between the BCTs are in the quantity and types of vehicles, configuration of medical equipment sets, and number of personnel assigned.

9-136. The treatment squad consists of two teams (treatment team alpha and team bravo). The treatment squad operates the aid station and provides Role 1 medical care and treatment (to include sick call, TCCC, and advance trauma management). Team alpha is clinically staffed with the battalion or squadron surgeon while team bravo is clinically staffed with the physician assistant.

9-137. Medical platoon ambulances provide medical evacuation and en route care from the Soldier's point of injury, the casualty collection point, or an ambulance exchange point to the aid station. The ambulance squad is four teams of two ambulances composed of one emergency care sergeant and two ambulance aide/drivers assigned to each ambulance.

9-138. Combat medics are normally allocated to the supported maneuver company and troop on a basis of one emergency care sergeant per company and troop plus one combat medic per platoon. The medical platoon's emergency care sergeants normally locate with, or near, the maneuver company commander or first sergeant to provide guidance and direction to the subordinate platoon combat medics. The platoon's combat medic locates with, or near, the platoon leader or platoon sergeant. (See ATP 4-02.3 for additional information.)

Note. BCT echelon specific ATPs address how each tactical echelon employs its organic medical resources.

MEDICAL COMPANY (OF THE BRIGADE SUPPORT BATTALION)

9-139. The mission of the medical company in the BSB, also referred to as the BSMC, is to provide Role 2 AHS support to supported battalions and squadron of the BCT with organic medical platoons. The medical company provides both Roles 1 and 2 medical treatment, on an area basis, to those units without organic medical assets operating in the BCT area of operations.

9-140. The medical company within the BCT is configured with a company headquarters, preventive medicine section, mental health section, medical treatment platoon (with a medical treatment squad, area support squad, medical treatment squad [area], and patient hold squad), and evacuation platoon. Differences of personnel, equipment, and vehicles may exist, based upon the BCT type, with the medical companies, however, the mission remains the same for all AHS units and elements and they execute their mission in a similar fashion.

9-141. The medical company headquarters provides command and control for the company and attached units. The headquarters provides unit-level administration, general supply, and CBRN defense support. The company headquarters is organized into a command element, a supply element, and CBRN operations element consisting of unit decontamination and CBRN defense.

9-142. The preventive medicine section provides advice and consultation in the area of health threat assessment, force health protection, environmental sanitation, epidemiology, sanitary engineering, and pest management. The mission of the mental health section is to support commanders in the prevention and control of combat and operational stress reaction through the BCT's behavioral health activities by the provision of advice and assistance in the areas of behavioral health and combat and operational stress control.

9-143. The medical treatment platoon receives, triages, treats, and determines the disposition of patients in the BCT area of operations. The platoon provides for advance trauma management, TCCC, general medicine, general dentistry, and physical therapy. In addition, the medical treatment platoon has limited radiology, medical laboratory, and patient holding capabilities. The medical treatment platoon is organized with a headquarters, a medical treatment squad, an area support squad, a medical treatment squad (area), and patient holding squad.

9-144.   The evacuation platoon performs ground evacuation and en route patient care for supported units. The evacuation platoon headquarters provides command and control for the evacuation squad (forward) and the evacuation squad (area). The platoon employs ten evacuation teams. The evacuation platoon provides ground medical evacuation support for the maneuver battalions and squadron, BEB, and field artillery battalion of the BCT. In addition, it provides ground medical evacuation support to units receiving area medical support from the medical company.

Note. The medical operations cell of the supporting combat aviation brigade (see chapter 4) provides assistance in planning and coordination for air ambulance employment and utilization. The medical operations cell assists with the synchronization of the air and ground medical evacuation plan. The medical operations officer and operations sergeant also manage medical treatment facility information from AHS support commands and surgeon cells from higher roles of care including combat support hospital locations and status (beds by type and number available), evacuation routes, casualty collection points, and ambulance exchange points. (See ATP 4-02.2, ATP 4-02.3, and FM 3-04 for additional information.)

9-145.   The medical company's supply element is the brigade medical supply office. This office provides brigade level, Role 2, class VIII coordination, synchronization, and execution of medical logistics support for the BSMC and supported BCT. Class VIII organizational assets in the BCT, are fixed and deploy with assigned AHS support units. Operational medical logistics support relies on the application of a class VIII supply chain that is agile, responsive, and swift and that possesses situational understanding of the supported organizations, the operational environment, mission, and the area of operations. During the initial deployment phase, the BSMC receives medical resupply mainly through preconfigured push packages, medical resupply sets from the supporting medical logistics company, or a higher logistics support activity (see ATP 4-02.1).

SECTION III – ECHELON SUPPORT

9-146.   How BCT support organizations, including external and attached organizations, array in echelon varies widely based upon METT-TC. The BSB, in support of the BCT's concept of support, plans and synchronizes *echelon support*—the method of supporting an organization arrayed within an area of operations (ATP 4-90). Current mission, task organization, command and control, concept of support, and terrain influence how support is echeloned.

ECHELON OF SUPPORT

9-147.   Echeloning support within the BCT is a carefully planned and executed process. The method employed to echelon support is a deliberate, collaborative decision based upon a thorough mission analysis within the MDMP. During this analysis, there must be an understanding at all levels of the capabilities of each support organization within and supporting the BCT. Commanders must understand that echeloned support will vary by BCT and each battalion or squadron. As the BCT's primary sustainment organization, the BSB's organization facilitates echeloned support. Common echelon of support at the lowest level of sustainment is executed at the battalion, squadron, company, battery, and troop echelons. Figure 9-2 on page 9-28 provides a notional concept of support for a BCT conducting offensive combat operations.

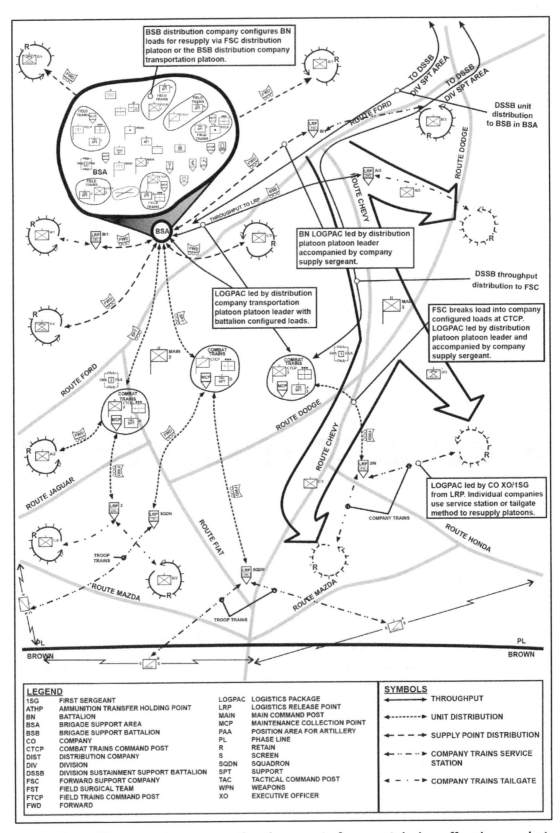

Figure 9-2. Brigade combat team notional concept of support during offensive combat operations

Note. (See ATP 3-21.20, appendix H for a notional concept of support scenario, used for discussion purposes, of an IBCT during the conduct of offensive combat operations.)

BATTALION AND SQUADRON ECHELONS

9-148. As discussed earlier, an FSC from the BSB supports each battalion and squadron in the BCT. The FSC performs the logistics function within the battalion or squadron echelon of support, referred to as unit trains in one location, or echeloned trains within an area of operations. Unit trains at the battalion or squadron level are appropriate when the unit is consolidated in an assembly area, during reconstitution, major movements, or when terrain or distances restrict movement causing the unit to depend on aerial resupply and evacuation for support. The BCT normally operates in echeloned trains where subordinate unit trains employ into multiple locations.

9-149. Echeloned trains at the battalion and squadron can be organized into combat trains and field trains. Battalion and squadron trains are used to array subordinate sustainment elements (unit personnel, vehicles, and equipment) including their designated FSC. The battalion or squadron commander and staff, the BSB commander and staff, and the FSC commander collaborate to determine the best method of employment commensurate with the BCT's concept of support and commander's guidance. Echeloning of support can include the battalion or squadron aid station, elements of the S-1 section and S-4 section, and elements of the FSC.

Combat Trains

9-150. Combat trains usually consist of elements of the battalion or squadron S-1 section, S-4 section, and aid station, the maintenance collection point and other selected elements of the FSC. The FSC typically positions its commander or first sergeant, field feeding section, portions of the distribution platoon, maintenance control officer, and portions of the maintenance platoon in the combat trains. The battalion and squadron commanders position key personnel, staff and subordinate company leaders, and assets in the trains based on the best location to support the mission. Commanders consider the mission variables of METT-TC when selecting the location for their combat trains.

9-151. When established, the combat trains command post (graphically depicted as the CTCP) plans and coordinates sustainment operations in support of the tactical operations. The combat trains command post serves as the focal point for all administrative and logistical functions for the battalion or squadron. The combat trains command post may serve as an alternate CP for the battalion or squadron main CP. The battalion or squadron S-4 usually serves as the combat trains CP sustainment officer in charge and the maintenance control officer usually serves as the maintenance collection point officer in charge. The headquarters and headquarters company (battery or troop) commander usually exercises command and control for their respective combat trains CP. The combat trains CP serves the following functions:

- Tracks the current battle.
- Controls sustainment support to the current operation.
- Provides sustainment representation to the main CP for planning and integration.
- Monitors supply routes and controls the sustainment flow of materiel and personnel.
- Coordinates evacuation of casualties, equipment, and detainees.

9-152. Units position the maintenance collection point where recovery vehicles have access, or where maintenance personnel perform major or difficult maintenance. The combat trains must be mobile enough to support frequent changes in location, time and terrain permitting, under the following conditions when— heavy use or traffic in the area may cause detection, area becomes worn by heavy use such as in wet and muddy conditions, or security is compromised.

Field Trains

9-153. Field trains are positioned based on METT-TC considerations and are often located in the BSA. The field trains include battalion or squadron sustainment assets not located with the combat trains. Field trains can provide direct coordination between the battalion or squadron and the BSB.

9-154. When established, the field trains usually consist of the elements of the headquarters and headquarters company (battery or troop) and the battalion or squadron S-1 and S-4 sections and may include FSC elements not located in the combat trains. Field trains personnel help facilitate the coordination and movement of support from the BSB to the battalion or squadron. The battalion or squadron S-4 coordinates all unit supply requests with the BCT S-4 and BSB. The BSB fills orders with on-hand stocked items through unit distribution to the FSC, typically located at the combat trains. Requests for items not on-hand in the BSA are forwarded to the BCT S-4.

9-155. The FSC typically places personnel in the field trains that can facilitate the resupply of rations, water, fuel, and ammunition. These FSC elements should also enable the flow of class IV, VIII, and IX. FSC elements in the field trains may consist of the FSC XO or first sergeant, ammunition handlers, field feeding Soldiers, fuel handlers, motor transport operators, and supply sergeant or other representatives from the FSC. The food operations noncommissioned officer may coordinate ration ordering and class I break bulk configuration for units and Soldiers in the field trains.

9-156. When established, the field trains CP (graphically depicted as the FTCP) serves as the battalion or squadron commander's primary direct coordination element with the supporting BSB in the BSA. The field trains CP usually consists of the headquarters and headquarters company (battery or troop) XO and first sergeant, an S-4 and S-1 representative, and supply sergeant or representative. The headquarters and headquarters company XO or designated representative can control the field trains CP. The field trains CP serves the following functions:

- Synchronizes and integrates the BCT concept of support.
- Coordinates logistics requirements with the BSB support operations.
- Configures LOGPACs tailored to support requirements.
- Coordinates with the BCT for personnel services and replacement operations.
- Forecasts and coordinates future sustainment requirements.
- Coordinates retrograde of equipment.
- Coordinates retrograde of personnel (casualty evacuation, personnel movement, and human remains).

9-157. Maneuver battalions and squadrons do not necessarily have to locate their field trains in the BSA. While it is common to have field trains co-located in the BSA, the mission variables of METT-TC can dictate the necessity to move the field trains forward closer to maneuver unit combat trains. With the field trains collocated in the BSA, the BCT will not utilize the distribution trucks and lift platforms of the distribution company fully as designed. As maneuver forces move forward, field trains may move forward outside the BSA in order to keep FSC assets closer and more responsive to the BCT's maneuver. The distribution company then distributes supplies to the field trains or further forward if the situation permits. If the FSC establishes field trains at the BSA, they can receive commodities at the BSA and push them forward to the combat trains using organic distribution assets, enabling the distribution company to use its assets to weight the main effort or perform unit distribution to units not collocated in the BSA.

COMPANY, BATTERY, AND TROOP ECHELONS

9-158. Echeloning of support begins at the company (battery or troop) level. Companies (batteries or troops) within the BCT have no organic logistics organizations. Echeloning support within these units, if required, must be done with internal personnel and equipment used to facilitate or expedite logistics support within these units.

9-159. The commander determines the composition of echeloned support, often referred to as company (battery or troop) trains, and may consist of the first sergeant, supply sergeant, and medic. Maintenance teams from the FSC may be included. This echeloned support expedites replenishment of subordinate elements using either the supply point distribution or the unit distribution method. The operation order must describe the method used.

9-160. Supply point distribution requires unit representatives to move to a supply point to pick up their supplies. Supply point distribution is commonly executed by means of an LRP. The LRP may be any place on the ground where unit vehicles return to pick up supplies and then take them forward to their unit. In unit distribution, supplies are configured in unit sets and delivered to one or more central locations. Depending on the distribution method used, the first sergeant may send unit personnel and vehicles to an LRP designated by the FSC (supply point distribution) or the first sergeant may coordinate for the FSC to deliver supplies to a location (unit distribution).

9-161.   Within the company (battery or troop), the first sergeant will replenish company elements using various techniques depending on the situation. Unit elements may move from their positions to the designated site to feed, resupply, or turn in damaged equipment. This is often referred to as a service station technique. This technique is normally used in assembly areas and when contact is not likely. This technique takes the least amount of time for the sustainment operators.

9-162.   Conversely, the first sergeant may use unit or support personnel and vehicles to go to each element to replenish them. Soldiers can remain in position when using this technique. This technique is the lengthiest resupply method and may compromise friendly positions. This is often referred to as the tailgate technique or the in-position resupply.

FORWARD LOGISTICS ELEMENT

9-163.   A *forward logistics element* is comprised of task-organized multifunctional logistics assets designed to support fast-moving offensive operations in the early phases of decisive action (ATP 4-90). The forward logistics element (FLE) operates out of a forward logistics base or support area. The FLE represents the BSB commander's ability to weight the effort for the operation by drawing on all sustainment assets across the BCT. Additionally, the BSB commander may coordinate with echelons above brigade to provide support capabilities to augment the FLE in the concept of support. This includes identifying and the positioning of echelons above brigade unit assets in proximity to geographically dispersed forces to extend operational reach and prolong endurance. The intent for employing an FLE is to minimize tactical pauses to the offensive plan and enable momentum for the commander.

9-164.   While the mission analysis dictates an FLE's composition, a BSB typically establish an FLE with fuel handlers, ammunition handlers, water and class I supplies, recovery assets, and medical personnel. Typically, there is limited requirement for maintenance capability in an FLE while the BSA displaces. The lack of maintenance required is a direct result of the BSB's field maintenance company not providing direct support to the BCT's maneuver battalions and squadron. FSCs continue to provide direct support to their assigned battalions and squadron using LRPs as required to support by their design. Security is also a concern for the FLE. Because the FLE is generally, a fixed node for an extended period it requires more significant security considerations to defend against a level I threat. An FLE requires more security planning and defense than an LRP, which is established for a limited duration of time at a location.

ECHELONS ABOVE BRIGADE SUSTAINMENT

9-165.   The objective of sustainment during decisive action is to provide support and services to ensure freedom of action, extend operational reach, and prolong endurance. (See paragraph 9-2.) Echelons above brigade sustainment organizations synchronize and execute sustainment operations under all conditions to assist the maneuver commander to seize, retain, and exploit the initiative and dominate in increasingly challenging and complex environments. For example, in the offense sustainment units supporting decisive action are focused on sustaining and maintaining the combat power necessary to defeat, destroy or dislocate enemy forces. Regardless of which element of decisive action (offense, defense, or stability) currently dominates, successful sustainment commanders and planners will act, rather than react. To support decisive action, sustainment forces at all echelons consider echeloning support assets to expedite replenishment of critical support.

ARMY ECHELONS AND SUSTAINMENT UNITS AND STAFFS

9-166.   Army echelons and sustainment units and staffs operate across the strategic, operational, and tactical levels; many are affiliated with either supported or supporting commands and operate under a variety of command relationships. The Army Service component command (ASCC) assigned to each combatant command is responsible for the preparation and administrative support of Army forces assigned or attached to the combatant command. The effectiveness of the sustainment warfighting function is dependent upon the actions of sustainment units and staffs at each echelon of support. Understanding Army echelons and sustainment unit and staff roles and capabilities are essential to conducting sustainment operations. Knowing the roles, responsibilities, and authorities of sustainment units is essential to planning, preparing, executing, and assessing sustainment operations. A critical element within the headquarters at each echelon is the sustainment cell at echelons above brigade and the sustainment staff (brigade and below) that works in conjunction with the supporting sustainment headquarters to plan and synchronize support during decisive action. Figure 9-3 depicts a notional area of responsibility (AOR) command and control structure of sustainment forces.

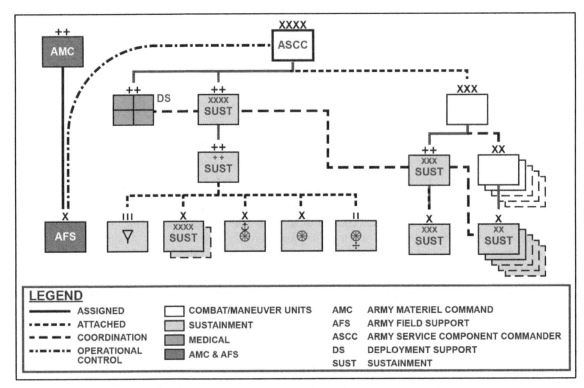

Figure 9-3. Notional area of responsibility command and control of sustainment forces

> ***Note.*** Theater ASCCs also support Army special operations forces when performing theater special operations missions. Support to other Services while executing assigned executive agent or lead Service responsibilities is commonly referred to as Army support to other Services. In both instances, the ASCC supports sustainment requirements through its designated theater sustainment command (TSC), expeditionary sustainment command (ESC), and MEDCOM (DS).

Theater Sustainment Command

9-167.   The TSC is the Army's command for the integration and synchronization of sustainment in the AOR. The MEDCOM (DS) is also assigned to the ASCC (see paragraph 9-177). It is the theater medical command that is responsible for command and control, integration, synchronization, and execution of AHS support within the AOR. The TSC connects strategic enablers to the tactical formations. The TSC commander also commands and task organizes attached ESCs, sustainment brigades, and additional sustainment units. The TSC executes the sustainment concept of support for planning and executing sustainment-related support to the AOR. TSCs execute sustainment operations through their assigned and attached units. The TSC integrates and synchronizes sustainment operations across the AOR from a home station command and control center or through a deployed CP. The TSC has four operational responsibilities to forces in theater: theater opening, theater distribution, sustainment, and theater closing. The task organized TSC is tailored to provide operational-level sustainment support within an assigned AOR. It integrates and synchronizes sustainment operations for an ASCC including all Army forces forward-stationed, transiting, or operating within the AOR. The TSC coordinates Title 10, Army support to other Services, Department of Defense executive agent, and lead service responsibilities across the entire theater. Figure 9-4 depicts a notional task organized TSC.

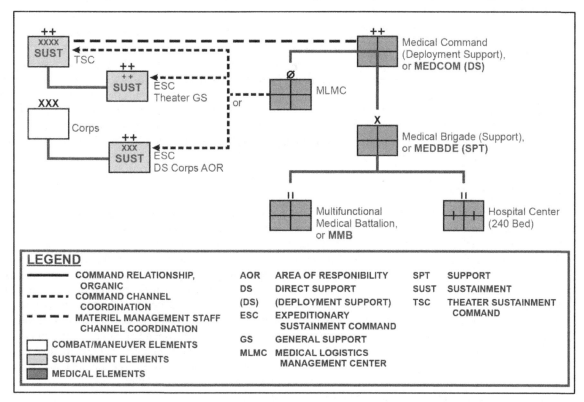

Figure 9-4. Notional task organized theater sustainment command

Expeditionary Sustainment Command

9-168. At the theater echelon, one or more ESCs are attached to a TSC. The ESC attached to a TSC commands and controls all assigned and attached units in an operational area as directed by the TSC commander. A task organized ESC attached to a TSC normally includes one or more sustainment brigades, a transportation brigade expeditionary, and a movement control battalion to support theater opening, theater distribution, and theater closing operations. The ESC plans for near term operations and synchronizes operational-level sustainment operations to meet the current and future operational requirements of the TSC. It may perform as a forward CP for the TSC if directed. The ESC attached to a TSC is dependent on the TSC staff for long-range planning capability and enabling capabilities like signal support. The ESC and its subordinate units must be able to move and displace at the pace of large-scale combat operations. (For more information on the ESC, see ATP 4-94. For more information on the sustainment brigade, see ATP 4-93.)

SUSTAINING THE CORPS

9-169. An ESC is assigned to the corps. The ESC is the corps' command for the integration and synchronization of sustainment in an operational area. The ESC assists the corps sustainment cell with planning and coordinating sustainment. The corps' ESC and its subordinate task organized functional and multifunctional sustainment units provide general support for all units in the corps area of operations as directed by the corps commander. A task organized ESC assigned to a corps normally includes enablers that include a corps logistics support element, petroleum group, movement control battalion, and one or more sustainment brigades task organized with combat sustainment support battalions (CSSBs) to support sustainment operations. The corps' echelon above brigade sustainment is dependent on the corps units for medical support, signal support, intelligence, long-range surveillance and reconnaissance, fires, protection (engineer support and route security), and strategic partner planning capability for field maintenance support.

The MEDCOM (DS) provides direct or general support to the corps through the MEDBDE (SPT) that will have hospital centers and medical battalions (multifunctional) attached.

Sustainment Brigade

9-170. Sustainment brigades can be attached to a corps ESC. The sustainment brigade attached to a corps ESC commands and controls all assigned and attached units in an operational area as directed by the corps commander providing general support logistics, financial management, and personnel services to forces operating in the corps area of operations. The corps commander determines the task organization for the sustainment brigade attached to a corps ESC. A task organized sustainment brigade attached to a corps ESC normally includes attached CSSBs, a petroleum battalion and motor transportation battalion to support tactical-level sustainment operations. The sustainment brigade coordinates and synchronizes tactical-level sustainment operations to meet the current and future operations. (See ATP 4-93 for more information on the sustainment brigade.)

Special Troops Battalion

9-171. The special troops battalion is organic to the sustainment brigade. The special troops battalion's role is to exercise command and control for all units assigned, attached, and OPCON to the sustainment brigade headquarters. The special troops battalion plans, prepares, executes, and assesses the internal support requirements for the sustainment brigade headquarters. Its core competencies are to establish a battalion CP, execute the operations process, and synchronize internal support operations in support of mission requirements. The battalion consists of a command group, unit ministry team, and coordinating staff. A battalion headquarters is organized to provide administrative support, life support, and communications for the sustainment brigade headquarters. Capable of operating at the tactical level throughout an operational area, it can command up to seven organizations. Organic to the special troops battalion is a headquarters company that includes a maintenance section, medical treatment team, and medical evacuation team. Assigned to the special troops battalion are a signal company, human resources company, and a financial management support unit. These units all support the special troops battalion, sustainment brigade headquarters, and all the organic, assigned, and attached units.

Combat Sustainment Support Battalion

9-172. The CSSB can be attached to sustainment brigades supporting a corps. The CSSB attached to sustainment brigades supporting the corps commands and controls all assigned and attached units in an operational area as directed by the sustainment brigade commander and conducts maintenance, transportation, supply, field services, and distribution. The corps commander determines the task organization for the CSSBs attached to sustainment brigades supporting the corps. Task organized CSSBs attached to sustainment brigades supporting the corps normally include a composite supply company, support maintenance company, modular ammunition company, palletized load system truck company and inland cargo transfer company, and a field feeding company. The CSSB synchronizes and executes logistics support to functional brigades and multifunctional support brigades attached to the corps. (See ATP 4-93.1 for more information on the CSSB.)

SUSTAINING THE DIVISION

9-173. A division will conduct operations with their assigned DSB, and the organic DSSB of the DSB. The DSB provides materiel management capability to the division. The division's assigned, task organized DSB provides general support for all units in or passing through their geographic area. The division and its subordinate units must be able to move and displace at the pace of large-scale combat operations. Divisions may have additional CSSBs attached to meet operational requirements. Multiple echelons above division sustainment units and elements of the MEDBDE (SPT) may be operating in the area alongside the DSB with its organic DSSB. Medical elements of the MEDBDE (SPT) are normally OPCON to the division commander and their parent medical organization retains administrative control. (See ATP 4-93, ATP 4-93.1, and FM 3-94 for additional information.)

Division Sustainment Brigade

9-174.   The DSB is assigned to a division. The DSB is a renamed sustainment brigade. The DSB commander is the primary senior advisor to the division commander and the deputy commanding general (support) for the sustainment warfighting function. The commander is responsible for the integration, synchronization, and execution of sustainment operations at echelon. The DSB employs sustainment capabilities to create desired effects in support of the division commander's objectives. Depending upon operational and mission variables, the DSB can command up to seven battalions. Figure 9-5 and figure 9-6 on page 9-36 depict notional task organized DSBs in support of an Infantry division and an Armored division respectively. The DSB and its subordinate units assigned to a division provides direct support to all assigned and attached units in an operational area as directed by the division commander. The DSB provides general support logistics, personnel services, and financial management to non-divisional forces operating in the division area of operations. A task organized DSB assigned to a division includes an organic division sustainment troops battalion (known as DSTB) and an organic DSSB to support tactical-level sustainment operations. The DSB coordinates and synchronizes tactical-level sustainment operations to meet current and future operations. The DSB is dependent on the division staff for long-range planning capability. The DSB and its subordinate units must be able to move and displace at the pace of large-scale combat operations. Additional modular CSSBs and companies may be attached to the DSB to sustain large-scale combat operations.

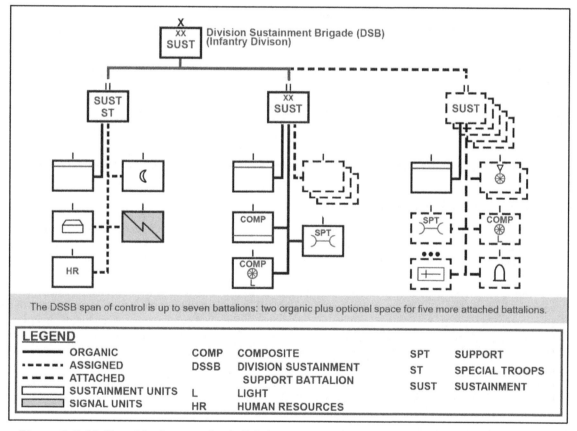

The DSSB span of control is up to seven battalions: two organic plus optional space for five more attached battalions.

LEGEND

———	ORGANIC	COMP	COMPOSITE	SPT	SUPPORT
·-·-·-·	ASSIGNED	DSSB	DIVISION SUSTAINMENT	ST	SPECIAL TROOPS
- - - -	ATTACHED		SUPPORT BATTALION	SUST	SUSTAINMENT
☐	SUSTAINMENT UNITS	L	LIGHT		
▨	SIGNAL UNITS	HR	HUMAN RESOURCES		

Figure 9-5. Notional task organized division sustainment brigade for an Infantry division

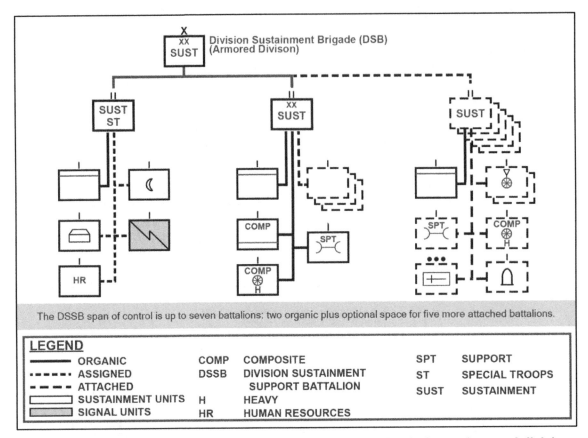

The DSSB span of control is up to seven battalions: two organic plus optional space for five more attached battalions.

LEGEND					
——— ORGANIC	COMP	COMPOSITE		SPT	SUPPORT
·····• ASSIGNED	DSSB	DIVISION SUSTAINMENT		ST	SPECIAL TROOPS
–––• ATTACHED		SUPPORT BATTALION		SUST	SUSTAINMENT
☐ SUSTAINMENT UNITS	H	HEAVY			
▨ SIGNAL UNITS	HR	HUMAN RESOURCES			

Figure 9-6. Notional task organized division sustainment brigade for an Armored division

Division Sustainment Troops Battalion

9-175. The DSTB is organic to DSBs. The battalion's role is to exercise command and control for all units assigned, attached, and OPCON to the DSB headquarters as shown in figure 9-5 on page 9-35 and figure 9-6 above. The battalion plans, prepares, executes, and assesses the internal support requirements for the DSB headquarters. Its core competencies are to establish a battalion CP, execute the operations process, and synchronize internal support operations in support of mission requirements. The DSTB consists of a command group, unit ministry team, and coordinating staff. It is a battalion headquarters organized to provide administrative support, life support, and communications for the DSB headquarters. Capable of operating at the tactical level throughout an operational area, it can command up to seven organizations. Organic to the DSTB is a headquarters company, which includes a maintenance section, medical treatment team, and medical evacuation team. Assigned to the DSTB are a signal company, human resources company, field feeding company, and a financial management support unit. These units all support the DSTB, DSB headquarters, and all the organic, assigned, and attached units.

Division Sustainment Support Battalion

9-176. The DSSB is employed using various task organizations as shown in figure 9-7. The DSSB is a renamed CSSB. The DSSB is organic to DSBs assigned to divisions. The DSSB and its subordinate units must be able to move and displace at the pace of large-scale combat operations. The DSSB commands and controls all organic, assigned, and attached units. As directed by the DSB commander, the DSSB conducts maintenance, transportation, supply, and distribution. DSSBs organic to DSBs supporting divisions have an organic composite supply company, composite truck company, and support maintenance company. Other capabilities are task organized by the division commander in accordance with requirements. The DSSB synchronizes and executes logistics support to BCTs and multifunctional support brigades attached to the division and non-divisional units operating in the division area of operations.

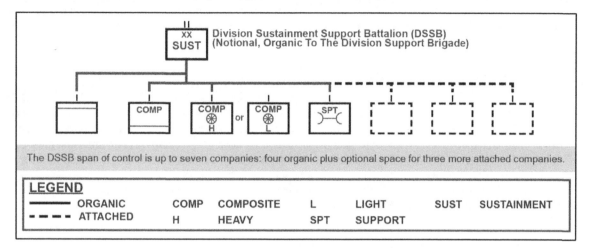

Figure 9-7. Notional division sustainment support battalion

MEDICAL SUPPORT ORGANIZATIONS

9-177. Battalion medical platoons and the BSMC provide health service support and force health protection to BCTs (see paragraphs 9-139 through 9-145). The theater Army has a MEDCOM (DS) for command and control of all medical units in a theater of operations at echelons above brigade (see paragraph 9-165). The MEDCOM (DS) provides subordinate medical organizations that operate under the MEDBDE or medical battalion (multifunctional). The MEDBDE provides a scalable expeditionary medical capability for assigned and attached medical organizations that are task organized to support BCTs and echelons above brigade. The medical battalion (multifunctional) also provides medical command and control, administrative assistance, logistical support, and technical supervision for assigned and attached companies and detachments. The medical battalion (multifunctional) is assigned to the MEDCOM (DS) or MEDBDE. The hospital center is a modular Role 3 medical treatment facility tailored to provide hospitalization support during decisive action and serves as the replacement for the current combat support hospital. The hospital center provides essential care within the theater evacuation policy to return patients either to duty or stabilization for further evacuation to a role 4 medical treatment facility in the continental United States or another safe haven. (See ATP 4-02.1 and FM 4-02 for additional information.)

BRIGADE SUPPORT AREA

9-178. The *brigade support area* is a designated area in which sustainment elements locate to provide support to a brigade (ATP 4-90). The BSA is the sustainment hub of the supported BCT. The BCT commander approves the location of the BSA based upon recommendations from the BSB commander and BCT staff. The BSA is a subset of the larger support area of the BCT, which can encompass a greater terrain footprint than the BSA. The support area of a BCT in a contiguous area of operation extends from its rear boundary forward to the rear boundary of its battalions. The support area of a BCT typically includes the BSA, airfields, lodgments, other battalion rear areas, the BCT alternate CP if required, the BEB, and may include areas of operations the BCT has not assigned to a maneuver battalion. In a noncontiguous area of operations, the BSA can be a base, base cluster, or sub-set inside the support area of the BCT.

OPERATIONAL OVERVIEW

9-179. The BSA is the sustainment (logistics, medical, personnel, and administrative) node for the BCT, and is the BSB's terrain from which to conduct sustainment operations. It consists of the BSB main CP (which can also serve as a BCT alternate CP if required), signal assets, and other sustainment units from echelons above brigade. The BSB commander is responsible for the command and control of all support organizations within the BSA for terrain management and security unless otherwise stated by the operations or fragmentary order. The BCT commander, with the support of the staff and upon the advice of the BSB commander determines the control exercised by the BSB commander in governing the authority and limitations of the BSB to execute area security within the BSA. Considerations used in determining the authority and limitations of the BSB commander to execute area security within the BSA are—threat levels and situation; utility of different locations; and civil considerations.

LOCATIONS FOR SUPPORT AREAS

9-180. The BSB commander, in coordination with the BCT S-3 and S-4, recommends to the BCT commander the layout of the BSA. The BSB commander coordinates land usage with the overall support area's terrain manager, typically the BEB commander. Support areas are located so that support to the BCT can be maintained but does not interfere with the tactical movement of BCT units or with units that must pass through the BCT area, while still maximizing security. The BSA's size varies with terrain and number of sustainment units. Usually the BSA is on a main supply route and ideally out of the range of the enemy's indirect-fire artillery. Position the BSA away from the enemy's likely avenues of approach and entry points into the BCT's main battle area (MBA).

9-181. In determining the location for the BSA, there is a constant balancing of support and security, which ultimately determines the best placement of support areas. The BSB commander balances constant support operations and security requirements for the BSB as it establishes and operates the BSA. The BSB commander integrates both activities to not degrade the BCT's combat effectiveness. The BSB commander ensures logistics missions and associated activities continue without restriction and that all units within or transiting the support area are capable of conducting self-protection against a level I threat.

9-182. Threats in the BCT and higher echelon support areas are categorized by the three levels of defense required to counter them. Any or all threat levels may exist simultaneously in these support areas. Emphasis on defense and security measures depends on the anticipated threat level. A *level I threat* is a small enemy force that can be defeated by those units normally operating in the echelon support area or by the perimeter defenses established by friendly bases and base clusters (ATP 3-91). A level I threat for an echelon support area or base camp consists of a squad-sized unit or smaller groups of enemy soldiers, agents, or terrorists. Typical objectives for a level I threat include supplying themselves from friendly supply stocks; disrupting friendly command and control nodes and logistics areas; and interdicting friendly lines of communication.

9-183. A *level II threat* is an enemy force or activities that can be defeated by a base or base cluster's defensive capabilities when augmented by a response force (ATP 3-91). A typical response force is a military police platoon (with appropriate supporting fires) for an echelon support area or base camp; however, it can be a combined arms maneuver element. Level II threats consist of enemy special operations teams, long-range reconnaissance units, mounted or dismounted combat reconnaissance teams, and partially attrited small combat units. Typical objectives for a level II threat include the destruction, as well as the disruption, of friendly command and control nodes and logistics and commercial facilities, and the interdiction of friendly lines of communications.

9-184. A *level III threat* is an enemy force or activities beyond the defensive capability of both the base and base cluster and any local reserve or response force (ATP 3-91). It consists of mobile enemy combat forces. Possible objectives for a level III threat include seizing key terrain, interfering with the movement and commitment of reserves and artillery, and destroying friendly combat forces. Its objectives could also include destroying friendly sustainment facilities, supply points, CP facilities, airfields, aviation assembly areas, arming and refueling points, and interdicting lines of communications and major supply routes. The response (usually task organized at the division or corps level) to a level III threat is a *tactical combat force*, a rapidly deployable, air-ground, mobile combat unit with appropriate combat support and combat service support assets assigned to, and capable of, defeating level III threats, including combined arms (JP 3-10). (See ATP 3-91 for additional information.)

9-185. Once positioned, units should not consider echelon support areas such as the trains or BSA as permanent or stationary. Support areas (specifically echeloned trains) must be mobile to support the units when they move and should change locations frequently depending on available time and terrain. A change of location may occur with a change of mission or change in a unit's area of operations. Movement to a new location may be required to avoid detection caused by heavy use or traffic in the area or an area becomes worn by heavy use (wet and muddy conditions). Echeloned trains locations may need to change when security becomes lax or complacent due to familiarity. (See ATP 4-90 for additional information.) Support area location considerations include the following:

- Cover and concealment (natural terrain or manmade structures).
- Room for dispersion.
- Level, firm ground to support vehicle traffic and sustainment operations.
- Suitable helicopter landing sites.
- Distance from known or templated enemy indirect fire assets.
- Good road or trail networks.
- Good routes in and out of the area (preferably separate routes going in and going out).
- Access to lateral routes.
- Good access or positioned along the main supply route.
- Positioned away from likely enemy avenues of approach.

PROTECTION OF SUPPORT AREAS

9-186. The BSB commander's responsibility for protection includes the BSA and extends to self-protection of BSB assets operating outside of the BSA, unless otherwise stated by the operation or fragmentary order. Forces engaged in area security protect the force, installation, route, area, or asset. Although vital to the success of military operations, area security is normally an economy-of-force mission, often designed to ensure the continued conduct of sustainment operations and to support decisive and shaping operations by generating and maintaining combat power. Area security may be the predominant method of protecting support areas that are necessary to facilitate the positioning, employment, and protection of resources required to sustain, enable, and control forces.

9-187. The BSB commander and staff must plan for and coordinate protection for subordinate units and detachments located within and away from the BSA. While the BSB S-3 is responsible overall for developing the BSA security plan, the BSB S-2 assists by developing the information collection plan to support intelligence operations, reconnaissance, surveillance, and security operations within the BSA. The BSB commander and staff use the intelligence preparation of the battlefield (IPB) to analyze the mission variables of enemy, terrain, weather, and civil considerations to determine their effect on sustainment operations.

9-188. Sustainment operations in noncontiguous area of operations require commanders to emphasize protection. Sustainment organizations are normally the least capable of self-defense against an enemy force and are often the target of enemy action. As the threat increases, supported commanders and sustainment unit commanders cannot decrease sustainment operations in favor of enhancing protection. The supported commander and the sustainment unit commander must discuss what risks are reasonable to accept and what risk mitigation measures they should implement based on requirements and priorities. The supported commander and sustainment unit commander carefully weigh and balance options and alternatives and derive solutions that both accomplish the mission for the supported BCT while minimizing or mitigating the risk of the sustainment units. Protection within the BSA includes terrain management, fire support coordination, airspace management, and other security and protection activities including node protection, lines of communications security, and checkpoints. Sustainment units plan for, train, and rehearse support area protection measures and immediate response actions against enemy threats. Additional operations can include convoy security (see ATP 3-91 and ATP 4-01.45), coordination of base camp and based cluster defense (see ATP 3-37.10), area damage control (see ATP 3-91), and response force operations (see FM 3-39 and ATP 3-37.10).

9-189. As the enemy may avoid maneuver forces, preferring to attack targets commonly found in sustainment areas. Sustainment elements must organize and prepare to defend themselves against ground or air attacks. The security of the support areas or trains at each echelon is the responsibility of the individual in charge of the support area or echeloned trains. All elements in, or transiting the support area, assist with forming and defending the area. Based on mission analyses, the BSB S-3 subdivides the area, and assigns subordinate and tenant units to those subdivided areas. When a subordinate or tenant unit receives a change of mission or can no longer occupy an assigned area, area adjustments are made to the support area by the BSB S-3. When a particular supply point is sufficiently large, it may be assigned its own area for defense, and a security force may be attached to provide security. Additional activities to enable BSA protection include—

- Select sites that use available cover, concealment, and camouflage.
- Use movement and positioning discipline, as well as noise and light discipline, to prevent detection.
- Establish area defenses.
- Establish engagement areas.
- Establish obstacles.
- Establish observation posts and conduct patrols.
- Position weapons (small arms, machine guns, and antitank weapons) for self-defense.
- Plan mutually supporting positions to dominate likely avenues of approach.
- Prepare fire support plans.
- Make area of operations sketches and identify sectors of fires.
- Emplace target reference points (TRPs) to control fires.
- Integrate available combat vehicles within the area into the plan and adjust the plan when vehicles depart.
- Conduct rehearsals.
- Establish rest plans.
- Identify alarms or warning systems to enable rapid execution of the defense plan.
- Designate a response force (see ATP 3-91) with appropriate fire support.
- Ensure the response force is equipped to perform its mission.
- Response force must be well-rehearsed or briefed on—
  - Unit assembly.
  - Friendly and threat force recognition.
  - Actions on contact.

SUPPLY ROUTES AND CONVOYS

9-190. The BCT S-4, in coordination with the BSB support operations officer and the BCT S-3, select supply routes between echeloned support areas. Main supply routes are designated within the BCT's area of operations. A main supply route is selected based on the terrain, friendly disposition, enemy situation, and scheme of maneuver. Alternate supply routes are planned if a main supply route is interdicted by the enemy or becomes too congested. In the event of CBRN contamination, either the primary or the alternate main supply route can be designated as the dirty main supply route to handle contaminated traffic. Alternate supply routes should meet the same criteria as the main supply route. Military police may assist with regulating traffic and the security of routes and convoys on those routes, and engineer units, if available, can maintain routes. Main supply route considerations include—

- Location and planned scheme of maneuver for subordinate units.
- Location and planned movements of other units moving through the BCT's area of operations.
- Route classification, width, obstructions, steep slopes, sharp curves, and roadway surface.
- Two way, all weather trafficability.
- Classification of bridges and culverts. Location and planned scheme of maneuver for subordinate units.
- Requirements for traffic control such as choke points, congested areas, confusing intersections, or through built up areas.
- Location and number of crossover routes from the main supply route to alternate supply routes.
- Requirements for repair, upgrade, or maintenance of the route, fording sites, and bridges.
- Route vulnerabilities that must be protected, such as bridges, fords, built up areas, and choke points.
- Enemy threats such as air attack, mines, ambushes, and CBRN attacks.

- Known or likely locations of enemy penetrations, attacks, CBRN attacks, or obstacles.
- Known or potential civilian and dislocated civilian movements that must be controlled or monitored.

9-191.   Security of supply routes in noncontiguous area of operations may require the BCT commander to commit combat units. The security and protection of supply routes along with lines of communications are critical to military operations since most support traffic moves along these routes. The security of supply routes presents one of the greatest security challenges in an area of operations. Route security operations are defensive in nature and are terrain oriented. A route security force may prevent an enemy or adversary force from impeding, harassing, or destroying traffic along a route or portions of a route by establishing a movement corridor. Units conduct synchronized operations (mobility and information collection) within the movement corridor. A movement corridor may be established in a high-risk area to facilitate the movement of a single element, or it may be an enduring operation. (See FM 3-90-2 for additional information.)

9-192.   A convoy security operation is a specialized kind of area security operation conducted to protect convoys. Units conduct convoy security operations anytime there are insufficient friendly forces to secure routes continuously in an area of operations and there is a significant danger of enemy or adversary ground action directed against the convoy. The commander may conduct convoy security operations in conjunction with route security operations. Planning includes designating units for convoy security; providing guidance on tactics, techniques, and procedures for units to provide for their own security during convoys; or establishing protection and security requirements for convoys carrying critical assets. Local or theater policy typically dictates when or which convoys receive security and protection. (See ATP 4-01.45 for additional information.)

This page intentionally left blank.

Source Notes

These are the sources used for historical examples cited and quoted in this manual. They are listed by paragraph number.

2-24   1. *Perceptions Are Reality: Historical Case Studies of Information Operations in Large-Scale Combat Operations*, Fort Leavenworth, Kansas: Army University Press [2018] Series.

2. Department of the Army, Field Manual (FM) 3-13, *Information Operations* (Washington, DC: 2016), 1-2.

3. Kaplan, Fred. *Dark Territory: The Secret History of Cyber War* (New York: Simon and Schuster, 2016), 31.

2-31   1. Center for Army Lessons Learned, *Strategic Landpower in Europe Special Study*, CALL website: http://call.army.mil.

2. 10th Combat Aviation Brigade, 10th Mountain Division Fact Sheet, 03 FEB 2017. Online at https://static.dvidshub.net/media/pubs/pdf_34851.pdf.

3. 3rd Armored Brigade Combat Team, 4th Infantry Division Fact Sheet, 04 JAN 2017. Online at https://static.dvidshub.net/media/pubs/pdf_34851.pdf.

4. Operation Atlantic Resolve, Online at https://www.eur.army.mil/Portals/19/Atlantic%20Resolve%20Fact%20Sheet.pdf?ver=jP-mMtPdOc4kvb_8r_uGkA%3d%3d, 2 NOV 2020.

2-58   1. *From Domination to Consolidation: at the Tactical Level in Future Large-Scale Combat Operations*, A U.S. Army Command and General Staff College Press Book Published by the Army University Press [2020] Series.

2. 82d Airborne Division, After Action Report 82d Airborne, December 1944-February 1945. (World War II Operational Documents, Combined Arms Research Digital Library, Fort Leavenworth, KS).

2-60   Lundy, Michael D. *Military Review: Large-Scale Combat Operations Special Edition*, Fort Leavenworth, Kansas: Army University Press [September –October 2018], Forward. Available online at https://www.armyupress.army.mil/Journals/Military-Review/English-Edition-Archives/September-October-2018/.

4-4   Wright, Donald P., Ph., ed. *16 Cases of Mission Command*. U.S. Army Combined Arms Center. Fort Leavenworth, KS: Combat Studies Institute Press, 2013. Available online at https://www.armyupress.army.mil/Portals/7/Primer-on-Urban-Operation/Documents/Sixteen-Cases-of-Mission-Command.pdf.

6-2   Salinger, Jerry, *Urban Warfare-The 2008 Battle for Sadr City*. Rand Research Brief, RAND Arroyo Center. Santa Monica, CA: 2012. Available online at http://www.rand.org/content/dam/rand/pubs/research_briefs/2012/RAND_RB9652.pdf.

7-2   *Kasserine Pass Battles. Readings, volume I*. United States Army Center of Military History, United States Army. Available online at https://history.army.mil/books/Staff-Rides/kasserine/kasserine.htm.

7-88   Esposito, Vincent, ed. *West Point Atlas of American Wars*. United States Military Academy, Department of Military Art and Engineering. New York: Praeger, 1972.

7-92   Mossman, Billy C. *Ebb and Flow, November 1950-July 1951*. U.S. Army History of the Korean War Series, volume 5. Washington, DC: Center of Military History, United States Army. Available online at https://history.army.mil/books/korea/ebb/fm.htm.

7-93   Appleman, Roy E., James M. Burns, Russell A. Gugeler, and John Stevens. *The War in the Pacific, Okinawa: The Last Battle*. Washington, DC: Historical Division, Department of the Army, 1048. Available online at https://history.army.mil/html/books/005/5-11-1/CMH_Pub_5-11-1.pdf.

Glossary

The glossary lists acronyms and terms with Army or joint definitions. Where Army and joint definitions differ, (Army) precedes the definition. Terms for which FM 3-96 is the proponent are marked with an asterisk (*). The proponent publication for other terms is listed in parentheses after the definition.

SECTION I – ACRONYMS AND ABBREVIATIONS

ABCT	Armored brigade combat team
ABE	assistant brigade engineer
ADAM	air defense airspace management
ADP	Army doctrine publication
ADRP	Army doctrine reference publication
AHS	Army Health System
AOR	area of responsibility
AR	Army regulation
ASCC	Army Service component command
ASCOPE	areas, structures, capabilities, organizations, people, and events
ATGM	antitank guided missile
ATP	Army techniques publication
ATTP	Army tactics, techniques, and procedures
BAE	brigade aviation element
BCT	brigade combat team
BEB	brigade engineer battalion
BSA	brigade support area
BSB	brigade support battalion
BSMC	brigade support medical company
CBRN	chemical, biological, radiological, and nuclear
CCIR	commander's critical information requirement
CEMA	cyberspace electromagnetic activities
CERF	cyber effects request format
CJCSM	Chairman of the Joint Chiefs of Staff Manual
COA	course of action
COL	colonel

COP	common operational picture
COR	contracting officer representative
CP	command post
CPCE	Command Post Computing Environment
CPOF	Command Post of the Future
CSSB	combat sustainment support battalion
CTCP	combat trains command post
CWMD	countering weapons of mass destruction
DA	Department of the Army
DCGS-A	Distributed Common Ground System-Army
DD	Department of Defense
DLIC	detachment left in contact
DODIN	Department of Defense information network
DODIN-A	Department of Defense information network-Army
DSB	division sustainment brigade
DSM	decision support matrix
DSSB	division sustainment support battalion
DST	decision support template
DSTB	division sustainment troops battalion
EA	engagement area
EEFI	essential element of friendly information
EMS	electromagnetic spectrum
EP	electromagnetic protection
ESC	expeditionary sustainment command
EW	electromagnetic warfare
EWO	electromagnetic warfare officer
FARP	forward arming and refueling point
FLE	forward logistics element
FM	field manual
FSC	forward support company
FSF	foreign security forces
FST	forward surgical team
GCSS-Army	Global Combat Support System-Army
GPS	Global Positioning System
GPW	Geneva Convention Relative to the Treatment of Prisoners of War
HPT	high-payoff target
HUMINT	human intelligence
HVT	high-value target
IBCT	Infantry brigade combat team
ID	Infantry division
IDAD	internal defense and development

IED	improvised explosive device
IPB	intelligence preparation of the battlefield
JAGIC	joint air-ground integration center
JBC-P	Joint Battle Command-Platform
JCR	Joint Capabilities Release
JCR-Log	Joint Capabilities Release-Logistics
JP	joint publication
JSTARS	Joint Surveillance Target Attack Radar System
JTAC	joint terminal attack controller
LOGCAP	logistics civil augmentation program
LOGPAC	logistics package
LOGSTAT	logistics status
LRP	logistics release point
LTC	lieutenant colonel
LTG	lieutenant general
MBA	main battle area
MC4	Medical Communications for Combat Casualty Care
MDMP	military decision-making process
MEDBDE (SPT)	medical brigade (support)
MEDCOM (DS)	medical command (deployment support)
METT-TC	mission, enemy, terrain and weather, troops and support available, time available, civil considerations [mission variables] (Army)
MG	major general
MGS	mobile gun system
mm	millimeter
MOE	measure of effectiveness
MOP	measure of performance
NAI	named area of interest
OAKOC	observation and fields of fire, avenues of approach, key terrain, obstacles, and cover and concealment [military aspects of terrain]
OPCON	operational control
ORP	objective rally point
OTERA-A	organize, train, equip, rebuild and build, advise and assist, and assess
PACE	primary, alternate, contingency, and emergency
PL	phase line
PMESII-PT	political, military, economic, social, information, infrastructure, physical environment, and time [operational variables]
POD	port of debarkation
POE	port of embarkation
POL	petroleum, oils, and lubricants
PSYOP	psychological operations
RETRANS	retransmission

RM	risk management
ROM	refuel on the move
S-1	battalion or brigade personnel staff officer
S-2	battalion or brigade intelligence staff officer
S-3	battalion or brigade operations staff officer
S-4	battalion or brigade logistics staff officer
S-6	battalion or brigade signal staff officer
S-8	battalion or brigade financial management staff officer
S-9	battalion or brigade civil affairs operations staff officer
SBCT	Stryker brigade combat team
SFA	security force assistance
SFAB	security force assistance brigade
SOP	standard operating procedure
SOSRA	suppress, obscure, secure, reduce, and assault (breaching fundamentals)
SSA	supply support activity
SSG	staff sergeant
TAC	tactical command post (TAC [graphic])
TACP	tactical air control party
TAI	target area of interest
TC	training circular
TCCC	tactical combat casualty care
TRP	target reference point
TSC	theater sustainment command
TUAS	tactical unmanned aircraft system
U.S.	United States
UAS	unmanned aircraft system
WMD	weapons of mass destruction
XO	executive officer

SECTION II – TERMS

actions on contact

A series of combat actions, often conducted simultaneously, taken on contact with the enemy to develop the situation. (ADP 3-90)

administrative movement

A movement in which troops and vehicles are arranged to expedite their movement and conserve time and energy when no enemy ground interference is anticipated. (ADP 3-90)

adversary

(DOD) A party acknowledged as potentially hostile to a friendly party and against which the use of force may be envisaged. (JP 3-0)

air assault

(DOD) The movement of friendly assault forces by rotary-wing or tiltrotor aircraft to engage and destroy enemy forces or to seize and hold key terrain. (JP 3-18)

air assault operation

(DOD) An operation in which assault forces, using the mobility of rotary-wing or tiltrotor aircraft and the total integration of available fires, maneuver under the control of a ground or air maneuver commander to engage enemy forces or to seize and hold key terrain. (JP 3-18)

air movements

(Army) Operations involving the use of utility and cargo rotary-wing assets for other than air assaults. (FM 3-90-2)

airborne assault

(DOD) The use of airborne forces to parachute into an objective area to attack and eliminate armed resistance and secure designated objectives. (JP 3-18)

airborne operation

(DOD) An operation involving the air movement into an objective area of combat forces and their logistic support for execution of a tactical, operational, or strategic mission. (JP 3-18)

airspace management

(DOD) The coordination, integration, and regulation of the use of airspace of defined dimensions. (JP 3-52)

all-source intelligence

(DOD) Intelligence products and/or organizations and activities that incorporate all sources of information in the production of finished intelligence. (JP 2-0)

all-source intelligence

(Army) The integration of intelligence and information from all relevant sources in order to analyze situations or conditions that impact operations. (ADP 2-0)

alternate position

A defensive position that the commander assigns to a unit or weapon system for occupation when the primary position becomes untenable or unsuitable for carrying out the assigned task. (ADP 3-90)

ambush

An attack by fire or other destructive means from concealed positions on a moving or temporarily halted enemy. (FM 3-90-1)

approach march

The advance of a combat unit when direct contact with the enemy is intended. (ADP 3-90)

area defense

A type of defensive operation that concentrates on denying enemy forces access to designated terrain for a specific time rather than destroying the enemy outright. (ADP 3-90)

area of influence

(DOD) A geographical area wherein a commander is directly capable of influencing operations by maneuver or fire support systems normally under the commander's command or control. (JP 3-0)

area of interest

(DOD) That area of concern to the commander, including the area of influence, areas adjacent thereto, and extending into enemy territory. (JP 3-0)

area of operations

(DOD) An operational area defined by a commander for land and maritime forces that should be large enough to accomplish their missions and protect their forces. (JP 3-0)

area reconnaissance

A type of reconnaissance operation that focuses on obtaining detailed information about the terrain or enemy activity within a prescribed area. (ADP 3-90)

area security

A type of security operation conducted to protect friendly forces, lines of communicaiton, and activities within a specific area. (ADP 3-90)

area support

A task assigned to a sustainment unit directing it to support units in or passing through a specified location. (ATP 4-90)

Army design methodology

A methodology for applying critical and creative thinking to understand, visualize, and describe problems and approaches to solving them. (ADP 5-0)

assailable flank

A flank exposed to attack or envelopment. (ADP 3-90)

assault position

A covered and concealed position short of the objective from which final preparations are made to assault the objective. (ADP 3-90)

assessment

(DOD) Determination of the progress toward accomplishing a task, creating a condition, or achieving an objective. (JP 3-0)

attack

A type of offensive operation that destroys or defeats enemy forces, seizes and secures terrain, or both. (ADP 3-90)

attack position

(Army) The last position an attacking force occupies or passes through before crossing the line of departure. (ADP 3-90)

avenue of approach

(Army) A path used by an attacking force leading to its objective or to key terrain. Avenues of approach exist in all domains. (ADP 3-90)

backbrief

A briefing by subordinates to the commander to review how subordinates intend to accomplish their mission. (FM 6-0)

battle drill

Rehearsed and well understood actions made in response to common battlefield occurrences. (ADP 3-90)

battle handover line

A designated phase line where responsibility transitions from the stationary force to the moving force and vice versa. (ADP 3-90)

battle position

A defensive location oriented on a likely enemy avenue of approach. (ADP 3-90)

battle rhythm

(Army) A deliberate daily cycle of command, staff, and unit activities intended to synchronize current and future operations. (FM 6-0)

breakout

An operation conducted by an encircled force to regain freedom of movement or contact with friendly units. (ADP 3-90)

brigade support area

A designated area in which sustainment elements locate to provide support to a brigade. (ATP 4-90)

bypass criteria

Measures established by higher echelon headquarters that specify the conditions and size under which enemy units and contact may be avoided. (ADP 3-90)

chemical, biological, radiological, and nuclear defense

(DOD) Measures taken to minimize or negate the vulnerabilities to, and/or effects of, a chemical, biological, radiological, or nuclear hazard or incident. (JP 3-11)

chemical, biological, radiological, and nuclear environment

(DOD) An operational environment that includes chemical, biological, radiological, and nuclear threats and hazards and their potential resulting effects. (JP 3-11)

chemical, biological, radiological, and nuclear protection

Measures taken to keep chemical, biological, radiological, and nuclear threats and hazards from having an adverse effect on personnel, equipment, and facilities. (ATP 3-11.32)

civil considerations

The influence of manmade infrastructure, civilian institutions, and attitudes and activities of the civilian leaders, populations, and organizations within an area of operations on the conduct of military operations. (ADP 6-0)

close area

The portion of a commander's area of operations where the majority of subordinate maneuver forces conduct close combat. (ADP 3-0)

close combat

Warfare carried out on land in a direct-fire fight, supported by direct and indirect fires, and other assets. (ADP 3-0)

collaborative planning

Two or more echelons planning together in real time, sharing information, perceptions, and ideas to develop their respective plans simultaneously. (ADP 5-0)

combat information

(DOD) Unevaluated data, gathered by or provided directly to the tactical commander which, due to its highly perishable nature or the criticality of the situation, cannot be processed into tactical intelligence in time to satisfy the user's tactical intelligence requirements. (JP 2-01)

combat outpost

A reinforced observation post capable of conducting limited combat operations. (FM 3-90-2)

combat power

(Army) The total means of destructive, constructive, and information capabilities that a military unit or formation can apply at a given time. (ADP 3-0)

combined arms

The synchronized and simultaneous application of arms to achieve an effect greater than if each element was used separately or sequentially. (ADP 3-0)

command

(DOD) The authority that a commander in the armed forces lawfully exercises over subordinates by virtue of rank or assignment. (JP 1)

command and control

(DOD) The exercise of authority and direction by a properly designated commander over assigned and attached forces in the accomplishment of mission. (JP 1)

command and control system

(Army) The arrangement of people, processes, networks, and command posts that enable commanders to conduct operations. (ADP 6-0)

command and control warfighting function

The related tasks and a system that enable commanders to synchronize and converge all elements of power. (ADP 3-0)

command post

A unit headquarters where the commander and staff perform their activities. (FM 6-0)

command post cell

A grouping of personnel and equipment organized by warfighting function or by planning horizon to facilitate the exercise of mission command. (FM 6-0)

commander's critical information requirements

(DOD) An information requirement identified by the commander as being critical to facilitating timely decision making. (JP 3-0)

commander's intent

(DOD) A clear and concise expression of the purpose of the operation and the desired military end state that supports mission command, provides focus to the staff, and helps subordinate and supporting commanders act to achieve the commander's desired results without further orders, even when the operation does not unfold as planned. (JP 3-0)

common operational picture

(Army) A display of relevant information within a commander's area of interest tailored to the user's requirements and based on common data and information shared by more than one command. (ADP 6-0)

complex terrain

A geographical area consisting of an urban center larger than a village and/or of two or more types of restrictive terrain or environmental conditions occupying the same space. (ATP 3-34.80)

*concealment

Protection from observation or surveillance.

concept of operations

(Army) A statement that directs the manner in which subordinate units cooperate to accomplish the mission and establishes the sequence of actions the force will use to achieve the end state. (ADP 5-0)

confirmation brief

A briefing subordinate leaders give to the higher commander immediately after the operation order is given to confirm understanding. (ADP 5-0)

consolidate gains

Activities to make enduring any temporary operational success and to set the conditions for a sustainable security environment, allowing for a transition of control to other legitimate authorities. (ADP 3-0)

consolidation

Organizing and strengthening a newly captured position so that it can be used against the enemy. (FM 3-90-1)

consolidation area

The portion of the land commander's area of operations that may be designated to facilitate freedom of action, consolidate gains through decisive action, and set conditions to transition the area of operations to follow on forces or other legitimate authorities. (ADP 3-0)

contamination mitigation

(DOD) The planning and actions taken to prepare for, respond to, and recover from contamination associated with all chemical, biological, radiological, and nuclear threats and hazards to continue military operations. (JP 3-11)

contiguous area of operations

Where all a commander's subordinate forces' areas of operations share one or more common boundary. (FM 3-90-1)

control

(Army) The regulation of forces and warfighting functions to accomplish the mission in accordance with the commander's intent. (ADP 6-0)

control measure

A means of regulating forces or warfighting functions. (ADP 6-0)

cordon and search

A technique of conducting a movement to contact that involves isolating a target area and searching suspected locations within that target area to capture or destroy possible enemy forces and contraband. (FM 3-90-1)

counterattack

Attack by part or all of a defending force against an enemy attacking force, for such specific purposes as regaining ground lost, or cutting off or destroying enemy advance units, and with the general objective of denying to the enemy the attainment of the enemy's purpose in attacking. In sustained defensive operations, it is undertaken to restore the battle position and is directed at limited objectives. (FM 3-90-1)

countering weapons of mass destruction

(DOD) Efforts against actors of concern to curtail the conceptualization, development, possession, proliferation, use, and effects of weapons of mass destruction, related expertise, materials, technologies, and means of delivery. (JP 3-40)

counterreconnaissance

A tactical mission task that encompasses all measures taken by a commander to counter enemy reconnaissance and surveillance efforts. Counterreconnaissance is not a distinct mission, but a component of all forms of security operations. (FM 3-90-1)

*cover

Protection from the effects of fires.

cover

(Army) A type of security operation done independent of the main body to protect them by fighting to gain time while preventing enemy ground observation of and direct fire against the main body. (ADP 3-90)

covering force

(Army) A self-contained force capable of operating independently of the main body, unlike a screen or guard force to conduct the cover task. (FM 3-90-2)

covering force area

The area forward of the forward edge of the battle area out to the forward positions initially assigned to the covering force. It is here that the covering force executes assigned tasks. (FM 3-90-2)

cross-domain fires

Fires executed in one domain to create effects in a different domain. (ADP 3-19)

cueing

The integration of one or more types of reconnaissance or surveillance systems to provide information that directs follow-on collection of more detailed information by another system. (FM 3-90-2)

cyberspace

(DOD) A global domain within the information environment consisting of the interdependent networks of information technology infrastructures and resident data, including the Internet, telecommunications networks, computer systems, and embedded processors and controllers. (JP 3-12)

cyberspace operations

(DOD) The employment of cyberspace capabilities where the primary purpose is to achieve objectives in or through cyberspace. (JP 3-0)

data

In the contect of decision making, unprocessed observations detected by a collector of any kind (human, mechanical, or electronic). (ADP 6-0)

decision point

(DOD) A point in space and time when the commander or staff anticipates making a key decision concerning a specific course of action. (JP 5-0)

decision support matrix

A written record of a war-gamed course of action that describes decision points and associated actions at those decision points. (ADP 5-0)

decision support template

(DOD) A combined intelligence and operations graphic based on the results of wargaming that depicts decision points, timelines associated with movement of forces and the flow of the operation, and other key items of information required to execute a specific friendly course of action. (JP 2-01.3)

decisive action

(Army) The continuous, simultaneous execution of offensive, defensive, and stability operations or defense support of civil authorities tasks. (ADP 3-0)

decisive operation

The operation that directly accomplishes the mission. (ADP 3-0)

decisive terrain

Key terrain whose seizure and retention is mandatory for successful mission accomplishment. (ADP 3-90)

deep area

Where the commander sets conditions for future success in close combat. (ADP 3-0)

defeat

To render a force incapable of achieving its objective. (ADP 3-0)

defeat mechanism

A method through which friendly forces accomplish their mission against enemy opposition. (ADP 3-0)

defensive operation

An operation to defeat an enemy attack, gain time, economize forces, and develop conditions favorable for offensive or stability operations. (ADP 3-0)

delay

When a force under pressure trades space for time by slowing down the enemy's momentum and inflicting maximum damage on enemy forces without becoming decisively engaged. (ADP 3-90)

deliberate operation

An operation in which the tactical situation allows the development and coordination of detailed plans, including multiple branches and sequels. (ADP 3-90)

demonstration

(DOD) In military deception, a show of force similar to a feint without actual contact with the adversary, in an area where a decision is not sought that is made to deceive an adversary. (JP 3-13.4)

Department of Defense information network-Army

An Army-operated enclave of the Department of Defense information network that encompasses all Army information capabilities that collect, process, store, display, disseminate, and protect information worldwide. (ATP 6-02.71)

deployment

(DOD) The movement of forces into and out of an operational area. (JP 3-35)

depth

The extension of operations in time, space, or purpose to achieve definitive results. (ADP 3-0)

destroy

A tactical mission task that physically renders an enemy force combat-ineffective until it is reconstituted. Alternatively, to destroy a combat system is to damage it so badly that it cannot perform any function or be restored to a usable condition without being entirely rebuilt. (FM 3-90-1)

detachment left in contact

An element left in contact as part of the previously designated (usually rear) security force while the main body conducts its withdrawal. (FM 3-90-1)

direct fire

(DOD) Fire delivered on a target using the target itself as a point of aim for either the weapon or the director. (JP 3-09.3)

disengage

A tactical mission task where a commander has the unit break contact with the enemy to allow the conduct of another mission or to avoid decisive engagement. (FM 3-90-1)

disengagement line

A phase line located on identifiable terrain that, when crossed by the enemy, signals to defending elements that it is time to displace to their next position. (ADP 3-90)

disintegrate

To disrupt the enemy's command and control system, degrading its ability to conduct operations while leading to a rapid collapse of the enemy's capabilities or will to fight. (ADP 3-0)

dislocate

To employ forces to obtain significant positional advantage, rendering the enemy's dispositions less valuable, perhaps even irrelevant. (ADP 3-0)

double envelopment

This results from simultaneous maneuvering around both flanks of a designated enemy force. (FM 3-90-1)

early-entry command post

A lead element of a headquarters designed to control operations until the remaining portions of the headquarters are deployed and operational. (FM 6-0)

echelon support

The method of supporting an organization arrayed within an area of operations. (ATP 4-90)

electromagnetic spectrum

(DOD) The range of frequencies of electromagnetic radiation from zero to infinity. It is divided into 26 alphabetically designated bands. (JP 3-85)

electromagnetic warfare

(DOD) Military action involving the use of electromagnetic and directed energy to control the electromagnetic spectrum or to attack the enemy. (JP 3-85)

employment

(DOD) The strategic, operational, or tactical use of forces. (JP 5-0)

encirclement operations

Operations where one force loses its freedom of maneuver because an opposing force is able to isolate it by controlling all ground lines of communication and reinforcement. (ADP 3-90)

end state

(DOD) The set of required conditions that defines achievement of the commander's objectives. (JP 3-0)

enemy

A party identified as hostile against which the use of force is authorized. (ADP 3-0)

engagement area

An area where the commander intends to contain and destroy an enemy force with the massed effects of all available weapons and supporting systems. (ADP 3-90)

engagement criteria

Protocols that specify those circumstances for initiating engement with an enemy force. (FM 3-90-1)

engagement priority

Specifies the order in which the unit engages enemy systems or functions. (FM 3-90-1)

envelopment

A form of maneuver in which an attacking force seeks to avoid the principal enemy defenses by seizing objectives behind those defenses that allow the targeted enemy force to be destroyed in their current positions. (FM 3-90-1)

essential element of friendly information

A critical aspect of a friendly operation that, if known by a threat would subsequently compromise, lead to failure, or limit success of the operation and therefore should be protected from enemy detection. (ADP 6-0)

evaluating

Using indicators to judge progress toward desired conditions and determining why the current degree of progress exists. (ADP 5-0)

execution

The act of putting a plan into action by applying combat power to accomplish the mission and adjusting operations based on changes in the situation. (ADP 5-0)

execution matrix

A visual representation of subordinate tasks in relationship to each other over time. (ADP 5-0)

exploitation

(Army) A type of offensive operation that usually follows a successful attack and is designed to disorganize the enemy in depth. (ADP 3-90)

feint

(DOD) In military deception, an offensive action involving contact with the adversary conducted for the purpose of deceiving the adversary as to the location and/or time of the actual main offensive action. (JP 3-13.4)

financial management

(Army) The sustainment of the U.S. Army and its unified action partners through the execution of Fund the Force, Banking and Disbursing, Accounting Support and Cost Management, Pay Support and Management Internal Controls. (FM 1-06)

*fire and movement

The concept of applying fires from all sources to suppress, neutralize, or destroy the enemy, and the tactical movement of combat forces in relation to the enemy (as components of maneuver applicable at all echelons). At the squad level, fire and movement entails a team placing suppressive fire on the enemy as another team moves against or around the enemy.

fire support

(DOD) Fires that directly support land, maritime, amphibious, space, cyberspace, and special operations forces to engage enemy forces, combat formations, and facilities in pursuit of tactical and operational objectives. (JP 3-09)

fires

(DOD) The use of weapons systems or other actions to create a specific lethal or nonlethal effects on a target. (JP 3-09)

fires warfighting function

The related tasks and systems that create and converge effects in all domains against the adversary or enemy to enable operations across the range of military operations. (ADP 3-0)

fixing force

A force designated to supplement the striking force by preventing the enemy from moving from a specific area for a specific time. (ADP 3-90)

flank attack

A form of offensive maneuver directed at the flank of an enemy. (FM 3-90-1)

force projection

(DOD) The ability to project the military instrument of national power from the United States or another theater, in response to requirements for military operations. (JP 3-0)

force tailoring

The process of determining the right mix of forces and the sequence of their deployment in support of a joint force commander. (ADP 3-0)

foreign internal defense

(DOD) Participation by civilian and military agencies of a government in any of the action programs taken by another government or other designated organization to free and protect its society from subversion, lawlessness, insurgency, terrorism, and other threats to its security. (JP 3-22)

forms of maneuver

Distinct tactical combinations of fire and movement with a unique set of doctrinal characteristics that differ primarily in the relationship between the maneuvering force and the enemy. (ADP 3-90)

forward logistics element

Comprised of task-organized multifunctional logistics assets designed to support fast-moving offensive operations in the early phases of decisive action. (ATP 4-90)

forward passage of lines

Occurs when a unit passes through another unit's positions while moving toward the enemy. (ADP 3-90)

free-fire area

(DOD) A specific region into which any weapon system may fire without additional coordination with the establishing headquarters. (JP 3-09)

friendly force information requirement

(DOD) Information the commander and staff need to understand the status of friendly and supporting capabilities. (JP 3-0)

fusion

    Consolidating, combining, and correlating information together. (ADP 2-0)

governance

    (DOD) The state's ability to serve the citizens through the rules, processes, and behavior by which interests are articulated, resources are managed, and power is exercised in a society. (JP 3-24)

graphic control measure

    A symbol used on maps and displays to regulate forces and warfighting functions. (ADP 6-0)

guard

    A type of security operation done to protect the main body by fighting to gain time while preventing enemy ground observation of and direct fire against the main body. (ADP 3-90)

hasty operation

    An operation in which a commander directs immediately available forces, using fragmentary orders, to perform tasks with minimal preparation, trading planning and preparation time for speed of execution. (ADP 3-90)

hazard

    (DOD) A condition with the potential to cause injury, illness, or death of personnel; damage to or loss of equipment or property; or mission degradation. (JP 3-33)

health service support

    (Army) Encompasses all support and services performed, provided, and arranged by the Army Medical Department to promote, improve, conserve, or restore the mental and physical well-being of personnel in the Army. Additionally, as directed, provide support to other Services, agencies, and organizations. This includes casualty care, (encompassing a number of Army Medical Department functions—organic and area medical support, hospitalization, the treatment aspects of dental care and behavioral/neuropsychiatric treatment, clinical laboratory services, and treatment of chemical, biological, radiological, and nuclear patients) medical evacuation, and medical logistics. (FM 4-02)

high-payoff target

    (DOD) A target whose loss to the enemy will significantly contribute to the success of the friendly course of action. (JP 3-60)

high-value target

    (DOD) A target the enemy commander requires for the successful completion of the mission. (JP 3-60)

human intelligence

    (Army) The collection by a trained human intelligence collector of foreign information from people and multimedia to identify elements, intentions, composition, strength, dispositions, tactics, equipment, and capabilities. (ADP 2-0)

hybrid threat

    The diverse and dynamic combination of regular forces, irregular forces, terrorist, or criminal elements acting in concert to achieve mutually benefitting effects. (ADP 3-0)

indicator

    (DOD) In the context of assessment, a specific piece of information that infers the condition, state, or existence of something, and provides a reliable means to ascertain performance or effectiveness. (JP 5-0)

infiltration

    (Army) A form of maneuver in which an attacking force conducts undetected movement through or into an area occupied by enemy forces to occupy a position of advantage behind those enemy positions while exposing only small elements to enemy defensive fires. (FM 3-90-1)

infiltration lane

A control measure that coordinates forward and lateral movement of infiltrating units and fixes fire planning responsibilities. (FM 3-90-1)

information

In the context of decision making, data that has been organized and processed in order to provide context for further analysis. (ADP 6-0)

information collection

An activity that synchronizes and integrates the planning and employment of sensors and assets as well as the processing, exploitation, and dissemination of systems in direct support of current and future operations. (FM 3-55)

information environment

(DOD) The aggregate of individuals, organizations, and systems that collect, process, disseminate, or act on information. (JP 3-13)

information management

(Army) The science of using procedures and information systems to collect, process, store, display, disseminate, and protect data, information, and knowledge products. (ADP 6-0)

information operations

(DOD) The integrated employment, during military operations, of information related capabilities in concert with other lines of operation to influence, disrupt, corrupt, or usurp the decision-making of adversaries and potential adversaries while protecting our own. (JP 3-13)

information-related capability

(DOD) A tool, technique, or activity employed within a dimension of the information environment that can be used to create effects and operationally desired conditions. (JP 3-13)

information requirements

(DOD) In intelligence usage, those items of information regarding the adversary and other relevant aspects of the operaitonal environment that need to be collected and processed in order to meet the intelligence requirements of a commander. (JP 2-0)

integration

(DOD) The arrangement of military forces and their actions to create a force that operates by engaging as a whole. (JP 1)

intelligence analysis

The process by which collected information is evaluated and integrated with existing information to facilitate intelligence production. (ADP 2-0)

intelligence operations

(Army) The tasks undertaken by military intelligence units through the intelligence disciplines to obtain information to satisfy validated requirements. (ADP 2-0)

intelligence preparation of the battlefield

(Army) The systematic process of analyzing the mission variables of enemy, terrain, weather, and civil considerations in an area of interest to determine their effect on operations. (ATP 2-01.3)

intelligence synchronization

The art of integrating information collection; intelligence processing, exploitation, and dissemination; and intelligence analysis with operations to effectively and efficiently fight for intelligence in support of decision making. (ADP 2-0)

intelligence warfighting function

The related tasks and systems that facilitate understanding the enemy, terrain, weather, civil considerations, and other significant aspects of the operational environment. (ADP 3-0)

interdiction

(DOD) An action to divert, disrupt, delay, or destroy the enemy's military surface capability before it can be used effectively against friendly forces, or to otherwise achieve objectives. (JP 3-03)

isolate

To separate a force from its sources of support in order to reduce its effectiveness and increase its vulnerability to defeat. (ADP 3-0)

key tasks

Those significant activities the force must perform as a whole to achieve the desired end state. (ADP 6-0)

key terrain

(Army) An identifiable characteristic whose seizure or retention affords a marked advantage to either combatant. (ADP 3-90)

knowledge

In the context of decision making, information that has been analyzed and evaluated for operational implications. (ADP 6-0)

knowledge management

The process of enabling knowledge flow to enhance shared understanding, learning, and decision making. (ADP 6-0)

large-scale combat operations

Extensive joint combat operations in terms of scope and size of forces committed, conducted as a campaign aimed at achieving operational and strategic objectives. (ADP 3-0)

large-scale ground combat operations

Sustained combat operations involving multiple corps and divisions. (ADP 3-0)

leadership

The activity of influencing people by providing purpose, direction, and motivation to accomplish the mission and improve the organization. (ADP 6-22)

Level I threat

A small enemy force that can be defeated by those units normally operating in the echelon support area or by the perimeter defenses established by friendly bases and base clusters. (ATP 3-91)

Level II threat

An enemy force or activities that can be defeated by a base or base cluster's defensive capabilities when augmented by a response force. (ATP 3-91)

Level III threat

An enemy force or activities beyond the defensive capability of both the base and base cluster and any local reserve or response force. (ATP 3-91)

limit of advance

A phase line used to control forward progress of the attack. (ADP 3-90)

line of contact

A general trace delineating the location where friendly and enemy forces are engaged. (ADP 3-90)

line of departure

(DOD) In land warfare, a line designated to coordinate the departure of attack elements. (JP 3-31)

local security

The low-level security activities conducted near a unit to prevent surprise by the enemy. (ADP 3-90)

logistics

(Army) Planning and executing the movement and support of forces. It includes those aspects of military operations that deal with: design and development, acquisition, storage, movement, distribution, maintenance, evacuation, and disposition of materiel; acquisition or construction, maintenance, operation, and disposition of facilities; and acquisition or furnishing of services. (ADP 4-0)

main battle area

The area where the commander intends to deploy the bulk of the unit's combat power and conduct decisive operations to defeat an attacking enemy. (ADP 3-90)

main command post

A facility containing the majority of the staff designed to control current operations, conduct detailed analysis, and plan future operations. (FM 6-0)

main effort

A designated subordinate unit whose mission at a given point in time is most critical to overall mission success. (ADP 3-0)

maneuver

(Army) Movement in conjunction with fires. (ADP 3-0)

massed fire

(DOD) Fire from a number of weapons directed at a single point or small area. (JP 3-02)

measure of effectiveness

(DOD) An indicator used to measure a current system state, with change indicated by comparing multiple observations over time. (JP 5-0)

measure of performance

(DOD) An indicator used to measure a friendly action that is tied to measuring task accomplishment. (JP 5-0)

meeting engagement

A combat action that occurs when a moving force, incompletely deployed for battle, engages an enemy at an unexpected time and place. (ADP 3-90)

military decision-making process

An iterative planning methodology to understand the situation and mission, develop a course of action, and produce an operation plan or order. (ADP 5-0)

mission

(DOD) The task, together with the purpose, that clearly indicates the action to be taken and the reason therefore. (JP 3-0)

mission command

(Army) The Army's approach to command and control that empowers subordinate decision making and decentralized execution appropriate to the situation. (ADP 6-0)

mission orders

Directives that emphasize to subordinates the results to be attained, not how they are to achieve them. (ADP 6-0)

mission statement

(DOD) A short sentence or paragraph that describes the organization's essential task(s), purpose, and action containing the elements of who, what, when, where, and why. (JP 5-0)

mission variables

Categories of specific information needed to conduct operations. (ADP 1-01)

mixing

Using two or more different assets to collect against the same intelligence requirement. (FM 3-90-2)

mobile defense

A type of defensive operation that concentrates on the destruction or defeat of the enemy through a decisive attack by a striking force. (ADP 3-90)

mobilization

(DOD) The process by which Armed Forces of the United States, or part of them, are brought to a state of readiness for war or other national emergency. (JP 4-05)

monitoring

Continuous observation of those conditions relevant to the current operation. (ADP 5-0)

movement

The positioning of combat power to establish the conditions for maneuver. (ADP 3-90)

movement and maneuver warfighting function

The related tasks and systems that move and employ forces to achieve a position of relative advantage over the enemy and other threats. (ADP 3-0)

movement formation

An ordered arrangement of forces for a specific purpose and describes the general configuration of a unit on the ground. (ADP 3-90)

movement to contact

(Army) A type of offensive operation designed to develop the situation and to establish or regain contact. (ADP 3-90)

multi-domain fires

Fires that converge effects from two or more domains against a target. (ADP 3-19)

mutual support

(DOD) That support which units render each other against an enemy, because of their assigned tasks, their position relative to each other and to the enemy, and their inherent capabilities. (JP 3-31)

named area of interest

(DOD) A geospatial area or systems node or link against which information that will satisfy a specific information requirement can be collected, usually to capture indications of adversary courses of action. (JP 2-01.3)

neutralize

(Army) A tactical mission task that results in rendering enemy personnel or materiel incapable of interfering with a particular operation. (FM 3-90-1)

no-fire area

(DOD) An area designated by the appropriate commander into which fires or their effects are prohibited. (JP 3-09.3)

noncontiguous area of operations

Where one or more of the commander's subordinate force's areas of operation do not share a common boundary. (FM 3-90-1)

objective

(Army) A location used to orient operations, phase operations, facilitate changes of direction, and provide for unity of effort. (ADP 3-90)

objective rally point

An easily identifiable point where all elements of the infiltrating unit assemble and prepare to attack the objective. (ADP 3-90)

offensive operation

An operation to defeat or destroy enemy forces and gain control of terrain, resources, and population centers. (ADP 3-0)

operational approach

(DOD) A broad description of the mission, operational concepts, tasks, and actions required to accomplish the mission. (JP 5-0)

operational environment

(DOD) A composite of the conditions, circumstances, and influences that affect the employment of capabilities and bear on the decisions of the commander. (JP 3-0)

operational framework

A cognitive tool used to assist commanders and staffs in clearly visualizing and describing the application of combat power in time, space, purpose, and resources in the concept of operations. (ADP 1-01)

operational variables

A comprehensive set of information categories used to describe an operational environment. (ADP 1-01)

operations in depth

The simultaneous application of combat power throughout an area of operations. (ADP 3-90)

operations process

The major command and control activities performed during operations: planning, preparing, executing, and continuously assessing the operation. (ADP 5-0)

parallel planning

Two or more echelons planning for the same operations nearly simulaneously facilitated by the use of warning orders by the higher headquarters. (ADP 5-0)

passage of lines

(DOD) An operation in which a force moves forward or rearward through another force's combat positions with the intention of moving into or out of contact with the enemy. (JP 3-18)

penetration

A form of maneuver in which an attacking force seeks to rupture enemy defenses on a narrow front to disrupt the defensive system. (FM 3-90-1)

personnel services

Sustainment functions that man and fund the force, maintain Soldier and Family readiness, promote the moral and ethical values of the nation, and enable the fighting qualities of the Army. (ADP 4-0)

phase

(Army) A planning and execution tool used to divide an operation in duration or activity. (ADP 3-0)

physical security

(DOD) That part of security concerned with physical measures designed to safeguard personnel; to prevent unauthorized access to equipment, installations, material, and documents; and to safeguard them against espionage, sabotage, damage, and theft. (JP 3-0)

planning

The art and science of understanding a situation, envisioning a desired future, and determining effective ways to bring that future about. (ADP 5-0)

planning horizon

A point in time that commanders use to focus the organization's planning efforts to shape future events. (ADP 5-0)

position of relative advantage

A location or the establishment of a favorable condition within the area of operations that provides the commander with temporary freedom of action to enhance combat power over an enemy or influence the enemy to accept risk and move to a position of disadvantage. (ADP 3-0)

positive control

(DOD) A method of airspace control that relies on positive identification, tracking, and direction of aircraft within an airspace, conducted with electronic means by an agency having the authority and responsibility therein. (JP 3-52)

preparation

Those activities performed by units and Soldiers to improve their ability to execute an operation. (ADP 5-0)

primary position

The position that covers the enemy's most likely avenue of approach into the area of operations. (ADP 3-90)

priority intelligence requirement

(DOD) An intelligence requirement that the commander and staff need to understand the threat and other aspects of the operational environment. (JP 2-01)

probable line of deployment

A phase line that designates the location where the commander intends to deploy the unit into assault formation before beginning the assault. (ADP 3-90)

procedural control

(DOD) A method of airspace control which relies on a combination of previously agreed and promulgated orders and procedures. (JP 3-52)

procedures

(DOD) Standard, detailed steps that prescribe how to perform specific tasks. (CJCSM 5120.01B)

processing, exploitation, and dissemination

The execution of the related functions that converts and refines collected data into usable information, distributes the information for further analysis, and, when appropriate, provides combat information to commanders and staffs. (ADP 2-0)

protection warfighting function

The related tasks and systems that preserve the force so the commander can apply maximum combat power to accomplish the mission. (ADP 3-0)

pursuit

A type of offensive operation designed to catch or cut off a hostile force attempting to escape, with the aim of destroying it. (ADP 3-90)

radio silence

The status on a radio network in which all stations are directed to continuously monitor without transmitting, except under established criteria. (ATP 6-02.53)

raid

(DOD) An operation to temporarily seize an area to secure information, confuse an adversary, capture personnel or equipment, or to destroy a capability culminating with a planned withdrawal. (JP 3-0)

rally point

An easily identifiable point on the ground at which units can reassemble and reorganize if they become dispersed. (ATP 3-21.20)

rearward passage of lines

Occurs when a unit passes through another unit's positions while moving away from the enemy. (ADP 3-90)

reconnaissance

(DOD) A mission undertaken to obtain, by visual observation or other detection methods, information about the activities and resources of an enemy or adversary, or to secure data concerning the meteorological, hydrographic, or geographic characteristics of a particular area. (JP 2-0)

reconnaissance handover

The action that occurs between two elements in order to coordinate the transfer of information and/or responsibility for observation of potential threat contact, or the transfer of an assigned area from one element to another. (FM 3-98)

reconnaissance handover line

A designated phase line on the ground where reconnaissance responsibility transitions from one element to another. (FM 3-98)

reconnaissance in force

A type of reconnaissance operation designed to discover or test the enemy's strength, dispositions, and reactions or to obtain other information. (ADP 3-90)

reconnaissance objective

A terrain feature, geographic area, enemy force, adversary, or other mission or operational variable about which the commander wants to obtain additional information. (ADP 3-90)

reconnaissance-pull

Reconnaissance that determines which routes are suitable for maneuver, where the enemy is strong and weak, and where gaps exist, thus pulling the main body toward and along the path of least resistance. This facilitates the commander's initiative and agility. (FM 3-90-2)

reconnaissance-push

Reconnaissance that refines the common operational picture, enabling the commander to finalize the plan and support shaping and decisive operations. It is normally used once the commander commits to a scheme of maneuver or course of action. (FM 3-90-2)

reconstitution

(Army) Actions that commanders plan and implement to restore units to a desired level of combat effectiveness commensurate with mission requirements and available resources. (ATP 3-21.20)

redeployment

(Army) The transfer of forces and materiel to home and/or demobilization stations for reintegration and/or out-processing. (ATP 3-35)

redundancy

Using two or more like assets to collect against the same intelligence requirement. (FM 3-90-2)

rehearsal

A session in which the commander and staff or unit practices expected actions to improve performance during execution. (ADP 5-0)

relevant information

All information of importance to the commander and staff in the exercise of command and control. (ADP 6-0)

relief in place

(DOD) An operation in which, by direction of higher authority, all or part of a unit is replaced in an area by the incoming unit and the responsibilities of the replaced elements for the mission and the assigned zone of operations are transferred to the incoming unit. (JP 3-07.3)

reorganization

All measures taken by the commander to maintain unit combat effectiveness or return it to a specified level of combat capability. (FM 3-90-1)

reserve

(Army) That portion of a body of troops that is withheld from action at the beginning of an engagement to be available for a decisive movement. (ADP 3-90)

restrictive fire line

(DOD) A specific boundary established between converging, friendly surgace forces that prohibits fires or their effects from crossing. (JP 3-09)

retirement

When a force out of contact moves away from the enemy. (ADP 3-90)

retrograde

(Army) A type of defensive operation that involves organized movement away from the enemy. (ADP 3-90)

risk management

(DOD) The process to identify, assess, and control risks and make decisions that balance risk cost with mission benefits. (JP 3-0)

route reconnaissance

A type of reconnaissance operation to obtain detailed information of a specified route and all terrain from which the enemy could influence movement along that route. (ADP 3-90)

rule of law

A principle under which all persons, institutions, and entities, public and private, including the state itself, are accountable to laws that are publicly promulgated, equally enforced, and independently adjudicated, and that are consistent with international human rights principles. (ADP 3-07)

running estimate

The continuous assessment of the current situation used to determine if the current operation is proceeding according to the commander's intent and if planned future operations are supportable. (ADP 5-0)

screen

A type of security operation that primarily provides early warning to the protected force. (ADP 3-90)

search and attack

A technique for conducting a movement to contact that shares many of the characteristics of an area security mission. (FM 3-90-1)

security area

That area occupied by a unit's security elements and includes the areas of influence of those security elements. (ADP 3-90)

security cooperation

(DOD) All Department of Defense interactions with foreign security establishments to build security relationships that promote specific United States security interests, develop allied and partner nation military and security capabilities for self-defense and multinational operations, and provide United States forces with peacetime and contingency access to allied and partner nations. (JP 3-20)

security force assistance

(DOD) The Department of Defense activities that support the development of the capacity and capability of foreign security forces and their supporting institutions. (JP 3-20)

security operations

Those operations performed by commanders to provide early and accurate warning of enemy operations, to provide the forces being protected with time and maneuver space within which to react to the enemy, and to develop the situation to allow commanders to effectively use their protected forces. (ADP 3-90)

security sector reform

(DOD) A comprehensive set of programs and activities undertaken by a host nation to improve the way it provides safety, security, and justice. (JP 3-07)

sequential relief in place

Occurs when each element within the relieved unit is relieved in succession, from right to left or left to right, depending on how it is deployed. (ADP 3-90)

shaping operation

An operation at any echelon that creates and presrves conditions for success of the decisive operation through effects on the enemy, other actors, and the terrain. (ADP 3-0)

simultaneity

The execution of related and mutually supporting tasks at the same time across multiple locations and domains. (ADP 3-0)

simultaneous relief in place

Occurs when all elements are relieved at the same time. (ADP 3-90)

single envelopment

A form of maneuver that results from maneuvering around one assailable flank of a designated enemy force. (FM 3-90-1)

situational understanding

The product of applying analysis and judgment to relevant information to determine the relationship among the operational and mission variables. (ADP 6-0)

special reconnaissance

(DOD) Reconnaissance and surveillance actions conducted as a special operation in hostile, denied, or diplomatically and/or politically sensitive environments to collect or verify information of strategic or operational significance, employing military capabilities not normally found in conventional forces. (JP 3-05)

spoiling attack

A tactical maneuver employed to seriously impair a hostile attack while the enemy is in the process of forming or assembling for an attack. (FM 3-90-1)

stability mechanism

The primary method through which friendly forces affect civilians in order to attain conditions that support establishing a lasting, stable peace. (ADP 3-0)

stability operation

An operation conducted outside the United States in coordination with other instruments of national power to establish or maintain a secure environment and provide essential governmental services, emergency infrastructure reconstruction, and humanitarian relief. (ADP 3-0)

staff section

A grouping of staff members by area of expertise under a coordinating, special, or personal staff officer. (FM 6-0)

staggered relief in place

Occurs when a commander relieves each element in a sequence determined by the tactical situation, not its geographical orientation. (ADP 3-90)

striking force

A dedicated counterattack force in a mobile defense constituted with the bulk of available combat power. (ADP 3-90)

strong point

A heavily fortified battle position tied to a natural or reinforcing obstacle to create an anchor for the defense or to deny the enemy decisive or key terrain. (ADP 3-90)

subsequent position

A position that a unit expects to move to during the course of battle. (ADP 3-90)

supplementary position

A defensive position located within a unit's assigned area of operations that provides the best sectors of fire and defensive terrain along an avenue of approach that is not the primary avenue where the enemy is expected to attack. (ADP 3-90)

supply point distribution

A method of distributing supplies to the receiving unit at a supply point. The receiving unit then moves the supplies to its own area using its own transportation. (FM 4-40)

support area

The portion of the commander's area of operations that is designated to facilitate the positioning, employment, and protection of base sustainment assets required to sustain, enable, and control operations. (ADP 3-0)

supporting distance

The distance between two units that can be traveled in time for one to come to the aid of the other and prevent its defeat by an enemy or ensure it regains control of a civil situation. (ADP 3-0)

supporting effort

A designated subordinate unit with a mission that supports the success of the main effort. (ADP 3-0)

supporting range

The distance one unit may be geographically separated from a second unit yet remain within the maximum range of the second unit's weapons systems. (ADP 3-0)

suppress

A tactical mission task that results in the temporary degradation of the performance of a force or weapon system below the level needed to accomplish its mission. (FM 3-90-1)

suppression

(DOD) Temporary or transient degradation by an opposing force of the performance of a weapons system below the level needed to fulfill its mission objectives. (JP 3-01)

surveillance

(DOD) The systematic observation of aerospace, cyberspace, surface, or subsurface areas, places, persons, or things by visual, aural, electronic, photographic, or other means. (JP 3-0)

survivability

(Army) A quality or capability of military forces which permits them to avoid or withstand hostile actions or environmental conditions while retaining the ability to fulfill their primary mission. (ATP 3-37.34)

survivability move

A move that involves rapidly displacing a unit, command post, or facility in response to direct and indirect fires, the approach of a threat or as a proactive measure based on intelligence, meteorological data, and risk assessment of enemy capabilities and intentions. (ADP 3-90)

survivability operations

(Army) Those protection activities that alter the physical environment to providing or improving cover, camouflage, and concealment. (ATP 3-37.34)

sustaining operation

An operation at any echelon that enables the decisive operation or shaping operations by generating and maintaining combat power. (ADP 3-0)

sustainment

(Army) The provision of logistics, financial management, personnel services, and health service support necessary to maintain operations until successful mission completion. (ADP 4-0)

sustainment preparation of the operational environment

The analysis to determine infrastructure, physical environment, and resources in the operational environment that will optimize or adversely impact friendly forces means for supporting and sustaining the commander's operations plan. (ADP 4-0)

sustainment warfighting function

The related tasks and systems that provide support and services to ensure freedom of action, extend operational reach, and prolong endurance. (ADP 3-0)

tactical combat force

(DOD) A rapidly deployable, air-ground mobile combat unit, with appropriate combat support and combat service support assets assigned to and capable of defeating Level III threats including combined arms. (JP 3-10)

tactical command post

A facility containing a tailored portion of a unit headquarters designed to control portions of an operation for a limited time. (FM 6-0)

tactical mobility

The ability of friendly forces to move and maneuver freely on the battlefield relative to the enemy. (ADP 3-90)

tactical road march

A rapid movement used to relocate units within an area of operations to prepare for combat operations. (ADP 3-90)

tactics

(Army) The employment, ordered arrangement, and directed actions of forces in relation to each other. (ADP 3-90)

target area of interest

(DOD) The geographical area where high-value targets can be acquired and engaged by friendly forces. (JP 2-01.3)

target reference point

(DOD) A predetermined point of reference, normally a permanent structure or terrain feature that can be used when describing a target location. (JP 3-09.3)

targeting

(DOD) The process of selecting and prioritizing targets and matching the appropriate response to them, considering operational requirements and capabilities. (JP 3-0)

task

(DOD) A clearly defined action or activity specifically assigned to an individual or organization that must be done as it is imposed by an appropriate authority. (JP 1)

task organization

(Army) A temporary grouping of forces designed to accomplish a particular mission. (ADP 5-0)

task-organizing

(Army) The act of designing a force, support staff, or sustainment package of specific size and composition to meet a unique task or mission. (ADP 3-0)

techniques

(DOD) Non-prescriptive ways or methods used to perform missions, functions, or tasks. (CJCSM 5120.01B)

tempo

The relative speed and rhythm of military operations over time with respect to the enemy. (ADP 3-0)

terrain management

The process of allocating terrain by establishing areas of operations, designating assembly areas, and specifying locations for units and activities to deconflict activities that might interfere with each other. (ADP 3-90)

threat

Any combination of actors, entities, or forces that have the capability and intent to harm United States forces, United States national interests, or the homeland. (ADP 3-0)

trigger line

A phase line located on identifiable terrain that crosses the engagement area—used to initiate and mass fires into an engagement area at a predetermined range for all or like weapon systems. (ATP 3-21.20)

troop movement

The movement of Soldiers and units from one place to another by any available means. (ADP 3-90)

turning movement

(Army) A form of maneuver in which the attacking force seeks to avoid the enemy's principle defensive positions by seizing objectives behind the enemy's current positions thereby causing the enemy force to move out of their current positions or divert major forces to meet the threat. (FM 3-90-1)

understanding

In the context of decision making, knowledge that has been synthesized and had judgment applied to comprehend the situation's inner relationships, enable decision making, and drive action. (ADP 6-0)

unified action

(DOD) The synchronization, coordination, and/or integration of the activities of governmental and nongovernmental entities with military operations to achieve unity of effort. (JP 1)

unified action partners

Those military forces, governmental and nongovernmental organizations, and elements of the private sector with whom Army forces plan, coordinate, synchronize, and integrate during the conduct of operations. (ADP 3-0)

unified land operations

The simultaneous execution of offense, defense, stability, and defense support of civil authorities across multiple domains to shape operational environments, prevent conflict, prevail in large-scale ground combat, and consolidate gains as part of unified action. (ADP 3-0)

unit distribution

A method of distributing supplies by which the receiving unit is issued supplies in its own area, with transportation furnished by the issuing agency. (FM 4-40)

unity of effort

(DOD) Coordination and cooperation toward common objectives, even if the participants are not necessarily part of the same command or organization, which is the product of successful unified action. (JP 1)

warfighting function

A group of tasks and systems united by a common purpose that commanders use to accomplish missions and training objectives. (ADP 3-0)

weapons of mass destruction

(DOD) Chemical, biological, radiological, or nuclear weapons capable of a high order of destruction or causing mass casualties, excluding the means of transporting or propelling the weapon where such means is a separable and divisible part from the weapon. (JP 3-40)

withdraw

To disengage from an enemy force and move in a direction away from the enemy. (ADP 3-90)

working group

(Army) A grouping of predetermined staff representatives who meet to provide analysis, coordinate, and provide recommendations for a particular purpose or function. (FM 6-0)

zone reconnaissance

A type of reconnaissance operation that involves a directed effort to obtain detailed information on all routes, obstacles, terrain, and enemy forces within a zone defined by boundaries. (ADP 3-90)

This page intentionally left blank.

References

URLs accessed on 6 January 2021.

REQUIRED PUBLICATIONS

DOD Dictionary of Military and Associated Terms. December 2020.

FM 1-02.1. *Operational Terms.* 21 November 2019.

FM 1-02.2. *Military Symbols.* 10 November 2020.

RELATED PUBLICATIONS

These documents contain relevant supplemental information.

JOINT PUBLICATIONS

Most joint publications are available online: https://www.jcs.mil/Doctrine

CJCSM 5120.01B. *Joint Doctrine Development Process.* 6 November 2020.

DODI 5000.68. *Security Force Assistance (SFA).* 27 October 2010. https://www.esd.whs.mil/dd/.

JP 1. *Doctrine for the Armed Forces of the United States.* 25 March 2013.

JP 2-0. *Joint Intelligence.* 22 October 2013.

JP 2-01. *Joint and National Intelligence Support to Military Operations.* 5 July 2017.

JP 2-01.3. *Joint Intelligence Preparation of the Operational Environment.* 21 May 2014.

JP 3-0. *Joint Operations.* 17 January 2017.

JP 3-01. *Countering Air and Missile Threats.* 21 April 2017.

JP 3-02. *Amphibious Operations.* 4 January 2019.

JP 3-03. *Joint Interdiction.* 9 September 2016.

JP 3-05. *Joint Doctrine for Special Operations.* 22 September 2020.

JP 3-07. *Stability.* 3 August 2016.

JP 3-07.3. *Peace Operations.* 1 March 2018.

JP 3-08. *Interorganizational Cooperation.* 12 October 2016.

JP 3-09. *Joint Fire Support.* 10 April 2019.

JP 3-09.3. *Close Air Support.* 10 June 2019.

JP 3-10. *Joint Security Operations in Theater.* 25 July 2019.

JP 3-11. *Operations in Chemical, Biological, Radiological, and Nuclear Environments.* 29 October 2018.

JP 3-12. *Cyberspace Operations.* 8 June 2018.

JP 3-13. *Information Operations.* 27 November 2012.

JP 3-13.3. *Operations Security.* 6 January 2016.

JP 3-13.4. *Military Deception.* 14 February 2017.

JP 3-16. *Multinational Operations.* 1 March 2019.

JP 3-18. *Joint Forcible Entry Operations.* 27 June 2018.

JP 3-20. *Security Cooperation.* 23 May 2017.

JP 3-22. *Foreign Internal Defense.* 17 August 2018.

JP 3-24. *Counterinsurgency*. 25 April 2018.

JP 3-31. *Joint Land Operations*. 3 October 2019.

JP 3-33. *Joint Task Force Headquarters*. 31 January 2018.

JP 3-35. *Deployment and Redeployment Operations*. 10 January 2018.

JP 3-40. *Joint Countering Weapons of Mass Destruction*. 27 November 2019.

JP 3-52. *Joint Airspace Control*. 13 November 2014.

JP 3-60. *Joint Targeting*. 28 September 2018.

JP 3-63. *Detainee Operations*. 13 November 2014.

JP 3-85. *Joint Electromagnetic Spectrum Operation*. 22 May 2020.

JP 4-05. *Joint Mobilization Planning*. 23 October 2018.

JP 5-0. *Joint Planning*. 1 December 2020.

ARMY PUBLICATIONS

Most Army publications are available online: https://armypubs.army.mil

ADP 1. *The Army*. 31 July 2019.

ADP 1-01. *Doctrine Primer*. 31 July 2019.

ADP 2-0. *Intelligence*. 31 July 2019.

ADP 3-0. *Operations*. 31 July 2019.

ADP 3-05. *Army Special Operations*. 31 July 2019.

ADP 3-07. *Stability*. 31 July 2019.

ADP 3-19. *Fires*. 31 July 2019.

ADP 3-28. *Defense Support of Civil Authorities*. 31 July 2019.

ADP 3-37. *Protection*. 31 July 2019.

ADP 3-90. *Offense and Defense*. 31 July 2019.

ADP 4-0. *Sustainment*. 31 July 2019.

ADP 5-0. *The Operations Process*. 31 July 2019.

ADP 6-0. *Mission Command: Command and Control of Army Forces*. 31 July 2019.

ADP 6-22. *Army Leadership and the Profession*. 31 July 2019.

ADRP 1-03. *The Army Universal Task List*. 2 October 2015.

AR 27-1. *Legal Services, Judge Advocate Legal Services*. 24 January 2017.

AR 385-10. *The Army Safety Program*. 24 February 2017.

AR 600-20. *Army Command Policy*. 24 July 2020.

AR 700-137. *Logistics Civil Augmentation Program*. 23 March 2017.

ATP 1-0.1. *G-1/AG and S-1 Operations*. 23 March 2015.

ATP 1-06.1. *Field Ordering Officer (FOO) and Pay Agent (PA) Operations*. 10 May 2013.

ATP 1-20. *Military History Operations*. 9 June 2014.

ATP 2-01. *Plan Requirements and Assess Collection*. 19 August 2014.

ATP 2-01.3. *Intelligence Preparation of the Battlefield*. 1 March 2019.

ATP 2-19.4. *Brigade Combat Team Intelligence Techniques*. 10 February 2015.

ATP 2-22.2-1. *Counterintelligence Volume I: Investigations, Analysis and Production, and Technical Services and Support Activities (U)*. 11 December 2015.

ATP 2-22.2-2. *(U) Counterintelligence Volume II: Operations and Collection Activities (S)*. 22 December 2016.

ATP 3-01.8. *Techniques for Combined Arms for Air Defense*. 29 July 2016.

ATP 3-01.15/MCTP 10-10B/NTTP 3-01.8/AFTTP 3-2.31. *Multi-Service Tactics, Techniques, and Procedures for Air and Missile Defense.* 14 March 2019.

ATP 3-01.50. *Air Defense and Airspace Management (ADAM) Cell Operation.* 5 April 2013.

ATP 3-01.60. *Counter-Rocket, Artillery, and Mortar Operations.* 10 May 2013.

ATP 3-04.1. *Aviation Tactical Employment.* 7 May 2020.

ATP 3-04.64/MCRP 3-42.1A/NTTP 3-55.14/AFTTP 3-2.64. *Multi-Service Tactics, Techniques, and Procedures for the Tactical Employment of Unmanned Aircraft Systems.* 22 January 2015.

ATP 3-05.2. *Foreign Internal Defense.* 19 August 2015.

ATP 3-05.40. *Special Operations Sustainment.* 3 May 2013.

ATP 3-06. *Urban Operations.* 7 December 2017.

ATP 3-07.5. *Stability Techniques.* 31 August 2012.

ATP 3-09.30. *Observed Fires.* 28 September 2017.

ATP 3-09.32/MCRP 3-31.6/NTTP 3-09.2/AFTTP 3-2.6. *JFIRE Multi-Service Tactics, Techniques, and Procedures for Joint Application of Firepower.* 18 October 2019.

ATP 3-09.42. *Fire Support for the Brigade Combat Team.* 1 March 2016.

ATP 3-11.23/MCWP 3-37.7/NTTP 3-11.35/AFTTP 3-2.71. *Multi-Service Tactics, Techniques, and Procedures for Weapons of Mass Destruction Elimination Operations.* 1 November 2013.

ATP 3-11.32/MCWP 10-10E.8/NTTP 3-11.37/AFTTP 3-2.46. *Multi-Service Tactics, Techniques, and Procedures for Chemical, Biological, Radiological, and Nuclear Passive Defense.* 13 May 2016.

ATP 3-11.36/MCRP 10-10E.1/NTTP 3-11.34/AFTTP 3-2.70. *Multi-Service Tactics, Techniques, and Procedures for Chemical, Biological, Radiological, and Nuclear Planning.* 24 September 2018.

ATP 3-11.37/MCWP 3-37.4/NTTP 3-11.29/AFTTP 3-2.44. *Multi-Service Tactics, Techniques, and Procedures for Chemical, Biological, Radiological, and Nuclear Reconnaissance and Surveillance.* 25 March 2013.

ATP 3-11.50. *Battlefield Obscuration.* 15 May 2014.

ATP 3-12.3. *Electronic Warfare Techniques.* 16 July 2019.

ATP 3-13.1. *The Conduct of Information Operations.* 4 October 2018.

ATP 3-20.15. *Tank Platoon.* 3 July 2019.

ATP 3-20.96. *Cavalry Squadron.* 12 May 2016.

ATP 3-20.97. *Cavalry Troop.* 1 September 2016.

ATP 3-20.98. *Scout Platoon.* 4 December 2019.

ATP 3-21.8. *Infantry Platoon and Squad.* 12 April 2016.

ATP 3-21.10. *Infantry Rifle Company.* 14 May 2018.

ATP 3-21.11. *SBCT Infantry Rifle Company.* 25 November 2020.

ATP 3-21.18. *Foot Marches.* 17 April 2017.

ATP 3-21.20. *Infantry Battalion.* 28 December 2017.

ATP 3-21.21. *SBCT Infantry Battalion.* 18 March 2016.

ATP 3-21.50. *Infantry Small-Unit Mountain and Cold Weather Operations.* 27 August 2020.

ATP 3-21.51. *Subterranean Operations.* 1 November 2019.

ATP 3-21.91. *Stryker Brigade Combat Team Weapons Troop.* 11 May 2017.

ATP 3-28.1/MCWP 3-36.2/NTTP 3-57.2/AFTTP 3-2.67. *Multi-Service Tactics, Techniques, and Procedures for Defense Support of Civil Authorities (DSCA).* 25 September 2015.

ATP 3-34.22. *Engineer Operations—Brigade Combat Team and Below.* 5 December 2014.

ATP 3-34.40. *General Engineering.* 25 February 2015.

ATP 3-34.80. *Geospatial Engineering.* 22 February 2017.

ATP 3-34.81/MCWP 3-17.4. *Engineer Reconnaissance*. 1 March 2016.

ATP 3-35. *Army Deployment and Redeployment*. 23 March 2015.

ATP 3-35.1. *Army Pre-Positioned Operations*. 27 October 2015.

ATP 3-37.2. *Antiterrorism*. 3 June 2014.

ATP 3-37.10. *Base Camps*. 27 January 2017.

ATP 3-37.11. *Chemical, Biological, Radiological, Nuclear, and Explosives Command*. 28 August 2018.

ATP 3-37.34. *Survivability Operations*. 16 April 2018.

ATP 3-39.10. *Police Operations*. 26 January 2015.

ATP 3-39.11. *Military Police Special Reaction Teams*. 26 November 2013.

ATP 3-39.12. *Law Enforcement Investigations*. 19 August 2013.

ATP 3-39.20. *Police Intelligence Operations*. 13 May 2019.

ATP 3-39.30. *Security and Mobility Support*. 21 May 2020.

ATP 3-39.32. *Physical Security*. 30 April 2014.

ATP 3-39.34. *Military Working Dogs*. 30 January 2015.

ATP 3-39.35. *Protective Services*. 31 May 2013.

ATP 3-52.1/MCRP 3-20F.4/NTTP 3-56.4/AFTTP 3-2.78. *Multi-Service Tactics, Techniques, and Procedures for Airspace Control*. 14 February 2019.

ATP 3-52.2/MCRP 3-20.1/NTTP 3-56.2/AFTTP 3-2.17. *Multi-Service Tactics, Techniques, and Procedures For The Theater Air-Ground System*. 21 May 2020.

ATP 3-55.4. *Techniques for Information Collection During Operations Among Populations*. 5 April 2016.

ATP 3-60. *Targeting*. 7 May 2015.

ATP 3-60.1/MCRP 3-16D/NTTP 3-60.1/AFTTP 3-2.3. *Dynamic Targeting, Multi-Service Tactics, Techniques, and Procedures for Dynamic Targeting*. 10 September 2015.

ATP 3-90.1. *Armor and Mechanized Infantry Company Team*. 27 January 2016.

ATP 3-90.4. *Combined Arms Mobility*. 8 March 2016.

ATP 3-90.5. *Combined Arms Battalion*. 5 February 2016.

ATP 3-90.8. *Combined Arms Countermobility Operations*. 17 September 2014.

ATP 3-90.15. *Site Exploitation*. 28 July 2015.

ATP 3-90.40. *Combined Arms Countering Weapons of Mass Destruction*. 29 June 2017.

ATP 3-90.90. *Army Tactical Standard Operating Procedures*. 1 November 2011.

ATP 3-90.97. *Mountain Warfare and Cold Weather Operations*. 29 April 2016.

ATP 3-90.98/MCTP 12-10C. *Jungle Operations*. 24 September 2020.

ATP 3-91. *Division Operations*. 17 October 2014.

ATP 3-91.1. *The Joint Air Ground Integration Center*. 17 April 2019.

ATP 3-92. *Corps Operations*. 7 April 2016.

ATP 3-96.1. *Security Force Assistance Brigade*. 2 September 2020.

ATP 4-01.45/MCRP 3-40F.7[MCRP 4-11.3H]/AFTTP 3-2.58. *Multi-Service Tactics, Techniques, and Procedures for Tactical Convoy Operations*. 22 February 2017.

ATP 4-02.1. *Army Medical Logistics*. 29 October 2015.

ATP 4-02.2. *Medical Evacuation*. 12 July 2019.

ATP 4-02.3. *Army Health System Support to Maneuver Forces*. 9 June 2014.

ATP 4-02.5. *Casualty Care*. 10 May 2013.

ATP 4-02.7/MCRP 4-11.1F/NTTP 4-02.7/AFTTP 3-42.3. *Multi-Service Tactics, Techniques, and Procedures for Health Service Support in a Chemical, Biological, Radiological, and Nuclear Environment.* 15 March 2016.

ATP 4-02.8. *Force Health Protection.* 9 March 2016.

ATP 4-02.55. *Army Health System Support Planning.* 30 March 2020.

ATP 4-10/MCRP 4-11H/NTTP 4-09.1/AFMAN 10-409-O. *Multi-Service Tactics, Techniques, and Procedures for Operational Contract Support.* 18 February 2016.

ATP 4-10.1. *Logistics Civil Augmentation Program Support to Unified Land Operations.* 1 August 2016.

ATP 4-16. *Movement Control.* 5 April 2013.

ATP 4-25.13. *Casualty Evacuation.* 15 February 2013.

ATP 4-32. *Explosive Ordnance Disposal (EOD) Operations.* 30 September 2013.

ATP 4-32.1. *Explosive Ordnance Disposal (EOD) Group and Battalion Headquarters Operations.* 24 January 2017.

ATP 4-32.2/MCRP 10-10D.1/NTTP 3-02.4.1/AFTTP 3-2.12. *Multi-service Tactics, Techniques, and Procedures for Explosive Ordnance.* 12 March 2020.

ATP 4-32.3. *Explosive Ordnance Disposal (EOD) Company, Platoon, and Team Operations.* 1 February 2017.

ATP 4-33. *Maintenance Operations.* 9 July 2019.

ATP 4-43. *Petroleum Supply Operations.* 6 August 2015.

ATP 4-46. *Contingency Fatality Operations.* 17 December 2014.

ATP 4-48. *Aerial Delivery.* 21 December 2016.

ATP 4-90. *Brigade Support Battalion.* 18 June 2020.

ATP 4-93. *Sustainment Brigade.* 11 April 2016.

ATP 4-93.1. *Combat Sustainment Support Battalion.* 19 June 2017.

ATP 4-94. *Theater Sustainment Command.* 28 June 2013.

ATP 5-0.1. *Army Design Methodology.* 1 July 2015.

ATP 5-19. *Risk Management.* 14 April 2014.

ATP 6-0.5. *Command Post Organization and Operations.* 1 March 2017.

ATP 6-01.1. *Techniques for Effective Knowledge Management.* 6 March 2015.

ATP 6-02.53. *Techniques for Tactical Radio Operations.* 13 February 2020.

ATP 6-02.54. *Techniques for Satellite Communications.* 5 November 2020.

ATP 6-02.60. *Tactical Networking Techniques for Corps and Below.* 9 August 2019.

ATP 6-02.70. *Techniques for Spectrum Management Operations.* 16 October 2019.

ATP 6-02.71. *Techniques for Department of Defense Information Network Operations.* 30 April 2019.

ATP 6-02.72. *Multi-Service Tactics, Techniques, and Procedures for Tactical Radios.* 19 May 2017.

ATP 6-02.75. *Techniques for Communications Security.* 18 May 2020.

ATTP 3-06.11. *Combined Arms Operations in Urban Terrain.* 10 June 2011.

FM 1-0. *Human Resources Support.* 1 April 2014.

FM 1-04. *Legal Support to Operations.* 8 June 2020.

FM 1-05. *Religious Support.* 21 January 2019.

FM 1-06. *Financial Management Operations.* 15 April 2014.

FM 2-0. *Intelligence.* 6 July 2018.

FM 2-22.3. *Human Intelligence Collector Operations.* 6 September 2006.

FM 3-0. *Operations.* 6 October 2017.

FM 3-04. *Army Aviation.* 6 April 2020.

FM 3-05. *Army Special Operations*. 9 January 2014.

FM 3-07. *Stability*. 2 June 2014.

FM 3-09. *Fire Support and Field Artillery Operations*. 30 April 2020.

FM 3-11. *Chemical, Biological, Radiological, and Nuclear Operations*. 23 May 2019.

FM 3-12. *Cyberspace and Electronic Warfare Operations*. 11 April 2017.

FM 3-13. *Information Operations*. 6 December 2016.

FM 3-14. *Army Space Operations*. 30 October 2019.

FM 3-18. *Special Forces Operations*. 28 May 2014.

FM 3-21.38. *Pathfinder Operations*. 25 April 2006.

FM 3-22. *Army Support to Security Cooperation*. 22 January 2013.

FM 3-24.2. *Tactics in Counterinsurgency*. 21 April 2009.

FM 3-34. *Engineer Operations*. 18 December 2020.

FM 3-39. *Military Police Operations*. 9 April 2019.

FM 3-50. *Army Personnel Recovery*. 2 September 2014.

FM 3-52. *Airspace Control*. 20 October 2016.

FM 3-53. *Military Information Support Operations*. 4 January 2013.

FM 3-55. *Information Collection*. 3 May 2013.

FM 3-57. *Civil Affairs Operations*. 17 April 2019.

FM 3-61. *Public Affairs Operations*. 1 April 2014.

FM 3-63. *Detainee Operations*. 2 January 2020.

FM 3-81. *Maneuver Enhancement Brigade*. 21 April 2014.

FM 3-90-1. *Offense and Defense, Volume 1*. 22 March 2013.

FM 3-90-2. *Reconnaissance, Security, and Tactical Enabling Tasks, Volume 2*. 22 March 2013.

FM 3-94. *Theater Army, Corps, and Division Operations*. 21 April 2014.

FM 3-98. *Reconnaissance and Security Operations*. 1 July 2015

FM 3-99. *Airborne and Air Assault Operations*. 6 March 2015.

FM 4-0. *Sustainment Operations*. 31 July 2019.

FM 4-01. *Army Transportation Operations*. 3 April 2014.

FM 4-02. *Army Health System*. 17 November 2020.

FM 4-40. *Quartermaster Operations*. 22 October 2013.

FM 6-0. *Commander and Staff Organization and Operations*. 5 May 2014.

FM 6-02. *Signal Support to Operations*. 13 September 2019.

FM 6-22. *Leader Development*. 30 June 2015.

FM 6-27/MCTP 11-10C. *The Commander's Handbook on the Law of Land Warfare*. 7 August 2019.

FM 7-0. *Train to Win in a Complex World*. 5 October 2016.

FM 90-3/FMFM 7-27. *Desert Operations*. 24 August 1993.

TC 2-91.4. *Intelligence Support to Urban Operations*. 23 December 2015.

TC 3-09.81. *Field Artillery Manual Cannon Gunnery*. 13 April 2016.

WEBSITES

3rd Armored Brigade Combat Team, 4th Infantry Division Fact Sheet. Available online at https://static.dvidshub.net/media/pubs/pdf_34851.pdf.

10th Combat Aviation Brigade, 10th Mountain Division Fact Sheet. Available online at https://static.dvidshub.net/media/pubs/pdf_34851.pdf.

16 Cases of Mission Command. U.S. Army Combined Arms Center. Fort Leavenworth, KS: Combat Studies Institute Press, 2013. Available online at https://www.armyupress.army.mil/Portals/7/Primer-on-Urban-Operation/Documents/Sixteen-Cases-of-Mission-Command.pdf.

82d Airborne Division, After Action Report 82d Airborne, December 1944-February 1945 (World War II Operational Documents, Combined Arms Research Digital Library, Fort Leavenworth, KS).

Dark Territory: The Secret History of Cyber War. New York: Simon and Schuster, 2016.

Ebb and Flow, November 1950-July 1951. U.S. Army History of the Korean War Series, volume 5. Washington, DC: Center of Military History, United States Army. Available online at https://history.army.mil/books/korea/ebb/fm.htm.

From Domination to Consolidation: at the Tactical Level in Future Large-Scale Combat Operations, A U.S. Army Command and General Staff College Press Book Published by the Army University Press [2020] Series.

Geneva Convention Relative to the Treatment of Prisoners of War of July 27, 1929. Available online at https://www.loc.gov/law/help/us-treaties/index.php.

Kasserine Pass Battles. Readings, volume I. United States Army Center of Military History, United States Army. Available online at https://history.army.mil/books/Staff-Rides/kasserine/kasserine.htm.

Military Review: Large-Scale Combat Operations Special Edition, Fort Leavenworth, Kansas: Army University Press. September –October 2018. Available online at https://www.armyupress.army.mil/Journals/Military-Review/English-Edition-Archives/September-October-2018/.

Operation Atlantic Resolve, Online at https://www.eur.army.mil/Portals/19/Atlantic%20Resolve%20Fact%20Sheet.pdf?ver=jP-mMtPdOc4kvb_8r_uGkA%3d%3d, 2 NOV 2020.

Perceptions Are Reality: Historical Case Studies of Information Operations in Large-Scale Combat Operations, Fort Leavenworth, Kansas: Army University Press [2018] Series.

Strategic Landpower in Europe Special Study. Center for Army Lessons Learned. Available online at http://call.army.mil.

The War in the Pacific, Okinawa: The Last Battle. Washington, DC: Historical Division, Department of the Army, 1948. Available online at https://history.army.mil/html/books/005/5-11-1/CMH_Pub_5-11-1.pdf.

Urban Warfare-The 2008 Battle for Sadr City. Rand Research Brief, RAND Arroyo Center. Santa Monica, CA: 2012. Available online at http://www.rand.org/content/dam/rand/pubs/research_briefs/2012/RAND_RB9652.pdf.

West Point Atlas of American Wars. United States Military Academy, Department of Military Art and Engineering. New York: 1972.

UNITED STATES LAW

Most acts and public law are available at: https://uscode.house.gov/.

Title 10. United States Code. *Armed Forces.*

PRESCRIBED FORMS

This section contains no entries.

REFERENCED FORMS

Unless otherwise indicated, DA forms are available on the Army Publishing Directorate website at https://armypubs.army.mil/. DD forms are available on the Executive Services Directorate website at https://www.esd.whs.mil/dd/.

DA Form 2028. *Recommended Changes to Publications and Blank Forms.*

DD Form 1380. *Tactical Combat Casualty Care (TCCC) Card.* (Available through normal supply channels.)

DD Form 1972. *Joint Tactical Air Strike Request.*

Index

Entries are by paragraph number.

Made in the USA
Coppell, TX
31 October 2023

23622293R00247